AF670117

Core Knowledge in Critical Care Medicine

Wolfgang Krüger • Andrew James Ludman

Core Knowledge in Critical Care Medicine

Wolfgang Krüger
Medizinische Universitätsklinik
Kantonsspital Aarau
Aarau
Switzerland

Andrew James Ludman
Department of Cardiology
The London Chest Hospital
London
UK

ISBN 978-3-642-54970-0 ISBN 978-3-642-54971-7 (eBook)
DOI 10.1007/978-3-642-54971-7
Springer Heidelberg New York Dordrecht London

Library of Congress Control Number: 2014944745

© Springer-Verlag Berlin Heidelberg 2014
This work is subject to copyright. All rights are reserved by the Publisher, whether the whole or part of the material is concerned, specifically the rights of translation, reprinting, reuse of illustrations, recitation, broadcasting, reproduction on microfilms or in any other physical way, and transmission or information storage and retrieval, electronic adaptation, computer software, or by similar or dissimilar methodology now known or hereafter developed. Exempted from this legal reservation are brief excerpts in connection with reviews or scholarly analysis or material supplied specifically for the purpose of being entered and executed on a computer system, for exclusive use by the purchaser of the work. Duplication of this publication or parts thereof is permitted only under the provisions of the Copyright Law of the Publisher's location, in its current version, and permission for use must always be obtained from Springer. Permissions for use may be obtained through RightsLink at the Copyright Clearance Center. Violations are liable to prosecution under the respective Copyright Law.
The use of general descriptive names, registered names, trademarks, service marks, etc. in this publication does not imply, even in the absence of a specific statement, that such names are exempt from the relevant protective laws and regulations and therefore free for general use.
While the advice and information in this book are believed to be true and accurate at the date of publication, neither the authors nor the editors nor the publisher can accept any legal responsibility for any errors or omissions that may be made. The publisher makes no warranty, express or implied, with respect to the material contained herein.

Printed on acid-free paper

Springer is part of Springer Science+Business Media (www.springer.com)

Special thanks appertains to Dr. M. Morgan, Christchurch Hospital, NZ, for his inspiring suggestions and particularly his editorial support.

I would like to thank my wife Manuela for her understanding and support during the writing of this book.

Wolfgang Krüger

Contents

Chapter 1
Mechanical Ventilation

1.1 Acute Respiratory Failure

Acute respiratory failure (ARF) is defined as the inability of the respiratory system to meet the oxygenation, ventilatory, or metabolic requirements of the patient [1].

Most authors divide respiratory failure based on the two gas exchange functions, oxygenation and elimination of carbon dioxide. Either, "only" oxygen replenishment may be compromised or a joint disruption occurs [2, 3]:

I. Hypoxaemic respiratory failure
II. Hypercapnic respiratory failure

Hypoxaemic respiratory failure refers to the failure of the lungs to oxygenate mixed venous blood sufficiently, P_aO_2 <60 mmHg- [4, 5], while hypercapnic respiratory failure indicates a blunted elimination of carbon dioxide resulting in respiratory acidaemia with a P_aCO_2 >50 mmHg in the presence of hypoxaemia [5–7].

Hypercapnic respiratory failure is called ventilatory failure as well [6], highlighting that the ventilatory part of the respiratory system – the "pump function" of the respiratory apparatus – has failed, mainly due to ventilatory muscle fatigue, rather than to the gas exchange element [5].

As such, hypercapnia is a hallmark of ventilatory failure [5, 8], and an acutely decompensated ventilatory failure is characterized by a respiratory acidosis (pH <7.35) in the presence of hypercapnia [5].

Accordingly, the majority of authors classify respiratory failure into two subtypes [7, 9, 10], although others may distinguish four types [11, 12]:

Type 1: Hypoxaemic respiratory failure

Shunting and a ventilation–perfusion mismatch (V/Q mismatch) are the most common underlying pathophysiological mechanisms causing hypoxaemia [10, 13, 14]. While V/Q mismatch will easily respond to oxygen delivery (chronic obstructive pulmonary disease is a typical example [11]), hypoxaemia due to shunting will persist even if oxygen is supplied [11, 15]. Pulmonary shunting can be interpreted as an extreme form of V/Q mismatch [15] which occurs in the

W. Krüger, A.J. Ludman, *Core Knowledge in Critical Care Medicine*,
DOI 10.1007/978-3-642-54971-7_1, © Springer-Verlag Berlin Heidelberg 2014

setting of alveolar hypoventilation or alveolar collapse related to atelectasis and/or alveolar flooding from infection, blood or fluid [16]. Typical examples are cardiogenic pulmonary oedema, non-cardiogenic pulmonary oedema (ARDS), pneumonia, lung haemorrhage, and atelectasis [5, 11, 17]. Other conditions which may produce hypoxaemic respiratory failure include alveolar hypoventilation due to high altitude (low FiO_2), diffusion abnormalities and low mixed venous oxygen content subsequent to increased peripheral uptake [10, 13, 14].

Type 2: Ventilatory, hypercapnic respiratory failure

Type 2 respiratory failure is attributed to alveolar hypoventilation as found in (a) central nervous system disturbance, e.g. anaesthesia, head injury, drug overdose;(b) neuromuscular diseases, e.g. myasthenia gravis, Guillain–Barre syndrome, spinal cord diseases, myopathies;(c) elevated breathing workload, e.g. COPD, asthma, pulmonary fibrosis; and (d) increased dead space, e.g. pulmonary embolism, hypovolaemia, poor cardiac output, alveolar distension and/or increased CO_2 production as in fever, sepsis or burns trauma [10, 11].

Respiratory acidosis in the setting of chronic ventilatory failure must be considered a potentially life-threatening situation, and early mechanically ventilatory support is warranted [5].

Some authors additionally discriminate two further types:

Type 3: Perioperative ventilatory failure

This is actually a subtype of type 1 and is especially common in the postoperative phase. The main pathophysiology is atelectasis resulting from decreased functional residual capacity (FRC), anaesthesia, upper abdominal incision, airway secretions, supine position, obesity and ascites [11].

Type 4: Respiratory failure in conditions of shock

Hypoperfusion may affect respiration attributed to either increased demand or compromised delivery, leading to ARF [11, 12]. In addition, the central respiratory drive may be blunted [11].

Accordingly, a wide variety of etiological conditions may cause ARF, often not directly affecting the lung tissue [7].

Causes of type I respiratory failure include [5, 17]:

- ARDS
- Acute asthma
- Pulmonary fibrosis
- Pneumonia
- Pulmonary embolism
- Pneumothorax
- Pulmonary oedema
- COPD

Conditions such as pneumothorax, asthma or particularly COPD may present initially as type I respiratory failure but become complicated by developing hypercapnia, related to worsening ventilation–perfusion mismatch and/or increasing work of breathing resulting in exhaustion of the respiratory muscles [5]. The latter is referred to as *ventilatory failure* [5, 6] (see above).

Causes of type II respiratory failure include [5, 17]:

- COPD
- Kyphoscoliosis
- Obstructive sleep apnoea
- Acute severe asthma
- Abdominal distension (ascites, blood, peritonitis, pancreatitis, etc.)
- Morbid obesity resulting in obesity hypoventilation syndrome
- Chest wall trauma, e.g. flail chest or pneumothorax
- Central nervous system with depression of central respiratory drive
  - Coma
  - Raised intracranial pressure
  - Drugs, i.e. opioids, sedative
- Neuromuscular diseases
  - Peripheral nervous system, e.g. Guillain–Barre syndrome, critical illness polyneuropathy
  - Neuromuscular junction, i.e. myasthenia gravis, muscle relaxants, organophosphate poisoning
  - Myopathy, e.g. muscular dystrophy
  - Cervical cord lesion, e.g. trauma, tumour, etc.

1.2 Epidemiology

A review by Esteban [18] revealed that the indications for positive pressure mechanical ventilation (PPMV) include:

- Acute respiratory failure 66 %
- Coma 15 %
- Acute exacerbation of COPD 13 %
- Neuromuscular disorders 5 %

This was largely confirmed in other surveys [19, 20], however all of these studies were also performed by Esteban and colleagues. Acute respiratory failure may be caused by a number of different disease entities such as pneumonia, postoperative conditions, acute heart failure, ARDS and sepsis being the most frequent ones (between 10 and 18 % each) [18–20]. Thirty three percent [20] to 46 % [21] of all patients admitted to an ICU need PPMV for more than 12–24 h and the overall median duration of PPMV is reported to be 3 days [20]. However, a wide range exists: from 2 days in postoperative cases [22], over 4 days in COPD [20], up to 11 days in pneumonia [23] and 6–15 days in ALI/ARDS [20]. Only a very small number (3 %) needed more than 3 weeks of ventilatory support [20].

Ventilator-associated pneumonia (VAP) occurs in 8–28 % of cases [24], and the incidence increases with the duration of PPMV [25]. Interestingly, the risk rate lowers the longer the patient is ventilated, as shown by Cook [26], with the risk rates of 3% on day 5, of 2% day 10 and of 1% on day 15. However, the cumulative risk is estimated to be 7 % at 10 days and 19 % at 20 days [27]. However, differences in the definitions require some caution when generalizing this data.

Published mortality rates vary widely [28, 29], and it is important to note that ventilator settings have changed fundamentally since the implementation of lung protective ventilation in 2000, specifically regarding rates of acute lung injury/acute respiratory distress syndrome (ALI/ARDS) [30]. In unselected populations of PPMV patients, the reported mortality rates range from 64 % in an older (1993) study [31] to 39 % in a publication in 2002 [20]. Ideally therefore, when comparing ventilator-related mortality rates, the disease-specific rates should be considered.

1.3 Ventilator Modes Nomenclature

There is a profusion of terms used to describe ventilator modes which may be inconsistent and confusing, and multiple different names may be used to describe the same function [32, 33]. However, there are a number of main general principles.

The mode of mechanical ventilation refers to the method of inspiratory support [34] and comprises three components: (a) the control variable which may be volume or pressure controlled, or dual switching from one mode to the other during one breath may be permitted; (b) the breath sequence which may be continuous mandatory, intermittent mandatory or spontaneous; and (c) the targeting scheme, feedback or type of control mechanism(s) referring to the programmed ventilator target settings, i.e. respiratory rate, tidal volume, minute volume or combined targets [32, 33]. The new generation of ventilators are equipped with adaptive features and use modelled algorithms which calculate how to achieve set goals. However, it is not possible to control both pressure and volume simultaneously [33, 35].

Based on these principles, eight ventilation modes can be identified [32, 36]:

Control variable	Breath sequence	Acronym
Volume	Continuous mandatory ventilation	VC – CMV (IPPV)
	Intermittent mandatory ventilation	VC – IMV
Pressure	Continuous mandatory ventilation	PC – CMV (IPPV)
	Intermittent mandatory ventilation	PC – IMV
	Continuous spontaneous ventilation	PC – CSV (PSV/ASB)
Dual	Continuous mandatory ventilation	DC – CMV (CMV, pressure limited)
	Intermittent mandatory ventilation	DC – IMV (IMV, pressure limited)
	Continuous spontaneous ventilation	DC – CSV (CSV, pressure limited)

PC pressure control, *VC* volume control, *DC* dual control, *IPPV* intermittent positive pressure ventilation, *PSV* pressure support ventilation, *ASB* assisted spontaneous breathing

Pressure support ventilation (PSV) or just pressure support (PS) amplifies the patient's own respiratory efforts on patient-initiated breaths [37, 38]. With contemporary ventilators, triggering requires the patient to create a small negative inspiratory flow of −1 to −2 cm H_2O [39] which will, if achieved, lead to the initiation of the set support pressure [40]. Compared to conventional pressure triggering, flow triggering probably decreases the work of breathing [41, 42] and minimizes the risk of development of auto-PEEP/gas trapping inherent to the ventilator settings [43, 44].

Synchronized intermittent mandatory ventilation (SIMV), or now just shortened to IMV [32], refers to a volume- or pressure-controlled mode where the patient is able to trigger set mandatory breaths. This is allowed to occur only during a limited, short phase at the end of expiration prior to the next mandatory breath being delivered, and in addition to a set rate of breaths per minute supplied by the ventilator (whether patient triggered or not), the patient may spontaneously activate further breaths. These will usually be pressure supported [32, 36, 40], in order to ensure an adequate tidal volume which otherwise may vary according to the patient's respiratory muscle capability [40].

The term dual control was coined by Branson [45]to describe the technical ability to switch from one control mode to the other during a single breath cycle, i.e. starting with volume control in order to achieve a set tidal volume (V_T) target but limiting the pressure automatically generated to meet that V_T [32]. The same intention is also known as volume target pressure control which, in line with the results from the ARDSnet group in 2000 [30], limits the pressure to match the lung protective ventilation settings [40]. In recent years the dual or adaptive pressure-control algorithms have become widely available and combine pressure-limiting and volume-cycling features. This is achieved either by regulating the pressure in a volume-controlled mode (PRVC) or by assuring a specific volume in a pressure-controlled manner. It is physically achieved by affecting the flow delivered over a variable time, and the pressure is held after flow has stopped [46]. However, this approach has its limits; a minimum V_T is guaranteed but will not be constant because V_T depends on a complex relationship between respiratory compliance, airway resistance and patient effort and the ventilator is unable to distinguish between changes in lung mechanical properties and improved patient effort [46].

1.4 Volume-Controlled (VC) Versus Pressure-Controlled (PC) Ventilation

Positive pressure mechanical ventilation (PPMV) either in VC or PC mode replaces the physiological negative pressure respiration by the exact opposite mechanism [40]. Indeed, negative-pressure ventilatory support was used in at the advent of mechanical ventilation, but following the results observed during the polio epidemics in the 1950s, where those patients ventilated with negative-pressure generating mechanical ventilators had worse outcomes compared to those treated with PPMV, the technology shifted completely towards the positive pressure variant [47, 48].

Mechanical ventilatory support mainly increases lung volume that decreases in various disease states due to altered lung mechanics, namely, diminished lung and/or chest wall compliance and elevated airway resistance. This considerably supports the work of breathing by unloading the exhausted respiratory muscles, allowing them to recover, and thus improves pulmonary gas exchange, the latter further improved by a revised ventilation–perfusion mismatch [49–52].

Volume-controlled ventilation is the most frequently used mode worldwide as physicians are more familiar with this type of ventilatory support rather than with pressure-controlled [18]. This may also represent a desire to safeguard tidal and minute volumes which have traditionally been considered the most important target of respiratory support, and high V_T may avoid formation of atelectasis by avoiding alveolar hypoventilation [53]. In VC, the tidal volume set is delivered; however, the pressure necessary to reach the desired V_T will vary and may exceed our limits of peak and plateau pressure [35, 54, 55]. In contrast, in PC mode the pressure set will definitely be generated, but the consecutive tidal volumes will vary as it follows a complex function (see below) largely determined by the respiratory mechanics [54, 55]. The pressure level has to be adjusted if compliance or resistance change, in order to avoid too low or too high V_Ts which may risk atelectasis or over-distension, respectively [46]. Although a large body of literature concludes that no significant differences exist between PC and basic modes of ventilation, in terms of outcome, duration of ventilation or ICU and hospital stay exist [56–58], PC may still have some advantages. Even in VC, the maximal pressures applied need to be limited as there is good evidence that plateau pressures higher than 26–28 cm H_2O [59–61] or a peak inspiratory pressure (PIP) higher than 30–35 cm H_2O [53, 62] may have detrimental effects and should be avoided [54]. Moreover, in PC mode a more favourable pressure distribution and dissemination of the airway pressures including a significant reduction in peak pressure are found compared to VC [63, 64]. Decelerating inspiratory flow patterns are applicable in PC but not in VC (only constant patterns) and are associated with improved air distribution in the lungs which have heterogeneous mechanical properties [65–67] and will facilitate gentle and tissue preserving airway pressure and air distribution conditions. Decelerating inspiratory flow is shown to be especially of value in lungs with poor compliance [40, 68]. At the least, PC may be more comfortable for patients due to a better interaction between patient and ventilator, particularly in obstructive lung diseases [69].

In PC the main disadvantage is potential hypoventilation due to varying lung and chest wall mechanics which allows variations in V_T, whereas in VC the high pressures applied may be harmful, VC cannot compensate for leaks [54, 69] and the fixed flow may lead to patient–ventilator dyssynchrony [46]. However, VC will control the ventilation, may better manage hypoventilation during the first phase of respiratory failure and exhibits the lowest degree of hyperinflation if high inspiratory flows are applied and long expiratory times allowed [69].

In order to better synchronize patient and ventilator, two very similar pressure-controlled methods of ventilation have been developed, the so-called bilevel, BiPAP or BiLevel ventilation, and airway pressure release ventilation, APRV. Biphasic positive airway pressure (BiPAP, BiLevel) and airway pressure release ventilation are modified pressure-controlled modes [40, 70]. APRV combines repetitive application of a constant high positive airway pressure (P_{HIGH} or high PEEP level), generating a tidal volume with intermittent pressure releases to a lower pressure level (P_{LOW} or low PEEP level) causing expiration [71]. As the inspiratory time is kept very prolonged there is only a short exhalation time in this pressure-controlled (IMV) ventilatory mode, potentially and intentionally creating auto-PEEP [70]. However, spontaneous breathing is possible at any time, making patient–ventilator interaction much more comfortable [32, 72]. Biphasic positive airway pressure ventilation (BiPAP) also allows, in principle, for unrestricted spontaneous ventilation (inspiration as well as expiration) at any time during the respiratory cycle, resulting in reduced sedation requirements and promoting weaning [54, 70]. This mode also applies two different pressure levels, P_{HIGH} and P_{LOW}, and is conceptually equal to AVPR [70] with the exception that the duration of the lower pressure level is far longer than in AVPR where it is by convention less than 1.5 s [33] and that BiPAP is supportive in spontaneous breaths [70]. AVPR represents a form of inversed ratio ventilation (IVR) which means that the inspiratory time is extremely prolonged in order to strongly support oxygen replenishment [33]. In BiPAP, I:E ratio can be determined by the physician and generally any ratio is available (e.g. inversed, 1:1, up to 1:4 [5]) [33]. In both types, the lower pressure level applies PEEP while with the change to the higher level, air will be inflated [33, 70]. Unfortunately, in AVPR the inversed ratio which may be necessary in severe hypoxaemia requires sedation or even paralysis [54]. However, as both techniques significantly improve oxygenation, attributed to alveolar recruitment and improved ventilation–perfusion matching [73], and as the more moderate settings in BiPAP allow spontaneous breathing, they are commonly applied in patients with hypoxaemic respiratory failure and BiPAP may routinely be the initial ventilation mode. Recommended settings are an initial fairly high-pressure level of 12–15 cm H_2O above set PEEP [74]. The potential for spontaneous breathing at any time is of very high value as it not only facilitates patient–ventilator interaction and synchrony but helps to avoid respiratory muscle fatigue and longer-term respirator dependency of which the main cause is diaphragm dysfunction and fatigue [75–77].

CPAP is a PC mode delivering a constant level of positive pressure throughout the respiratory cycle [40]. It may be applied in a broad range of causes of respiratory failure as by increasing mean airway pressure, collapsed and hypoventilated lung units will be reinflated and kept open during expiration with consecutive increase in functional residual capacity (FRC). This results in improved gas exchange and oxygenation [78] and so this technique is particularly indicated in hypoxaemic respiratory failure [40]. Moreover, as CPAP will improve the lung compliance as well, it reduces the work of breathing and thus may avert the development of overt muscle or ventilatory failure [79–81].

1.5 Indications for Intubation and Mechanical Ventilation

The decision to intubate a patient is a complex assessment process requiring the consideration and integration of numerous aspects and facts, but remains largely a clinical judgement, and in daily practice is essentially concurrent with the determination to apply PPMV [5, 17, 35, 82–84].

A review by Esteban revealed that the indications for PPMV include [18]:

- Acute respiratory failure 66 %
- Coma 15 %
- Acute exacerbation of COPD 13 %

- Neuromuscular disorders 5 %

Frequently designated/specified indications for PPMV [85]:

- Acute respiratory arrest
- Apnoea and impending respiratory arrest
- Acute hypoxaemic respiratory failure
- Coma and acute neuromuscular diseases
- Acute exacerbation of COPD
- Heart failure and cardiogenic shock
- Cardiac arrest
- Acute severe asthma
- Acute brain injury
- Flail chest

The physiological consequences of a sustained pH >7.65 or <7.10 are considered dangerous in itself if not quickly reversible and thus may require mechanical ventilation [35]. Within this range from pH 7.1 to 7.65, the clinical condition is seminal in how to approach the patient [86]. Some indicators in the setting of respiratory dysfunction and distress which support the initiation of PPMV are [50]:

Respiratory rate	>35/min
Tidal volume, spontaneously	<5 ml/kg
Vital capacity	<10 ml/kg
Rise in P_aCO_2 from baseline	>10 mmHg
Negative inspiratory force	<−25 cm H_2O
P_aO_2 with supplemented	O_2 <55 mmHg
Alveolar–arterial gradient (on $FiO_2=1.0$)	>450 mmHg
P_aO_2/P_AO_2 ratio	<0.15
GCS	<8

A GCS less than eight comes with the risk of not protecting the airway. Intubation should generally be considered and may be mandatory in head injuries [87]; however, a GCS <8 is otherwise not an absolute indication for intubation [88].

Aside from the clinical assessment, other features may help in making the decision whether to intubate or not [17]:

- Initiation of PPMV necessary and needs facilitation
- Protection from aspiration, particularly in patients not able to protect their airway, which is generally the case in altered mental status as indicated by a GCS <8
- Facilitation of tracheobronchial suction
- Relief of upper airway obstruction

Further, specific indications for ventilatory support and/or intubation are depicted in the paragraphs on ventilation in patients suffering from non-ARDS ventilator failure, COPD, asthma and the separate chapter on ALI/ARDS.

1.6 Patient–Ventilator Interaction

PPMV is applied to patients struggling with substantial respiratory difficulties in order to largely unload the respiratory muscles by taking over or sharing the work of breathing, facilitating lung inflation and gas exchange, thus reducing dyspnoea [89–91]. Spontaneous breaths and breathing efforts may be initially replaced by the ventilator, but as passive mechanical ventilation will lead to considerable respiratory muscle dysfunction and atrophy [92, 93], its timely withdrawal is of pivotal importance [89]. In order to facilitate this, the ventilator actions/responses must synchronize with patients' spontaneous breathing efforts and demands [89–91, 94, 95].

When considering why a patient is combating the ventilators, multiple factors may be contributing; these include underlying lung functional abnormalities, the ventilator settings set by the clinician, the specific ventilator functions, the patient–ventilator interface, and not at least the patients' own airway responses [89, 91, 96, 97]. NIV intolerance is most clearly related to asynchrony [94]. Of particular importance and interest are trigger asynchronies [89, 90, 94, 95], reported to be found in up to 58 % of all patients [94]. Asynchronies (patient–ventilator interactions) in general are associated with adverse outcomes, prolonged duration of PPMV and higher rate of tracheostomy [98–100] due to ineffective ventilation with increased work of breathing, lung over-distension, impaired gas exchange and patient discomfort [89]. Anxiety and dyspnoea often result from dyssynchronous interactions [90]. Particularly predisposed to mismatch between patients' request (ventilatory drive and muscular effort) and the machine's reply (airflow and pressure delivered) are patients with COPD and ALI/ARDS [89].

Trigger asynchronies comprise ineffective trigger efforts, auto-triggering, delayed triggering and premature and delayed release of flow and pressure–volume [89, 91] with the most common problem being ineffective or wasted efforts [101]. A trigger effort, in the vast majority occurring during the expiratory period, indicated by an abrupt decrease in airway pressure of >0.5 cm H_2O is ineffective if not resulting in an assisted breath from the ventilator (if the trigger occurs during expiration, accompanied by a decrease expiratory flow) [99]. Ineffective efforts, also known as ineffective triggering, untriggered breaths or trigger asynchrony, may occur as well during inspiration, indicated by an abrupt increase in inspiratory flow (in PC mode) or transient abrupt decrease in airway pressure (VC mode) [89, 95]. Dynamic hyperinflation, limited respiratory drive, weakness of respiratory muscles and insensitive trigger settings are causally underlying ineffective efforts [102]. Since resolving a dys-synchrony in one area often facilitates other adverse interactions as well [90], the most common problem is discussed in detail below.

Although the specific analysis of disadvantageous patient–ventilator interactions and how to tackle them have been more and more recognized in recent years, the first systematic approach of how to analyse and manage trigger asynchronies has been recently done by Sassoon [95]. In brief, low PEEP (5 cm H_2O are very common) should be applied, but adjusted in case of measured or suggested intrinsic PEEP (PEEPi) to 75–80 % of PEEPi as with Nava [103]. If there is still an asynchrony index of >10 %, increase PEEP by steps of 1 cm H_2O up to max 8 cm H_2O. If still wasted efforts are present, adjust V_T to 6–8 ml/kg PBW [104]. Thereafter, increases in inspiratory flow rate (VC) or pressure (PC) are recommended. Further steps and details, see Fig. 1.1 below:

[Def asynchrony index (AI) = number of wasted efforts/number of wasted efforts plus triggered breaths during a period of 2 min [105] in percent [99]]

Keep in mind that some degree of asynchrony may always be present [95].

1.7 Basics of Respiratory Physiology and Pathophysiological Issues

To breathe and thus inflate the lungs, a pressure gradient between the nose/mouth (atmosphere) and the lungs (alveoli) is needed as air, like fluid, moves from the higher pressure level towards the lower one [106, 107]. During inspiration the contraction of the respiratory muscles, diaphragm and external intercostal muscles enlarges the thoracic cage. Due to the (elastic) recoil properties of the lungs, adhered to the chest wall by a thin layer of fluid, the enlarging thoracic cavity generates a negative intrapleural pressure which is accompanied by a subatmospheric, negative alveolar pressure, establishing a pressure gradient within the airways, called the distending or transpulmonary pressure [106, 108, 109]. Mathematically this distending or transpulmonary pressure (gradient) being the driving force of airflow can be described and is defined by

$$P_{\mathrm{TRANS}} = P_{\mathrm{AL}} - P_{\mathrm{PL}} \left(\text{formula A}\right)$$

with P_{AL} being the pressure within the alveoli, and P_{PL} represents the pressure in the intrapleural space [106, 108, 110]. In mechanically ventilated patients, the airway pressure (P_{AW}) can be measured and equals alveolar pressure (P_{AL}) if there is no airflow [111, 112]. This is usually determined at end inspiration, as the end-inspiratory pressure is considered as being the most critical one since it distends the alveoli/alveolar units [113], but is otherwise influenced by lung and chest specific factors and properties

Fig. 1.1 Algorithm to improve patient–ventilator synchrony (With permission from Sassoon [95]). Start up to fine-tune PEEP as depicted. If $PEEP_i$ (intrinsic PEEP) cannot be measured or is ≤5 cm H_2O, go ahead with PEEP of 5 cm H_2O! With persisting asynchrony index (number of wasted efforts/number of wasted efforts + triggered breaths during a period of 2 min [%]) >10 %, increase applied PEEP by 1 cm H_2O steps up to a max of 8 cm H_2O, if measurable $PEEP_i$ apply 75–85 % of that. If after PEEP adjustment still a relevant asynchrony (asynchrony index >10 %) remains, adopt V_T: 6–8 ml/kg IBW. With ongoing asynchrony, increase respiratory flow rate in case of VCV or pressurization rate if the patient is on PCV. In the end, in case of time-cycled pressure target settings decrease inspiratory time, while in pressure support-ventilation, adjustment of the flow-cycling threshold is recommended: upward in case of prolonged expiration, otherwise downward

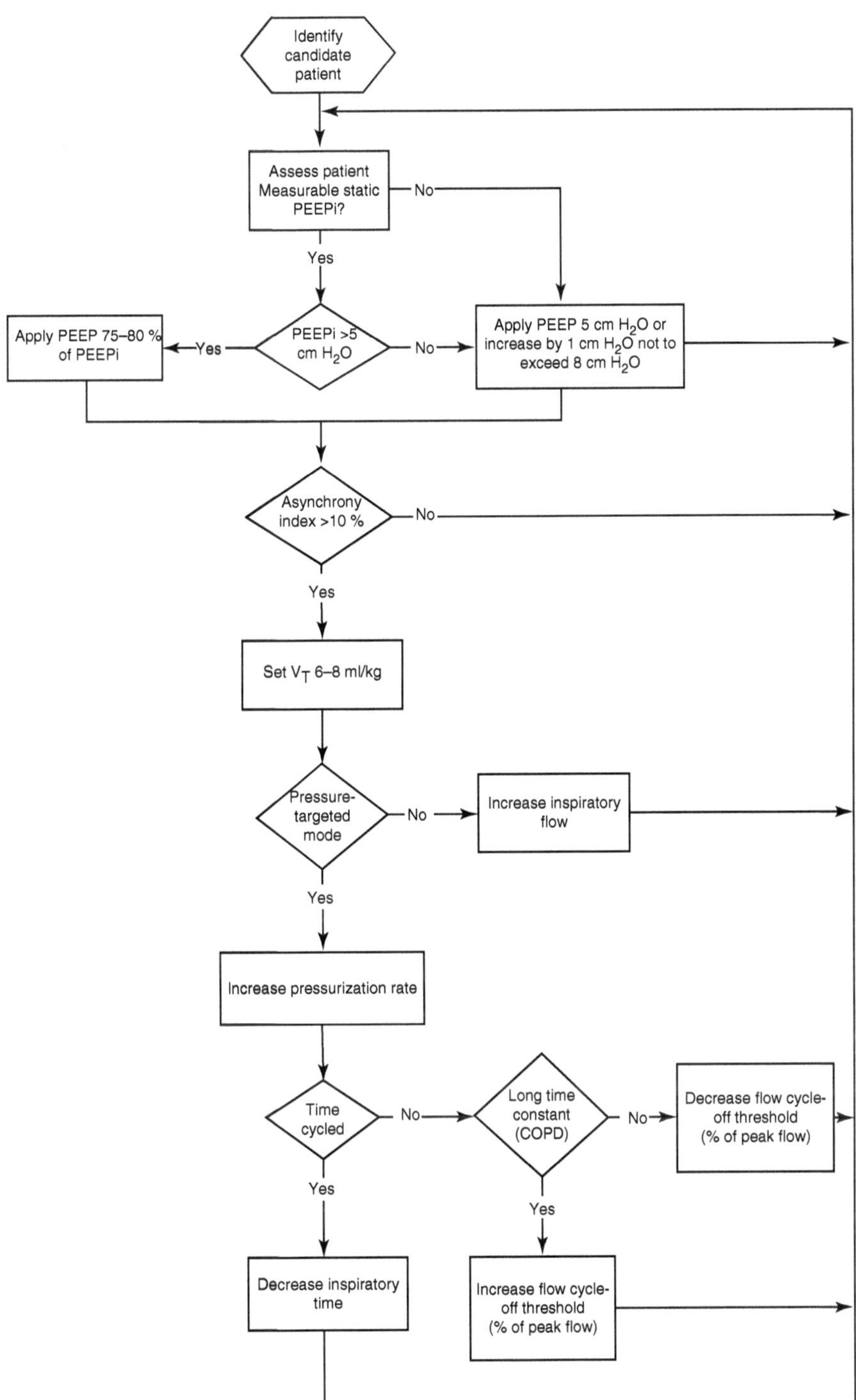
Identify candidate patient
Assess patient Measurable static PEEPi?
No
Yes
PEEPi >5 cm H_2O
Apply PEEP 75–80 % of PEEPi
Yes
No
Apply PEEP 5 cm H_2O or increase by 1 cm H_2O not to exceed 8 cm H_2O
Asynchrony index >10 %
No
Yes
Set V_T 6–8 ml/kg
Pressure-targeted mode
No
Increase inspiratory flow
Yes
Increase pressurization rate
Time cycled
No
Long time constant (COPD)
No
Decrease flow cycle-off threshold (% of peak flow)
Yes
Yes
Decrease inspiratory time
Increase flow cycle-off threshold (% of peak flow)

[110, 114]. Accordingly, the airway pressure needed to inflate the lungs either during spontaneous breathing or by ventilator is affected by the compliance of the respiratory system (C_{RS}), the airway resistance (R), the volume (V) inhaled (tidal volume, V_T) and the airflow (Q) [110], depicted by the following relationship:

$$P_{AW} = V / C_{RES} + R / Q \qquad [110]$$

In contrast, P_{TRANS} measured at end inspiration is free from such influences, allowing estimation of the actual true distending pressure in the passive lungs [113, 114]. The transpulmonary pressure is negative during spontaneous inspiration, zero at the functional residual capacity (FRC) of the lungs where the opposing forces of lungs and chest wall are equal and opposite to each other [110], and at which the P_{PL} is −5 cm H_2O as most authors mention [108], and positive during expiration [110].

Of those features mentioned above affecting the airway pressure, the compliance is of special interest and relevance.

The compliance of the respiratory system is made up by the compliance of the lungs (C_L) and the compliance of the chest wall (C_W), related by $1/C_{RS} = 1/C_L + 1/C_{CW}$ (or alternatively $C_{RS} = (C_{CW} \times C_L)/(C_{CW} + C_L)$) as both components are arranged in series [108–110]. The elastic properties of the respiratory system, C_{Rs}, correlate well with the amount of aerated lung tissue in patients with acute lung injury and ARDS [115]. Compliance is the inverse of elastance ($E_{TOT} = E_L + E_{CW}$ [112]-formula (B)), with E_{TOT} indicating the elastance of the whole respiratory system (RS) as counterpart to its inverse (C_{RS}) and a measure of the distensibility of the tissues, in this case an estimate of the ease of the lungs **and** the chest wall to distend [108, 110]. As compliance per definition equals the change in volume (usually V_T) per cm H_2O change in pressure, $C = \Delta V / \Delta P$ [106, 108, 109], the chest wall elastance can be separated by pleural pressure and rearrangement of formula (A) and (B):

$P_{PL} = P_{AW} \times E_{CW}/E_{TOT}$ [112], at which P_{AW} represents P_{AL} since literally the airway opening (mouth/tube) pressure in end-inspiratory hold is assessed, also called plateau pressure (see below); thus in a static setting after proximal airway pressure and distal, alveolar pressures have equalized [116]. As Barberis et al. [117] nicely show, in the clinical setting this happens 0.5 s after the onset of occlusion. In physiological circumstances $E_{CW}/E_{TOT} \approx 0.4$ to 0.5 at FRC level [112]. Consecutively, transpulmonary pressure reflecting the alveolar distending forces can be calculated by $P_{TRANS} = P_{AW} \times E_L/E_{TOT}$ [112].

The compliances of lung tissue and chest wall are depicted in Fig. 1.3, as obviously, their lung volumes are markedly different at recoil pressure zero and any change in either compliance will affect total elastance; see Fig. 1.2 by Kacmarek [110].

Pulmonary (lung) and chest compliance may be substantially different, depending on the disease and underlying medical conditions [118–120], and as such may considerably affect the transpulmonary pressure with potentially marked clinical implications [112, 116, 121]. The transpulmonary pressure most closely represents the actual alveolar distending pressure [114] responsible for and considered as being the main cause of ventilator-induced lung injury [65, 112, 122–124].

Impaired chest wall compliance is common and may be present in ALI/ARDS patients with abdominal diseases (extrapulmonary ALI/ARDS) associated with increased intra-abdominal pressure such as bowel distension, ascites, sepsis, pancreatitis, (pre-) eclampsia, multi-trauma or peritonitis [112, 118, 119], or even obesity

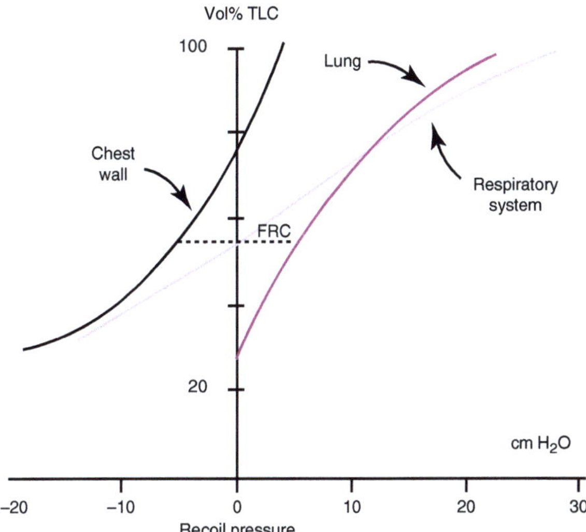

Fig. 1.2 Depicted are the compliance curves of the lungs, thorax and the total respiratory system. As can be derived from the curve progression, any change in either or both lung and thoracic compliance affects the total compliance of the respiratory system. *TLC* total lung capacity, *RFC* functional residual capacity (With permission from Kacmarek [110])

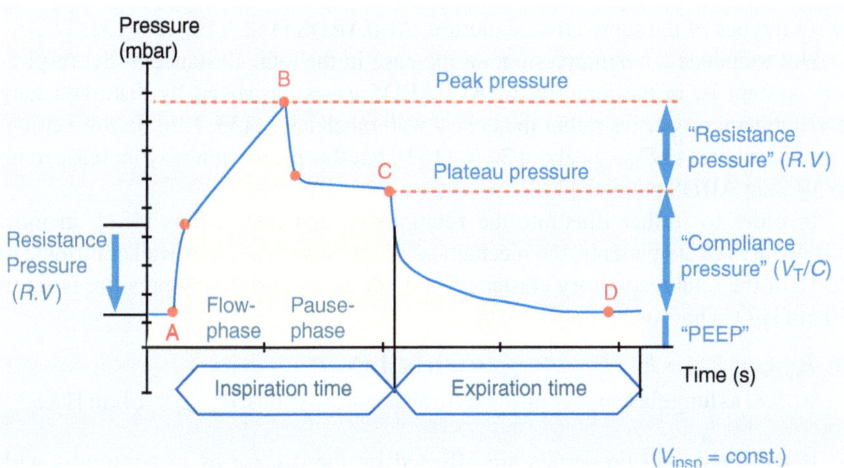

Fig. 1.3 Depicts the course of airway pressures in chronological sequence during mechanical ventilation. Pressure – time diagram of a volume controlled, constant flow positive pressure mechanical ventilation mode to explain pressure behaviour and ventilatory parameters. *R* airway resistance, V_T tidal volume, *C* compliance of the lungs and the chest – total compliance and of the mechanically ventilator system (like tubes etc.), *V* flow. *A* beginning of inspiration (mechanical application of air), *B* peak pressure, *C* plateau pressure, *D* end-expiratory pressure, if positive called PEEP (Modified with permission from Rittner and Döring [146])

[112, 125]. Gattinoni et al [118] established a linear relationship between increases in intra-abdominal pressures and chest wall elastance:

$$E_{CW} = 0.47 \times \text{intra} - \text{abdominal pressure}\left(\text{cm}\,H_2O\right) + 1.43 \qquad [118, 126].$$

Furthermore, pleural effusions which are often due to a positive fluid balance [127], obesity, sedation and paralysis in anaesthetized patients, and anatomical chest abnormalities all may, at least to some degree, cause an increase in chest wall elastance [120, 128–130]. Thus, for any given applied airway pressure, with increasing chest wall elastance, the pleural pressure will increase, while the transpulmonary distending pressure will drop [112, 131]. However, in patients with pulmonary ARDS, as may develop in the setting of diffuse pneumonia, aspiration, inhalation-trauma or multi-localized pulmonary embolism, the lung elastance may be considerably altered while the chest wall elastance is not affected, causing the transpulmonary pressure to increase but leaving the pleural pressure unchanged/normal [118, 119, 121, 132] as alveolar pressure is not transmitted [112]. Hence, impaired chest wall compliance (E_{CW}) associated with consecutively altered E_{CW} to E_{TOT} ratio and abnormally increased pleural pressure as found in a remarkable number (up to 30 %) [133, 134] of patients suffering from extrapulmonary ALI/ARDS commonly due to abdominal hypertension or compartment syndrome [133, 134] induces for any given airway pressure a reduction in P_{TRANS}. While patients with pulmonary ALI/ARDS, although holding the same total elastance of the respiratory system (E_{TOT}), exhibit considerably altered lung mechanics (E_L) and a concomitant rise of the E_L to E_{TOT} fraction, displaying a significant increase in the transpulmonary lung parenchyma distending pressure P_{TRANS}, but leaving the pleural pressures unchanged as alveolar pressures are not transmitted [112], accordingly characterizing two subtypes of the same clinical picture, ALI/ARDS [112, 119, 127, 131, 132].

Not to induce a misimpression, an increase in the total elastance of the respiratory system is, in the majority of ALI/ARDS cases, provoked by disturbed lung parenchymal properties rather than chest wall mechanics [135, 136]. In this setting, E_{CW} contributes to E_{TOT} by about 20 % [132], but this proportion may increase to up to 50 % in ARDS patients [112].

In order to further illustrate the relationship and interactions [131], in most pulmonary ARDS patients, the mechanical properties of the chest wall contribute to 20 % of the total respiratory elastance, and, let us say the airway pressure is set to 30 cm H_2O. Thus, using the formula

$$P_{TRANS} = P_{AW} \times E_L / E_{TOT}, P_{TRANS} = 30\,\text{cm}\,H_2O$$
$$\times\ 0.8\left(\text{as lung elastance contributes to } 80\%\text{ to } E_{TOT}\right); \text{thus, } P_{TRANS} = 24\,\text{cm}\,H_2O.$$

If the chest wall properties are affected by the disease as in peritonitis with increased intra-abdominal pressure, the chest wall may contribute to 50 % to E_{TOT}.

Hence, $P_{TRANS} = P_{AW} \times E_L/E_{TOT} = 30$ cm $H_2O \times 0.5 = 15$ cm H_2O, quite a highly significant difference and certainly will affect management.

However, the described differences in mechanical and morphological properties and the consecutive behaviour of the respiratory system in patients suffering from pulmonary and extrapulmonary ALI/ARDS imply different therapeutic approaches [112, 114, 131, 137]. As the real distending and thus potentially injurious pressure (P_{TRANS}) for lung tissue is, due to altered chest wall mechanics, markedly lower in patients suffering from extrapulmonary ALI/ARDS [118, 119, 131, 132], the application of higher airway pressures intending to increase/adjust V_T or PEEP may be advisable:

1. Adequate V_Ts are not only possible but may even be necessary to avoid atelectasis following too low tidal volumes [131, 138] as a relative large amount of the applied pressure will dissipate against the stiff chest [114, 131].

2. Specifically a higher PEEP will contribute to avoid cyclic reopening and collapse of the diffuse localized and unstable alveolar units [126, 139, 140] typically found as a result of the diffuse pulmonary oedema and the inflammatory cascade which originates outside the lung [141] in extrapulmonary ALI/ARDS.

The opposite conclusions could be drawn in cases of pulmonary ALI/ARDS [112]. Furthermore, based on the diverse chest wall and lung mechanics in both subtypes, recruitment attempts are more successful in extrapulmonary rather than in pulmonary ALI/ARDS [127, 132, 142]. Extrapulmonary ARDS patients demonstrate a bigger improvement in oxygenation when put into a prone position [143] attributed to regional changes in transpulmonary pressure resulting in lung density redistributions; however, in pulmonary ARDS, a more even distribution of ventilation will take a beneficial effect as well [144].

Of note, higher pleural pressures may compromise venous return and thus cardiac filling, hence resulting in lower cardiac output [145].

1.8 Pressures

Plateau pressure (P_{PLAT}) reflects the applied airway pressure during the end-inspiratory hold after inflation has finished and before exhalations starts [147], also called static elastic recoil pressure, as it largely reflects elastic and resistive properties of the respiratory system in ARF [117, 148]. The mechanical properties of the respiratory system are determined by its two main components, lung and chest wall, which are arranged in series, and their interactions [149]. Thus plateau pressure represents the sum of pressures required to inflate the lungs and to expand the chest wall [124]; hence the amount of tidal volume applied depends on P_{PLAT} [150]; on the other hand P_{PLAT} may be decisively influenced by chest wall mechanic properties [124, 151], as already described in detail above. Nevertheless, it is taken as an estimate of end-inspiratory lung distension [117], as such alveolar stretch is reflected by P_{PLAT} [152].

Increases in P_{PLAT} are associated with declines in respiratory system compliance (C_{RS}) and vice versa [153]. If comparing P_{PLAT} with peak airway pressure (P_{PEAK}) (see Fig. 1.3) in normal lungs, P_{PEAK} is found to be only slightly above P_{PLAT} [147, 154]. P_{PEAK} indicates the resistive characteristics of the respiratory system, specifically the airways during inspiratory flow [155]. If respiratory compliance decreases or V_T increases, P_{PEAK} and P_{PLAT} rise proportionately [147, 154]. Situations where P_{PLAT} remains unchanged while P_{PEAK} increases are indicative for increased total inspiratory resistance which includes tube and airways resistance and should lead to a check for airway obstruction [153, 154].

Of course, the real distending force of the lungs and hence alveoli is determined by the pressure difference between the alveolar pressure (P_{AL}) and, more generally expressed, the surrounding pressure (P_{SUR}), called transpulmonary pressure (P_{TRANS}), and thus is defined as [106, 108, 112, 156]:

$$P_{TRANS} = P_{AL} - P_{SUR}$$

at which the surrounding pressure is traditionally reflected by the intra-pleural pressure (also called intrathoracic pressure [157]), but esophageal pressure may, in turn, represent P_{SUR} as well [156]. Furthermore, we have to recognize that the primary determinant of pressure change is the change in lung volume, thus the amount of air volume inflated [158].

Unfortunately, as already described above, P_{PLAT} is dependent on chest wall and pulmonary elastic properties [124] and as such does not indicate the true alveolar distending force and strain resting on the alveolar units. P_{TRANS}, is a marker of ventilator-induced lung injury as it causes strain to the lung tissue [159]. Furthermore, Chiumello et al. analysed imaging data to show that the plateau pressure and tidal volume are imprecise markers to assess lung tissue stress and strain [124]. It is suggested that the anecdotally common sense upper limit of 30 cm H_2O P_{PLAT} which is thought to avoid relevant lung injury [67, 160–163] may be misleading, because alveolar distension may be overestimated and hyperinflation not recognized as chest and lung mechanical properties are not considered [114]. Thus, P_{PLAT} is challenged as being an inaccurate surrogate for lung distension and may be an improper guide to ventilator settings, particularly to avoid ventilator-induced lung injury (VILI) [119, 124, 126].

In order to use P_{TRANS} as the true distending pressure and hence proper indicator for stress and strain upon the small airways and alveolar units, a surrogate of the surrounding pressure needs to be determined and measured, and the proportional relation between lung and chest compliance/elastance has to be measured/calculated.

Alveolar pressure is reflected by the airway pressure (P_{AW}) and the total elastance (E_{TOT}) (sum of lung (E_L) and chest wall (E_{CW}) elastance), giving

$$P_{PL} = P_{AW} \times E_{CW} / E_{TOT} \; (1), \text{and thus}\, P_{TRANS} = P_{AW} \times E_L / E_{TOT} \quad [112].$$

E_{TOT} as the inverse of compliance can be obtained by calculating static compliance at end inspiration $C_{STAT}/C_{TOT} = V_T/P_{PLAT}$ –PEEP [ml/cm H_2O] [164], giving $E_{TOT} = (P_{PLAT}–\text{PEEP})/V_T$.

$P_{AW} = P_{PLAT}$ and can be measured, as well as pleural pressure. Unfortunately, P_{PL} is not routinely measured because the measurement is invasive [156]. Furthermore, pleural pressure is markedly affected by the pressure applied to the airways and by the proportion of chest wall elastance to total respiratory elastance [112]. Furthermore, pleural pressure shows regional differences during positive pressure application and will have the lowest increase at the diaphragm [165] and in addition a hydrostatic pressure gradient from posterior to anterior surface is recognized [156, 166]. As such, the measured P_{PL} will represent only regional conditions which might be quite different even in the near vicinity [156] and thus are not satisfactory to guide overall ventilator settings.

To overcome those shortcomings and particularly to adapt management to the true alveolar distending pressure indicated by P_{TRANS}, oesophageal pressures can be a reliable estimate of the P_{PL} [167]. Oesophageal pressures (P_{ES}) are generally accepted to reasonably estimate pleural pressure variations during the respiratory cycle [114, 168, 169], and Agostoni already verified in the 1970s a solid correlation between tidal changes in P_{ES} and P_{PL} applied to the lung surface and in consequence is a valid estimation of P_{TRANS} [170, 171]. However, there are a couple of constraints. The absolute P_{ES} values are less well defined and are influenced by positioning and abdominal pressure, e.g. obesity [172, 173]. The ability to gather global average changes are limited in the supine position and in asymmetrical lung disease [140, 174, 175] which is frequent in ALI/ARDS. P_{ES} reflects P_{PL} only at one locus of the pleural space, overestimating the pressure in non-dependent areas while dependent regions will be underestimated [176]. In addition, technical use with correct application, tracking and monitoring is quite challenging and requires specific training [175]. Hence, based on these problems and uncertainties and also some unexplained results of the Talmor study [167], the reliability of using measured oesophageal pressure to estimate pleural pressures upon which to base therapeutic considerations, particularly PEEP adjustment, has been heavily challenged [173, 177]. As the proportioning of E_{TOT} between E_L and E_{CW} is based on valid P_{ES} values, a considerable amount of uncertainty exists and has to be taken into therapeutic considerations. P_{ES} is not indicative of local gradients often attributed to abdominal pressure changes [135, 178] and because it reflects more the regional environment rather than the average intrathoracic (intrapleural) pressure (ITP) [114] (as assumed for calculation of chest wall to respiratory system elastance ratio [179]), it is insensitive to diverse lung volumes at end expiration [178, 180]. Furthermore, even tidal volumes may be subject to lung and chest wall elastance resulting in completely different P_{TRANS} influenced by the different and disease-related uneven elastances of lungs and chest wall [112, 114, 175, 181, 182].

For the P_{TRANS}, a physiological upper limit of 25 cm H_2O which optimizes alveolar recruitment is postulated [119, 127, 183]; however, Grasso recently favoured 27 cm H_2O [184] demonstrating that at present there is no evidence-based safe limit. However, although transpulmonary pressure as a surrogate of P_{PLAT} might be the more physiological marker of alveolar distension and with which to adapt respirator settings [114], the plateau pressure represents the inspiratory alveolar pressure at end inspiration (measured during an inspiratory pause (occlusion manoeuvre) if there is no air/gas flow [113]) and as such is considered as an estimate of the end-inspiratory lung distending forces which put stress on the alveolar units potentially causing parenchymal injury [117]. The magnitude of P_{PLAT} depends on both, compliance and V_T [146]. Furthermore, plateau pressure monitoring seems to be indispensable. Recent study results explicitly prefer P_{PLAT} as an indicator to guide ventilator settings, since P_{PLAT} may provide integrated information about lung and heart function and their interdependence [59, 61, 185, 186]. As such, P_{PLAT} emerged as being a more than reasonable parameter to coordinate artificial ventilation, lung protection and associated circulatory features, especially right heart function. Accordingly, it is frequently suggested that a P_{PLAT} of $\leq$26 cm H_2O should be aimed for in order to achieve the currently best therapeutic outcome [59, 61, 185, 187]. The pressure limit of $\leq$30 (−35) cm H_2O as recommended until recently has been more or less arbitrarily [188] defined [20, 49, 189]. Remarkably, even quite different clinical settings and conditions, such as asthma compared to ALJ/ARDS, with a different management approach achieved exactly the same cut-off level at 26 cm H_2O [190].

Of course, an individual approach is warranted, e.g. in case of significant abdominal hypertension or compartment syndrome, higher levels are acceptable [112, 132, 133]. Optimum PEEP application and titration is still a highly controversially area [191–193], and individualized management is achievable if a coordinated, well-balanced approach is taken, integrating both respiratory and circulatory requirements, using a max plateau pressure of less than 26 cm H_2O. This is a valuable and reliable indicator, even more if additionally supported by haemodynamic information, preferably obtained by echocardiography [194, 195]; see further details below.

1.9 Ventilator-Induced Lung Injury (VILI)

Ventilator-induced injury (VILI) evolves due to lung strain characterized by a change in lung volume from resting volume which adversely stretches and insults lung tissue [124, 196]. This may be further specified by:

(a) Barotrauma is attributed to excessively high airway pressures commonly applied in the early days of mechanical ventilation [197, 198] but far less common now due to pressure limitation, typically presents by causing air leaks with air entry into the pleural, soft tissues like the subcutis and infrequently mediastinum [199, 200].
(b) Volutrauma occurs as the result of excess stress at end inspiration. Regional overdistension/inflation of the most compliant alveolar units is attributed to too high volumes applied and mismatched air distribution in a highly heterogeneous lung parenchyma due to variable regional compliance [201]. This causes the lung/areas of the lung to deform above its resting volume exerting stress on alveolar epithelial and adjunctive capillary endothelial cells [65, 122, 124, 202, 203].
(c) Atelectrauma occurs following cyclic and repetitive closing at end expiration and reopenings during inflation [204, 205], termed recruitment–derecruitment. Unstable alveolar units emerging in injured lungs [206] are subject to substantial shear forces on the alveolar tissue resulting in functional and structural alterations [199, 201, 207, 208]. This can be attenuated or even prevented by setting an adequate PEEP [65, 209, 210].
(d) Biotrauma is associated with the release of inflammatory agents triggered by mechanical injury. This may aggravate local tissue damage and contribute to pulmonary and remote organ dysfunction [65, 122, 204, 211–213].

Artificial ventilation, applying pressure negative or positive, to the airways [40], is associated with a couple of potentially adverse effects termed ventilator-induced lung injury, VILI [188, 214]. Due to the complexity of the interaction between externally applied inflation and the body, we are still facing significant problems understanding it in principle [137, 215], but particularly facing difficulties in managing individual cases [167, 173].

Already in healthy, spontaneously breathing humans, physiological gas volume distribution is complex with significant regional variations [216]. This is even more complex in diseased lungs which show a highly inhomogeneous tissue anatomy with areas of normal lung tissue in close vicinity with regions of highly altered morphological structure and hence physiology [123, 201, 217–220]. Furthermore, the pathophysiology varies throughout the disease sequence and needs to be taken into account [137].

Injurious ventilation provokes markedly diffuse structural alveolar damage with the development of pulmonary oedema from initially perivascular cuffing progressing to interstitial oedema and finally alveolar flooding [198, 221]. This is largely following increased microvascular/alveolar permeability due to high volumes inflated [202, 222–224] and is a hallmark of VILI [188, 202]. It is recognized that endothelial cells will be focally separated from their basement membrane, and regions with disrupted epithelial cells and destroyed type 1 pneumocytes exist [65, 202]. Moreover, the associated activation and recruitment of inflammatory cells, the immune response, with the initial local production of inflammatory mediators and their consequent systemic overspill, complete the picture [225–230]. This results in quite heterogeneous morphology and lungs with a total reduction in available lung parenchyma and termed "baby lung" by Gattinoni [220]. There are atelectatic areas in very close neighbourhood to still open alveolar units. Functionally this is depicted by regionally diverse and overall substantially impaired lung compliance, huge pressure differences in nearby units [198], ventilation–perfusion mismatch, shunting and dead space ventilation [123, 201, 211, 231], and there may be systemic multiorgan failure promoted by the biotraumatic response [228–230, 232, 233]. The described morphological and functional features are not specific for VILI and are recognized in other entities such as ALI/ARDS as well [215, 234].

Barotrauma was the first mechanism discovered causing harm during mechanical ventilation [197, 214], referring to the association between airway pressures applied and air leak [235, 236]. As adjacent structures, alveoli and terminal bronchioles share common walls (known as interdependence), high pressure gradients between alveolus and bronchovascular sheath will provoke ruptures and thus air leaks at those junctions enabling air to pass along the bronchovascular sheath. This can penetrate into the mediastinum and subcutaneous tissue and further into the pulmonary interstitium causing pulmonary interstitial emphysema or into the pleural space with pneumothorax being the most common form [200]. Occasionally, pericardial or peritoneal space air leaks occur either by tracking or if ruptured there [197, 214, 215]. During those early years of long-term mechanical ventilation, up to 39 % of all ventilated patients suffered from barotrauma [237]. Pressures applied now have been significantly lower for a long time already, although air leaks are still reported in up to 8–14 % of ventilated patients [30, 238, 239]. Importantly, more recent studies from the end of the 1990s could only demonstrate a weak relationship between airway pressure/tidal volume and air leaks [30, 215, 239, 240]. Although today the applied pressures are clearly lower (recommended P_{PLAT} <30(−35) cm H_2O [49, 188, 241]), locally very high pressures at greater than 140 cm H_2O can be generated by the interdependence of traction forces of neighbouring expanded units and the walls of collapsed alveoli (re-expansion pressures [215]) and as such promote barotrauma [199, 240].

The correlation between airway pressure and consecutive tidal volume (with increasing airway pressure applied, tidal volume should normally increase) and air leaks/barotrauma is disputed and often challenged [240]. The seminal study by Dreyfuss [202] confirmed [65, 222, 242, 243] that it is the degree of lung inflation and thereby the distending stress generated/excessive strain that affect the alveoli and alveolar ducts causing tissue injury rather than the airway pressure per se [202, 211, 214, 215]. Thus, it became evident that inappropriate volume rather than pressure may injure lung tissue in mechanical ventilation, a mechanism termed *volutrauma* [215, 228, 244]. Alveolar epithelial and vascular endothelial cells of the alveolar–capillary unit are exposed to cyclic stretch by the air volume inflated during mechanical ventilation [245–248]. Accordingly, alveolar over-distension and overinflation following high volume expansion during mechanical ventilation, particularly of the still healthy alveoli which are preferentially aerated as the most compliant alveoli and lung areas [201], are the contributing insults of VILI [203, 205, 249]. Even relatively low tidal volume ventilation may cause significant regional over-distension in the heterogeneously diseased lungs (e.g. baby lungs) [115, 250]. It is important to recognize that microvascular alterations (endothelial cell injury and altered transmural microvascular pressures) at the alveolar–capillary interface result from overinflation and are significantly involved in the disease process, e.g. disturbed permeability and activation of inflammation [222, 248, 251–253]. Thus, as large volume rather than high intrathoracic pressure per se may result in VILI [254], and as airway pressure is not a determinant of pulmonary oedema [202], some authors suggested to replace the term barotrauma by the more accurate term volutrauma; however, both are closely related in daily practice [211, 215, 255]. Alveolar distension is determined by the pressure gradient across the alveoli and not by the alveolar pressure alone [188, 215], which can be estimated by the transpulmonary pressure [215]. The absolute transpulmonary pressure is responsible for injury [126, 136], and might differ substantially depending on the contribution of chest wall properties to total elastance of the respiratory system as described above.

Too low lung volumes, more precise absolute lung volumes described as ventilation with relatively low V_T and/or insufficient PEEP [256], rather than low tidal volumes tend to provoke atelectasis [256] and are demonstrated in animal models as being potentially injurious [256, 257]. However, those results have never been confirmed in humans; moreover, recent publications are quite encouraging to use lower than the recommended 6 ml/kg (PBW) tidal volumes in the most severely respiratory compromised and difficult to oxygenate patients. Providing a sufficient PEEP is applied and thus attenuating a too low total lung volume associated with cyclic collapse of alveolar units and consecutive recruitment–derecruitment [215, 254, 258].

Atelectasis characterized by a state of airless alveoli due to prolonged collapse [206] is a hallmark of injured lungs [259], and atelectrauma arises in mechanically ventilated injured lungs as the result of repeated cyclic closure and reopening of those unstable, collapsible alveolar units triggered by shear stress injury [204–206]. Due to its high prevalence, atelectrauma appears to be a critical feature in VILI [205].

The atelectatic or partly atelectatic units may open up during inspiration or the inspiratory hold (plateau phase); unfortunately, most of those cyclic reopened alveoli will collapse during/after expiration again, if not prevented by applying PEEP [188]. Reopening or recruitment of collapsed alveoli which thereafter re-collapse (derecruitment) while exhaling/at the end of expiration has been demonstrated exerting marked shear forces on the tissues, causing alveolar epithelial disruption and loss of barrier function or atelectrauma [199, 201, 207, 208, 260]. As in mechanical ventilation, repetitive identical ventilation induces more damaging shear stress than time and volume variable ventilation [261, 262]. Compromised surfactant factor production and function [201] as found in injured lungs [263] substantially contributes to the collapsibility of the alveolar units, and measures promoting surfactant production consequently, like the application of PEEP, improve alveolar stability by ameliorating surface tension of the alveolar unit [201]. Moreover, avoidance of repeated collapse and reopenings by setting an adequate PEEP are shown to be protective against VILI [65, 198, 209, 210].

Recruitment manoeuvres (application of higher pressures to open up collapsed and stiff alveolar units and lung areas) have been proposed and applied for a long time [127, 264–266]. Unfortunately, no prospective randomized trial could verify any survival benefit following the application of recruitment manoeuvres in patients with ARDS [267]. Consequently, there is no consensus as to when and how to do so [206], and their validity is only based on experimental models [268].

Mechanical ventilation is associated with the release of biologically active mediators, with the inflammatory ones of special interest and probably critical importance in the disease process [212, 214, 229, 269]. Mechanical ventilation may expose the alveolar units to shear stress attributed to cyclic collapse and reopening of atelectatic areas and unstable alveoli [207, 208] as does volume inflation by putting strain on the terminal airways and gas exchange elements during inspiration [211, 214]; both processes disrupt pulmonary architecture leading to proinflammatory signalling by lung cells [122, 229, 269]. Mechanical strain, such as overdistension [270, 271] and cyclic strain [272, 273], is shown to cause the release of a variety of mediators from cells of the alveolar unit including alveolar epithelial cells, macrophages, smooth muscles cells and endothelial cells of the alveolar–capillary network [212, 272, 274–276]. To specify this association between injurious mechanical ventilation and molecular- and cellular-mediated [214, 277, 278] release of a range of highly active, largely inflammatory mediators, the term biotrauma was coined [228]. Clinical trials applying lung protective ventilation provided clear evidence of attenuated mediator release [279, 280], particularly cytokines, less organ dysfunction [280, 281]and lower mortality [161, 280], confirming the role and contribution of biotrauma in VILI [214, 269]. These mediators released may moreover aggravate lung injury and may promote extrapulmonary multiorgan dysfunction [214, 232, 233].

Results from trials in ventilated animals are highly suspicious for conventional ventilator settings, high V_T and low PEEP used until the early 2000s, promoting the release of proinflammatory mediators locally into the airspaces and even into the circulation [212, 227, 276], potentially aggravating VILI. Further evidence confirming their impact on the development/course of VILI was revealed by several experimental studies [282–284]. Otto et al. demonstrated that lung protective ventilation attenuates regional distribution of inflammation [285], and de Prost reported a reduction in and heterogeneity of regional pulmonary inflammatory cell metabolic activity in early ALI if protective ventilation is applied [286]. Hence, protective mechanical ventilation facilitates keeping inflammation locally/regionally compartmentalized [287]. All in all, ventilator settings can beyond doubt aggravate VILI as depicted in detail above.

It is not only commonly accepted [271] that over-distension of the alveoli may cause damage of the alveolar lining cells, potentially resulting in a local and systemic inflammatory response (SIR). Even if the lungs are not primarily infected [280], we have more evidence to suggest that mechanical ventilation may promote MODS. As demonstrated by several authors, a disrupted alveolar–capillary barrier with consecutively enlarged permeability allows proinflammatory mediators to leak from the alveolar space (compartmentalized) into the circulatory system which an intact alveolar–capillary interface would have prevented, and thus releasing the normally compartmentalized inflammatory substances into the bloodstream [288, 289]. In ARDS, compromised alveolar–capillary barrier function is the consequence and a typical functional shortcoming of the severe diffuse alveolar damage seen in that disease [290, 291]. Limited V_T (6 ml/kg PBW) ventilation accompanied by appropriate PEEP can localize proteinaceous pulmonary oedema and bacteria [292–294], while traditional ventilator settings (high tidal volumes and no or improper PEEP) may promote contralateral bacterial spreading or even translocation from airspace into the bloodstream [292, 293] as well dispersion of alveolar fluids within the lungs [294, 295]. In addition, mechanical ventilation is possibly shown to affect remote organs; increased ileal permeability [296], systemic capillary leak [233, 297] and increased renal and small intestine apoptosis [298], as well as changes in peripheral immune response and host susceptibility to infection [214], are reported in the literature.

There is evidence that modern, protective ventilator settings do not necessarily cause a systemic inflammatory response [226] and furthermore injurious mechanical ventilation might not per se deploy a strong inflammatory response with systemic overspill and consequent multiorgan dysfunction. Some authors consider two or more coexisting or consecutively occurring insults, one being mechanical ventilation and the other the need for mechanical ventilation sensitizing the lungs (the initial injury), a second hit hypothesis initiating MODS [213, 299].

Inspiratory flow may also have an impact on VILI development as most recent literature suggests. High inspiratory flow is believed to reinforce transmission of kinetic energy to lung tissue to distort lung parenchyma and bronchial epithelial cells and can augment shear stress albeit parallel to airway and alveolar walls [300]. As such, high inspiratory flow was shown to promote morphological damage [301], which may worsen gas exchange [302], whereas if it is limited, it mitigates VILI

[303]. Further clarification is warranted as currently high, but decelerating flow delivery is often applied [53, 304, 305].

1.10 PEEP

Positive end-expiratory pressure, PEEP, refers to a pressure at the end of the passive expiration still operating on the airways, keeping the transpulmonary pressure above atmospheric level, as physiologically airway pressure will equilibrate with atmospheric pressure at end expiration. PEEP may be applied externally (PEEPe, usually termed PEEP) during mechanical ventilation [306], but may arise internally (called intrinsic or auto-PEEP (PEEPi)), reflecting incomplete emptying of the lungs [110] still leaving a fraction of the precedent tidal volume in the airways not reaching the static equilibrium volume (FRC) before/while the next inspiration occurs [307–309]. This exerts a positive recoil force on the airways [310–312] as may be provoked by interrupted or unfinished expiration before its natural end [306, 312], too short expiration times as observed in tachypnoea and obstructive lung diseases [313, 314] or applied by the ventilator settings (too high respiratory rate, inverse ratio ventilation) [306, 311, 312]. The inability of the lungs to return to FRC level at end expiration, leaving air absorbed or trapped, in airways and alveolar units is termed dynamic hyperinflation [311, 315]. Dynamic hyperinflation is influenced by V_T, expiratory time, resistance and compliance [316, 317]. Due to this trapped air, the physiological equilibration between airway pressure and the atmosphere, e.g. mouth/nose or tube at end expiration, cannot occur [318, 319]; a pressure gradient at the end of that incomplete expiration persists, called intrinsic PEEP [314, 320]. Auto-PEEP is particularly to be found in patients with high airway resistance to flow (airway obstruction) following further airway narrowing attributed to worsened airway oedema, inflammation and/or bronchospasm and excessive secretions during acute disease exacerbations accompanied by altered elastic recoil and a high tendency of the airways to collapse [313, 319, 321] as in COPD or asthma [53, 312, 318, 322, 323]. If with each respiratory cycle a small amount of air volume is not exhaled, the volume left will continuously grow and the lung gets progressively hyperinflated, commonly found in patients with obstructive lung disease [312, 319]. PEEP, as does positive pressure mechanical ventilation [324, 325], will increase intrathoracic and thus pleural pressure [326, 327].

The application of PEEP in mechanical ventilation is well established [328, 329], exerting several beneficial effects on the lungs, and thus is a powerful tool in the management of acute and chronic respiratory failure.

Beneficial effects of PEEP may include

Increases/restores FRC [330–332] – proportional to increasing PEEP level [331] by decreasing collapse of the small airways and limits atelectasis [333]

Increases static lung compliance [331, 334] – but only with lower (5–8 cm H_2O) PEEP levels [335]

Recruits and stabilizes lung units (insulted alveoli) – and thus eliminates/reduces cyclic recruitment –derecruitment of alveoli [198, 202, 210] and the amount of atelectatic areas/units [217, 287]

Attenuates pulmonary shunting [328, 336]

Shifts the compliance curve – the actual working point of pressure–volume curve to its linear, steeper part where volumes can be generated for less pressure applied [337]

Reduces work of breathing [79, 338]

Redistributes fluids – within the alveoli to the interstitial space although without lowering the total amount of lung water[1] [339, 340]

May reduce chest wall stiffness – if PEEP is high enough [131]

Attenuates the release of inflammatory mediators – (see biotrauma below) [281, 295]

Mitigates lung injury – both in low and high PEEP setting [214]

Overall, improved arterial oxygenation will be accomplished due to contributions from several factors depicted above including enhanced FRC resulting from recruitment of collapsed alveoli [334, 350] and mitigation of alveolar oedema [351], reduction in shunting [217], stabilizing of damaged alveoli [217], by modifications in surfactant [352] and alterations in alveolar microhaemodynamics [222, 353].

PEEP in general leads to an increase in interstitial fluid by facilitating peripheral transcapillary fluid flux and it also affects microcirculation [354–356]. Further, unfavourable pulmonary effects of PEEP include:

Increase of dead space – the PEEP set may contribute to barotrauma by over-distending healthy, compliant alveoli [328, 334].

Parenchymal lung injury commonly is heterogeneous – suitable PEEP level required to achieve the above-mentioned positive effects may differ markedly from region to region, even in the near vicinity; thus the PEEP set may be adequate for one area but deleterious (e.g. over-distension) for another, and optimizing PEEP is very challenging [216, 357].

Relative contraindications to the use of PEEP exist. It should not be applied in lobar disease, unless essential for oxygenation [137]. Furthermore, PEEP might be contraindicated in hypovolaemic patients as it may further compromise blood pressure [358] and may decrease cerebral perfusion in raised intracranial pressure [359]. In pulmonary embolism, PEEP may worsen the ventilation–perfusion ratio [360] and is contraindicated in pneumothorax in the absence of a drain, and it should be avoided in bronchopleural fistula [328].

[1]Positive pressure ventilation and PEEP both affect the lung lymphatic drainage [201] by compressing the thin-walled lymphatic vessels/ducts, impeding flow [341, 342]. In expiration with decreasing pressure, recovery of flow occurs but will be attenuated by PEEP [343]; further the elevated CVP following increased intrathoracic pressure provides a hydrostatic barrier to lymphatic flow [344], leading to an increase in total lung water [345–348]. PEEP merely contributes to expel fluid from the alveoli to the interstitium where it will be retained due to the reduction of drainage by the thoracic duct [346, 347, 349].

There are haemodynamically relevant effects caused by PEEP application to be discussed; these can be positive or negative and depend on the overall situation:

Increases in intrathoracic (pleural) pressure may impede venous return and hence cardiac filling [328, 361], potentially provoking a fall in CO, but diluted venous return is not the only mechanism involved in the possible reduction in CO.

Increases pulmonary vascular resistance [362] and decreases pulmonary blood flow [363] due to the increase in lung inflation following the increased alveolar pressure resulting from PEEP application [252].

Increases RV outflow impedance [324, 364] and thus RV afterload [195, 365–367], but may decrease elevated RV outflow impedance following recruitment of atelectatic lung tissue [156, 368]; read more below (cardiovascular effect of positive pressure ventilation).

Decreases LV afterload [195, 337, 369, 370], a desired effect in LV failure [328]. However, LV filling may be compromised following the rise in RV afterload as with increasing PEEP levels and concomitant shift of the interventricular septum to the left; LV filling will be diminished and consecutively CO [186, 371].

Diminishes RV compliance [328, 372]; the effect on LV compliance is inconclusive; the majority of studies demonstrated a reduction [373, 374] while others found no affect [375, 376] or confirmed an affect only in circumstances of very high PEEP levels and/or coexisting RV ischaemia [377]. Potentially the assessment methods are failing and not precise [378].

Contributes and may promote the development of hypotension and hypoperfusion in hypovolaemic patients [328].

Impairs hepato-splanchnic perfusion [201, 379, 380] on the one hand, jeopardizing gut barrier function allowing for bacterial translocation and the second hit [381, 382] and the other a blunted hepatic metabolic function [383].

Impairs renal perfusion, putting the kidneys at risk for renal failure [384].

Reduces cerebral perfusion [334].

Since pleural pressure is directly transmitted to the pericardial space [385], every increase in transpulmonary pressure following PEEP application (or positive pressure ventilation [386, 387]) is accompanied by a parallel rise in pericardial pressure (PP), potentially causing diastolic ventricular interaction (DVI). Sharing the interventricular septum and being enclosed by the same pericardium, interactions between the right and left ventricle occur [388, 389]. DVI describes the direct influence of changes in volume of one ventricle on the other due to the influence on the compliance of one ventricle by changes in volume, intraventricular pressure and compliance (or a combination) of the other ventricle [390–392]. As such, an increase in PP will exert a constraining effect on both RV and, to a lesser extent, LV [186], resulting in RV dilation, pulmonary hypertension and even acute cor pulmonale, accompanied by LV dysfunction probably resulting in low cardiac output and circulatory failure [186, 393–396]. However, relatively low PEEP levels ($\leq$8–10 cm H_2O) are demonstrated having beneficial effects on pulmonary haemodynamics and do not compromise RV function significantly even though PP has increased consecutively to an increase in pleural pressure [397, 398]. In contrast, PEEP levels above 10–12 cm H_2O will generate some degree of DVI and thus RV dysfunction [186]. Interestingly, in LV failure, PEEP by decreasing LV afterload has a well-established beneficial effect on LV function and is a very useful good therapeutic option (CPAP) [399–402].

1.11 Cardiovascular Effects of Positive Pressure Mechanical Ventilation (PPMV)

Positive pressure mechanical ventilation (PPMV) is a non-physiological fashion to respirate [368] and may, by changing lung volume and intrathoracic pressures when positive pressure is periodically or continuously applied to the airways [403], considerably affect the cardiopulmonary haemodynamics and the systemic circulation [195, 368, 404, 405]. While lung volume will increase during inspiration in both spontaneous respiration and PPMV, the intrathoracic pressure (ITP) decreases during spontaneous breathing due to contraction of the respiratory muscles [368]. Conversely ITP will increase during mechanical ventilation due to passive lung expansion following increases in airway pressure [195, 358, 368]. Hence a simultaneous increase in lung volume and transpulmonary pressure (transpulmonary pressure = alveolar (airway) pressure – extra-alveolar pressure [106], at which the latter is equivalent to ITP [368]) is seen in PPMV [195, 358]. However, the increase in lung volume during inflation rather than the airway pressure determines the change of ITP in PPMV [406]. The affliction of the heart and the circulation, preferentially the pulmonary circulation, may be even a relevant factor in mortality as the results of the ARMA study suggest [60, 407].

The lungs and the heart share one cavity, located and linked in series within the circulation. Both being distensible and elastic, they are able to transmit forces to each other and their surroundings, respectively responding to changes of the other organ. Thus they are closely cross-linked and interact with each other, this is termed the heart–lung interaction [368, 408]. During spontaneous inhalation, the pressure gradient, being the driving force of the venous blood return, between the right atrium (RA) and the central abdominal venous system increases due to the even more negative ITP created by the expansion of the thoracic cavity. This lowers the RA pressure, and there is an increase in the abdominal pressure following contraction of the diaphragm [368, 409, 410]. The rise in that extra- to intrathoracic pressure gradient causes an inspiratory increase in venous return and thus RV preload, RV filling and consequently RV size [368]. As the right and the left ventricles among themselves are very closely interconnected and cross-linked, changes of the volume of one ventricle will directly influence the volume of the other, referred to as ventricular interdependence [390]; thus filling of the left ventricle decreases [368, 410]. Subsequently, LV–SV and CO may decrease during inspiration [368]. The reverse is seen during expiration [410], and the inspiratory increase in ITP may augment LV afterload [409].

On the airway side, in spontaneous breathing the negative pressure gradient between mouth and alveoli creates the airflow [106, 108], while during PPMV, a positive pressure higher than the actual end-expiratory alveolar pressure is applied to the airways during inspiration, generating a gradient between tube and alveoli necessary to facilitate airflow [368]. Thus, the pressure conditions are entirely opposite to the ones in spontaneous breathing [195], and the heart–lung interactions may readily become apparent in PPMV [408].

It is general physiology that the amount of air volume effectively inflated upon the pressure acting on the alveolar units, the transpulmonary pressure [368], is largely a function of pulmonary and chest wall compliance and the resistance to airflow [110, 156, 368]. Hence, the size of tidal volume applied in PPMV and the associated increase in ITP are to some extent related to the compliance of the respiratory system and the resistance to airflow [156]. Accordingly, as the increase in ITP with subsequent reduction in venous return is, in part, dependent on total lung volume and thereby on the change in lung volume during inspiration and as the RV preload is widely determined by ITP, the circulatory alterations will, to some extent, rely upon the lung volume applied [156, 406]. This is a very important interaction to be aware of and has to be considered when choosing ventilator volume and pressure (airway pressure and PEEP applied) settings [60, 368]; see below.

In PPMV as *an effect* of the *heart–lung interaction* (following the cyclic increases in ITP and lung volume [406, 411]), which is much more pronounced due to the overall complete change in pressure conditions during mechanical ventilation compared to spontaneous breathing [195], total pressure levels are significantly higher, and *marked reductions* in RV filling (preload, ↓ RVEDV) predominantly related to the reduction in venous return are found during inspiration [326, 375, 412, 413]. Whether the main underlying mechanism is a reduced pressure gradient between central venous vessels and the RA, as still is our basic view [156, 368], or an increase in venous resistance and venous vascular tone, and as such a reduced conductance compromising the venous blood return, remains disputed [409, 414, 415]. Furthermore, compression of the heart in the cardiac fossa by the mechanically ventilated lungs has been demonstrated to lower RV and LV filling and output [368, 416]. Thus PPMV with its cyclic positive airway pressures and/or continuous PEEP may cause an increase in ITP, impeding venous blood flow into the thorax with consecutive reduction of intrathoracic blood volume and pulmonary blood flow [156, 363, 372, 417, 418]. This reduced filling, in turn, will result in a diminished RV–SV following the Frank–Starling mechanism [419, 420]. A significant increase in RV output impedance found in a considerable amount of patients may aggravate the reduction of impaired RV output [324, 364]. However, this inspiratory reduction in RV–SV affects the LV after a few heart beats, generating a reduction in LV preload (↓ LVEDV) [421–423]. Consequently, LV–SV is reduced [421, 423, 424], and this takes effect during expiration. As the ITP will be markedly elevated – compared to spontaneous breathing – during PPMV, LV–EF will be diminished despite an attenuated LV afterload [425].

Conversely, during expiration, the opposite occurs: increased venous return will result in increased RV and few beats later, LV filling with slightly meliorated CO [422, 424, 426]. The dynamic swing in LV–SV and thereby arterial pressure variations during PPMV are currently gold standard in predicting response to fluid administration and may guide fluid management [427, 428], as described in detail in Chap 3 (shock) and Chap. 5 (AKI).

PPMV or continuous PEEP directly affects RV outflow impedance due to alterations in pulmonary vascular resistance (PVR) subject and related to changes in transpulmonary pressure [362, 372, 429–431]. Transpulmonary pressure directly correlates with RV afterload [324, 432], and since transpulmonary pressure rises with PPMV (and increasing tidal volumes [324, 432]) and PEEP application, consecutively RV outflow impedance will increase [324, 325, 421], potentially promoting the development of RV dysfunction or even failure [393, 396, 431]. Furthermore, with increasing tidal volumes, the RV has to generate more pressure to eject blood into the pulmonary circulation [432] since the raise in RV impedance, an essential determinant of RV afterload [368], correlates with the extent of tidal volume applied by PPMV [324]. Again, it is the lung volume rather than the pressure applied, thereby the change in lung volume affecting heart and pulmonary circulation [368, 404].

Moreover, PVR is confirmed as being related to lung volume (dependent upon the actual transpulmonary pressure [368]), in a bimodal fashion with the lowest resistance found at the level of the functional residual capacity (FCR), exponentially increasing with increasing over-distension but also exponentially increasing with reductions in lung parenchyma, e.g. atelectasis, available for ventilation [362, 433]. The latter is commonly found in the "baby lung" of ALI/ARDS [395]. This specific behaviour is hypothesized as being addressed to two different types of intra-parenchymal vessels: Extra-alveolar vessels are linked to the traction forces of expanding lung areas during inflation and thus are compressed by atelectatic and collapsed alveoli, while the intra-alveolar capillaries are, in turn, compressed and stretched by lung volume over-distension in compliant alveolar units; both effects create an increase in PVR [156, 368]. As long as the available lung volume lies beyond the FRC as in severe lung injury, extra-alveolar vessels will predominantly determine the PVR; in case lung volume exceeds FRC as in hyperinflation or in primarily regional over-distension, intra-alveolar vessels will preferably determine the level of PVR [364]. The static compliance in normal lungs correlates well with the pulmonary vascular resistance index [434].

The former common application of high tidal volumes (12 ml/kg) often requiring high positive airway pressures, with associated disproportional rises in transpulmonary pressure, caused considerable increases in PVR [368]; consecutively in up to 60 % of cases, acute RV dysfunction or failure had been recognized in ARDS patients [365]. Other factors such as imbalances in pulmonary vascular tone regulation [365, 435, 436], endothelial injury [437], hypoxic vasoconstriction (Euler–Liljestrand reflex) [438, 439] and microthrombi [440, 441] may in addition influence the PVR and thus RV afterload and are frequently found in critically ill patients [365]. In contrast, in spontaneous breathing humans, the changes of PVR throughout the respiratory cycle are small, less than 15 % and do not affect the RV performance [442].

The haemodynamic consequences attributed to the described increase in PVR following PPMV and PEEP are mediated by ventricular interdependence. As superficial myocardial fibres are surrounding both RV and LV, and both ventricles are sharing the pericardium and the interventricular septum, they are effectively arranged in series [387, 443]. Interactions occurring are referred to as to diastolic ventricular interaction (DVI), as the interaction functionally mainly occurs in diastole, or ventricular interdependence [388–390]. DVI describes the influence of one ventricle on the other, as such changes in the end-diastolic volume of one ventricle will directly influence the volume, intraventricular pressure and the compliance of the other ventricle [390, 444]. Particularly acute changes may affect the other ventricle dramatically [388, 445]. Under physiological circumstances similar end-diastolic volumes are seen in the RV and LV [194, 446]. However, due to the constraining effects of the pericardium an increase in the cross-sectional area of one ventricle, e.g. due to volume loading or enlargement, necessarily reduces the area of the opposite ventricle, resulting in less filling volume there [447, 448], while the total cardiac filling will remain unchanged [447, 448]. Simultaneously an increase in pericardial pressure is demonstrated, particularly when the right ventricle is primarily affected [447, 449]. Increasing pericardial pressures (PP) due to enlargement of the right ventricle related to raised afterload in the setting of an increased ITP as in PPMV [324, 386] (with the ITP directly transmitted to the PP [385]) will progressively exert restraining effects on the ventricles [447]. The thin-walled RV, not being able to substantially counteract this external pressure as the LV can to some extent, will be more affected than the LV; hence a disproportionally higher increase in RVEDP than in LVEDP is shown [444, 450]. Thus, the pericardium plays a key role in the ventricular loading conditions and that particularly in the acute situation [445, 451]. An increase in RVEDV as found in case of elevated RV afterload (acute pressure load) and the accompanied marked rise in RVEDP shifts the interventricular septum towards the LV cavity during diastole [396]. This is subject to the restrictions imposed by the acutely non-distensible pericardium as RV cavity size increases [450, 452]. Although the venous return may be impaired as illustrated above, the substantial rise in RV afterload will inevitably result in acute RV dilation [152, 396, 453] as the thin-walled RV may only by cavity enlargement (via Frank–Starling mechanism) match the required pressure increase necessary to eject blood against the augmented resistance in the pulmonary circulation [365, 396, 453]. Further, due to the high resistance of the pulmonary vasculature, the RV may not empty as much as before; hence end-systolic right ventricular volume will increase [454], contributing to the compromised haemodynamics. Accordingly, the consecutive leftward push of the interventricular septum (typically of a flattened or even concave silhouette at end diastole [61]), and (predominantely) the constraining effects of the pericardium, reduce the transmural LVEDP,[2] the effective LV distending pressure, and elicit a substantial decrease in end-diastolic LV filling and LV size [450, 452, 455]. This in turn results in an extensively reduced LV–SV [419, 456]. The altered LV compliance attributed to the reported changes and mechanisms (mainly RV enlargement, septal shift and increasing PP [156]) and thus the changed diastolic LV properties [394, 457] may contribute to the attenuated LV filling and in consequence diminished LV output and compromised systemic circulation [452, 457, 458]. Therefore, RV dysfunction/failure potentially displayed in PPMV may have marked consequences on LV function and the systemic circulation [455, 459].

---

[2] Transmural LVEDP = $LVEDP_{Intracavitary}$ – surrounding pressure [454], in this case PP, reflected by the RAP [455] with PP considerably elevated under the described circumstances.

Volume loading and fluid administration under those circumstances may further increase RVEDV, and consecutively RVEDP disproportionately more than LVEDP, concomitantly a reduction in LV–SV occurs [449, 460, 461]. This is a real challenge as any further increase in RVEDV, e.g. by fluids, may cause more deterioration in haemodynamics [449, 460, 461]. However, both hypovolaemia and hypervolaemia may reduce CO [462–464]; volume administration may at least in part attenuate the reduction in preload [186]. An initial fluid bolus may be beneficial in acute myocardial infarction with significant involvement of the right ventricle [465] or in acute pulmonary embolism [466].

Anyway, accurate monitoring essential [467, 468].

PEEP, particularly if inappropriate, may critically contribute to/aggravate the reduction in RV preload and the intrathoracic blood volume [145, 156, 414, 469] and to the elevation of RV afterload [156, 195, 324, 470]. As such, even in expiration the intrathoracic venous return may still be inhibited [201] and venous pulmonary flow can still be blunted as a PEEP above 15 cm H_2O may exceed alveolar–capillary pressure and thus impede pulmonary blood flow [471]. Furthermore, as cyclic tidal ventilation or continuous PEEP precipitate an increase in ITP and hence pleural pressure resulting in a reduction in effective RV compliance impeding RV filling, RV filling may be even more limited if PEEP is applied [372]. PEEP is shown to potentially and considerably enhance the afterload of the RV [156, 324, 395, 421, 472], in addition to the effect of positive pressure during PPMV which causes a rise in RV outflow impedance [431]. However, study results regarding the effect of PEEP on RV afterload are inconsistent, as even with increasing PEEP levels reductions in RV afterload have been demonstrated as well [395, 434, 473, 474].

PEEP is basically applied to avoid cyclic alveolar recruitment and derecruitment of disease-related collapsed alveolar units and small airways as repetitive closure and reopening of those atelectatic alveoli throughout the respiratory cycle has been identified as a mechanism of ventilator-induced lung injury (atelectrauma) [65, 204]. Recruitment of atelectatic lung parenchyma by temporarily applying high airway pressures in order to counteract the oedematous, inflamed and infiltrated lung tissue aiming to resume a normal functional residual capacity (FRC) [475] and thereafter keeping the lung open with a high PEEP is referred to as the "open lung concept" coined by Lachmann [476]. With re-recruitment of atelectatic, perfused but not ventilated lung areas (shunting) [368], the amount of parenchyma participating in ventilation increases, resulting in an enlargement of the disease-associated reduced functional residual capacity as well as a reduction in PVR [156, 330, 332, 368]. In collapsed alveoli, functionally not contributing to gas exchange (dead space) displaying hypoxic units (P_AO_2 <60 mmHg), a regional increase in pulmonary vasomotor tone applies, reducing local blood flow [477–479] and increasing PVR [478, 480]. This is known as hypoxic pulmonary vasoconstriction, the Euler–Liljestrand reflex [477, 481]. Consequently, re-recruitment of those areas not only results in an increased functional lung parenchyma (FRC) [330, 332] with improved lung compliance [331, 334], and a lower ITP [395], of equal importance is a reduction in PVR, as the regional pO_2 will increase, thereby blunting hypoxic pulmonary vasoconstriction [434, 478]. The resultant reduction in pulmonary vasomotor tone accompanied by off loading of the RV outflow impedance and improvement of RV–EF will prevent RV dysfunction or improve an already established dysfunction [434, 478, 482]. Furthermore, a second mechanism contributes to the load removal. As described above, PVR is related to lung volume in a bimodal fashion and is lowest around FCR [362, 433]. While recruiting collapsed lung tissue, the disease-associated reduced FCR will be enlarged and shifted towards the normal, physiological level, thereby unburdening the extra-alveolar vessels from compression and reducing PVR [156, 364, 368]. This explains why PEEP application may be beneficial for both the amount of ventilated lung tissue and cardiopulmonary function [368].

However, even in spontaneous breathing humans gas volume distribution shows regional variations [179]. Diseased lungs often have morphologically and functionally severely inhomogeneous parenchyma with atelectatic and over-distended regions in proximity [368], and thus marked regional differences in lung compliance exist [395]. During PPMV, the most compliant, still "healthy" alveolar units being preferably aerated [156, 201]. The applied gas volume, even if low [156], as the force causing stress on the alveoli and determining the height of the regional airway pressures [403, 483] may in turn overinflate those functioning units [250, 484]. Further generation of dead space (ventilated, but not perfused lung regions) and subsequent hypercapnia is seen [368], while also compressing the intra-alveolar capillaries, reducing their diameter and increasing their length leading to an *increase in PVR* [152, 252, 362, 485], and concomitantly RV outflow impedance and afterload [152, 372, 430, 486]. Interestingly, both hypoxia and hypercapnia facilitate an increase in RV afterload, and thus relevant hypercapnia should be avoided or at least the effect on RV afterload should be taken into account when allowing permissive hypercapnia [480, 487].

As such, PEEP may modify PVR significantly depending on its impact on lung recruitment and its distending force on lung units with their respective regional compliances [156, 368]: if by PEEP application collapsed lung tissue can be recruited, with at least partial recovery of the disease-related reduced FCR, PVR will fall and thus the impact of PPMV on the central haemodynamics, particularly RV afterload will be diminished, due to the improvement in lung compliance with consecutive fall in ITP [395]. If the PEEP applied predominantly leads to overinflation of normal, compliant alveolar units, RV afterload will increase, compromising circulation even further [156, 368].

Furthermore, both cyclic opening and closure of alveoli as well as (too) excessive PEEP with regional over-distension may provoke a local inflammatory response with release of inflammatory mediators potentially with systemic overspill [195, 201, 249, 488, 489] and the resultant haemodynamic effects as described in the sepsis and shock chapter. Once again the close relationship between respiratory issues and cardiopulmonary, especially heart–lung interactions, are apparent (Fig. 1.4).

It is important to consider a few more aspects related to the impact of PPMV including PEEP on LV performance: the main interference is arguably mediated by the diastolic ventricular interaction [156, 186, 423, 490, 491]. PPMV as well (increasing) PEEP will provoke an increase in ITP, resulting in a reduction of the transmural pressure of the heart and hence LV wall stress and LV afterload:

$$\left.\begin{array}{l}\text{PPMV / PEEP}\uparrow\ \rightarrow\ \text{ITP}\uparrow\ \rightarrow\ \text{transmural p}\downarrow\\ \rightarrow\ \text{LV wall stress}\downarrow\ \rightarrow\ \text{LV afterload}\ \downarrow\end{array}\right]\ [186, 387, 492].$$

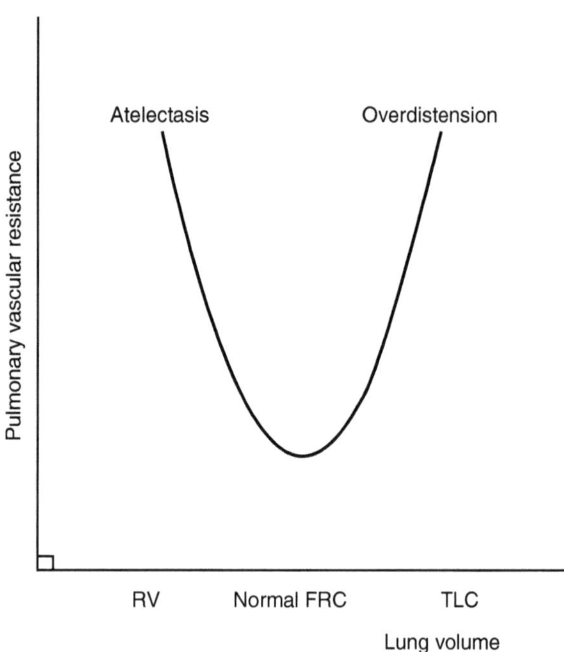

Fig. 1.4 Atelectatic and over-distended alveoli both cause an increase in pulmonary vascular resistance (PVR) attributed to compression of the "extra" (atelectasis) or "intra" (over-distension) alveolar capillaries constituting a bifashioned relation between PVR and lung (gas) volume with a nadir near-normal FRC. As PVR is the major determinant of right ventricular afterload, this relation decisively affects and designates right ventricular function *TLC* total lung capacity, *RV* residual volume, *FRC* functional residual capacity (Reprinted with permission from Cherpanath et al. [368])

While there is consensus that PEEP reduces LV afterload [337, 370, 425], the underlying mechanism is controversial. Is it really an unloading or simply a reduced mean arterial pressure in those situations [156], or even other changes in LV properties: some authors found a diminished LV compliance [373, 374], others not [376, 378] and some reported of unchanged LV intrinsic myocardial contractility [493, 494].

The most common cause of an increased RV outflow impedance is an increased PVR, the latter being the major determinant of RV afterload [365, 368, 386, 454, 495], which may consequently precipitate RV dysfunction or even acute right heart failure [396]:

$$\text{PVR} \uparrow \to \uparrow \text{RV afterload} \to \text{potentially RV} - \text{dysfunction} / \text{ACP}$$

"Right heart failure in the setting of acutely elevated PVR due to a primarily pulmonary disease is referred as to acute cor pulmonale (ACP)" [194, 496]. The main disease entities underlying ACP in the critical care setting are [194, 365, 386, 449, 452, 497] acute respiratory failure (especially pneumonia, exacerbated COPD (AECOPD) and ARDS), mechanical ventilation, pulmonary embolism, sepsis, acute compromised/acutely decompensated left heart failure decisively affecting the RV and RV malfunction in case of RV myocardial infarction and fluid overload – not all strictly follow the definition of ACP, but clinically present as acute right heart failure. Increases in pulmonary vascular resistance maybe of a functional ("reactive" hypoxic pulmonary vasoconstriction, vascular engorgement, vascular compression, PPMV, volume overload, inflammation [431, 482, 498]) or structural basis (vascular remodelling, i.e. obliteration, inflammation, toxic, micro-/macroemboli [499–501]). However, only those originally affecting the respiratory–pulmonary system may be literally termed cor pulmonale [194, 496]. Acute RV failure mainly provoked by PPMV is commonly defined as a primary pulmonary disease [61, 156, 395, 398].

PPMV and PEEP by inflating the lungs cause an increase in transpulmonary pressure [60, 195, 328, 368] with that increase being steeper in diseased lungs than in normal ones and disproportionately steeper than the increase in ITP [368]. As the transpulmonary pressure is the main determinant of PVR [324], it consequently precipitates RV afterloading [368, 404]:

$$\text{PPMV / PEEP} \rightarrow \uparrow \text{transpulmonary pressure} \rightarrow \uparrow \text{PVR} \rightarrow \\ \uparrow \text{RV afterload} \rightarrow \text{potentially RV dysfunction / ACP}$$

The severity of outflow resistance is strongly related to RV dysfunction/ACP and mortality [502].

Additional features may contribute to the elevation and sometimes considerable increase in RV afterload, such as hypoxic pulmonary vasoconstriction [478] and hypercapnia [480, 487], collapsed alveoli and hyperinflated relatively healthy areas [364, 478, 482], imbalances of inflammatory mediators [435, 436], endothelial injury [437] and microthrombosis [441]. Furthermore, in many instances aggravating factors are present and contributing such as sepsis, anaemia, trauma, surgery, pregnancy, arrhythmias and right ventricular ischaemia [503]. The elevated pulmonary resistance to blood flow generally puts the RV at risk of myocardial ischaemia [386, 504], with and without pre-existing coronary artery disease [505]. Furthermore, intrathoracic volume depletion with reduced venous return [358, 442] or fluid overload may cause an increase in PVR [396, 506].

Afterloading the RV attributed to an increase in PVR will result in a mismatch between RV elastance, largely a load independent index of RV contractility, and pulmonary arterial elastance, representing RV outflow impedance [507–509]. Accordingly, a progressive disruption of normal ventriculo-pulmonary coupling [510, 511], a major determinant of RV function [512, 513] will potentially lead to RV dysfunction or even RV failure [507, 510]. As the right ventricle is poorly able to adopt to sudden increases in afterload [396, 453, 514], a significant decrease in PA elastance following increased resistance to pulmonary blood flow whatever reason and mechanism underlying, the contractile properties of the thin-walled RV are rapidly overwhelmed resulting in acute RV dysfunction, indicated by RV distension [396, 453], impaired RV–EF [515] or even haemodynamic collapse [496, 510, 516, 517]. A vicious cycle may aggravate the situation since acute RV dilation consecutively requires an increased oxygen delivery and will even further diminish the already overstretched RV contractility [518]. Particularly sudden and significant increases in PVR acutely after loading the right ventricle provoke acute right heart failure [61, 152, 195, 519, 520]. ACP and RV dysfunction, mediated via DVI, may have marked consequences on LV function and systemic circulation [455, 459]. If pulmonary hypertension develops more gradually, RV will adapt [521], but it takes at least 96 h [522] to achieve a reasonable functional adjustment [521, 523].

Mechanical ventilation, at least in the former days of aggressive ventilator settings using high tidal volumes, has been a significant factor causing ACP [498, 524]. In recent years, RV dysfunction (characterized by RV dilatation, reduced RV-EF but still normal RV-SV) or failure/ACP (where the SV will be reduced as well [365, 515, 525]), occurs more in settings in which PPMV is associated with other conditioning factors. This is especially true if PPMV is accompanied by hypovolaemia, or in cases of severely diminished respiratory compliance, e.g. substantially increased intra-abdominal pressure as in peritonitis, comprehensive pneumonia, pneumothorax and auto-PEEP [526, 527]. Since lung protective ventilation was established, RV dysfunction and ACP directly associated with the mechanical ventilation and the ventilator settings probably accountable for the acute RV decompensation are significantly less often observed: in ARDS, where formerly patients have been ventilated most aggressively, a reduction of the ACP incidence from 60 % [502] to 10–25 % [61] is reported. However, recognition of PH and RV dysfunction is important as they have significant prognostic implications [365].

To summarize, positive airway pressure always constitutes a supplementary and separate load for the right ventricle [60, 395, 431]. But it is not the pressure applied per se affecting the haemodynamics. The change in lung volume [358, 403, 486], and thus the amount of air inflated rather than the airway pressure, per se may substantially affect the right heart and consecutively overall cardiac performance and systemic circulation [403–405]. The reduction in venous return [156, 486], following changes in venous conductance [414, 415] associated with PPMV is largely propagated by the heart–lung interactions [156, 406, 408, 410], while the marked systemic circulatory compromise including relevant LV dysfunction mainly results from substantial RV dysfunction or ACP. This right heart malfunction is predominantly attributed to a suddenly altered RV afterload [324, 432, 486] caused by a substantial rise in PVR [365, 372, 429], a result of the changes in lung volumes associated with PPMV and PEEP [372, 520]. Lung volumes and their changes are related to PVR in a bimodal fashion [362, 433], and its subsequent impact on RV function is predominantly mediated by heart-lung interaction [156, 423, 490]. Hence, "lung inflation primarily affects cardiac function and cardiac output by altering right ventricular preload and afterload" [486].

The basic physiological interactions and subtle harmony between heart and lung function become more apparent under pathological conditions, e.g. PPMV, and has been illustrated by a broad body of experimental [362, 433] and clinical evidence [59–61, 185–187]:

(a) RV outflow impedance correlates with the extent of tidal volume applied [324].
(b) The amount of air inflated upon actual transpulmonary pressure largely depends on total respiratory compliance, C_{RES} [368]. P_{TRANS}, a surrogate of P_{PLAT} [114], is the major determinant of PVR [324], and as the changes in lung volume inflated depend on P_{TRANS} [368], so does PVR and, in consequence, RV afterload correlates with lung volume. P_{TRANS} is responsible for RV afterload [324, 368].

(c) Whittenberger [362] established a bi-fashioned relationship between PVR/RV afterload (and consequently cardiocirculatory function) and lung volume/changes in lung volume.

By changing lung volume, atelectatic areas may be recruited, and if consecutively total lung volume returns towards a more normal FCR, a reduction in PVR results following the relation between PVR and lung volume established by Whittenberger [362] and later confirmed by Shekerdemian [433]. This is due to the attenuation of increased pulmonary vascular tone following hypoxic pulmonary vaso-constriction and the relief of compression of extra-alveolar capillaries [156, 368]. On the other hand, if compliant, healthy alveoli units are overinflated/overdistended by PEEP application, PVR will increase [156, 362, 368]. Thus, enhancement of PEEP is useless if it does not recruit lung tissue [405].

Eichacker [185] in a meta-analysis comparing the outcome of ventilated patients using high and low tidal volume strategies could demonstrate a U-shaped relationship between mortality and plateau pressure, very similar to the relation between PVR and lung volume shown by Whittenberger [362] in the 1960s. This analogy is not only suggestive for an interaction between airway pressure, or accurately plateau pressure (thereby lung volume changes) applied, and the circulatory effects of PPMV [60], but may also be indicative that a tidal volume too low in some settings may be disadvantageous. However, if following the concept of baby lung by Gattinoni [220], even further tidal volume adaptations beyond volumes of 6 ml/kg predicted body weight, thereby significantly lowering the plateau pressure [60], may be necessary, as predicted body weight will not reflect the size of severely damaged lungs which is not related to body weight [152]. Thus, the applied tidal volumes need to be based on plateau pressure, directly reflecting the size of the lungs as plateau pressure represents the value of the compliance of the respiratory system [60]. This change addressing the respiratory requirements in daily practice can be obtained by limiting P_{PLAT} to 26 cm [60, 528].

Importantly, Eichacker's results support other recent studies substantiating plateau pressure as a reasonable, practical and easy achievable parameter to guide mechanical ventilation [185], since it obviously integrates and mediates information about respiratory/pulmonary and haemodynamic conditions (the heart–lung interactions). Despite the shortcomings regarding the plateau pressure not reflecting true alveolar distending pressure (as transpulmonary pressure does, the physiological surrogate of lung injury [114]), P_{PLAT} apparently represents pulmonary mechanical and cardiocirculatory features, particularly right heart functional issues associated with PPMV and thus heart–lung interactions, at least well enough for daily clinical use, as several studies have confirmed [59, 61, 186, 187]. A strong relationship between the incidence of ACP and the height of the plateau pressure has been established [60].

Based on the currently available data, a plateau pressure of ideally ≤26 cm H_2O, at least <27–28 cm H_2O [59–61, 187], should be aimed for in order to provide good evidence-based respiratory support coordinating respiratory and circulatory needs. As particularly PEEP modifies RV afterload, and tidal volume ranges are more or less well defined, plateau pressure can be used to individually titrate a suitable

PEEP [156, 368, 395]. High PEEP not recruiting lung tissue may induce hypercapnia and impairs RV function [529]. Echocardiography is reported to provide additional useful haemodynamic information and is shown to substantially support individual PEEP setting [60, 194, 195] as RV function in part reflects lung status [405].

"The aim of lung protective ventilation is to prevent alveolar over-distension in open lung parts albeit concurrently avoiding repetitive opening and closing of alveoli in closed lung parts" [530], thereby "titrating PEEP to its lowest value while maintaining an open lung" [398]. Accordingly, the incidence of ACP will potentially be reduced or even prevented by lung protective ventilation with low V_T [194] and particular low PEEP [405]. Canada writes, "only one lung volume in normal and diseased lungs is optimal for ventilation, blood flow and transport of oxygen to tissues" [434].

1.12 Conclusion for Overall Respirator Settings

Very strong evidence and currently established facts are

1. *Tidal volume* applied 6 *ml/kg* of predicted body weight (PBW), a lung protective strategy [30, 67, 161, 241, 531]
   [calculation PBW: PBW=50+0.91 (height −152.4 cm) for men, PBW=45+0.91 (height −152.4 cm) for women [30, 167]]
2. *Plateau pressure* $\leq$*26 cm* H_2O (<27–28 cm H_2O) [59–61, 187, 528]
3. *PEEP adjustment by assessment of RV function* – avoid RV dysfunction. "Lung inflation primarily affects cardiac function and cardiac output by altering RV preload and afterload" [486]. Particularly PEEP will determine RV afterload and thus potentially promote RV dysfunction or even ACP if not recruiting lung tissue, resulting in severe circulatory malfunction [156, 362, 368]. Lower PEEP levels (10–12 [15] cm H_2O [405, 529, 532]) may be more suitable and are an RV protective approach [405].

The currently generally accepted oxygenation target is a pO_2 55–80 mmHg, SaO_2 $\geq$92 % [533, 534].

In case of very severe ARDS, if following the baby lung concept [220], even lower V_Ts may be advisable [60].

1.13 Ventilation of Nonobstructive Acute Respiratory Failure Patients Not Suffering from ALI/ARDS

A substantial and generally accepted body of literature provides convincing evidence that lung protective ventilation, basically characterized by applying a low tidal volume (6 ml/kg PBW) and a restricted pulmonary distension pressure (traditionally accomplished by a plateau pressure of less than 30 cm H_2O or a transpulmonary pressure of less than 25 cm H_2O), is currently the best approach to treating patients suffering from ALI/ARDS [30, 67, 161, 162, 239, 241, 531]. No doubt, a suitable PEEP is crucial and completes current standard of care [30, 181, 211, 241]. A subgroup analysis of the classic, seminal prospective ARDS network trail published in 2000 [30] with a multivariable logistic regression analysis confirmed that lower tidal volumes precipitating lower airway pressures indeed result in a lower mortality [59]. Remarkably, literature on how to deal with nonobstructive patients without ALI/ARDS suffering from pneumonia or cardiogenic oedema and requiring mechanical ventilation is lacking [123]. However, assessing the currently available evidence, to suggest a causal relationship between high tidal volumes applied at the onset of mechanical ventilation and the incidence of ALI/ARDS later in the course, is quite reasonable [535].

Indeed, physiologically, humans inhale about 6 ml/kg tidal volume at rest with each breath [536], and mechanical ventilation may cause the development of ALI/ARDS [537], particularly in the so-called "pre-primed" lungs resulting from an initial threat such as pneumonia, sepsis, blood transfusion, injurious mechanical ventilation or surgical procedures, especially cardiac surgery due to cardiopulmonary bypass [537, 538]. A few studies strongly suggest that lower tidal volumes (compared to conventionally applied 12ml/kg body weight) used in longer-term mechanical ventilation may also be beneficial in patients with acute respiratory failure but not with ALI/ARDS. Gajic reported two studies in medical–surgical ICU patients where a lower incidence of ALI occurrence could be found if tidal volumes of <6 ml/kg (PBW) or tidal volumes of less than 700 ml were applied [537, 539]. Moreover, in a variety of clinical conditions including the lungs of potential organ donors and high-risk surgical patients awaiting abdominal, lung and oesophageal surgery, patients who were ventilated with lower tidal volumes (administrating between 5 and 10–12 ml/kg PBW, the latter potential organ donors with healthy lungs), showed a trend or even developed significantly less often ALI [540–543]. The IMPROVE study evaluated 400 patients undergoing elective abdominal surgery, revealing low tidal volume ventilation (6–8 ml/kg PBW) compared to conventional V_T (V_T 10–12 ml/kg PBW) during the operation procedure elicits significant (10.5 % vs. 27.5 %) less major complications (death, sepsis, pneumonia, need for non-invasive/invasive ventilation due to acute respiratory failure in the postoperative period) [544]. However, some studies in high-risk surgical patients could not establish any correlation between V_T and the development of ALI/ARDS [545, 546].

Inflammatory mediators like cytokines and interleukins demonstrated in blood plasma and bronchoalveolar lavage fluid are considered to reflect the pulmonary inflammatory response caused by threats such as pneumonia and pulmonary oedema and by improper mechanical ventilation [229, 279, 547]. While in animal models high tidal volumes per se were shown to precipitate lung injury and hence the release of inflammatory agents [273, 548, 549] could just reveal that a high tidal volume ventilation in the already injured lungs elicits the investigations in humans release of detectable inflammatory mediators, at which low tidal volume application display significantly lower levels of inflammatory substances and the change from high to low V_T attenuates the elevated levels [30, 279, 547]. Short-term mechanical ventilation in otherwise healthy patients undergoing surgery is reported not to be associated with the increasing levels of cytokines independent of ventilator settings [382, 550]. Protective ventilation at least does not worsen pre-existing lung injury in the vast majority of patients [229], and in humans, detectable levels of inflammatory mediators in blood plasma and/or bronchoalveolar lavage fluid obviously only arise in pre-primed or already injured lungs, the two hit model where pulmonary inflammation is already present to some degree but may be aggravated by a second hit/insult (injurious mechanical ventilation) [218, 382, 535]. This allows locally produced mediators to disseminate from the lung compartment into the systemic circulation [551] and is supported by several experimental studies [212, 276, 284]. However, very recently, a dissertation by Wolthuis could give evidence that even short-term mechanical ventilation is a proinflammatory stimulus in non-injured lungs [551], though surgery and general anaesthesia are associated with inflammation measurable in blood plasma, making causation difficult to ascertain [552, 553].

Patients undergoing cardiac surgery are quite well examined as circulatory bypass may prime the lungs and thus put the patients at special risk [554, 555]. A number of studies, particularly a large trial including more than 3,400 cardiac surgical patients [556], revealed in principle lower levels of cytokines in the bronchial lavage and blood plasma if lower tidal volumes (8 ml/kg vs. 10–12 ml/kg, 6 ml/kg vs. 10 ml/kg or <10 ml/kg vs. >12 ml/kg) were supplied. Conflicting results were shown regarding issues such as shorter ICU stay, more stable haemodynamics or shorter duration of mechanical ventilation and less frequent reintubation. However, none of the trials found a significant better outcome in the lower tidal volume group [556–560]. Comparable results are demonstrated in a wide range of clinical settings [30, 547, 561, 562], and most recently even in patients without pre-existing ALI/ARDS, but may be primed by other threats like sepsis or transfusions [563].

The clinical implications are quite obvious. The main objective of putting a lung protective ventilator strategy (comprising low tidal volumes and limited plateau pressure) into practice in patients not primarily suffering from ALI/ARDS but requirung mechanical ventilation is to minimize regional end-inspiratory stretch thereby avoiding, at least substantially reducing, diffuse alveolar damage [535]. Not only based on physiological grounds [211] but suggested by observational studies, one randomized trial [563, 564], supported by a systematic literature review [565] and a recent meta-analysis from 2012 [566], a low tidal volume strategy is pivotal. Lower than traditionally used, more physiological tidal volumes substantially reduce the risk to injure lung parenchyma during mechanical ventilation in both primary lung diseases and initially healthy lungs (extra-pulmonary maladies), and thus the progression to ALI/ARDS [123, 211, 535, 567]. Furthermore, as recommended, peak inspiratory (<35 cm H_2O) or more meaningful plateau pressure (<30 cm H_2O [30, 162, 241, 568]) has to be limited – although the height is more or less arbitrarily set [188] and recent study results give substantial evidence that P_{PLAT} should be ≤26 cm H_2O [59–61, 185, 187]. Combining low V_T with a suitable, relatively high PEEP is essential in preventing disease progression and development of ARDS [30, 123, 211, 569]. However, we still do not know how to properly determine the optimal PEEP level needed [211, 569]. Using P_{PLAT} limitation [156, 368, 395] supported by echocardiographic assessment of cardiac function, particularly exclusion of haemodynamic relevant pulmonary hypertension compromising RV function

[194, 195] is the currently best evidence-based way to individually titrate PEEP. PEEP is not only the appropriate measure to avoid cyclic closure and reopening of atelectatic areas [65, 198, 202, 210], but due to its recruitment effects, it may allow to turn down oxygen content in the inspired air. FiO_2 levels above 60 % are believed to cause denitrogenation atelectasis and precipitate atelectasis development [151, 570, 571]. As in ARDS it is suggested to apply higher PEEP levels (up to 24 cm H_2O) to the most severely ill/lung-injured patients [572]. Lipes et al. suggest to start in non-ARDS patients with 8 cm H_2O, titrating PEEP in regard to FiO_2 required and haemodynamic conditions [123] while monitoring P_{PLAT} and echocardiography. Other authors principally suggest initial PEEP settings in non-ALI/ARDS patients between 5 and 10 cm H_2O [156, 434].

Somewhat surprising, the same authors recommend a tidal volume of 6–8 ml/kg PBW in patients at risk for ALI/ARDS; in those without risk factors for lung injury, they feel a $V_T \leq 10$ ml/kg being advisable [123]. We think, their suggestions do not really match the study results they are citing in their article (Determann used V_T 6.4 ± 1.0 ml/kg [563], in the systematic review, trials included used V_Ts between 7.0 and 9.5 ml/kg, on average 8.8 ml/kg) and advise lower values, 6–8 ml/kg. Further consider possible risk factors such as multiple transfusions, sepsis, shock, primary airway infection, acidaemia, surgical procedures, etc. [535]; when determing ventilator settings.

1.13.1 Summary, Invasive Mechanical Ventilation, Initial Settings in Non-ALI/ARDS Patients

- Mode: VC (and pressure limited) or PC [46, 54, 56], it may be that BiPAP is the most useful and comfortable initial mode [54, 70, 74]
- V_T: 6–8 *ml/kg PBW* [123, 540, 556, 562], the more severe the respiratory compromise, the closer to 6 ml/kg PBW [535, 563]
- Respiratory rate: 12–20*/min* [40, 123]
- PEEP: 8 *cm* H_2O (5–10 [12] cm H_2O) [123, 434, 540, 543]
- Plateau pressure: ≤26 *cm* H_2O [59–61, 190, 528]
- Inspiratory–expiratory: *ratio,* 1:2 to 1:1 [573]
- FiO_2: *up to* 1.0, but as soon as possible, reduce to <0.6 maintaining a paO_2 >60 mmHg [123]
- Peak inspiratory pressure: <30–35 *cm* H_2O [162, 241, 568]

A few aspects and concerns need to be addressed. Dyspnoea may arise during small tidal volume ventilation while aiming for high minute volume by setting high respiratory frequencies [137]. High frequencies may induce auto-PEEP, at least in ARDS patients, due to short time constants of some of the alveolar units and low lung compliance [307]. Reduction of respiratory rates set below 30/min minimizes the risk of intrinsic PEEP [307]. Whereas reduction in desired minute volume (reduction in the intensity of ventilation), even allowing for some degree of permissive hypercapnia, particularly enabling the patient to participate and contribute to ventilation, is a suitable approach [137]. Dyssynchrony may occur in cases where volume or flow delivered does not meet the patient's demand or their neural drive to breathe [574] and will result in intense anxiety and may be dyspnoea [575]. Unfortunately, muscle fatigue attributed to increased work of breathing promoted by this dyssynchrony may be deleterious [576, 577]. Adjustment of inspiratory flow by assessing the patient and analysing the waveforms usually applying 80 l/min is helpful [123]. Further, a decelerating flow waveform is superior to other types like square waveforms because high flows with relatively long inspiratory times can be administered while avoiding double triggering by the patient [577]. If this manoeuvre is not successful, patients may be allowed a V_T of 8 ml/kg while keeping P_{PLAT} strictly ≤26 cm H_2O [60, 187, 395, 576]; this modification is associated with a reduction in the work of breathing [576].

It is important to discuss ventilator settings in primarily brain-injured patients requiring mechanical ventilation. Roughly 20 % of those patients need intubation [578], and as brain injury triggers an inflammatory reaction, the extracerebral remote organs, particularly the lungs (even more so if artificially ventilated), are at risk of getting affected [579]. The brain itself has a high risk being affected by second hits [580, 581]. Notably if mechanical ventilation is required, pulmonary complications such as acute lung injury/ARDS are seen in 10–30 %, and between 20 and 45 % of patients with traumatic brain injury and subarachnoid haemorrhage may develop ventilator-associated pneumonia [580, 581].

The new lung protective strategy with low tidal volume, higher PEEP levels, allowing for hypercapnia, and low airway pressures is quite contradictive to the previously common usage of V_T 10 ml/kg, high FiO_2 and low or zero PEEP [580]. Lower tidal volumes are certainly acceptable as long as hypercapnia precipitating or even aggravating intracranial hypertension can be avoided [582, 583]. PEEP may exhibit adverse neurologic effects under special circumstances as PEEP is associated with increased intracranial (ICP) and intrathoracic pressure (ITP), the latter potentially compromising CO and MAP, consequently reducing cerebral perfusion pressure [584]. However, with reasonable levels of PEEP, these effects remain clinically modest [580]. Since hypercapnia leads to dilation of the cerebrovasculature and consecutively increased cerebral blood flow resulting in increased intra-cerebral blood volume and hence ICP with all its harmful effects on the brain, normocarbia should be aimed for [582, 585].

1.13.2 Non-invasive Positive Pressure Ventilation (NIV) in Non-ALI/ARDS Patients

As most of the complications seen in PPMV are attributed to intubation (see ventilator-associated pneumonia) [586], NIV may be preferred instead of invasive respiratory support, avoiding most of the complications while offering in principal the same degree of efficacy [587]in patients with non-ALI-/non-ARDS-related respiratory failure [588]. However, obviously the successful application of NIV and patients' benefit decisively depend on the underlying conditions/disease. There is very strong evidence to apply NIV in acute exacerbations of COPD (AECOPD) [589–592], in patients with cardiogenic pulmonary oedema [402, 593–595], weaning of AECOPD [596–598], and in immunocompromised patients with respiratory failure requiring ventilator support [592, 599, 600]. Less but still convincing results are demonstrated in NIV used in postoperative patients needing respiratory support [601–603]. At least some data support NIV in patients with asthma exacerbations [604–607], pneumonia (see below), critical ill patients with acute hypoxaemic respiratory problems [608–610] and patients with weaning failure of other than AECOPD aetiology [592]. Unfortunately in pneumonia, the results found in the studies are inconclusive and disappointingly not indicative of a beneficial effect of NIV in this condition [610–612]. However, although the routine use is not recommended [613] due to two quite promising studies, a trial of NIV may nevertheless be considered [588, 614, 615].

1.13.2.1 Initial Respirator Settings in NIV

- PC mode with pressure support (PS) and PEEP, thus bilevel ventilation (PC mode is nearly exclusively delivered; nevertheless VC mode may be applied [616]) [588, 617] BiPAP [54, 70].
- PEEP 3–5 *cm* H_2O [338, 617–619] gradually increased up to [620] 5–10 *cm* H_2O [617, 618].
- PS 8–10 *cm* H_2O [617, 619, 621] increasing in steps of 2–3 cm H_2O [620][3] *up to* 15 *cm* H_2O [617, 621]. Targeting a V_T >*300 ml* and particularly a respiratory rate <25–30 *breaths/min* [617, 618].
- Keep PIP ≤20–30 *cm* H_2O [617, 618]; others recommend <30 *cm* H_2O [620, 621].
- F_IO_2 to maintain SaO_2 ≥90 % but aim for F_IO_2 <0.6 as soon as possible [618].

Sometimes the nomenclature can vary; PEEP is termed EPAP (end-expiratory positive airway pressure), PS (pressure support) and IPAP (inspiratory positive airway pressure), with IPAP = PS + PEEP [622]. Yeow and Santanilla [618] describe a high-pressure approach starting with a PS level of 18–23 cm H_2O, titrating gradually down, or a low-pressure approach starting with a PS of 5–8 cm H_2O, titrating gradually up until the targets like reduction in respiratory rate, alleviation of dyspnoea and increase in tidal volume and the patient's comfort is achieved.

1.14 Mechanical Ventilation in COPD and Asthma

Acute exacerbations of COPD (AECOPD) are in the vast majority (70–80 %) caused by infections of the respiratory tract [623, 624]. While traditionally bacterial organisms have been considered as the predominant cause [625, 626], infections by viruses are increasing [627, 628]. In 20–30 % of patients, an unknown aetiology is

[3] Until F_IO_2 <0.6 [620].

described in the literature [629, 630]; conditions like environmental pollution, acute heart failure, myocardial ischaemia, aspirations, arrhythmias and pulmonary embolism are associated with acute deteriorations [631, 632]. Risk factors to develop AECOPD are advanced age, productive cough, duration of COPD, history of antibiotic therapy, chronic mucus hypersecretion, hospital stay within the previous year and comorbidities such as chronic heart failure, ischaemic heart disease, diabetes and possible gastroesophageal reflux disease [624, 633–636].

The preeminent pathophysiological feature in COPD is the marked limitation of expiratory airflow [629, 637, 638]. A substantial increase in airway resistance to airflow is mainly due to considerable peripheral airway narrowing [637, 639] related to inflammation and destruction of the airway walls and tissues such as wall thickening/swelling, vascular congestion, smooth muscle contraction and production of thickened secretions [319, 629, 640]. This results in a marked airflow obstruction, impeding expiration; during the disease process of COPD, the airway obstruction is slowly progressive and in acute states may suddenly exacerbate [319, 629]. The considerably impaired expiratory airflow entails that expiration is still persisting, has not finished and will be interrupted while the next inspiration ensues, leaving air from the preceding inspiration trapped in airways and alveolar units. This is referred to as dynamic hyperinflation [311, 312, 315, 641]. As no pressure equilibrium between airways and atmosphere can be achieved, there will still be a pressure gradient, auto- or intrinsic PEEP exists [314, 320]; see described above in the Sect. 1.10. Dynamic hyperinflation is a hallmark of (AE)COPD [312, 639, 642, 643]. Dynamic hyperinflation will be enhanced and exacerbated by:

(a) Small airway collapse during tidal expiration. When in COPD patients lung deflates, dynamic airway compression occurs [644].
(b) Loss of lung recoil in severe COPD exacerbates hyperinflation as well. The loss of elastic recoil favours airway collapse at low lung volumes and entails that the expiratory airflow depends only on the passive recoil elastance and no longer on the expiratory muscle effort as well. Thus, if the COPD patient contracts his/her expiratory muscles, the pleural pressure increases, and this compresses the airways, limiting the expiratory flow even more [318, 644].
(c) In cases of acute exacerbations, patients may adopt rapid shallow breathing pattern, further limiting expiration time and thus aggravating dynamic hyperinflation [640].

Functional residual capacity (FRC) will substantially increase [319], static lung compliance is reduced (indicated by an elevation of P_{PLAT} [319]) and the respiratory muscles being distended by the persisting end-expiratory volume causing inspiratory muscles to start their contraction on a higher than resting level to initiate inspiration, as the intrinsic PEEP has to be counterbalanced before the inspiratory airflow can begin [637, 638]. This forces the muscles to operate on a higher, suboptimal proportion of their pressure–volume curve, increasing work of breathing and potentially tiring muscle function [314, 322, 645–647]. The breathing workload for the muscles will be further compromised as the patients try by help of inspiratory muscle force to exhale the inspired air [314, 648].

An acute, severe asthma attack is similar to AECOPD characterized by a critical limitation of the expiratory airflow [649, 650], but unlike in COPD, in asthma, it is largely the central, non-collapsible airways that are severely affected and narrowed [651]. This is attributed to an increased bronchomotor tone ("bronchospasm") and a considerable inflammatory infiltration causing a stiffening of the airway walls while decreasing their collapsibility, nevertheless resulting in a substantial reduction of the airway calibre [637, 650, 652]. Even the subglottic aperture may be narrowed during expiration [653]. Consequently, expiration is considerably prolonged and incomplete when the next inspiration ensues, leaving the end-expiratory pressure out of equilibrium well above atmospheric pressure (auto-PEEP) [310, 311], precipitating gas trapping and dynamic hyperinflation [53, 650, 654, 655]. This is further promoted by an abnormally low elastic recoil of unknown reason found in asthma [321, 656] and by a high outward recoil of the chest wall related to an inappropriate (and into the expiration phase) prolonged activity of the inspiratory muscles [657]. In necroscopic examinations, all parts of the airway walls reveal severe inflammatory changes and thus swelling due to oedema [658, 659]. Further, massive mucus plugging of bronchi [660] may cause marked narrowing of or even blocking of the respiratory tree and hence exert a substantial resistance to airflow [661].

As predominantly the central airways are affected, the peripheral, collapsible parts of the airways (comprising bronchi and bronchioli) remain open during expiration [651]. This implies that in terms of passive expiration there is no flow limitation. Further characteristics include a severely reduced lung compliance related to hyperinflation and airway obstruction [318, 662], a markedly increased FRC [663], and an extremely uneven air distribution [664] with mechanically normal areas in near vicinity to highly abnormally behaving ones [53] resembling conditions described as "baby lungs" in ARDS patients [115]. This implies that in acute asthma, over-distensions of normally behaving alveolar units are highly likely and, as such, the ventilator settings chosen have to consider that only a small, tiny lung is going to be ventilated [53]. However, in contrary to COPD, the airflow obstruction is variable and principally reversible [318, 650], while in COPD fixed airflow obstruction will progressively get worse [650].

The consequences of the mechanical derangements are abnormalities in intrapulmonary gas exchange, in particular a severe ventilation to perfusion mismatch with an increased dead space to tidal volume ratio causing hypoxaemia and hypercapnia [665]. Intrinsic PEEP predisposes for barotrauma [311, 322, 666]; however, of note, if the P_{PLAT} <26 cm H_2O, only 4 % of patients may develop barotrauma as has been reported by Leatherman [190].

P_{PEAK} principally will be increased due to the high airflow resistance, while an augmented P_{PLAT} is under the described conditions often indicative for auto-PEEP [319]. As alveolar pressure increases with increasing lung volume due to air trapping, P_{PLAT} should reflect gas trapping [314] and is, in general, considered valid to indicate increasing air trapping in the discussed settings [667]. However, a reduction in the static respiratory system compliance may lead to a parallel increase of both pressures [153, 154]. Static auto-PEEP can be calculated, measuring total PEEP (auto-PEEP+external PEEP) by an end-expiratory hold manoeuvre in patients without spontaneous respiratory effort [640]. Most of the pressure applied will dissipate to overcome airflow resistance and as such will not over-distend alveoli [668]. However, as there is frequently a considerable heterogeneity in air distribution in obstructive lung diseases, less diseased alveolar units will, at least temporarily, be overinflated and thus potentially injured; hence a high peak airway pressure should be avoided [668]. The main requirement for ventilator settings in both, COPD and asthma, is to attenuate and minimize dynamic hyperinflation [669, 670]. As the main determinants of hyperinflation are tidal volume, minute ventilation, expiratory time and the severity of obstruction, ventilator settings should address those parameters, of which minute volume turned out to be the most important determinant [671]. The following measures are most effective to attenuate gas trapping and dynamic hyperinflation [319, 640, 671, 672]:

- Provision of a suitable extension of the expiratory time (primary method) [319]
- Reduction of minute volume and respiratory support, first and foremost a reduction in tidal volume
- Reduction of airflow obstruction (bronchodilators, corticosteroids)
- Shortening of inspiratory time by increasing airflow to up to 60–80 l/min [671, 672] or even up to 100 l/min [54, 672, 673]

To prolong expiration time is best accomplished by a reduction of the respiratory rate [319]. Further, inspiration to expiration ratio (I/E ratio) should be altered with an I/E ratio = 1:3, or even 1:4 being appropriate [318]. The main ventilator setting influencing the degree of dynamic hyperinflation [671] is determined by expiratory tidal volume and the duration of expiration (expiration time); Leatherman demonstrated that an expiratory tidal volume ≤10 l/min and an expiration time somewhere around 4 s are optimal in case of severe airway obstruction [573]. Thus, a simple adjustment of inspiration to expiration time is not sufficient. The absolute expiration time and the adaption of minute volume, the latter being the most important determinant of hyperinflation [671], are of substantial impact in reducing dynamic hyperinflation [640]. Reductions of tidal volume bear some risk of progressive increase of the dead space fraction; hence too low V_T has to be avoided [185, 672]. High inspiratory airflows of square-wave pattern (up to 60 l/min) are very rapidly flooding the airways and thus by reducing the inspiratory time will contribute to reduce hyperinflation [673, 674]. Moreover, high airway flows are shown as bearing a very low risk of barotrauma [655, 675]. Tuxen, based on results by McFadden [663], revealed inspiratory airflows of 60–80 l/min being optimal in decreasing air trapping and thus intrinsic PEEP [671]. Furthermore, decelerating inspiratory flow wave patterns are demonstrated to exhibit a significant lower P_{PEAK} [304]. In escalated inspiratory flows, up to 100 l/min as recommended by other authors decelerating flow patterns are preferred [672], resulting in even more improved gas exchanges due to less air trapping [673]. Moreover, high inspiratory flow rates are thought to attenuate the high respiratory drive and effort typically found in patients with respiratory failure due to obstructive pulmonary disease [318, 676]. Importantly, an insufficient inspiratory airflow may augment pulmonary and ventilator impedance increasing the work of breathing [677, 678].

1.14.1 Respiratory Support in COPD Patients

1. External PEEP
   As auto-PEEP has to be balanced before an inspiratory airflow can begin at all or the patient can trigger the respirator putting an additional load on respiratory muscles [640, 679], external PEEP applied is able to reduce the existing end-expiratory pressure gradient as explained by the stream of water down a waterfall with less drop height [309, 680]. This unloads the muscles and as such reduces the work of breathing alleviating the patient's respiratory efforts [309, 338, 641]. This also results in a stabilization of the collapsible small airways, and thus improvement of expiratory flow is reported [681], but recently challenged by Gladwin [320]. However, this only works as long as $PEEP_{EXT}$ is below the pressure necessary for the airways to constrict; pressures above this beneficial threshold will turn into detrimental consequences increasing upstream pressure and potentially exaggerating dynamic hyperinflation [309, 640]. As several authors documented, in order to stay on the safe side, external PEEPs should not be higher than 75–85 % of measured auto-PEEP [641, 666, 682]. Determining the level of auto-PEEP is difficult and the exact value can often not be defined [319, 683], and due to a non-linear pressure–volume relationships, the effect of external PEEP may be unpredictable [682]. As such it is suggested that an external PEEP of 4–8 cm H_2O may be chosen pragmatically [318, 587].
2. Oxygen application in COPD patients is tricky as too high levels may worsen hypercapnia [684, 685]. The suggested underlying pathomechanism is increased dead space ventilation with consecutive hypoxic vasoconstriction; further the respiratory drive may be impeded [318, 684, 686]. However, during mechanical ventilation the conditions may be quite different as F_IO_2 amounts from 0.3 to 0.7 did neither show any relevant increase in pCO_2 nor any effect on respiratory drive [687]. F_IO_2 levels between 0.24 and 0.4 seem to be noncritical regarding the pCO_2 ensuring an oxygen saturation of about 90 % and pO_2 about 60 mmHg in the vast majority of patients [688].
3. Non-invasive ventilation (NIV) in COPD
   There is more than ample evidence verifying non-invasive ventilator support (NIV) being not only beneficial compared to standard medical treatment but superior to invasive mechanical ventilation [589, 591, 689–691] and is in COPD patients the first choice of respiratory support [690, 692–696]. Improved outcomes, reduced need for intubation, decreased complications and both shortened length of ITU and hospital stay are demonstrated [589, 591, 689–691]. NIV is the recommended first-line respiratory support therapy in AECOPD [693, 697]. NIV is indicated in mild-to-moderate exacerbations reflected by a pH of not less than 7.3 [591]. In more severe forms, the rate of NIV failure (up to 62 %, even 70 % [698]) is inversely related to the severity of respiratory acidosis, and NIV should be withheld and invasive ventilation considered, if pH <7.25 [689, 698, 699]. Furthermore, a GCS <11, respiratory rate ≥30/min and an APACHE score ≥29 on admission were found to be markers of failure of NIV [698].

Nevertheless, in severe cases a non-invasive attempt may be tried to preventing loss of consciousness and to alleviating dyspnoea [590, 699]. In patients with altered consciousness related to hypercapnic acute respiratory failure, traditionally a contraindication for NIV (loss of airway control, risk of aspiration) [587], some studies describe a positive experience if NIV is applied [700, 701]. This is mainly due to a lower incidence of nosocomial pneumonia, shorter durations of ventilatory support and length of hospital stay [702], with early initiation of NIV associated with better outcomes [703]. Relative contraindications to NIV use include excessive airway secretions, agitated and non-cooperative patient, unable to fit the mask, recent upper abdominal surgery, medically unstable patients or an unsecure airway, e.g. GCS <8 [704].

NIV settings: The patient's comfort and cooperation is central; thus the setting has to be adapted to the patient's compliance and needs [705]. However, it is common practice to apply a PS of 10–15 cm H_2O and a PEEP between 4 and 8 cm H_2O, [624], with no mandatory breaths supplied by the respirator [78, 318, 592]. The extrinsic PEEP will counterbalance at least in part $PEEP_I$ and thus unload muscles and reduce the work of breathing [706]. The level of pressure support provided should aim to reduce the very high breathing rates of the patient, targeting the respiratory rate in the still spontaneously breathing patient to 25–30/min, as clinical research suggests [707]. However, the inter-individual range is broad; hence markedly higher and lower values are possible [318]. Due to the possibly considerable additive elastic load and work of breathing [322], trigger threshold needs to be carefully chosen as non-triggering (respiratory effort without ventilator response) will waste patient's energy, further tiring the respiratory muscles [103, 708]. This patient–ventilator asynchrony may be avoided by choosing a lower trigger sensitivity [322]. Oxygen saturation SaO_2 ≥88 % (90 %) should be acceptable and a pCO_2 target should be based on individual levels when stable; hypercapnia should be tolerable [318].

- PS mode [588, 617] – however, both, pressure and volume controlled are providing appropriate respiratory muscle unload [709].
- PS: 5–7 [8] *cm* H_2O [591, 622, 710], increases every 10 min by 2–5 *cm* H_2O steps until the target, a spontaneous respiratory rate of 25–30*/min* is achieved [318, 676, 707]. Usually levels between 15 and 20 *cm* H_2O are needed; however continue with lowest pressure support level for comfortable breathing with a rate 25–30*/min* [318, 707]; note there is a wide individual variation [710].
- PEEP external: 3–5 *cm* H_2O [591, 622, 624, 710]. Keep in mind, recommended are 66–85 % of auto-PEEP, which is most often not known. External PEEP stays as long beyond auto-PEEP as its application does not increase peak and/or plateau pressure [666, 682, 711].
- Inspiratory airflow: 60–90 [98] *l/min* [40, 53, 318, 671], at least in case of flow rates >80 l/min a decelerating flow pattern is recommended [304, 672].
- Low sensitivity to trigger the ventilator [318, 712]. Flow triggering due to less patient effort is necessary and preferred to pressure triggering [713].
- Max PIP: 20–25 *cm* H_2O [710, 714], FiO_2 usually max *0.4,* SaO_2 ≥88 % (90 %) [318, 715].

4. Indications for intubation, largely a clinical decision, include [596, 640, 716]:
   - Deterioration of gas exchange despite medical treatment
   - Cardiorespiratory arrest
   - Altered level of consciousness
   - Severe respiratory distress (indicated by tachypnoea, use of accessory muscles, diaphoresis, pulsus paradoxus, recession of suprasternal and intercostal spaces)
   - Haemodynamically and electrically unstable patients
   - Life-threatening hypoxia
   - Severe, progressive muscular exhaustion [53]
   - No improvement with trial of NIV [53]

   In case of severe muscular exhaustion, some authors recommend controlled ventilation for the patients over the next 24 h in order to allow the respiratory muscles to recover [92, 93].

5. Invasive mechanical ventilation, settings in COPD
   - Mode: volume controlled (and pressure limited) [53, 54, 318]or pressure controlled [717], BiPAP [54, 70]
   - V_T: *5 (6)–7 ml/kg* predicted body weight [54, 318]
   - Respiratory rate: *10–15/min* [318, 640], may be even less than *8–10/min* [54]
   - Plateau pressure: *≤26 cm H_2O* [59–61, 190, 528]
   - Inspiratory flow rate: 60–80 *l/min* [53, 318, 671],may be up to 100 *l/min* [54, 318, 640, 673]
   - Inspiratory wave form: decelerating [304, 318]
   - Inspiratory–expiratory ratio: I:E = 1:3 or even 1:4 [318, 319], absolute expiratory time ~4 s [573]
   - PEEP (external): 0 [311, 640, 680], max 75–85 % of $PEEP_I$ [641, 666, 682]
   - Expiratory minute volume: <10 l/min [53, 573]
   - FiO_2: 0.4–0.7, aiming for a saturation of about 90 % [318]
   - Peak pressure: <35 *cm H_2O* [53])

   In order to give the respiratory muscles a change to recover, the patients should be completely passively ventilated in the first 24 h after intubation [92, 93] as the necessity to intubate COPD patients is always associated with muscle fatigue/failure.

6. Medications
   β_2-agonists and anticholinergic bronchodilators are very well-established elements in the medical therapy of acutely exacerbated COPD, and numerous studies have verified their effect and value [650, 718–720]. Combined administration has been shown to provide some additional effect [721–723] with ipratropium bromide, particularly reducing bronchoconstriction of the central airways [650]; however, not all studies found a synergism [724, 725]. Intermittent nebulizations are reported to be more effective than continuous application [726]. Anti-inflammatory treatment (corticosteroids) in case of acute exacerbation of COPD is supported by quite a substantial number of studies, mitigating obstruction, lesser

treatment failures, shorter hospitalization rates and more rapid recovery of arterial oxygenation are demonstrated [727–729]. However, not all studies could verify beneficial effects [730]. Furthermore, the dosing is not determined, 100 mg iv hydrocortisone four times a day over 4 days followed by 40 mg prednisone once daily (OD) for another 4 days is described [731]; in the largest study 125 mg iv methylprednisolone six hourly over a 72-h period followed by 60 mg oral prednisone OD, titrated down during a fortnight has been used [728]. A more recent publication found a 5-day course of steroids (40 mg prednisolone applied in the morning) being not inferior to the conventional fortnight regimen in acute exacerbations of COPD [732].

Although not a consistent finding, routine administration of antibiotics is demonstrated in several studies to shorten the length or the severity of AECOPD [668], even in "milder" infections as a recent publication demonstrated. Llor et al. showed that patients with mild to moderate, indicated by a CRP-level of >40 mg/l on admission, acute exacerbations of COPD clearly benefitted from antibiotic therapy (amoxicillin/clavulanic acid 500/125 mg eight hourly) [733]. An older meta-analysis (1995) showed a statistically significant effect favouring routine antibiotic use [734]. However, this issue remains controversial [727]. Nevertheless, antibiotics should be given to moderate or severe exacerbations [640, 668, 735]; otherwise the study results by Anthonisen [736] and clinical features issued by Stockley [737] may be followed; hence patients with green sputum probably have a high bacterial load and may benefit in any case from antibiotics [737].

1.14.2 Respiratory Support in Asthma Patients

1. External PEEP
   External, ventilator-applied PEEP in sedated, controlled ventilated asthmatic patients without any spontaneous breathing activity is not recommended at all [672]. A respirator-applied PEEP will extend end-expiratory pressure all the way down to the periphery, thus the small airways and the alveoli principally not involved in the disease process, usually not collapsing [651]. External PEEP means just expansion of the end-expiratory lung volume and the lungs will be hyperinflated [311, 680]. However, in spontaneously breathing patients and during weaning, low PEEP levels (~5 cm H_2O) will mitigate the pressure gradient between the central airways and mouth/atmosphere, unloading the inspiratory muscles and thus substantially blunting work of breathing [287].

2. NIV in asthma patients
   In asthma, the evidence to apply non-invasive ventilation is less strong, not at least because randomized controlled studies confirming the benefit of non-invasive support/ventilation are lacking [738]. Drawing analogous conclusions from results in the COPD studies are not admissible due to the pathophysiological differences between COPD and asthma [738]. Nevertheless, there are a few studies showing beneficial effects compared to medical and/or invasive treatment

approaches when offering non-invasive respiratory support to patients suffering from moderate to severe acute asthma episodes [604, 739, 740]. Positive end-expiratory pressure applied (i.e. CPAP) will counteract subglottic/glottis constriction [653]. However, although the body of literature is small, a recent Cochrane analysis of the available data concludes that although the hitherto published results are controversial, they are promising [607]. Nevertheless, remember to consider the need for intubation to avoid severe complications: the appropriate intubation should not be delayed [318].

Proposed ventilator settings for an attempt of NIV in asthma patients [606]:

(a) Pressure-controlled ventilation
(b) PEEP: *3 cm* H_2O initially, to be increased up to max. *5 cm* H_2O *by 1 cm* H_2O steps every 15 min.
(c) Initial PS: *8 cm* H_2O, to be increased every 15 min to a max of *15 cm* H_2O or until respiratory rate $< 25/min$.

The currently best documented evidence on how to apply invasive mechanical ventilation in asthma is published in a review article based on most recent evidence-based data by Brenner et al. [672]. Although about two million patients per year visit, the emergency department of US hospitals for acute asthma [741] where of 2–20 % have been admitted to an ICU [742], only 5 (out of 41 randomized controlled trials and 6 meta-analyses) met the (strong) criteria being included in that review [672]; the data on this issue is still scarce. Nevertheless, the authors called their recommendations at the end strong. In addition, we considered other literature for our summary on how to set initial ventilator settings.

3. Indications for intubation in asthma patients
   The decision to intubate patients with an acute asthma attack is still largely a clinical one [672, 743]; however, some typical reasons to intubate a patient are described and are of common sense [744]:

   - Cardiac arrest
   - Respiratory arrest
   - Progressive physical exhaustion
   - Altered mental status
   - A pH <7.2, a pCO_2 increasing by more than 5 mmHg/h or values ≥55 to mmHg, or hypoxaemia with pO_2 <60 mmHg on 100 % oxygen
   - Silent chest on auscultation, respiratory rate >40/min, barotrauma
   - Therapy-resistant acidosis [745].

   It needs to be noted that even minor airway manipulation may promote laryngospasm or may worsen bronchospasm [672]; hence good sedation, rapid sequence induction and hydrocortisone iv are advocated [672].

4. Invasive mechanical ventilation in asthma
   Initial respirator settings in asthmatic patients [53, 318, 672]

   - Mode: volume controlled (VC) and pressure limited [53, 54, 318], is the generally worldwide preferred practice [746], but PC is of equal value [151, 717], BiPAP [54, 70]

- V_T: 7–8 *ml/kg* PBW [655, 672], range from 5–7 *ml/kg* PBW [661] to 6–10 *ml/kg* PBW [53, 652]
- Respiratory rate: ~10/*min* [655, 672], range from 8 to 14/*min* [53, 314, 318, 661]
- Minute volume (due to the required V_{Tex} <10 l/min): <10 *l/min* [573, 747]
- Plateau pressure: ≤26 *cm* H_2O [59–61, 190, 528]
- Inspiration to expiration ratio: I:E = 1:3–1:4 [654, 661]
- Expiration time: somewhat about 4 *s*, but in any case <5 *s* [53, 573]
- Inspiratory flow rate: 60–80 *l/min* [53, 655, 672], some recommend up to *100 l/min* [314]
- Inspiratory wave form: decelerating if flow rate is 80–100 *l/min* [53, 655, 672], may be constant/square from pattern at lower rates 60 *l/min* [655, 672]
- PEEP (external): 0[4] *to* <5 *cm* H_2O in case of spontaneous breaths or spontaneously breathing patients who are on ASB or BiPAP [287, 654, 661, 672]
- F_IO_2: up to 1.0, initially *1.0* to maintain PaO_2 >60 *mmHg* [655, 661, 672]
- Peak pressure: *<35 cm* H_2O [53]

  Permissive hypercapnia is probably the safest way to ventilate patients with acute asthma attacks [669, 670, 748], and reductions of respirator rate may lead to hypercapnia [672]. Fortunately, even p_aCO_2 values of 90 mmHg are frequently well tolerated [672]. The contraindications to permissive hypercapnia are an increased intracranial pressure, pre-existing epilepsy and severe myocardial dysfunction [748].

5. Additional (medical) treatment options in severe asthma and in therapy-resistant patients:
   Key elements are still β_2-agonists, ipratropium and steroids [53, 318, 650, 672]. Systemic corticosteroids are essential to diminish bronchospasm (which may continue after intubation [672]) and to attenuate inflammation [650, 654, 672, 675], and are considered gold standard in intubated patients [672]. The dosing is not as well investigated, e.g. 200 mg initially of hydrocortisone, thereafter a 100 mg every 6 h [654]; however, a Cochrane meta-analysis by Manser concluded 40 mg methylprednisolone (or equivalent) four times a day to be sufficient [749]. Nebulized β_2-agonists are fixed component in the therapeutic approach [650, 654, 672] as well. The most effective dosing is unclear [672]; in daily practice short-acting agents like salbutamol (5 mg) [654] may be given repeatedly or as continuous nebulization (salbutamol 10 mg/h) [654]. As nebulized ipratropium bromide elicits an even stronger bronchodilation and works synergistically to β_2-agonists, it should be administered as well, 500 μg every 4 h [650, 654].

Magnesium is very effective in severe asthma attacks and may be considered in life-threatening, therapy-resistant cases [750, 751]. By inhibition of bronchial smooth muscle contraction [650], magnesium has bronchodilating properties [751, 752], is

[4] PEEP is not recommended in sedated and paralysed patients [311, 672, 680].

shown to improve respiratory conditions in asthma attacks [751] and has evidence as being a good adjunct to β_2-agonists [753]. Two grams magnesium i.v. infusion over 20–30 min can be given and repeated to a maximum of 10 g/day [650, 654, 752, 754].

Heliox, a combination of helium and oxygen (60–70 % helium and 30–40 % O_2), may be applied in cases of absolute refractory disease [53, 314, 650]. Heliox is reported to increase peak expiratory flow rate, thereby alleviating dynamic hyperinflation, and to reduce peak inspiratory pressure and airway resistance; further oxygenation will improve and the carbon dioxide concentration will fall [755–758]. A clinical effect with less dyspnoea will ensue after 1 h time [759]. Unfortunately, clinical study results are conflicting [760–762].

1.15 Ventilator-Associated Pneumonia (VAP)

VAP is defined as a complication of mechanical ventilation arising ≥48–72 h after the onset of invasive, endotracheal intubation and mechanical ventilation determined by newly emerged pulmonary infiltrates on chest x-ray and fever, and/or leukocytosis, and/or purulent trachea–bronchial secretions [763, 764].

The endotracheal tube (ETT) constitutes a kind of bridge linking trachea and oropharynx and thus allowing bacterial loaded secretions to be drained and/or bacteria to move towards the lower airways triggering pneumonia [763]. Furthermore, the cough reflex is impaired, mucociliary flow is blocked by the inflated ETT cuff and, probably of special meaning, the same the same bacteria colonized in the oropharynx or stomach are often confirmed in the distal airways of patients with VAP; these are obviously significant factors in the disease development and progress [763, 765]. As such, drainage of saliva and/or gastric fluid is suggested to entail colonization of tracheal and lower airway mucosa, potentially provoking pneumonia [763]. It is not only the drainage of bacterial-loaded secretions along the trachea allowed by a not completely lining and blocking cuff, rather than the contamination of the inner luminal side of the tube with bacteria that acts as another potential bacterial source likely to instigate pneumonia [763, 766]. Despite regular suctioning, over a short time a layer of mucus and cells, called a biofilm, will cover the ETT lumen being an optimum condition for bacterial growth, and unfortunately this is hardly penetrated by antibiotics [767]. Up to 70 % of all intubated patients show the same bacterial flora in their tracheal secretion as can be analysed in their endotracheal tube's biofilm [768].

Thus, it is not the artificial ventilation per se causing ventilator-associated pneumonia rather the consequences of intubation and the concurrent circumstances [763]. The risk to develop VAP correlates with prolonged intubation, and reintubation [769], and even early tracheotomy cannot significantly inhibit the development of VAP. This indicates that the formation of VAP is multifactorial rather than mainly associated to the link (ETT) between oropharynx and trachea [770]. This is supported by the results of recent studies suggestive that bundle care is the way to markedly reduce the incidence of VAP appearance [771]. Established measures with protective potential include oral care [771], airway care, tube type and care [772, 773], cuff care [774], positioning [771], kinetic therapy [775], probiotics [776] and early enteral feeding [763]. In addition, sedation interruption and weaning protocols are also beneficial [771]:

I. To routinely rinse the oropharynx with chlorhexidine is an established procedure to markedly reduce tracheal colonization and the incidence of VAP in intubated patients [777].
II. Proper hand washing of the staff is a confirmed measure to reduce inoculation of bacteria within the ETT lumen [778].
III. Intermittent drainage of the subglottic secretions accumulating above the cuff by using a special tube equipped with a drainage system above the cuff is shown to be able to diminish without any relevant side effects the rate of VAP cases, length of ventilation and ICU stay [779]. Silver-coated ETTs with bacteriostatic properties of silver penetrating into the microbial membrane interfering with bacterial replication have been shown in the NASCENT study (of 2,000 patients) to cause substantially less VAP cases, particularly late-onset VAPs [773]. Furthermore, if mechanical ventilator support is required for more than 48 h, this tube is even cost-effective [780].
IV. Adjusted and sustained cuff pressure above 20 cm H_2O, but less than 30 cm H_2O, is paramount to effectively prevent microaspiration of bacterial-loaded subglottic secretions [763, 781, 782]. Unfortunately, despite close follow-ups, cuff pressure variations are common, promoting the development of VAP [781]. Importantly, pressures over 30 cm H_2O are associated with tracheal mucosal afflictions [782]. Devices keeping the cuff pressure constant are on the market with the first clinical results, encouraging for reducing the incidence of VAP [774].
V. There is no doubt that head-up positioning of the patients is beneficial in avoiding microaspirations of subglottic secretions compared to supine positioning, and a lower incidence of VAP occurrences are demonstrated in several studies [783, 784]. As such, 45° semirecumbent position is usually recommended [763]. However, recent pilot studies suggest lateral–horizontal positioning may be even more effective [785, 786]; a clinical trial investigating this approach in daily practice is under way since June 2012 [787].
VI. Since a recently published meta-analysis verified a lower incidence of VAP in mechanically ventilated patients receiving probiotics [776], their administration may have an ongoing position in our management [763]. Due to immune-modulatory properties and obviously their competition with the VAP-inducing microbes, they influence the colonization and reduce *Pseudomonas aeruginosa* growth and settlement in the respiratory tract [776].
VII. Enteral feeding is suspected as being a risk factor for aspiration, particularly in ventilated patients and thus for the development of VAP [788, 789]. However, the literature is inconsistent [790] with early enteral feeding (initiation within the first 48 h after admission), decreasing ICU and hospital mortality [788, 791]. Three risk factors have been identified in a recent large meta-analysis [792]: feeding method (continuous–intermittent), feeding site (gastric–small intestinal)

and timing (early–late); unfortunately again the results to each of these are equivocal and not decisively helpful in daily practice. Furthermore, large gastric volumes are an established factor provoking tracheobronchial aspiration in ventilated patients [793]; hence a feeding protocol controlling gastric filling is vital. Feeding is pivotal as malnutrition increases mortality [794, 795] and enteral feeding is clearly preferred to parenteral feeding which is used as an adjunct if insufficient calories are provided [796–798].

1.16 Weaning

The term weaning as defined by Thomason and Ely "implies the safe and yet rapid removal from mechanical ventilations as early as possible after identifying that the patient is able to breathe spontaneously" [799]. It is well known and documented that both prolonged ventilation and premature termination of PPMV are associated with increased mortality [800–803]. Unfortunately intensivists are generally poor in liberating their patients from PPMV, resulting in weaning time accounting for 2/3 of total ventilator time [19, 20]. Using just single, not really evaluated tests and the physician's clinical impression to evaluate whether a patient is ready to be successfully discontinued form PPMV is highly inaccurate; a positive predictive value of only 50% (a negative of 67 %) is reported [804]. Concurrently, 50 % of all patients who accidentally extubate themselves prematurely do not need re-intubation within the following 24 h [805, 806]. However, re-intubations are reported to increase the mortality rate [807]. Meanwhile several trails could establish evidence that daily assessment of the patient's readiness to be weaned and the use of standardized protocols significantly reduce respirator days and increase the accuracy of extubation [20, 808, 809]. Nevertheless, still 10–20 % of the patients need re-intubating, defined as the necessity to be re-intubated within 48–72 h post extubation [810–812].

Weaning from mechanical ventilation starts with the reduction of respiratory support and should begin as soon as possible [813, 814]. There is broad evidence that extubation can be achieved rapidly in the vast majority (75–80 %) of patients once the causal reason of the respiratory failure leading to PPMV has substantially improved or even resolved [810, 811, 815, 816]. Furthermore, this should happen in a standardized setting, and the method of weaning influences the duration of the weaning period and the success rate [817, 818]. A key element of care has become the daily interruption of sedation [819], also termed sedation hold, as the continuous administration of sedative agents in order to improve the patients' comfort [820, 821] will on the other hand prolong PPMV and hence ICU/hospital stay due to accumulation of the sedatives delaying mental recovery, particularly alertness of the patient [822]. The patient needs to be awake enough to sufficiently trigger the ventilator and to initiate adequate breaths, recover their cough reflex and protect the airway [816]. The clinician must be able to distinguish between disease-related issues, side effects especially of the sedative medication and mental alterations attributed to insufficient spontaneous breathing. The patient's respiratory capabilities and her/his reaction and behaviour during spontaneous breathing needs to be closely observed and monitored [819]. Furthermore, a clearly structured protocol has turned out in most trials to be preferred and superior to all other methods of care [816–818, 823]. It is highly suggested that a structured approach rather than a specific method to evaluate readiness, type of ventilator settings (see beyond) or criteria when to disconnect is important for the reduction of length of PPMV and early extubation [692, 816].

Quite different respirator settings are used to wean a patient. A typical, traditional and still the most common approach is PSV ventilation (or termed assisted spontaneous breathing mode, ASB), gradually reducing the pressure support assisting the spontaneous breaths of the patient [815, 816], e.g. 2–4 cm H_2O steps twice a day, until a pressure support level of 5–8 cm H_2O is reached or the criteria for a spontaneous breathing test are met [815]. However, other techniques are possible and in use, e.g. direct spontaneous breathing via T-piece over a prior determined time or as long as the patient feels comfortable and does not fatigue or develop unfavourable haemodynamic, respiratory or gas exchange patterns combined with phases to rest by connecting the patient back to the ventilator [810, 816].

Esteban [810] and Brochard [811] evaluated different ventilator modes for weaning a patient. Esteban found that intermittent spontaneous breathing trials over a T-tube facilitate extubation three times quicker than SIMV and twice as quick as PSV. The technique involved applying a low PEEP, at least twice a day, or once a day for a period for about two hours with PEEP of ≤5 cm H_2O. It was essential to avoid exhaustion and haemodynamic compromise while maintaining the patient in predefined respiratory and clinical limits [810]. In Brochard's study the PSV method performed significant better, with nearly 80 % of patients liberated from PPMV by PSV, while by SIMV or T-piece it was only 60 % ($p=0.05$) [811]. Thus, SIMV does not have sufficient evidence for efficient weaning and should not be used any longer [810, 811].

"Standardized weaning protocols offer the opportunity for safe and rapid removal from mechanical ventilation by identifying patients who are able to breathe spontaneously appropriate and thus are ready to wean" [824, 825].

A. Screening criteria identifying patients ready for spontaneous breathing trial (SBT):

- Evidence of relevant reversal of the underlying cause of respiratory failure [801, 824, 825]
- Adequate respiratory patterns [801, 812, 824, 826, 827]
  - FiO_2 0.4–0.5; p_aO_2/FiO_2– >150–200
  - PEEP ≤5–8 cm H_2O and pH ≥7.25
- Absence of clinically significant hypotension (MAP ≥60 mmHg) without vasopressors/inotropes or only in a minimal amount – dopamine or dobutamine <5 µg/kg/min
- Appropriate level of sedation/mentation [819, 822, 824, 826] (GCS ≥13)
- Ability to breath spontaneously [810, 812, 825]

- No evidence of increased intracranial pressure [799, 826]
- Active cough when suctioned [692, 826, 827]
- Other factors, but of minor importance like being afebrile [692], having minor amounts of secretion [828] and a stable metabolic status, e.g. blood glucose, electrolytes, etc. [692]

B. Failure of SBT, if *any* of the mentioned criterion occurs, and the test has to be stopped

- Respiratory rate (RR) >35/min [29, 692] or <8/min [824, 826]
- SaO_2 ≤88 % (≤90 %) [692, 799, 826], P_aO_2 ≤6.6–8.0 kpa [692],
- P_aCO_2 increase ≥1.3 kpa [692]
- Change in mental status, especially sustained anxiety, delirium, somnolence or coma [692, 824, 826]
- Marked use of accessory muscles [692, 799]
- Abdominal paradox [692, 799]
- Diaphoresis [692, 824, 826]
- Marked subjective dyspnoea [692, 799]
- Systolic BP >180–200 mmHg or <90 mmHg or change in SBP of more than 20 % [692, 824, 826]
- HR >120–140/min or change >20 % since SBT started [692, 824, 826]

One of the most common parameters used during the first minutes of SBT in order to predict weaning outcome is the ratio of respiratory rate (*f*) to tidal volume (V_T) [829], the so-called rapid shallow breathing index. A ratio <105 was said to predict a high probability of successful weaning [829]. A recently published re-evaluation by Khamiees [828] and unpublished data by Epstein (presented at the American Thoracic Society Meeting in May 2001) concluded that patients are safe and should be allowed to undergo SBT/weaning, even if f/V_T >105. A very recently published, newly defined discrimination factor, combining load/force balance with f/V_T might provide an even better predictive safety [830] but needs confirmation, and it may be that both parameters can be helpful in difficult situations.

C. Spontaneous Breathing Trial

- The reason causing acute respiratory failure has resolved to a major extent; the clinical status has clearly stabilized. This includes having COPD patients intubated and completely passively ventilated for at least 48–72 h,[5] definite improvement in case of pneumonia or ARDS, and furthermore the patient must pass all screening criteria; see A [692, 824, 825, 831].

[5] If a patient, suffering from exacerbated COPD is exhausted and has to be intubated and mechanically ventilated, a period of 48–72 h of exclusive mechanical ventilation is required for appropriate recovery of the diaphragm and other respiratory muscles. Only after such a period of controlled PPMV can we expect the respiratory pump to have recovered. Therefore, it makes no sense to start an attempt of spontaneous breathing in that patient population earlier than 48 h (72 h) after initiation of PPMV [714, 834].

- Respirator settings during SBT:
  The mandatory respiratory rate should be set at zero, with the patient breathing spontaneously either by T-tube or on pressure support of 0–7 cm H_2O [692, 799, 832]. The endotracheal tube imposes a resistive load on the respiratory muscles, and thus physicians recommend a low pressure support to offset this imposed load [832], termed tube compensation.
  In patients with significant auto-PEEP (COPD patients), PEEP or CPAP may conceivably enhance triggering [338, 641]. When these patients are spontaneously breathing, external PEEP is needed to counterbalance auto-PEEP. External PEEP will counterbalance the elastic mechanical load induced by auto-PEEP (intrinsic PEEP) and thus reduce the inspiratory muscle effort lowering the respiratory work [338] as discussed above. Most authors recommend applying up to 85 % of the auto-PEEP; if unknown, a PEEP between 3 and 5 cm H_2O is suitable [666, 682].
  It is also worth discussing some measures that facilitate respiration and reduce the respiratory workload of the patients. As such, the patients should be placed in an upright position [596], and on a comfortable, safe, and well-monitored mode of ventilation" [825].
- If detrimental effects occur it is often early during the trail [812, 829, 833]. Therefore, in the initial 5 min of the SBT the patient has to be monitored closely. Although some authors suggest, minor differences in regard to the failure criteria, (see B), maybe tolerable [692, 799], principally only those patients should continue the SBT, who pass all of the criteria mentioned (see B) [692, 799].
- The SBT should be continue for 30 min but not >2 h if tolerated [338, 692, 799, 812]. While most authors recommend the test should last for 2 h [810–812, 834], there is evidence that a duration of 30 min of sufficient spontaneous breathing is conclusive evidence to decide whether to extubate the patient or not [835].
- When passing all criteria of the SBT over a period of 30–120 min, the patient should be extubated immediately [692, 799, 810, 812, 835]. If any of the criteria is missed out, the patient has to be put back to the ventilator [692, 799, 810, 812, 835]. Successful completion of a 2 h SBT indicates a 85–90 % change of staying off mechanical ventilation for 48 h [811, 835–837]. However, Esteban [835] found in an SBT lasting just 30 min a similar range of reintubations (13 % in 48 h).

D. Reasons for failure of SBT

- Reasons for weaning failure comprise either incomplete resolution of the illness initially causing the need for mechanical ventilation or new problems [832].
- About 25 % of all patients passing the initial screening criteria (ready for SBT) fail to be weaned. This occurs particularly in patients suffering from COPD and/or left ventricular heart disease [811].

- A failure of weaning will increase the

  I. Duration of mechanical ventilation
  II. The mortality and the morbidity in those critically ill patients [812].

- The most frequent reasons for weaning failure are respiratory ones [692, 832, 838]:

Hyperinflation (COPD)	Bronchospasm
Obesity	Airway oedema, secretions
Abdominal distension	Upper airway obstruction
Secretions encrustation	Pleural effusion/ascites
Atelectasis	Interstitial inflammation/oedema

- Other relevant factors causing weaning failure [49, 692, 832, 838, 839] are:

LV heart failure	Drug overdose	Brain stem lesion
Acute myocardial ischaemia	**Muscle fatigue**	Sepsis
Critical illness myopathy	Hypothyroidism	Malnutrition/starvation
Drugs, e.g. steroids, Aminoglycosides		

- Hyperinflation can have a pivotal role in weaning failure in all patients with chronic obstructive lung disease [832]. Measurement of intrinsic (auto) PEEP is valuable in every COPD patient who fails to wean [832]. Appropriate treatment comprises of:

  1. Maximizing bronchodilator application
  2. Adjustment of ventilator settings to allow as much time as possible for expiration
  3. Addition of external PEEP [641, 666, 840, 841] to approximately 85 % of auto-PEEP [666, 682]

  When the ventilator settings are not optimal, the patient will have to generate more energy while breathing on the ventilator than when breathing spontaneously [678]. Optimal patterns of trigger sensitivity [832] and inspiratory flow rate [832] are the key factors which primarily determine the workload of breathing during mechanically supported ventilation [832]. Bonmarch and [842] showed in COPD patients that a high inspiratory peak flow, when using the pressure support modes, will reduce the respiratory work of the patient, while lengthening of the time of pressure rise will increase the respiratory work markedly. Thus, a high inspiratory flow rate (60–100 l/min), a PEEP of max 85 % of auto-PEEP and a low trigger sensitivity are necessary.

- A major issue before weaning is started is to allow respiratory muscles to recover from fatigue, in particular in COPD patients who are admitted with severe exhaustion of their respiratory muscles [832]. Importantly, critical illness itself induces muscle weakness, making it difficult to wean patients [843]. So there are a proportion of patients with marked weakness of their respiratory muscles due to critical illness and/or unresolved fatigue due to exhaustion as a result of their acute severe illness and unphysiological high demand of respiratory work.

Mechanical ventilation (with complete resting of the respiratory muscles) is the only way to treat muscle fatigue, in particular in the acute phase [711, 844]. This fatigue can last up to 72 h in the acute phase [711]; thus those patients have to be exclusively ventilated in the first 48 [70] hours [711, 844]. In the further course of the disease, controlled training might be effective. Skeletal muscles respond to training regimens when repetitive training periods of infra-maximal levels alternate with periods of rest [843, 845]. On the other hand, strength training differs conceptually and in practice from endurance training [843, 846]. Although the respiratory muscles should behave in a similar way as all other skeletal muscle groups, no definite studies examining especially the behaviour of the respiratory muscles while trained are done as yet [843]. There are suggestions that the effect of training might be at least partly genetically determined [847–849], and endurance training seems to be the superior way to train respiratory muscles [849]. For the purpose of weaning difficult-to-wean patients, increasing ventilator support during night-time is highly recommended [49]. Thus, the respiratory muscles must have had a proper period of rest; thereafter appropriate muscle training in the weaning phase (slow reduction of PS or training periods of infra-maximal intensity in turns with rest periods with PS guaranteeing respiratory rates <25/min) and increased pressure support during night-time are necessary.

- LV heart failure is another major cause of weaning failure [838], in particular if the patient suffers from both COPD and left heart failure [850]. Switching from positive pressure ventilation to spontaneous breathing is reported to be associated with pulmonary oedema, myocardial ischaemia [851, 852] and tachycardias [404]. The possible pathophysiological mechanisms inducing left heart dysfunction, in particular in patients with established coronary artery disease and chronic heart disease [853, 854], are:

  (a) Increased venous return due to abrupt shift from positive to negative intrathoracic pressure and increased sympathetic activity when switching from respirator-supported ventilation to spontaneous breathing resulting in enhanced RV and LV preload [855] and hence right ventricular as well as left ventricular end-diastolic volume and filling pressures (RVEDP and LVEDP) [850, 856].
  (b) Reduction of LV compliance related to myocardial ischaemia as weaning is a cardiac stress test [358, 850] with increased myocardial oxygen demand, often in a setting of pre-existing coronary artery disease [857–861]. Diastolic ventricular interdependence due to increased right ventricular filling and filling pressures may occur and compromise LV performance [862].
  (c) Abrupt increase of LV afterload due to negative intrathoracic pressures when merging to spontaneous breathing [363, 863].

The increased filling pressures have to be seen as the major determinant of the cardio-haemodynamic features, further affecting the airway resistance by narrowing the small airways and the work of breathing facilitating muscle fatigue of the just-recovering muscles, thus may precipitate a vicious cycle [850].

Accordingly, a typical setting of central fluid redistribution (redistribution of blood/volume to the thoracic circulation) and an abrupt increase in LV outflow impedance causing an afterload mismatch, often accompanied by diastolic dysfunction (as represented by altered LV compliance), is described by Lemaire investigating the causation of left ventricular dysfunction during unsuccessful weaning [850]. Meanwhile, central fluid redistribution and afterload mismatch have been identified playing a fundamental role in the pathobiology of acute heart failure and are encroached upon the modern and current concept of acute heart failure pathophysiology launched by Cotter [864, 865].

As demonstrated in a couple of recent studies, patients with altered diastolic properties are at higher risk of developing diastolic dysfunction or failure during weaning resulting in weaning failure [866–868]. It may be that "LV diastolic dysfunction is a key player" in unsuccessful weaning [868]. A simple echocardiographic examination as described by Papanikolaou [868] done immediately before SBT, and during the first minutes of weaning may be valuable to recognize diastolic dysfunction or the onset of failure developing in the weaning process as Voga recommends [869]. Therefore, an extended cardiologic diagnostic examination including at least an ECG and echocardiogram are essential in weaning failure. Remember, it is often difficult, and maybe impossible, to distinguish clinical symptoms resulting from respiratory or cardiac failure [838].

- Abdominal distension and obesity alter the mechanical conditions of the thorax. Increased intra-abdominal pressure (abdominal compartment syndrome in its worst case) considerably affects the chest wall elastance and thus the transpulmonary pressure. Intra-abdominal hypertension, e.g. acute abdomen, is always associated with a markedly reduced chest wall compliance [119, 870] and thus has relevant effects on the mechanics of the respiratory system [139, 871].

E. Weaning of patients who failed SBT (failed to be extubated)

- Patients who have failed an SBT should receive a stable, non-exhausting, comfortable form of ventilatory support [692]. Some patients have to continue mechanical ventilation; the vast majority can be supported by a reasonable level of pressure support (PSV) [692, 799, 832]. The lowest pressure support level providing comfortable breathing should be chosen [318]. Clinical research data suggest that the most appropriate pressure support is achieved when the patient's respiratory rate ranges between 25/min and 30/min, higher than common practice would expect [596]. A pressure support level of 15–20 cm H_2O will commonly match this goal, but there is a wide variation from patient to patient [318]. It is recommended to focus during the first 24 h after failed SBT on:

(a) Maintaining adequate muscle unloading
(b) Optimizing comfort (i.e. sedation)
(c) Avoiding complications
(d) Identification of the reasons which caused SBT failure and their proper treatment [692, 832, 838, 839].

An aggressive ventilatory support reduction must be avoided [692].

- The reasons why the patient failed the SBT should be accurately analysed [692, 832]. Due to the fact that failure to wean is usually multifactorial [832], all factors described above have to be evaluated, including the performance of an echocardiogram. Check in particular for hyperinflation (COPD), LV heart failure, muscle fatigue, abdominal distension, acute myocardial ischaemia, secretions encrustation, critical illness weakness, obesity and adequate consciousness. Appropriate treatment of identified reasons for SBT failure is essential for successful weaning.
- Special strategy and ventilator settings:
There are a number of ventilator modes available providing substantial ventilatory support and offering simultaneously the opportunity for partial reduction of ventilator assistance. We know that the former widely used SIMV technique (reducing the rate of ventilator supplied breaths/min) is the least efficient one and should not be used any longer [801, 810, 811, 832]. Esteban [810] found in a study comprising 546 medical–surgical patients that the success rate in weaning the patients was significantly higher when training the patient using a constant alternation between multiple T-piece trials daily (avoiding exhaustion of the patient) and phases of adequate pressure support in between. Adequate PS levels resulting in a comfortable patient and respiratory rates of 25–30/min compared to the method of continuous reduction of the level of pressure support in accordance with the patients improvements have minimal and inconsistent data [692, 832]. Potentially, especially if taking the training effects into account, COPD patients might benefit from stable phases of rest with comfortable breathing support and low respiratory rates. Thus favouring the daily T-piece training periods and proper rest periods on comfortable pressure support with high inspiratory peak flow for muscle recovery and avoidance of hyperinflation in between [318, 641, 666, 810, 832, 840, 842, 843].
The ACCP consensus conference on mechanical ventilation states that when using the PSV weaning strategy (reducing the PS stepwise and continuously until the screening criteria are met), the reduction of the PS should be orientated more on the respiratory rate and should not exceed 30/min rather than on the VT [49].
Switching to non-invasive mechanical ventilation during the weaning process is not extensively evaluated yet (no studies in postoperative

patients, patients with altered neurologic status and severe concomitant diseases), but there are some encouraging study results, especially in COPD patients [596, 872, 873]. Therefore a swap from invasive weaning with the patient still intubated to non-invasive ventilation in the weaning phase should be seriously considered on an individual basis, in particular in COPD patients.

F. Summary [692, 809–812, 832, 835]

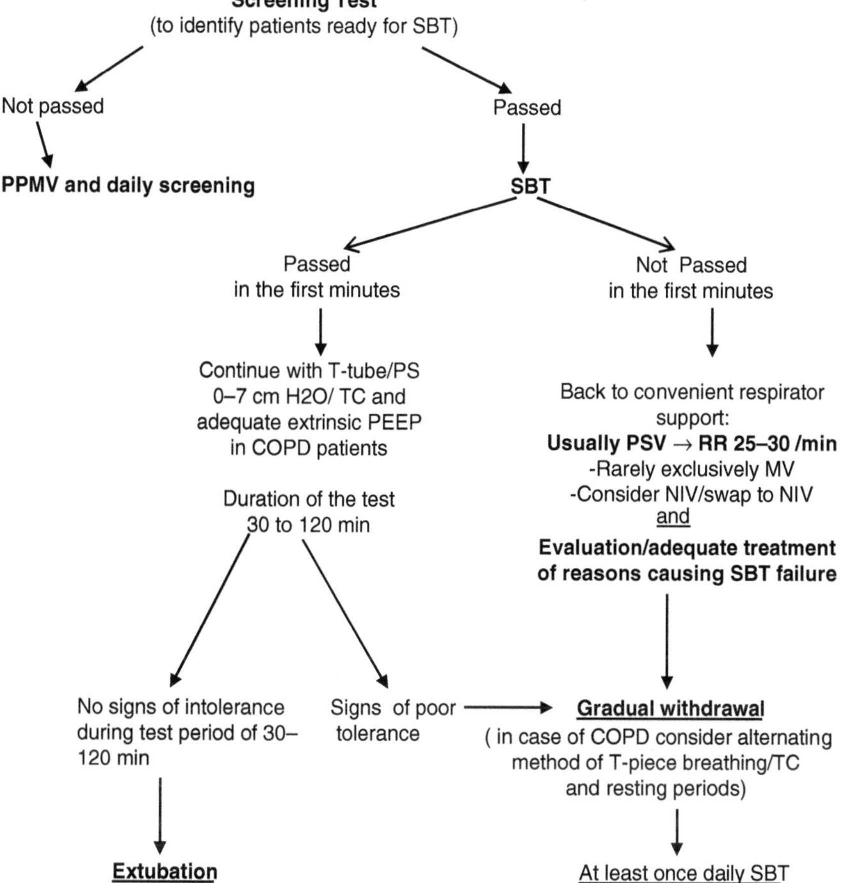

References

1. Green KE, Peters JI. Pathophysiology of acute respiratory failure. Clin Chest Med. 1994;15:1.
2. Grippi MA. Respiratory failure: an overview. In: Elias JA, Fishman JA, Fishman PA, editors. Fishman's pulmonary diseases and disorders. 3rd ed. New York: McGraw-Hill; 1998. p. 2525–35.
3. Wood LD. The pathophysiology and differential diagnosis of acute respiratory failure. In: Schmidt GA, Wood LDH, Hall JB, editors. Principles of intensive care. New Dehli: McGraw-Hill; 2005. p. 417–22.

4. Tisi GM. Pulmonary pathophysiology in clinical medicine. 2nd ed. Baltimore: Williams and Wilkens; 1983.
5. Chakrabarti B, Calverley PMA. Management of acute ventilatory failure. Postgrad Med J. 2006;82:438–45.
6. Weinberger SE, Schwartzstein RM, Weiss JW. Hypercapnia. N Engl J Med. 1989;321:1223–31.
7. Balk RA, Bone RG. Acute respiratory failure. Med Clin North Am. 1983;67:351.
8. Campbell EJM. Respiratory failure. BMJ. 1965;1:1451–60.
9. Gurka DP, Balk RA. Acute respiratory failure, including acute lung injury and ARDS. In: Dellinger RP, Parrillo JE, editors. Critical care medicine: principles of diagnosis and management in the adult. St Louis: Mosby; 2002. p. 705.
10. Roussos C, Koutsoukou A. Respiratory failure. Eur Respir J. 2003;22 Suppl 47:s3–14.
11. Melanson P. McGill University, Critical Care Med, [Online] 18 Oct 2011. http://www.mcgill.ca/criticalcare/teaching/files/acute. Cited 10 Aug 2013.
12. Suri HS, Li G, Gajic O. Epidemiology of acute respiratory failure and mechanical ventilation. In: Vincent JL, editor. Yearbook of intensive care and emergency medicine. Berlin/Heidelberg/New York: Springer; 2008. p. 183.
13. Marini JJ. Initial management of acute hypoxemia. In: Ayres SM, Holbrook PR, Grenvik A, editors. Textbook of critical care. Philadelphia: Saunders; 2000. p. 1404–19.
14. Hall JB, Schmidt GA, Wood LD. Acute hypoxemic respiratory failure. In: Nadel JA, Murray JF, editors. Textbook of respiratory medicine. Philadelphia: Saunders; 2000. p. 2413–42.
15. Glenny RW. Teaching ventilation/perfusion relationships in the lung. Adv Physiol Educ. 2008;32:192–5.
16. Weinberger SE, Cockrill BA, Mandel J. Pulmonary anatomy and physiology- the basics. In: Principles of pulmonary medicine. 5th ed. Philadelphia: Saunders Elsevier; 2008. p. 1–18.
17. Gunning EEJ. Pathophysiology of respiratory failure and indications for respiratory support. Surgery (Oxford). 2003;21:72–6.
18. Esteban A, Anzueto A, Alia I. How is mechanical ventilation employed in the intensive care unit? An international utilization review. Am J Respir Crit Care Med. 2000;161:1450–8.
19. Esteban A, Alia I, Ibanez J. The Spanish Lung Failure Collaborative Group, modes of mechanical ventilation and weaning. Chest. 1994;106:1188–93.
20. Esteban A, Anzueto A, Frutos F. Characteristics and outcomes in adult patients receiving mechanical ventilation. A 28-day international study. JAMA. 2002;287:345–55.
21. Esteban A, Alia I, Ibanez J. Modes of mechanical ventilation and weaning. A national survey of Spanish hospitals. Chest. 1994;106:1188–93.
22. Troche G, Moine P. Is the duration of mechanical ventilation predictable? Chest. 1997;112: 745–51.
23. Stauffer JL, Fayter NA, Graves B. Survival following mechanical ventilation for acute respiratory failure in adlut men. Chest. 1993;104:1222–9.
24. Chastre J, Fagon JY. Ventilator-associated pneumonia. Am J Respir Crit Care Med. 2002;165:1618–23.
25. Langer M, Mosconi P, Cigada M. Long-term respiratory support and risk of pneumonia in critically ill patients. Am Rev Respir Dis. 1989;140:302–5.
26. Cook DJ, Walter SD, Cook RJ. Incidence of and risks factors for ventilator-associated pneumonia in critically ill patients. Ann Intern Med. 1998;129:433–40.
27. Fagon JY, Chastre J, Domart Y. Nosocomial pneumonia in patients receiving continuous mechanical ventilation. Prospective analysis of 52 episodes with use of a protected specimen brush and quantitative culture techniques. Am Rev Respir Dis. 1989;139:877–84.
28. Esteban A, Ferguson ND, Frutos-Vivar F. Mechanical ventilation: epidemiology and outcomes. In: Anzueto A, Cook DJ, Esteban A, editors. Evidence-based management of patients with respiratory failure. Berlin/Heidelberg: Springer; 2004. p. 5.
29. Frutos-vivar F, Ferguson ND, Esteban A. The epidemiology of mechanical ventilation. In: Brochard L, Slutsky AS, editors. Mechanical ventilation. Berlin/Heidelberg: Springer; 2005. p. 11.

30. Brower RG, Matthay MA, Morris A. Ventilation with lower tidal volumes as compared with traditional tidal volumes for acute lung injury and respiratory distress syndrome. N Engl J Med. 2000;342:1301–8.
31. Papadakis MA, Lee KK, Brower WS. Prognosis of mechanically ventilated patients. West J Med. 1993;159:659–64.
32. Chatburn RL. Classification of ventilator modes: update and proposal for implementation. Respir Care. 2007;52:301–23.
33. Mireles-Cabodevila E, Diaz-Guzman E, Heresi GA. Alternative modes of mechanical ventilation: a review for the hospitalist. Cleve Clin J Med. 2009;76:417–30.
34. Hyzy RC, Parsons PE, Wilson KC. Modes of mechanical ventilation. UpToDate. [Online] 30 Apr 2010. http://www.uptodate.com/contents/modes-of-mechanical-ventilation. Cited 28 Oct 2013.
35. Dries DJ, Marini JJ. Mechanical ventilation, types of invasive ventilation. In: Bellomo R, Kellum JA, Ronco C, editors. Critical care nephrology. Philadelphia: Saunders; 2009. p. 22–31.
36. Bahns E. Das Beatmungsbuch – Spontanatmung in maschineller Beatmung. Lübeck: Drägerwerk AG; 2010. www.draeger.com.
37. MacIntyre NR. Respiratory function during pressure support ventilation. Chest. 1986;89:677.
38. Brochard L, Pluskwa F, Lemaire F. Improved efficacy of spontaneous breathing with inspiratory pressure support. Am Rev Respir Dis. 1987;136:411.
39. Sassoon C, Gruer S. Characteristics of the ventilator pressure- and flow- trigger variables. Intensive Care Med. 1995;21:159.
40. Gali B, Goyal DG. Positive pressure mechanical ventilation. Emerg Med Clin North Am. 2003;21:453–73.
41. Branson RD, Campbell RS, Jr Davis K. Comparison of pressure and flow triggering systems during continuous positive airway pressure. Chest. 1994;106:540.
42. Sassoon CS, Del Rosario N, Fei R. Influence of pressure- and flow-triggered synchronous intermittent mandatory ventilation on inspiratory muscle work. Crit Care Med. 1994;22:1933.
43. Gurevitch MJ, Gelmont D. Importance of trigger sensitivity to ventilator response delay in advanced chronic obstructive pulmonary disease with respiratory failure. Crit Care Med. 1989;17:354.
44. Sassoon CS, Lodia R, Rheeman CH. Inspiratory muscle work of breathing during flow-by, demand-flow, and continous-flow systems in patients with chronic obstructive pulmonary disease. Am Rev Respir Dis. 1992;145:1219.
45. Branson R, Chatburn RL, Hess D. Respiratory care equipment. 2nd ed. Philadelphia: Lippincott; 1998.
46. Branson RD, Chatburn RL. Should adaptive pressure control modes be utilized for virtually all patients receiving mechanical ventilation? Respir Care. 2007;52:478–85.
47. Morch E. History of mechanical ventilation. In: Banner M, Downs J, Kirby R, editors. Clinical applications of ventilatory support. New York: Churchill-Livingstone Inc; 1990.
48. Ralfs F. Protective ventilation booklet. Lübeck: Dräger Werke; 2011. p. 8, chapter 2.
49. Slutsky AS. Mechanical ventilation. American College of Chest Physician's Consensus Conference. Chest. 1993;104:1833–59.
50. Courey AJ, Parsons PE, Wilson KC. Overview of mechanical ventilation. UpToDate. [Online] 29 Aug 2013. http://www.uptodate.com/contents/overview-of-mechanical-ventilation. Cited 30 Oct 2013.
51. Gil A, Carrizosa F, Herrero A. Influence of mechanical ventilation on blood lactate in patients with acute respiratory failure. Intensive Care Med. 1998;24:924.
52. Cohen CA, Zagenlbaum G, Gross D. Clinical manifestations of inspiratory muscle fatigue. Am J Med. 1982;73:308.
53. Oddo M, Feihl F, Schaller MD. Management of mechanical ventilation in acute severe asthma: practical aspects. Intensive Care Med. 2006;32:501–10.
54. Santanilla JI, Daniel B, Yeow M-E. Mechanical ventilation. Emerg Med Clin North Am. 2008;26:849–62.

55. Frutos-Vivar F, Ferguson ND, Esteban A. Ventilator modes: which do we use and how should we use them? In: Anzueto A, Cook D, Esteban A, editors. Evidence-based management of patients with respiratory failure. Berlin/Heidelberg: Springer; 2004. p. 15–20.
56. Rappaport SH, Shpiner R, Yoshihara G. Randomised, prospective trail of pressure-limited versus volume-controlled ventilation in severe respiratory failure. Crit Care Med. 1994;22:22–32.
57. Campbell RS, Davis SR. Pressure-controlled vs volume-controlled ventilation: does it matter? Respir Care. 2002;47:416–24.
58. Chacko B, Peter JV, Tharyan P. Pressure-controlled versus volume-controlled ventilation for acute respiratory failure due to acute lung injury (ALI) or acute respiratory distress syndrome (ARDS). Cochrane Database Syste Rev. 2010;(11):CD008807. doi:10.1002/14651858.CD008807.
59. Hager DN, Krishman JA, Hayden DL. Tidal volume reduction in patients with acute lung injury when plateau pressures are not high. Am J Respir Crit Care Med. 2005;172:1241–5.
60. Jardin F, Vieillard-Baron A. Is there a safe plateau pressure in ARDS? The right heart only knows. Intensive Care Med. 2007;33:444–7.
61. Vieillard-Baron A, Schmitt JM, Augarde R. Acute cor pulmonale in acute respiratory distress syndrome submitted to protective ventilation: incidence, clinical implications and prognosis. Crit Care Med. 2001;29:1551–5.
62. Boussarsar M, Thierry G, Jaber S. Relationship between ventilatory settings and barotrauma in the acute respiratory distress syndrome. Intensive Care Med. 2002;28:406–13.
63. Marini JJ, Crooke III PS, Truwit JD. Determinants and limits of pressure-preset ventilation: a mathematical model of pressure control. J Appl Physiol. 1989;67:1081–92.
64. Hubmayr RD, Gay PC, Tayyab M. Respiratory system mechanics in ventilated patients: techniques and indications. Mayo Clin Proc. 1987;62:358–68.
65. Dreyfuss D, Saumon G. Ventilator- induced lung injury: lessons from experimental studies. Am J Respir Crit Care Med. 1998;157:294–323.
66. Artigas A, Bernard GR, Carlet J. The American-European Consensus Conference on ARDS, part2: Ventilatory, pharmacologic, supportive therapy, study design strategies, and issues related to recovery and remodeling. Acute respiratory distress syndrome. Am J Respir Crit Care Med. 1998;157:1332–47.
67. Brochard L, Roudot-Thoraval F, Roupie E. Tidal volume reduction for prevention of ventilator-induced lung injury in acute respiratory distress syndrome: The Multicenter Trail Group on Tidal Volume reduction in ARDS. Am J Respir Crit Care Med. 1998;158:1831–8.
68. Baker AB, Restall R, Clark BW. Effects of varying inspiratory flow waveform and time in intermittent positive pressure ventilation: emphysema. Br J Anaesth. 1982;54:547–54.
69. Branson RD, Campbell RS. Modes of ventilator operation. In: Branson RD, MacIntyre NR, editors. Mechanical ventilation. Philadelphia: WB Saunders; 2001. p. 55.
70. Modrykamien A, Chatburn RL, Ashton RW. Airway pressure release ventilation: an alternative mode of mechanical ventilation in acute respiratory distress syndrome. Cleve Clin J Med. 2011;78:101–10.
71. Stock MC, Downs JB, Frolicher DA. Airway pressure release ventilation. Crit Care Med. 1987;15:462–6.
72. Martin LD, Wetzel RC. Optimal release time during airway pressure release ventilation in neonatal sheep. Crit Care Med. 1994;22:486–93.
73. Putensen C, Rasanen J, Lopez FA. Ventilation- perfusion distributions during mechanical ventilation with superimposed spontaneous breathing in canine lung injury. Am J Respir Crit Care Med. 1994;150:101–8.
74. Hörmann CH, Baum M, Putensen CH. Biphasic positive airway pressure (BIPAP XE "BIPAP") – a new mode of augmented ventilation. Eur J Anaesthesiol. 1994;11:37–42.
75. Vassilakopoulos T, Petrof B. Ventilator-induced diaphragmatic dysfunction. Am J Respir Crit Care Med. 2004;169:336–41.
76. Petrof B, Jaber S, Matecki S. Ventilator-induced diaphragmatic dysfunction. Curr Opin Crit Care. 2010;16:19–25.

77. Callaghan LA. Invited editorial on 'acquired respiratory muscle weakness in critically ill patients: what is the role of mechanical ventilation- induced diaphragm dysfunction?'. J Appl Physiol. 2009;106:360–1.
78. Mehta S, Hill NS. Non invasive ventilation. State of the art. Am J Respir Crit Care Med. 2001;163:540–77.
79. Katz JA, Marks JD. Inspiratory work with and without continous positive airway pressure in patients with acute respiratory failure. Anesthesiology. 1985;63:598–607.
80. Carry Z, Gottfried SB, Levy RD. Ventilatory muscle support in respiratory failure with nasal positive pressure ventilation. Chest. 1990;97:150.
81. Renston JP, DiMarco AF, Supinski GS. Respiratory muscle rest using nasal BiPAP ventilation in patients with stable severe COPD. Chest. 1994;105:1053.
82. Pierson DJ. Conference proceedings: indications for mechanical ventilation in adults with acute respiratory failure. Respir Care. 2002;47:249–62.
83. Kovacs G, Law AJ, Ross J. Acute airway management in the emergency department by non-anesthesiologists. Can J Anaesth. 2004;51:174–80.
84. Byrd RP, Mosenifar Z. Mechanical ventilation. [Online] medscape: emedicine, 26 Apr 2012. http://emedicine.medscape.com/article/304068-overview#aw2aab6b5. Cited 01 Nov 2013.
85. Gunning KEJ, Pierson DJ. Indications for mechanical ventilation in adults with acute respiratory failure. Respir Care. 2002;47:249–62.
86. Marini JJ, Wheeler AP. Indications and options for mechanical ventilation. Critical care medicine: the essentials. 3rd ed. Philadelphia: Lippincott Williams & Wilkins; 2006. p. 120–38.
87. Restrepo RD. Neurologic assessment. In: Dexter JR, Heuer A Wilkins RC, editors. The netter collection of medical illustrations, vol 7, 6th ed. Clinical assessment in respiratory care. Elsevier Health B 978 - 1- 4160-5923-3.0006-7; 00006; 2010.
88. Duncan R, Thahore S. Decreased GCS score does not mandate endotracheal intubation in emergency department. J Emerg Med. 2009;37:451–5.
89. Pierson DJ. Patient-ventilator interaction. Respir Care. 2011;56:214–28.
90. MacIntyre N. Patient-ventilator trigger dyssynchrony: a common phenomenon with important implications. Crit Care. 2013;17:157.
91. Epstein SK. How often does patient-ventilator asynchrony occur and what are the consequences? Respir Care. 2011;56:25–38.
92. Sassoon CS, Caiozzo VJ, Manka A. Altered diaphragm contractile properties with controlled mechanical ventilation. J Appl Physiol. 2002;92:2585–95.
93. Le Bourdelles G, Viires N, Boczkowski J. Effects of mechanical ventilation on diaphragmatic contractile properties in rates. Am J Respir Crit Care Med. 1994;149:1539–44.
94. Carlucci A, Pisani L, Ceriana P. Patient-ventilator asynchronies: may the respiratory mechanics play a role? Crit Care. 2013;17:R54.
95. Sassoon CSH. Triggering of the ventilator in patient-ventilator interactions. Respir Care. 2011;56:39–48.
96. Kondili E, Prinianakis G, Georgopoulos D. Patient-ventilator interaction. Br J Anaesth. 2003;91:106–19.
97. Dick CR, Sassoon CS. Patient-ventilator interactions. Clin Chest Med. 1996;17:423–38.
98. Vignaux L, Vargas F, Roeseler J. Patient-ventilator asynchrony during non-invasive ventilation for acute respiratory failure: a multicenter study. Intensive Care Med. 2009;35:840–6.
99. Thille A, Rodriguez P, Cabello B. Patient-ventilator asynchrony during assisted mechanical ventilation. Intensive Care Med. 2006;32:1515–22.
100. Epstein SK. Optimizing patient-ventilator synchrony. Semin Respir Crit Care Med. 2001;22:137–52.
101. de Witt M, Miller KB. Ineffective triggering predicts increased duration of mechanical ventilation. Crit Care Med. 2009;37:2740–5.
102. Racca F, Squadrone V, Ranieri VM. Patient-ventilator interaction during the triggering phase. Respir Care Clin N Am. 2005;11:225–45.
103. Nava S, Bruschi C, Rubini F. Respiratory response and inspiratory effort during pressure support ventilation in COPD patients. Intensive Care Med. 1995;21:871–9.

104. Thille AW, Cabello B, Galia F. Reduction of patient-ventilator asynchrony by reducing tidal volume during pressure –support ventilation. Intensive Care Med. 2008;34:1477–86.
105. Chao DC, Scheinborn DJ, Stearn-Hassenpflug M. Patient-ventilator trigger asynchrony in prolonged mechanical ventilation. Chest. 1997;112:1592–9.
106. Macklen PT, Mead J. The respiratory system, vol III, parts 1 and 2: Mechanics of breathing. In: Macklem PT, Mead J, Geiger SR, Fishman AP, editors. Handbook of physiology: section 3, revisited edition. Oxford: Oxford University Press; 1988.
107. Silverthorn DU. Das respiratorische System: Atemmechanik (Kapitel 17). In: Human physiology: an integrated approach. 4th ed. München: Pearson Education Deutschland GmbH, German Translation; 2009.
108. Osborne S. Physiology of ventilation. Vancouver: Department of Cellular and Physiological Sciences, University of Columbia; 2012. www.sallyosborne.com/Med%20Lecture%20-%20P.
109. Schwartzstein RM, Parker MJ. Respiratory physiology: a clinical approach. Baltimore: Lippincott Williams and Wilkins; 2006. Chapter, 1,2 and 3.
110. Kacmarek RM. Respiratory physiology/pulmonary gas exchange. In: Steingrub JS, Kacmarek RM, Stoller JK, Higgins TL, editors. Cardiopulmonary critical care. Oxford: BIOS Scientific Publishers ltd; 2002.
111. Lexington Pulmonary and Critical Care. Lung mechanics and mechanical ventilation by DR Elton. [Online] 2006. http://www.lexingtonpulmonary.com/education/lungmech/lungmech.html. Cited 25 Aug 2013.
112. Gattinoni L, Chiumello D, Carlesso E. Bench-to-bedside review: chest wall elastance in acute lung injury/acute respiratory distress syndrome patients. Crit Care. 2004;8:350–5.
113. Zanella A, Bellani G, Pesenti A. Airway pressure and flow monitoring. Curr Opin Crit Care. 2010;16:255–60.
114. Richard J-C, Marini JJ. Transpulmonary pressure as a surrogate of plateau pressure for lung protective strategy: not perfect but more physiologic. Intensive Care Med. 2012;38:339–41.
115. Gattinoni L, Pesenti A, Avalli L. Pressure volume curve of total respiratory system in acute respiratory failure: computed tomographic scan study. Am Rev Respir Dis. 1987;136:730–6.
116. Sarge T, Talmor D. Targeting pulmonary pressure to prevent ventilatory induced lung injury. Minerva Anestesiol. 2009;75:293–9.
117. Barberis L, Manno E, Guerin C. Effect of end-inspiratory pause duration on plateau pressure in mechanically ventilated patients. Intensive Care Med. 2003;29:130–4.
118. Gattinoni L, Pelosi P, Suter PM. Acute respiratory distress syndrome caused by pulmonary and extrapulmonary disease. Am J Respir Crit Care Med. 1998;158:3–11.
119. Ranieri VM, Brienza N, Santostasi S. Impairment of lung and chest wall mechanics in patients with acute respiratory distress syndrome. Am J Respir Crit Care Med. 1997;156:1082–91.
120. Pelosi P, Croci M, Ravagnan I. The effect of body mass on lung volumes, respiratory mechanics, and gas exchange during general anesthesia. Anesth Analg. 1998;87:654–60.
121. Oczenski W. Chapter 10: ARDS. In: Atmen – Atemhilfen: Atemphysiologie und Beatmungstechnik. 9th ed. Stuttgart: Thieme Verlag; 2008.
122. Dos Santos CC, Slutsky AS. Invited review: mechanisms of ventilator-induced lung injury: a perspective. J Appl Physiol. 2000;89:1645–55.
123. Lipes J, Bojmehrani A, Lellouche F. Low tidal volume ventilation in patients without respiratory distress syndrome: a paradigm shift in mechanical ventilation. Crit Care Res Pract. 2012:416862, 12 pages. doi:10.1155/2012/416862.
124. Chiumello D, Carlesso E, Cadringher P. Lung stress and strain during mechanical ventilation for acute respiratory distress syndrome. Am J Respir Crit Care Med. 2008;178:346–55.
125. Salome CM, King GG, Berend N. Physiology of obesity and effects on lung function. J Appl Physiol. 2010;108:206–11.
126. Talmor D, Sarge T, O'Donnell CR. Esophageal and transpulmonary pressures in acute respiratory failure. Crit Care Med. 2006;34:1389–94.

127. Grasso S, Mascia L, Del Turco M. Effects of recruiting maneuvers in patients with acute respiratory distress syndrome ventilated with protective ventilatory strategy. Anesthesiology. 2002;96:795–802.
128. Pelosi P, Ravagnan I, Giurati G. Positive end expiratory pressure improves respiratory function in obese patients but not in normal subjects during anesthesia and paralysis. Anesthesiology. 1999;91:1221–31.
129. Sousa AS, Moll RJ, Pontes CF. Mechanical and morphometrical changes in progressive bilateral pneumothorax and pleural effusion in normal rats. Eur Respir J. 1995;8:99–104.
130. Ladosky W, Botelho MA, Albuquerque JP. Chest mechanics in morbidly obese non-hypoventilated patients. Respir Med. 2001;95:281–6.
131. Grasso S, Terragni P, Birocco A. ECMO criteria for influenza A (H1N1)-associated ARDS: role of transpulmonary pressure. Intensive Care Med. 2012 doi:10.1007/s00134-012-2490-7.
132. Gattinoni L, Pelosi P, Suter PM. Acute respiratory distress syndrome caused by pulmonary and extrapulmonary disease. Different syndromes? Am J Respir Crit Care Med. 1998;158:3–11.
133. Malbrain M. Abdominal pressure in the critically ill. Curr Opin Crit Care. 2000;6:1–21.
134. Sugrue M, Hilman KM. Intra-abdominal hypertension and intensive care. In: Vincent JL, editor. Yearbook of intensive care and emergency medicine. Berlin: Springer; 1998. p. 667–76.
135. Hubmayr RD, Rodarte JR, Walters BJ. Regional ventilation during spontaneous breathing and mechanical ventilation in dogs. J Appl Physiol. 1987;63:2467–75.
136. Brander L, Ranieri VM, Slutsky AS. Esophageal and transpulmonary pressure help optimize mechanical ventilation in patients with acute lung injury. Crit Care Med. 2006;34:1556–8.
137. Marini JJ. Mechanical ventilation: past lessons and the near future. Crit Care. 2013;17 Suppl 1:S1.
138. Muscedere JG, Mullen JB, Gan K. Tidal ventilation at low airway pressures can augment lung injury. Am J Respir Crit Care Med. 1994;149:1327–34.
139. Pelosi P, Cereda M, Foti G. Alterations of lung and chest wall mechanics in patients with acute lung injury: effects of positive end-expiratory pressure. Am J Respir Crit Care Med. 1995;152:531–7.
140. Washko GR, O'Donnell CR, Loring SH. Volume-related and volume-independent effects of posture on esophageal and transpulmonary pressures in healthy subjects. J Appl Physiol. 2006;100:753–8.
141. Pelosi P, D'Onofrio D, Chiumello D. Pulmonary and extrapulmonary acute respiratory distress syndrome are different. Eur Respir J. 2003;22:48s–56.
142. Pelosi P, Cadringher P, Bottino N. Sigh in acute respiratory distress syndrome. Am J Respir Crit Care Med. 1999;159:872–80.
143. Lim CM, Kim EK, Lee JS. Comparison of the response to the prone position between pulmonary and extrapulmonary acute respiratory distress syndrome. Intensive Care Med. 2001;27:477–85.
144. Gattinoni L, Pelosi P, Valenza F. Patient positioning in acute respiratory failure. In: Tobin M, editor. Principles and practice of mechanical ventilation. New York: McGraw-Hill; 1994. p. 1067–77.
145. Pinsky M, Desmet JM, Vincent JL. Effect of positive end expiratory pressure on right ventricular function in humans. Am Rev Respir Dis. 1991;143:25–31.
146. Rittner F, Döring M. Kurven und Loops in der Beatmung Dräger, Technik für das Leben. Lübeck: Dräger; 2013.
147. Moon R, Camporesi E. Respiratory monitoring. In: Miller R, editor. Miller's anesthesia. 6th ed. New York: Elsevier; 2005. p. 1466.
148. Kirby RR, Banner MJ, Downs JB. Clinical applications of ventilator support. New York: Churchill Livingstone; 1990. p. 415.
149. Agostoni E, Hyatt RE. Static behavior of the respiratory system. In: Geiger SR, editor. Handbook of physiology, section3: the respiratory system, volume III, part 1. Bethesda: American Physiological Society; 1986. p. 113–30.
150. Fanelli V, Ranieri VM. When pressure does not mean volume? Body mass index may account for dissociation. Crit Care. 2011;15:143.

151. Aboab J, Jonson B, Kouatchet A. Effect of inspired oxygen fraction on alveolar derecruitment in acute respiratory distress syndrome. Intensive Care Med. 2006;32:1979–86.
152. Jardin F, Vieillard-Baron A. Acute cor pulmonale. Curr Opin Crit Care. 2009;15:67–70.
153. Al-Rawas N, Banner MJ, Euliano NR. Expiratory time constant for determination of plateau pressure, respiratory system compliance, and total resistance. Crit Care. 2013;17:R23.
154. Morgan EG, Maged M, Murry M. Clinical anesthesiology. 4th ed. Philadelphia: McGraw-Hill; 2006. p. 82.
155. Churchill Livingstone 2000. Section 3: Anaesthesia management, Part B – Monitoring, Chapter 33. Churchill Livingstone 2000. web.squ.edu.om/…/anesthesia/…/040101r00.HTM.
156. Luecke T, Pelosi P. Clinical review: positive end-expiratory pressure and cardiac output. Crit Care. 2005;9:607–21.
157. Bloom JA. Monitoring of respiration and circulation. Oxford: Taylor & Francis; 2003. p. 7.
158. O'Qiun RJ, Marini JJ, Culver BH. Transmission of airway pressure to pleural space during lung edema and chest wall restriction. J Appl Physiol. 1985;59:1171–7.
159. Gattinoni L, Carlesso E, Cadringher P. Physical and biological triggers of ventilator induced lung injury and its prevention. Eur Respir J. 2003;47:15s–25.
160. Ricard JD, Dreyfuss D, Saumon G. Ventilator-induced lung injury. Eur Respir J. 2003;42:2s–9.
161. Amato MBP, Barbas CSV, Medeiros MD. Effect of a protective- ventilation strategy on mortality in the acute respiratory distress syndrome. N Engl J Med. 1998;338:347–54.
162. Brower RG, Shanholtz CB, Fessler HE. Prospective, randomized, controlled clinical trial comparing traditional versus reduced tidal volume ventilation in acute respiratory distress syndrome patients. Crit Care Med. 1999;27:1492–8.
163. Kolobow T, Moretti MP, Fumagalli R. Severe impairment in lung function induced by high peak airway pressure during mechanical ventilation. An experimental study. Am Rev Respir Dis. 1987;135:312–5.
164. Benito S, Subirana M, Gracia JM. What is the utility of monitoring pulmonary mechanics in the treatment of patients with acute respiratory failure? In: Anzuero A, Cook D, Esteban A, editors. Evidence-based management of patients with respiratory failure. Berlin/Heidelberg: Springer; 2004. p. 29.
165. Novak R, Matuschak GM, Pinsky MR. Effect of positive- pressure ventilatory frequency on regional pleural pressure. J Appl Physiol. 1988;65:1314–23.
166. Lai-Fook SJ, Beck KC, Sothorn PA. Pleural liquid pressure measurements by micropipettes in rabbits. J Appl Physiol. 1984;56:1633–9.
167. Talmor D, Sarge T, Malhotra A. Mechanical ventilation guided by esophageal pressure in acute lung injury. N Engl J Med. 2008;359:2095–104.
168. Suter PM. Assessment of respiratory mechanics in ARDS. In: Falke K, Zapol WM, editors. Acute respiratory failure. New York: Marcel Dekker; 1985.
169. Baydur A, Cha EJ, Saaaoon CSH. Validation of esophageal balloon technique at different lung volumes and postures. J Appl Physiol. 1987;62:315–21.
170. Agostoni E, Miserocchi G. Vertical gradient of transpulmonary pressure with active and artificial lung expansion. J Appl Physiol. 1970;29:705–12.
171. Agostoni E, D'Angelo E, Bonanni MV. Topography of pleural surface pressure above resting volume in relaxed animals. J Appl Physiol. 1970;29:297–306.
172. Owens RL, Campana LM, Hess L. Sitting and supine esophageal pressures in overweight and obese subjects. Obesity (Silver Spring). 2012;20:2354–60.
173. Cortes GA, Marini JJ. Two steps forward in bedside monitoring of lung mechanics: transpulmonary pressure and lung volume. Crit Care. 2013;17:219.
174. Graf J, Formenti P, Santos A. Pleural effusion complicates monitoring of respiratory mechanics. Crit Care Med. 2011;39:2294–9.
175. Talomr DS, Fessler HE. Are esophageal pressure measurements important in clinical decision-making in mechanically ventilated patients? Respir Care. 2010;55:162–72.
176. Hubmary RD, Walters BJ, Chevalier PA. Topographical distribution of regional lung volume in anesthetized dogs. J Appl Physiol. 1983;54:1048–56.

177. Vieillard-Baron A, Jardin F. Esophageal pressure in acute lung injury. N Engl J Med. 2009;360:832.
178. Valenza F, Chevallard G, Porro GA, Gattinoni L. Static and dynamic components of esophageal and central venous pressure during intra- abdominal hypertension. Crit Care Med. 2007;35:1575–81.
179. Milic-Emili J, Mead J, Turner JM. Improved technique for estimating pleural pressure from esophageal balloons. J Appl Physiol. 1964;19:207–11.
180. Rehder K, Abboud N, Rodarte JR. Positive airway pressure and vertical transpulmonary pressure gradient in man. J Appl Physiol. 1975;38:896–9.
181. Gattinoni L, Pelosi P, Crotti S. Effects of positive end-expiratory pressure on regional distribution of tidal volume and recruitment in adult respiratory distress syndrome. Am J Respir Crit Care Med. 1995;151:1807–14.
182. Gattinoni L, Vagginelli F, Chiumello D. Physiologic rationale for ventilator setting in acute lung injury/acute respiratory distress syndrome patients. Crit Care Med. 2003;31(4 Suppl):S300–4.
183. Colebatch HJ, Greaves IA, Ng CK. Exponential analysis of elastic recoil and aging in healthy males and females. J Appl Physiol. 1979;47:683–91.
184. Grasso S, Terragni P, Birocco A. ECMO criteria for influenza A (H1N1) – associated ARDS: role of transpulmonary pressure. Intensive Care Med. 2012. doi:10.1007/s00134-012-2490-7.
185. Eichacker P, Gerstenberger E, Banks S. Meta-analysis of scute lung injury and acute respiratory distress syndrome trails testing low tidal volumes. Am J Respir Crit Care Med. 2002;166:1510–4.
186. Jardin F, Farcot JC, Boisante L. Influence of positive end-expiratory pressure on left ventricular performance. N Engl J Med. 1981;304:387–92.
187. Jardin F. To the editor: tidal volume reduction in patients with acute lung injury when plateau pressures are not high. Am J Respir Crit Care Med. 2006;173:685–6.
188. Haitsma JJ. Physiology of mechanical ventilation. Crit Care Clin. 2007;23:117–34.
189. Dellinger RP, Levy MM, Carlet JM. Surviving sepsis campaign: international guidelines for the management of severe sepsis and septic shock. Crit Care Med. 2008;36:296–327.
190. Leatherman J. Life-threatening asthma. Clin Chest Med. 1994;15:453–9.
191. The National Heart, Lung, and Blood Institute ARDS Clinical Trials Network. Higher versus lower positive end-expiratory pressures in patients with the acute respiratory distress syndrome. N Engl J Med. 2004;351:327–36.
192. Gattinoni L, Caironi P, Cressoni M. Lung recruitment in patients with the acute respiratory distress syndrome. N Engl J Med. 2006;354:1775–86.
193. Maede O, Cook DJ, Guyatt GH. Ventilation strategy using Low tidal volumes, recruitment maneuvers, and high positive end-expiratory pressure for acute lung injury and acute respiratory distress syndrome. JAMA. 2008;299:637–45.
194. Vieillard-Baron A, Prin S, Chergui K. Echo–Doppler demonstration of acute cor pulmonale at the bedside in the medical intensive care unit. Am J Respir Crit Care Med. 2002;166:1310–9.
195. Smeding L, Lust E, Plötz FB. Clinical complications of heart-lung interactions during mechanical ventilation: an update. Neth J Med. 2010;68:56–61.
196. Protti A, Cressonig M, Santini A. Lung stress and strain during mechanical ventilation: any safe threshold? Am J Respir Crit Care Med. 2011;183:1354–62.
197. Macklin CC. Transport of air along sheaths of pulmonic blood vessels from alveoli to mediastinum. Arch Intern Med. 1939;64:913–26.
198. Webb HH, Tierney DF. Experimental pulmonary edema due to intermittent positive pressure ventilation with high inflation pressures. Protection by positive end-expiratory pressure. Am Rev Respir Dis. 1974;110:556–65.
199. Mead J, Takishima T, Leith D. Stress distribution in lungs: a model of pulmonary elasticity. J Appl Physiol. 1970;28:596–608.
200. Pingleton SK. Complications of acute respiratory failure (State of the Art). Am Rev Respir Dis. 1988;137:1463–93.

201. Soni N, Williams P. Positive pressure ventilation: what is the real cost? Br J Anaesth. 2008;101:446–57.
202. Dreyfuss D, Soler P, Basset G. High inflation pressure pulmonary edema. Respective effects of high airway pressure, high tidal volume, and positive end-expiratory pressure. Am Rev Respir Dis. 1988;137:1159–64.
203. Lionetti V, Recchia FA, Ranieri VM. Overview of ventilator-induced lung injury mechanisms. Curr Opin Crit Care. 2005;11:82–6.
204. Slutsky AS. Lung injury caused by mechanical ventilation. Chest. 1999;116(1 Suppl): 9S–16.
205. Carny D, DiRocco J, Nieman G. Dynamic alveolar mechanics and ventilator-induced lung injury. Crit Care Med. 2005;33(Suppl):S 122–8.
206. Mols G, Priebe H-J, Guttmann J. Alveolar recruitment in acute lung injury. Br J Anaesth. 2006;96:156–66.
207. Steinberg JM, Schiller HJ, Halter JM. Alveolar instability causes early ventilator-induced lung injury independent of neutrophils. Am J Respir Crit Care Med. 2004;169:57–63.
208. Gajic O, Lee J, Doerr CH. Ventilator-induced cell wounding and repair in the intact lung. Am J Respir Crit Care Med. 2003;167:1057–63.
209. Grasso S, Terragni P, Mascia L. Airway pressure-time curve profile (stress index) detects tidal recruitment/hyperinflation in experimental acute lung injury. Crit Care Med. 2004;32:1018–27.
210. Verbrugge SJ, Bohm SH, Gommers D. Surfactant impairment after mechanical ventilation with large alveolar surface area changes and effects of positive end-expiratory pressure. Br J Anaesth. 1998;80:360–4.
211. de Prost N, Ricard J-D, Saumon G. Ventilator-induced lung injury: historical perspectives and clinical implications. Ann Intensive Care. 2011;1:28.
212. Tremblay L, Valenza F, Ribeiro SP. Injurious ventilatory strategies increase cytokines and c-fos m-RNA expression in an isolated rat lung model. J Clin Invest. 1997;99:944–52.
213. Bouadma L, Schortgen F, Ricard J. Ventilation strategy affects cytokine release after mesenteric ischemia-reperfusion in rats. Crit Care Med. 2004;32:1563–9.
214. Tremblay LN, Slutsky AS. Ventilator-induced lung injury: from the bench to the bedside. Intensive Care Med. 2006;32:24–33.
215. Whitehead T, Slutsky AS. The pulmonary physician in critical care. 7: Ventilator induced lung injury. Thorax. 2002;57:635–42.
216. Milic-Emili J, Henderson JAN, Dolovich MB. Regional distribution of inhaled gas in the lung. J Appl Physiol. 1966;21:749–59.
217. Piantadosi CA, Schwartz DA. The acute respiratory distress syndrome. Ann Intern Med. 2004;141:460–70.
218. Gattinoni L, Bombino M, Pelosi P. Lung structure and function in different stages of severe adult respiratory distress syndrome. JAMA. 1994;271:1772–9.
219. Puybasset L, Cluzel P, Chao N. A computed tomography scan assessment of regional lung volume in acute lung injury. The CT Scan ARDS Study Group. Am J Respir Crit Care Med. 1998;158(5 Pt 1):1644–55.
220. Gattinoni L, Pesenti A. The concept of 'baby lung'. Intensive Care Med. 2005;31:776–84.
221. Dreyfuss D, Basset G, Soler P. Intermittent positive-pressure hyperventilation with high inflation pressures produces pulmonary microvascular injury in rats. Am Rev Respir Dis. 1985;132:880–4.
222. Carlton DP, Cummings JJ, Scheerer RG. Lung overexpansion increases pulmonary microvascular protein permeability in young lambs. J Appl Physiol. 1990;69:577–83.
223. Egan EA. Lung inflation, lung solute permeability, and alveolar edema. J Appl Physiol. 1982;53:121–5.
224. Parker JC, Townsley MI, Rippe B. Increased microvascular permeability in dog lungs due to high airway pressures. J Appl Physiol. 1984;57:1809–16.
225. Sugiura M, McCulloch PR, Wren S. Ventilator pattern influences neutrophil influx and activation in atelectasis-prone rabbit lung. J Appl Physiol. 1994;77:1355–65.

226. Ricard JD, Dreyfuss D, Saumon G. Production of inflammatory cytokines in ventilator-induced lung injury: a reappraisal. Am J Respir Crit Care Med. 2001;163:1176–80.
227. Haitsma JJ, Uhlig S, Goggel R. Ventilator-induced lung injury leads to loss of alveolar and systemic compartmentalization of tumor necrosis factor alpha. Intensive Care Med. 2000;26:1515–22.
228. Tremblay LN, Slutsky AS. Ventilator-induced injury: from barotrauma to biotrauma. Proc Assoc Am Physicians. 1998;110:482–8.
229. Cepkova M, Brady S, Sapru A. Biological markers of lung injury before and after the institution of positive pressure ventilation in patients with acute lung injury. Crit Care. 2006;10:R126.
230. Puhin J. Molecular mechanisms of lung cell activation induced by cyclic stretch. Crit Care Med. 2003;31(4 Suppl):S 200–6.
231. Granto JS, Slutsky AS. Ventilator-induced lung injury. In: Schmidt G, Wood L, Hall J, editors. Principles of critical care. Philadelphia: McGraw-Hill; 2005. p. 499–514.
232. Slutsky A, Tremblay LN. Multiple system organ failure: is mechanical ventilation a contributing factor? Am J Respir Crit Care Med. 1998;157:1721–5.
233. Plotz FB, Slutsky AS, van Vught AJ. Ventilator-induced lung injury and multiple system organ failure: a critical review of facts and hypotheses. Intensive Care Med. 2004;30:1865–72.
234. Evans T, Cranshaw J. Lung injury. In: Hayes M, Soni M, Fink N, editors. Classic papers in critical care. Oxfordshire: Bladon Medical Publishing; 2003. p. 31–58.
235. Macklin MT, Macklin CC. Malignant interstitial emphysema of the lungs and mediastinum as an important occult complication in many respiratory diseases and other conditions: an interpretation of the clinical literature in the light of laboratory experiment. Medicine. 1944;23:281–352.
236. Petersen GW, Baier H. Incidence of pulmonary barotrauma in a medical ICU. Crit Care Med. 1983;11:67–9.
237. Downs JB, Chapman Jr RL. Treatment of bronchopleural fistula during continuous positive pressure ventilation. Chest. 1967;69:363–6.
238. Weg JG, Anzueto A, Balk RA. The relation of pneumothorax and other air leaks to mortality in acute respiratory distress syndrome. N Engl J Med. 1998;338:341–6.
239. Stewart TE, Meade MO, Cook DJ. Evaluation of a ventilation strategy to prevent barotrauma in patients at high risk for acute respiratory distress syndrome: pressure- and volume – limited ventilation strategy group. N Engl J Med. 1998;338:355–61.
240. Amato MBP, Marini JJ. Barotrauma, volutrauma, and ventilation of acute lung injury. In: Slutsky AS, Marini JJ, editors. Physiological basis of ventilatory support. New York: Marcel Dekker; 1998. p. 1187–245.
241. Haas CF. Mechanical ventilation with lung protective strategies: what works? Crit Care Clin. 2011;27:469–86.
242. Bouhuys A. Physiology and musical instruments. Nature. 1969;221:1199–204.
243. Hernandez LA, Peevy KJ, Moise A. Chest wall restriction limits high airway pressure – induced lung injury in young rabbits. J Appl Physiol. 1989;66:2364–8.
244. Dreyfuss D, Saumon G. Barotrauma is volutrauma, but which volume is the one responsible? Intensive Care Med. 1992;18:139–41.
245. Waters CM, Sporn PH, Liu M. Cellular biomechanics in the lung. Am J Physiol Lung Cell Mol Physiol. 2002;283:L503–9.
246. Fisher JL, Margulies SS. Na(+)-K(+)-ATP activity in alveolar epithelial cells increases with cyclic stretch. Am J Physiol Lung Cell Mol Physiol. 2002;283:L737–46.
247. Kumar A, Lnu S, Malya R. Mechanical stretch activates nuclear factro-kappaB, activator protein-1, and mitogen-activated protein kinases in lung parenchyma: implications in asthma. FASEB J. 2003;17:1800–11.
248. Orfanos SE, Mavrommati I, Korovesi I. Pulmonary endothelium in acute lung injury: from basic science to the critically ill. Intensive Care Med. 2004;30:1702–14.
249. Chu EK, Whitehead T, Slutsky AS. Effects of cyclic opening and closing at low- and high – volume ventilation on bronchoalveolar lavage cytokines. Crit Care Med. 2004;32:168–74.

250. Gattinoni L, Mascheroni D, Torresin A. Morphological response to positive end expiratory pressure in acute respiratory failure. Computerized tomography study. Intensive Care Med. 1986;12:137–42.
251. Czarny M, Schnitzer JE. Neutral sphingomyelinase inhibitor scyphostatin prevents and ceramide mimics mechanotransduction in vascular endothelium. Am J Physiol Heart Circ Physiol. 2004;287:H1344–52.
252. Howell JBL, Permutt S, Proctor DF. Effect of inflation of the lung on different parts of pulmonary vascular bed. J Appl Physiol. 1961;16:71–6.
253. Parker JC, Hernandez LA, Longenecker GL. Lung edema caused by high peal inspiratory pressures in dogs. Role of increased microvascular filtration pressure and permeability. Am Rev Respir Dis. 1990;142:312–28.
254. Ricard JD, Dreyfuss D, Saumon G. VILI: physiological evidence. In: Brochard L, Slutsky AS, editors. Mechanical ventilation. New York/Berlin/Heidelberg: Springer; 2005. p. 243.
255. Dreyfuss D, Soler P, Saumon G. Spontaneous resolution of pulmonary edema caused by short periods of cyclic overinflation. J Appl Physiol. 1992;72:2081–9.
256. Muscadere JG, Mullen JBM, Gan K. Tidal ventilation at low airway pressures can augment lung injury. Am J Respir Crit Care Med. 1994;149:1327–34.
257. Robertson B. Lung surfactant. In: Van Golde L, Batenburg J, Robertson B, editors. Pulmonary surfactant. Amsterdam: Elsevier; 1984.
258. Retamal J, Libuy J, Jimenez M. Preliminary study of ventilation with 4 ml/kg tidal volume in acute respiratory distress syndrome: feasibility and effects on cyclic recruitment – derecruitment and hyperinflation. Crit Care. 2013;17:R16.
259. Matute-Bello G, Downey G, Moore BB. An official American Thoracic Society workshop report: features and measurements of experimental acute lung injury in animals. Am J Respir Cell Mol Biol. 2011;44:725–38.
260. Hubmayr RD. Perspective on lung injury and recruitment. A skeptical look at the opening and collapse story. Am J Respir Crit Care Med. 2002;165:1647–53.
261. Boker A, Haberman CJ, Girling L. Variable ventilation improves perioperative lung function in patients undergoing abdominal aortic aneurysmectomy. Anesthesiology. 2004;100:608–16.
262. Mutch WA, Eschun GM, Kowalski SE. Biologically variable ventilation prevents deterioration of gas exchange during prolonged anaesthesia. Br J Anaesth. 2000;84:197–203.
263. Panda AK, Nag K, Harbottle RR. Effect of acute lung injury on structure and function of pulmonary surfactant films. Am J Respir Cell Mol Biol. 2004;30:614–50.
264. Bond DM, McAloon J, Froese AB. Sustained inflations improve respiratory compliance during high -frequency oscillatory ventilation but not during large tidal volume positive pressure ventilation in rabbits. Crit Care Med. 1994;22:1269–77.
265. Kacmarek RM, Schwartz DR. Lung recruitment. Respir Care Clin N Am. 2000;6:597–623.
266. Lapinsky SE, Aubin M, Mehta S. Safety and efficacy of a sustained inflation for alveolar recruitment in adults with respiratory failure. Intensive Care Med. 1999;25:1297–301.
267. Hess DR, Bigatello LM. Lung recruitment: the role of recruitment manoeuvers. Respir Care. 2002;47:308–17.
268. van der Kloot TE, Blanch L, Youngblood AM. Recruitment maneuvers in three experimental models of lung injury: effects on lung volume and gas exchange. Am J Respir Crit Care Med. 2000;161:1485–94.
269. Imai Y, Slutsky AS. Systemic effects of mechanical ventilation. In: Brochard L, Slutsky AS, editors. Mechanical ventilation. Berlin/Heidelberg: Springer; 2005. p. 259.
270. International Consensus conferences in Intensive Care Medicine. Ventilator-associated Lung Injury in ARDS. Am J Respir Crit Care Med. 1999;160:1301–08.
271. Pop-Vicas A, Opal SM. Over-distension of the airways by mechanical ventilation in the elderly: adding insult to injury. Crit Care. 2013;17:165.
272. Vlahakis NE, Schroeder MA, Limper AH. Stretch induces cytokine release by alveolar epithelial cells in vitro. Am J Physiol. 1999;277:L167–73.
273. Pugin J, Dunn I, Jolliet P. Activation of human macrophages by mechanical ventilation in vitro. Am J Physiol. 1998;275:L1040–50.

274. Grembowicz KP, Sprague D, McNeil PL. Temporary disruption of the plasma membrane is required for c-fos expression in response to mechanical stress. Mol Biol Cell. 1999;10:1247–57.
275. Quinn D, Tager A, Joseph PM. Stretch-induced mitogen-activated protein kinase activation and interleukin-8 production in type II alveolar cells. Chest. 1999;116:89S–90.
276. Chiumello D, Pristine G, Slutsky AS. Mechanical ventilation affects local and systemic cytokines in an animal model of acute respiratory distress syndrome. Am J Respir Crit Care Med. 1999;160:109–16.
277. Parker JC, Ivey CL, Tucker JA. Gadolinium prevents high airway pressure -induced permeability increases in isolated rat lungs. J Appl Physiol. 1998;84:1113–8.
278. Pugin J. Molecular mechanisms of lung cell activation induced by cyclic stretch. Crit Care Med. 2003;31(4 Suppl):S200–6.
279. Ranieri VM, Suter PM, Tortorella C. Effect of mechanical ventilation on inflammatory mediators in patients with acute respiratory distress syndrome: a randomized controlled trial. JAMA. 1999;282:54–61.
280. International Consensus Conference in Intensive Care Medicine. Ventilator-associated Lung Injury and ARDS: Official conference report of Am Thoracic Society (ATS), European Society of Intensive Care Medicine, and the Societe of Reanimation de Langue Francaise, and approved by the ATS Board of Directors. Am J Respir Crit Care Med. 1999;60:2118–24.
281. Ranieri VM, Giunta F, Sutter PM. Mechanical ventilation as a mediator of multisystem organ failure in acute respiratory distress syndrome. JAMA. 2000;284:43–4.
282. Imai Y, Kawano T, Iwamoto S. Intratracheal anti-tumor necrosis factor-alpha antibody attenuates ventilator – induced lung injury in rabbits. J Appl Physiol. 1999;87:510–5.
283. Narimanbekov IO, Rozycki HJ. Effect of IL-1 blockade on inflammatory manifestations of acute ventilator-induced lung injury in a rabbit model. Exp Lung Res. 1995;21:239–54.
284. Held HD, Boettcher S, Hamann L. Ventilation-induced chemokine and cytokine release is associated with activation of nuclear factor- kappa B and is blocked by steroids. Am J Respir Crit Care Med. 2001;163:711–6.
285. Otto CM, Markstaller K, Kajikawa O. Spatial and temporal heterogenity of ventilator – associated lung injury after surfactant depletion. J Appl Physiol. 2008;104:1485–94.
286. de Prost D, Costa EL, Wellman T. Effects of ventilation strategy on distribution of lung inflammatory cell activity. Crit Care. 2013;17:R175.
287. Tobin MJ. Advances in mechanical ventilation. N Engl J Med. 2001;344:1986–96.
288. Tutor JD, Mason CM, Dobard E. Loss of compartmentalization of alveolar tumor necrosis factor after lung injury. Am J Respir Crit Care Med. 1994;149:1107–11.
289. Kurahashi K, Kajikawa O, Sawa T. Pathogenesis of septic shock in Pseudomonas aeroginosa pneumonia. J Clin Invest. 1999;104:743–50.
290. Greene KE, Wright JR, Steinberg KP. Serial changes in surfactant – associated proteins in lung and serum before and after onset of ARDS. Am J Respir Crit Care Med. 1999;160:1843–50.
291. Doyle IR, Bersten AD, Nicholas TE. Surfactant proteins-A and -B are elevated in plasma of patients with acute respiratory failure. Am J Respir Crit Care Med. 1997;156:1217–29.
292. Nahum A, Hoyt J, Schmitz L. Effects of mechanical ventilation strategy on dissemination of intratracheally instilled Escherichia coli in dogs. Crit Care Med. 1997;25:1733–43.
293. Verbrugge S, Sorn V, Van't Veen A. Lung overinflation without positive end-expiratory pressure promotes bacteremia after experimental Klebsiella pneumoniae inoculation. Intensive Care Med. 1998;24:172–7.
294. Schortgen F, Bouadma L, Joly-Guillou ML. Infectious and inflammatory dissimination are affected by ventilation strategy in rats with unilateral pneumonia. Intensive Care Med. 2004;30:693–701.
295. de Prost N, Roux D, Ricard JD. Alveolar edema dispersion and alveolar protein permeability during high volume ventilation: effect of positive end-expiratory pressure. Intensive Care Med. 2007;33:711–7.

296. Guery BP, Welsh DA, Viget NB. Ventilation-induced lung injury is associated with an increase in gut permeability. Shock. 2003;19:559–63.
297. Choi WI, Quinn DA, Park MM. Systemic microvascular leak in an in vivo rat model of ventilator-induced lung injury. Am J Respir Crit Care Med. 2003;167:1627–32.
298. Imai Y, Parodo J, Kajikawa O. Injurious mechanical ventilation and end-organ epithelial cell apoptosis and organ dysfunction in an experimental model of acute respiratory distress syndrome. JAMA. 2003;289:2104–12.
299. Bouadma L, Dreyfuss D, Ricard JD. Mechanical ventilation and hemorrhagic shock-resuscitation interact to increase inflammatory cytokine release in rats. Crit Care Med. 2007;35:2601–6.
300. Kotani M, Kotani T, Li Z. Reduced inspiratory flow attenuated IL-8 release and MAPK activation of lung overstretch. Eur Respir J. 2004;24:238–46.
301. Garcia C, Abreu SC, Soares RML. Pulmonary morphofunctional effects of mechanical ventilation with high inspiratory air flow. Crit Care Med. 2008;36:232–9.
302. Maeda Y, Fujino Y, Uchiyama A. Effects of peak inspiratory flow on development of ventilator-induced lung injury flow in rabbits. Anesthesiology. 2004;101:722–8.
303. Rich PB, Reickert CA, Sawada S. Effect of rate and inspiratory flow on ventilator-induced lung injury. J Trauma. 2000;49:903–11.
304. Wong PW, Nygard S, Sogoloff H. He effect of varying inspiratory flow waveforms on pulmonary mechanics in critically ill patients. J Crit Care. 2000;15:133–6.
305. Cheifetz IM. Invasive and noninvasive pediatric mechanical ventilation. Respir Care. 2003;48:442–53; discussion 453–8.
306. Brochard L. Intrinsic (or auto-) PEEP during controlled mechanical ventilation. In: Brochard L, Mancebo J, Pinsky MR, editors. Applied physiology in intensive care medicine. Berlin/Heidelberg: Springer; 2006. p. 3–6.
307. Richard JC, Brochard L, Breton L. Influence of respiratory rate on gas trapping during low volume ventilation of patients with acute lung injury. Intensive Care Med. 2002;28:1078–83.
308. Koutsoukou A, Armaganidis A, Stavrakaki-Kallergi C. Expiratory flow limitation and intrinsic positive end-expiratory pressure at zero positive end-expiratory pressure in patients with adult respiratory distress syndrome. Am J Respir Crit Care Med. 2000;161:1590–6.
309. Tobin MJ, Lodato RF. PEEP, auto-PEEP, and waterfalls (comment). Chest. 1989;96:449–51.
310. Ranieri VM, Grasso S, Fiore T. Auto-positive end-expiratory pressure and dynamic hyperinflation. Clin Chest Med. 1996;17:379–94.
311. Pepe PE, Marini JJ. Occult positive end-expiratory pressure in mechanically ventilated patients with airflow obstruction: the auto-PEEP effect. Am Rev Respir Dis. 1982;126:166–70.
312. Rossi A, Polese G, Brandi G. Intrinsic positive end-expiratory pressure (PEEPi). Intensive Care Med. 1995;21:522–36.
313. Wise RA, Liu MC. Obstructive airway diseases: asthma and COPD. In: Barker R, editor. Principles of ambulatory medicine. 6th ed. Philadelphia: Lippincott Williams & Wilkins; 2003. p. 808–41.
314. Stather DR, Stewart T. Clinical review: mechanical ventilation in severe asthma. Crit Care. 2005;9:581–7.
315. Stewart TE, Slutsky AS. Occult, occult auto-PEEP in status asthmaticus. Crit Care Med. 1996;24:379–80.
316. Kondili E, Alexopoulou C, Prinianakis G. Pattern of lung emptying and expiratory resistance in mechanically ventilated patients with chronic obstructive pulmonary disease. Am Rev Respir Dis. 2004;30:1311–8.
317. Fleury B, Murciano D, Talamo C. Work of breathing in patients with chronic obstructive pulmonary disease in acute respiratory failure. Am Rev Respir Dis. 1985;131:822–7.
318. Mancebo J. Physiological rationale for ventilation of patients with obstructive diseases. In: Brochard L, Slutsky AS, Vincent J-L, editors. Mechanical ventilation, update in intensive care medicine. New York/Berlin/Heidelberg: Springer; 2005. p. 97–112.
319. Ward NS, Dushay KM. Clinical concise review: mechanical ventilation of patients with chronic obstructive pulmonary disease. Crit Care Med. 2008;36:1614–9.

320. Gladwin MT, Pierson DJ. Mechanical ventilation of the patient with severe chronic obstructive pulmonary disease. Intensive Care Med. 1998;24:898–910.
321. MCCarthy DS, Sigurdson M. Lung elastic recoil and reduced air flow in clinically stable asthma. Thorax. 1980;35:298–302.
322. Dhand R. Ventilator graphics and respiratory mechanics in the patient with obstructive lung disease. Respir Care. 2005;50:246–59.
323. Nadel JA. Mucus and mucus-secreting cells. In: Nacnee W, Voelkel NF, editors. Chronic obstructive lung diseases. Hamilton: Dekker; 2002. p. 161–74.
324. Vieillard-Baron A, Loubieres Y, Schmitt JM. Cyclic changes in right ventricular output impedance during mechanical ventilation. J Appl Physiol. 1999;87:1644–50.
325. Parker JC, Hernandez LA, Peevy KJ. Mechanisms of ventilator-induced lung-injury. Crit Care Med. 1993;21:131–43.
326. Jellinek H, Krafft P, Fitzgerald RD. Right atrial pressure predicts hemodynamic response to apneic positive airway pressure. Crit Care Med. 2000;28:628–78.
327. Sessler CN. Mechanical ventilation of patients with acute lung injury. Crit Care Clin. 1998;14:707–29.
328. Acosta P, Santisbon E, Varon J. The use of positive end-expiratory pressure in mechanical ventilation. Crit Care Clin. 2007;23:251–61.
329. MacIntyre NR. Setting the positive expiratory –end pressure – FiO 2 in acute lung injury/acute respiratory distress syndrome. Respir Care Clin N Am. 2004;10:301–8.
330. McIntyre RW, Laws AK, Ramachandran RP. Positive expiratory pressure plateau: improved gas exchange during mechanical ventilation. Can Anaesth Soc J. 1969;16:477.
331. Falke KJ, Pontoppidan H, Kumar A. Ventilation with end-expiratory pressure in acute lung disease. Chest. 1969;61:376.
332. Suter P, Fairley B, Isenberg M. Optimum end-expiratory airway pressure in patients with acute pulmonary failure. N Engl J Med. 1975;292:284–9.
333. East TD, In't Veen JC, Pace NL. Functional residual capacity as a noninvasive indicator of optimal positive end-expiratory pressure. J Clin Monit. 1988;4:91–8.
334. Marini JJ, Wheeler AP. Critical care medicine: the essentials; actions of PEEP in acute hypoxic respiratoty failure. Philadelphia: Lippincott Williams & Wilkins; 2010. p. 176.
335. Kumar A, Falke KJ, Lowenstein E. Continous positive pressure ventilation in acute respiratory failure. N Engl J Med. 1970;283:1430–6.
336. Frumin MJ, Bergman NA, Holaday DA. Alveolar-arterial O2 differences during artificial respiration in man. J Appl Physiol. 1959;14:694–700.
337. Zaky A, Lang JD. The use of intraoperative positive end expiratory pressure. J Anesth Clin Res. 2011;S3:002. doi:10.4172/2155-6148.
338. Smith TC, Marini JJ. Impact of PEEP on lung mechanics and work of breathing in severe airflow obstruction. J Appl Physiol. 1988;65:1488–99.
339. Bashour CA, Stoller JK. The acute respiratory distress syndrome. In: Steingrub JS, Kacmarek RM, Stoller JK, Higgins TL, editors. Cardiopulmonary critical care. Oxford: BIOS Scientific Publishers; 2002. p. 221, chapter 14.
340. Hopewell PC. Failure of positive end-expiratory pressure to decrease lung water content in alloxan-induced pulmonary edema. Am Rev Respir Dis. 1979;120:813.
341. Moriondo A, Mukenge S, Negrini D. Transmural pressure in rat initial subpleural lymphatics during spontaneous or mechanical ventilation. Am J Physiol Heart Circ Physiol. 2005;289:H263–9.
342. Pearse DB, Searcy RM, Mitzner W. Effects of tidal volume and respiratory frequency on lung lymph flow. J Appl Physiol. 2005;99:556–63.
343. Schmid-Schonbein GW. Microlymphatics and lymph flow. Physiol Rev. 1990;70:987–1028.
344. Laine GA, Allen SJ, Katz J. Effect of systemic venous pressure elevation on lymph flow during lung edema formation. J Appl Physiol. 1986;61:1634–8.
345. Allen SJ, Drake RE, Laine GA. Effect of thoracic duct drainage on hydrostatic pulmonary edema and pleural effusion in sheep. J Appl Physiol. 1991;71:314–6.
346. Frostell C, Blomqvist H, Hedenstierna G. Thoracic and abdominal lymph drainage in relation to mechanical ventilation and PEEP. Acta Anaesthesiol Scand. 1987;31:405–12.

347. Maybauer DM, Talke PO, Westphal M. Positive end-expiratory pressure ventilation increases extravascular lung water due to a decrease in lung lymph flow. Anaesth Intensive Care. 2006;34:329–33.
348. van der Zee H, Cooper JA, Hakim TS. Alterations in pulmonary fluid balance induced by positive end-expiratory pressure. Respir Physiol. 1986;64:125–33.
349. Mallick A, Bodenham AR. Disorders of the lymph circulation: their relevance to anaesthesia and intensive care. Br J Anaesth. 2003;91:265–72.
350. Falke KJ, Pontoppidan H, Kumar A. Ventilation with end-expiratory pressure in acute lung disease. J Clin Invest. 1972;51:2315–23.
351. Dreyfuss D, Saumon G. Role of tidal volume, FRC, and end-inspiratory volume in the development of pulmonary edema following mechanical ventilation. Am Rev Respir Dis. 1993;148:1194–203.
352. Bilek AM, Dee KC, Gaver DP. Mechanisms of surface -tension-induced epithelial cell damage in a model of pulmonary airway reopening. J Appl Physiol. 2003;94:770–83.
353. de Prost N, Dreyfuss D, Saumon G. Evaluation of two-way protein fluxes across the alveolo-capillary membrane by scintigraphy in rates: effect of lung inflation. J Appl Physiol. 2007;102:794–802.
354. Bersten AD, Gnidec AA, Rutledge FS. Hyperdynamic sepsis modifies a PEEP-mediated redistribution in organ blood flows. Am Rev Respir Dis. 1990;141:1198–208.
355. Lehtipalo S, Biber B, Frojse R. Effects of positive end-expiratory pressure on intestinal circulation during graded mesenteric artery occlusion. Acta Anaesthesiol Scand. 2001;45:875–84.
356. Mutlu GM, Mutlu EA, Factor P. GI complications in patients receiving mechanical ventilation. Chest. 2001;119:1222–41.
357. Benito S, Lamaire F. Pulmonary pressure – volume relationship in acute respiratory distress syndrome in adults: role of positive end expiratory pressure. J Crit Care. 1990;5:27–34.
358. Pinsky MR. Cardiovascular issues in respiratory care. Crit Care. 2005;9:566–72.
359. Marik P, Chen K, Varon J. Management of acute intracranial hypertension: a review for clinicians. J Emerg Med. 1999;17:711–9.
360. Wheeler AP, Bernard GR. Minimal positive end-expiratory pressure: a safe, effective, defensible practice. Clin Intensive Care. 1990;1:175–9.
361. Fessler HE, Brower RG, Boisante L. Effects of end-expiratory pressure on the gradient for venous return. Am Rev Respir Dis. 1991;143:677–83.
362. Whittenberger J, McGregor M, Berglund E. Influence of state of inflation of the lung on pulmonary vascular resistance. J Appl Physiol. 1960;15:878–82.
363. Morgan B, Martin W, Hornbein T. Hemodynamic effects of intermittent positive pressure respiration. Anesthesiology. 1966;27:584–90.
364. Burton AC, Patel DJ. Effect on pulmonary vascular resistance of inflation of the rabbit lungs. J Appl Physiol. 1958;12:239–46.
365. Price LC. Pulmonary vascular and right ventricular dysfunction in adult critical care: current and emerging options of management, a systematic literature review. Crit Care. 2010;14:R169.
366. Groeneveld ABJ, Berendsen RR, Schneider AJ. Effect of the mechanical ventilatory cycle on thermodilution right ventricular volumes and cardiac output. J Appl Physiol. 2000;89:89–96.
367. Culver BH, Marini JJ, Butler J. Lung volume and pleural pressure effects on ventricular function. J Appl Physiol. 1981;50:630–5.
368. Cherpanath TGV, Lagrand WK, Schultz MJ. Cardiopulmonary interactions during mechanical ventilation in critically ill patients. Neth Heart J. 2013;21:166–72.
369. Bernard GR, Pou NA, Coggeshall JW. Comparison of the pulmonary dysfunction caused by cardiogenic noncardiogenic pulmonary edema. Chest. 1995;108:798–803.
370. Fessler HE, Brower RG, Wise RA. Mechanisms of reduced LA afterload by systolic and diastolic positive pleural pressure. J Appl Physiol. 1988;65:1244–50.
371. van den Berg PC, Grimbergen CA, Spaan JA. Positive pressure inspiration differently affects right and left ventricular outputs in postoperative cardiac surgery patients. J Crit Care. 1997;12:56–65.

372. Jardin F, Vieillard-Baron A. Right ventricular function and positive pressure ventilation in clinical practice: from hemodynamic subsets to respirator settings. Intensive Care Med. 2003;29:1426–34.
373. Haynes JB, Carlson SD, Whitney WP. Positive end-expiratory pressure shifts left ventricular diastolic pressure-area curves. J Appl Physiol. 1980;48:670–6.
374. Lloyd Jr JT. Mechanical cardiopulmonary interdependence. J Appl Physiol. 1982;52:333–9.
375. Fewell JE, Abendschein DR, Carlson CJ. Continous positive-pressure ventilation decreases right and left ventricular end-diastolic volumes in dog. Circ Res. 1980;46:125–32.
376. Fewell JE, Abendschein DR, Carlson CJ. Continous positive pressure ventilation does not alter ventricular pressure-volume relationship. Am J Physiol. 1981;240:H821–6.
377. Schulman DS, Biondi JW, Zohgbi S. Left ventricular diastolic function during positive end-expiratory pressure. Impact of right ventricular ischemia and ventricular interaction. Am Rev Respir Dis. 1992;145:515–21.
378. Dhainut JF, Devaux JY, Monsallier JF. Mechanisms of decreased left ventricular preload during continuous positive pressure ventilation in ARDS. Chest. 1986;90:74–80.
379. Dorinsky PM, Hamlin RL, Gadek JE. Alterations in regional blood flow during positive end-expiratory pressure ventilation. Crit Care Med. 1987;15:106–13.
380. Matuschak GM, Pinsky MR, Rogers RM. Effects of positive end-expiratory pressure on hepatic blood flow and performance. J Appl Physiol. 1987;62:1377–83.
381. Fujita Y. Effects of PEEP on splanchnic hemodynamics and blood volume. Acta Anaesthesiol Scand. 1993;37:427–31.
382. Wrigge H, Uhlig U, Zinserling J. The effects of different ventilatory settings on pulmonary and systemic inflammatory responses during major surgery. Anesth Analg. 2004;98:775–81.
383. Schricker T, Kugler B, Schywalsky M. Effects of PEEP ventilation on liver metabolism. Infusionsther Transfusionsmed. 1995;22:168–74.
384. Priebe HJ, Heiman JC, Hedley-Whyte J. Mechanisms of renal dysfunction during positive end-expiratory pressure ventilation. J Appl Physiol. 1981;50:643–9.
385. Morgan BC, Guntheroth WG, Dillard DH. Relationship of pericardial to pleural pressure during quiet respiration and cardiac tamponade. Circ Res. 1965;16:493–8.
386. Bhorade S, Christenson J, O'Connor M. Response to inhaled nitric oxide in patients with acute right heart syndrome. Am J Respir Crit Care Med. 1999;159:571–9.
387. Pinsky MR. Recent advances in the clinical application of heart-lung interactions. Curr Opin Crit Care. 2002;8:26–31.
388. Santamore WP, Lynch PR, Meier G. Myocardial interaction between the ventricles. J Appl Physiol. 1976;41:362–8.
389. Bemis CE, Serur JR, Borkenhagen D. Influence of right ventricular filling pressure on left ventricular pressure and dimension. Circ Res. 1974;34:498–504.
390. Atherton JJ, Moore TD, Lele SS. Diastolic ventricular interaction in chronic heart failure. Lancet. 1997;349:1720–4.
391. Laks MM, Garner D, Swan HJC. Volumes and compliances measured simultaneously in the right and left ventricles of the Dog. Circ Res. 1967;20:565–9.
392. Badke FR. Left ventricular dimensions and function during right ventricular pressure overload. Am J Physiol Heart Circ Physiol. 1982;242:H611–8.
393. Jardin F, Brun-Ney D, Hardy A. Combined thermodilution and two- dimensional echocardiographic evaluation of right ventricular function during respiratory support with PEEP. Chest. 1991;99:162–8.
394. Louie EK, Lin SS, Reynertson SI. Pressure and volume loading of the right ventricle have opposite effects on left ventricular ejection fraction. Circulation. 1995;92:819–24.
395. Gordo-Vidal F, Enciso-Calderon V. Acute respiratory distress syndrome, mechanical ventilation and right ventricular function. Med Intensiva. 2012;36:138–42.
396. Piazza G, Goldhaber SZ. The acutely decompensated right: pathways for diagnosis and management. Chest. 2005;128:1836–52.
397. Vieillard-Baron A. Right ventricular function and positive pressure ventilation in clinical practice: from hemodynamic subsets to respirator settings. In: Brochard L, Mancebo J,

Pinsky MR, editors. Applied physiology in intensive care medicine. Berlin: Springer; 2006. p. 207.
398. Schmitt JM, Vieillard-Baron A, Roch A. Positive end-expiratory pressure titration in acute respiratory distress syndrome patients: impact on right ventricular outflow impedance evaluated by pulmonary artery Doppler flow velocity measurements. Crit Care Med. 2001;29:1154–8.
399. Permutt S. Circulatory effects of weaning from mechanical ventilation: the importance of transdiaphragmatic pressure. Anesthesiology. 1988;69:157–60.
400. Peters J, Fraser C, Stuart RS. Negative intrathoracic pressure decreases independently left ventricular filling and emptying. Am J Physiol Heart Circ Physiol. 1989;257:H120.
401. Nieminen MS, Boehm M, Cowie MR. Executive summary of the guidelines on the diagnosis and treatment of acute heart failure. The Task Force on Acute Heart Failure of the European Society of Cardiology. Eur Heart J. 2005;26:384–416.
402. Peter JC, Moran JL, Phillips-Hughes J. Effect of non-invasive positive pressure ventilation (NIPPV) on mortality in patients with acute cardiogenic pulmonary oedema: a meta-analysis. Lancet. 2006;367:1155–63.
403. Pinsky MR. The hemodynamic consequences of mechanical ventilation: an evolving story. Intensive Care Med. 1997;23:493–503.
404. Pinsky MR. Cardiovascular issues in respiratory care. Chest. 2005;128:592S–7.
405. Bouferrache K, Vieillard-Baron A. Acute respiratory distress syndrome, mechanical ventilation, and right heart function. Curr Opin Crit Care. 2011;17:30–5.
406. Romand JA, Shi W, Pinsky MR. Cardiopulmonary effects of positive pressure ventilation during acute lung injury. Chest. 1995;108:1041–8.
407. Pinsky MR. Hemodynamic effects of artificial ventilation. In: Ayres SN, Grenvik A, Shoemaker WC, editors. Textbook of critical care. Philadelphia: WB Saunders; 1995. p. 911–22.
408. Fessler HE. Heart lung interactions: application in the critically ill. Eur Respir J. 1997;10:26–237.
409. Pinsky MR. Instantaneous venous return curves in an intact canine preparation. J Appl Physiol. 1984;56:765–71.
410. Laurant F, Montaudon M. Physiological interactions. In: Remy-Jardin M, Remey J, Baert AL, editors. Integrated cardiothoracic imaging with MDCT. Berlin/Heidelberg: Springer; 2009. p. 109–30.
411. Bendjelid K, Romand J-A. Fluid responsiveness in mechanically ventilated patients: a review of indices used in intensive care. In: Brochard L, Mancebo J, Pinsky MR, editors. Applied physiology in intensive care medicine. Berlin/Heidelberg: Springer; 2006. p. 95.
412. Van den Berg PCM, Jansen JRC, Pinsky MR. Effect of positive pressure on venous return in volume-loaded cardiac surgical patients. J Appl Physiol. 2002;92:1223–31.
413. Braunwald E, Binion JT, Morgan WL. Alterations in central blood volume and cardiac output induced by positive pressure breathing and counteracted by metaraminol (aramine). Circ Res. 1957;5:670–5.
414. Fessler H, Brower R, Wise R. Effects of positive end-expiratory pressure on the canine venous return curve. Am Rev Respir Dis. 1992;146:4–10.
415. Jellinek H, Krenn H, Oczenski W. Influence of positive airway pressure on the pressure gradient for venous return in humans. J Appl Physiol. 2000;88:926–32.
416. Groeneveld ABJ, Schreuder WO, Vermeij CG. Effect of incremental positive end-expiratory pressure ventilation on right ventricular function in anaesthetized pigs: a thermodilution study. J Crit Care. 1990;5:218–27.
417. Guyton AC. The effect on cardiac output of respiration, opening the chest, and cardiac tamponade. In: Circulatory physiology: cardiac output and its regulation. Philadelphia: WB Saunders; 1963. p. 378–86.
418. Cournand A, Motley HL, Werko L. Physiological studies of the effects of intermittent positive pressure breathing on cardiac output in man. Am J Physiol. 1948;152:162–74.
419. Frank O. On the dynamics of cardiac muscle. Am Heart J. 1959;58:282–317.

420. Michard F, Boussat S, Chemla D. Relation between respiratory changes in arterial pulse pressure and fluid responsiveness in septic patients with acute circulatory failure. Am J Respir Crit Care Med. 2000;162:134–8.
421. Scharf SM, Brown N, Saunders N. Hemodynamic effects of positive pressure inflation. J Appl Physiol. 1980;49:124–31.
422. Jardin F, Farcot JC, Gueret P. Cyclic changes in arterial pulse during respiratory support. Circulation. 1983;68:266–74.
423. Robotham JL, Cherry D, Mitzner W. Re-evaluation of the hemodynamic consequences of intermittent positive pressure ventilation. Crit Care Med. 1983;11:783.
424. Vieillard- Baron A, Chergui K, Augarde R. Cyclic changes in arterial pulse during respiratory support revisited by Doppler echocardiography. Am J Respir Crit Care Med. 2003;168:671–6.
425. Pinsky MR, Summer WR, Wise RA. Augmentation of cardiac function by elevation of intrathoracic pressure. J Appl Physiol. 1983;54:950–5.
426. Pinsky MR. Hemodynamic evaluation and monitoring in the ICU. Chest. 2007;132:2020–9.
427. Boulain T, Achard J-M, Teboul J-L. Changes in BP induced by passive leg raising predict response to fluid loading in critically ill patients. Chest. 2002;121:1245–52.
428. Preisman S, Kogan S, Berkenstadt H. Predicting fluid responsiveness in patients undergoing cardiac surgery: functional haemodynamic parameters including respiratory systolic variation test and static preload indicators. Br J Anaesth. 2005;95:746–55.
429. Permutt S, Bromberger-Barena B, Bane H. Alveolar pressure, pulmonary venous pressure, and the vascular waterfall. Med Thorac. 1962;19:239–60.
430. Hakim TS, Michel RP, Minami H. Site of pulmonary hypoxic vasoconstriction studied with arterial and venous occlusion. J Appl Physiol. 1983;54:1298–302.
431. Jardin F, Delorme G, Hardy A. Reevaluation of hemodynamic consequences of positive pressure ventilation: emphasis on cyclic right ventricular afterloading by mechanical lung inflation. Anesthesiology. 1990;72:966–70.
432. Jardin F, Brun-Ney D, Cazaux P. Relation between transpulmonary pressure and right ventricular isovolumetric pressure change during respiratory support. Cathet Cardiovasc Diagn. 1989;16:215–20.
433. Shekerdemian L, Bohn D. Cardiovascular effects of mechanical ventilation. Arch Dis Child. 1999;80:475–80.
434. Canada E, Benumof IL, Tousdale FR. Pulmonary vascular resistance correlates in intact normal and abnormal canine lungs. Crit Care Med. 1982;10:719–21.
435. Gillis CN, Pitt PR, Wiedemann HP. Depressed prostaglandin E1 and 5-hydroxytryptamine removal in patients with adult respiratory distress syndrome. Am Rev Respir Dis. 1986;134:739–44.
436. Wort SJ, Evans TW. The role of the endothelium in modulating vascular control in sepsis and related conditions. Br Med Bull. 1999;55:30–48.
437. Nuckton TJ, Alonso JA, Kallet RH. Pulmonary dead space fraction as a risk factor for death in the acute respiratory distress syndrome. N Engl J Med. 2002;346:1281–6.
438. Rossi P, Persson B, Boels PJ. Endotoxemic pulmonary hypertension is largely mediated by endothelin-induced venous constriction. Intensive Care Med. 2008;34:873–80.
439. Mrashall BE, Marshall C, Frasch F. Role of hypoxic pulmonary vasoconstriction in pulmonary gas exchange and blood floe distribution. 1. Physiologic concepts. Intensive Care Med. 1994;20:291–7.
440. Zapol WM, Kobayashi K, Snider MT. Vascular obstruction causes pulmonary hypertension in severe acute respiratory failure. Chest. 1977;71:306–7.
441. Tomashefski Jr JF, Davies P, Boggis C. The pulmonary vascular lesions of the adult respiratory distress syndrome. Am J Pathol. 1983;112:112–26.
442. Pinsky MR. Determinates of pulmonary arterial flow variation during respiration. J Appl Physiol. 1984;56:1237–45.
443. Stojnic BB, Brecker SJ, Xiao HB. Left ventricular filling characteristics in pulmonary hypertension: a new mode of ventricular interaction. Br Heart J. 1992;68:16–20.

444. Belenkie I, Dani R, Smith ER. Effects of volume loading during experimental acute pulmonary embolism. Circulation. 1989;80:178–88.
445. Janicki JS, Weber KT. The pericardium and ventricular interaction, distensibility and function. Am J Physiol Heart Circ Physiol. 1980;238:H494–503.
446. Heywood JT, Grimm J, Hess OM. Right ventricular diastolic function-during exercise: effect of ischemia. J Am Coll Cardiol. 1990;16:611–22.
447. Jardin F, Dubourg O, Gueret P. Quantitative two-dimensional echocardiography in massive pulmonary embolism: emphasis on ventricular interdependence and leftward septal displacement. J Am Coll Cardiol. 1987;10:1201–6.
448. Hoffman EA, Ritman EL. Invariant total heart volume in the intact thorax. Am J Physiol Heart Circ Physiol. 1985;249:H883–90.
449. Moore TD, Frenneaux MP, Sas R. Ventricular interaction and external constraint account for decreased stroke work during volume loading in CHF. Am J Physiol Heart Circ Physiol. 2001;281:H2385–91.
450. Jardin F, Gueret P, Prost JF. Two-dimensional echocardiographic assessment of the left ventricular function in chronic obstructive pulmonary disease. Am Rev Respir Dis. 1984;129:135–42.
451. Ross Jr J. Acute displacement of the diastolic pressure- volume curve of the left ventricle: role of the pericardium and the right ventricle. Circulation. 1979;59:32–7.
452. Mebazaa A, Karpati P, Renaud E. Acute right ventricular failure – from pathophysiology to new treatments. Intensive Care Med. 2004;30:185–96.
453. McIntyre KM, Sasahara AA. Determinants of right ventricular function and hemodynamics after pulmonary embolism. Chest. 1974;65:534–43.
454. Maughan WL, Shoukas AA, Sagawa K. Instantaneous volume-pressure relationship of the canine right ventricle. Circ Res. 1979;44:309–14.
455. Pinsky MR. Heart-lung-interactions. Curr Opin Crit Care. 2007;13:528–31.
456. Straling EH. The Linacre lecture on the Law of the heart. New York: Longmans Green & CO; 1918.
457. Pinsky MR. Pulmonary artery occlusion pressure. Intensive Care Med. 2003;29:19–22.
458. Galie N, Hinderliter AL, Torbicki A. Effects of the oral endothelin-receptor antagonist bosentan on echocardiographic doppler measures in patients with pulmonary arterial hypertension. J Am Coll Cardiol. 2003;41:1380–6.
459. Sibbald WJ, Driedger AA. Right ventricular function in acute disease states: pathophysiologic considerations. Crit Care Med. 1983;11:339–45.
460. Belenki I, Sas R, Mitchell J. Opening the pericardium during pulmonary artery constriction improves cardiac function. J Appl Physiol. 2004;96:917–22.
461. Bleasdale RA, Turner MS, Mumford CE. Left ventricular pacing minimizes diastolic ventricular interaction, allowing improved preload-dependent systolic performance. Circulation. 2004;110:2395–400.
462. Lunaldi JC, Goldhaber SZ. Right ventricular dysfunction after acute pulmonary embolism: pathophysiologic factors detection, and therapeutic implications. Am Heart J. 1995; 130:1276–82.
463. Naeije R, Vachiery JL. Medical therapy of pulmonary hypertension conventional therapies. Clin Chest Med. 2001;22:517–27.
464. Layish DT, Tapson VF. Pharmacologic hemodynamic support in massive pulmonary embolism. Chest. 1997;111:218–24.
465. Jacobs AK, Leopold JA, Bates E. Cardiogenic shock caused by right ventricular infarction: a report from the SHOCK registry. J Am Coll Cardiol. 2003;41:1273–9.
466. Mercat A, Diehl J-L, Meyer G. Hemodynamic effects of fluid loading in acute massive pulmonary embolism. Crit Care Med. 1999;27:540–54.
467. Goldhaber SZ. The approach to massive pulmonary embolism. Semin Respir Crit Care Med. 2000;21:555–61.
468. Forrest P. Anaesthesia and right ventricular failure. Anaesth Intensive Care. 2009;37:370–85.
469. Tyberg JV, Taichman GC, Smith ER. The relationship between pericardial pressure and the right atrial pressure: an intraoperative study. Circulation. 1986;73:428–32.

470. Robotham JL, Lixfeld W, Holland L. The effects of positive end-expiratory pressure on right and left ventricular performance. Am Rev Respir Dis. 1980;121:677–83.
471. Nieman GF, Pashanik AM, Bredenbeurg CE. Effect of positive end-expiratory pressure on alveolar capillary perfusion. J Thorac Cardiovasc Surg. 1988;95:712–6.
472. Tulevski II, Romkes H, Dodge-Khatami A. Quantitative assessment of the pressure and volume overloaded right ventricle: imaging is a real challenge. Int J Cardiovasc Imaging. 2002;18:41–51.
473. Miranda DR, Klompe L, Mekel J. Open lung ventilation does not increase right ventricular outflow impedance: an echo-Doppler study. Crit Care Med. 2006;34:2555–60.
474. Gernoth C, Wagner G, Pelosi P. Respiratory and hemodynamic changes during decremental open lung positive end-expiratory pressure titration in patients with acute respiratory distress syndrome. Crit Care Med. 2009;13:R59.
475. Maggiore SM, Jonson B, Richard JC. Alveolar derecruitment at decremental positive end-expiratory pressure levels in acute lung injury: comparison with the lower inflection point, oxygenation and compliance. Am J Respir Crit Care Med. 2001;164:795–801.
476. Lachmann B. Open up the lung and keep the lung open. Intensive Care Med. 1992;18:19–21.
477. Madden JA, Dawson CA, Harder DR. Hypoxia-induced activation in small isolated pulmonary arteries from the cat. J Appl Physiol. 1985;59:113–8.
478. Marshall BE, Marshall C, Frasch F. Roel of hypoxic pulmonary vasoconstriction in pulmonary gas exchange and blood flow distribution. 1. Physiologic concepts. Intensive Care Med. 1994;20:291–7.
479. Weir EK, Lopez-Barneo J, Buckler KJ. Mechanisms of disease: acute oxygen-sensing mechanisms. N Engl J Med. 2005;353:2042–55.
480. Moudgil R, Michelakis ED, Archer SL. Hypoxic pulmonary vasoconstriction. J Appl Physiol. 2005;98:390–403.
481. von Euler US, Liljestrand G. Observations on pulmonary arterial blood pressure in the cat. Acta Physiol Scand. 1946;12:301–20.
482. Marshall BE, Hanson CW, Frasch F. Role of hypoxic pulmonary vasoconstriction in pulmonary gas exchange and blood flow distribution. 2. Pathophysiology. Intensive Care Med. 1994;20:379–89.
483. O'Quin RJ, Marini JJ, Cluver BH. Transmission of airway pressure to pleural space during lung edema and chest wall restriction. J Appl Physiol. 1985;59:1171–7.
484. Marini JJ, Gattinoni L. Ventilatory management of acute respiratory distress syndrome: a consensus of two. Crit Care Med. 2004;32:250–5.
485. Hakim TS, Michel RP, Chang HK. Effect of lung inflation on pulmonary vascular resistance by arterial and venous occlusion. J Appl Physiol. 1982;53:1110–5.
486. Luce JM. The cardiovascular effects of mechanical ventilation and positive end-expiratory pressure. JAMA. 1984;252:807–11.
487. Dessap AM, Charron C, Devaquet J. Impact of acute hypercapnia and augmented positive end-expiratory pressure on right ventricle function in severe acute respiratory distress syndrome. Intensive Care Med. 2009;35:1850–8.
488. Grindlinger GA, Manny J, Justice R. Presence of negative inotropic agents in canine plasma during positive end-expiratory pressure. Circ Res. 1979;45:460–7.
489. Slutsky AS, Imai Y. Ventilator-induced lung injury, cytokines, PEEP, and mortality: implications for practice and for clinical trails. Intensive Care Med. 2003;29:1218–21.
490. Marini JJ, Culver BH, Butler J. Mechanical effect of lung distension with positive pressure on cardiac function. Am Rev Respir Dis. 1981;124:382–6.
491. Szold A, Pizov R, Segal E. The effect of tidal volume and intravascular volume state on systolic pressure variation in ventilated dogs. Intensive Care Med. 1989;15:46–53.
492. Armaganidis A. Heart lung interactions during mechanical interaction: effects on loading conditions. Intensivmedizin und Notfallmedizin. 1997;34:696–705.
493. Crottogini AJ, Willshaw P, Barra JG. End-systolic pressure-volume relationships in dogs during ventilation with PEEP. Am J Physiol. 1988;254:H664–70.

494. Johnston WE, Vinten-Johansen J, Santamoore WP. Mechanisms of reduced cardiac output during positive end-expiratory pressure in the dog. Am Rev Respir Dis. 1989;140:1257–64.
495. Schulman DS, Matthay RA. The right ventricle in pulmonary disease. Cardiol Clin. 1992;10:111–35.
496. Jardin F, Dubourg O, Bourdarias J-P. Echocardiographic pattern of acute cor pulmonale. Chest. 1997;111:209–17.
497. Kannel WB, Belanger AJ. Epidemiology of heart failure. Am Heart J. 1992;121:951–7.
498. Zapol W, Snider M. Pulmonary hypertension in severe acute respiratory failure. N Engl J Med. 1977;296:476–80.
499. Rubin LJ. Primary pulmonary hypertension. N Engl J Med. 1997;336:111–7.
500. Gaine S. Pulmonary hypertension. JAMA. 2000;284:160–8.
501. Zapol W, Jones R. Vascular components of ARDS. Clinical pulmonary hemodynamics and morphology. Am Rev Respir Dis. 1987;136:471–4.
502. Jardin F, Gueret P, Dubourg O. Two-dimensional echocardiographic evaluation of right ventricular size and contractility in acute respiratory failure. Crit Care Med. 1985;13:952–6.
503. SZtrymf B, Souza R, Bertoletti L. Prognostic factors of acute heart failure in patients with pulmonary arterial hypertension. Eur Respir J. 2010;35:1286–93.
504. Ghio S, Gavazzi A, Campana C. Independent and additive prognostic value of right ventricular systolic function and pulmonary artery pressure in patients with chronic heart failure. J Am Coll Cardiol. 2001;37:183–8.
505. Torbicki A, Kurzyna M, Kuca P. Detectable serum cardiac troponin T as a marker of poor prognosis among patients with chronic precapillary pulmonary hypertension. Circulation. 2003;108:844–8.
506. Kimchi A, Ellroth AG, Berman DS. Right ventricular performance in septic shock: a combined radionuclide and hemodynamic study. J Am Coll Cardiol. 1984;4:945–51.
507. Kasper W, Konstantinides S, Geibel A. Prognostic significance of right ventricular afterload stress detected by echocardiography in patients with clinically suspected pulmonary embolism. Heart. 1997;77:346–9.
508. Monchi M, Bellenfant F, Cariou A. Early predictive factors of survival in acute respiratory distress syndrome. A multivariate analysis. Am J Respir Crit Care Med. 1998;158:1076–81.
509. Clowes Jr GH, Farrington GH, Zuschneid W. Circulating factors in the etiology of pulmonary insufficiency and right heart failure accompanying severe sepsis (peritonitis). Ann Surg. 1970;171:663–78.
510. Bronicki RA, Baden HP. Pathophysiology of right ventricular failure in pulmonary hypertension. Pediatr Crit Care Med. 2010;11:S15–22.
511. Brimioulle S, Wauthy P, Ewalenko P. Single-beat estimation of right ventricular end systolic pressure–volume relationship. Am J Physiol Heart Circ Physiol. 2003;284:H1625–30.
512. Fourie PR, Coetzee AR, Bolliger CT. Pulmonary artery compliance: its role in right ventricular–arterial coupling. Cardiovasc Res. 1992;26:839–44.
513. Champion HC, Michelakis ED, Hassoun PM. Comprehensive invasive and noninvasive approach to the right ventricle–pulmonary circulation unit: state of the art and clinical and research implications. Circulation. 2009;120:992–1007.
514. Hoeper MM, Granton J. Intensive care unit management of patients with severe pulmonary hypertension and right heart failure. Am J Respir Crit Care Med. 2011;184:1114–24.
515. Vincent JL. What is right ventricular (dys)function? Crit Care Med. 1994;22:2014.
516. Osman D, Monnet X, Castelain V. Incidence and prognostic value of right ventricular failure in acute respiratory distress syndrome. Intensive Care Med. 2009;35:69–76.
517. Redington AN, Rigby ML, Shinebourne EA. Changes in the pressure–volume relation of the right ventricle when its loading conditions are modified. Br Heart J. 1990;63:45–9.
518. Kerbaul F, Rondelet B, Motte S. Effects of norepinephrine and dobutamine on pressure load-induced right ventricular failure. Crit Care Med. 2004;32:1035–40.
519. Block AJ, Boysen PG, Wynne JW. The origins of cor pulmonale: a hypothesis. Chest. 1979;75:109–10.
520. Via G, Braschi A. Pathophysiology of severe pulmonary hypertension in the critically ill patient. Minerva Anestesiol. 2004;70:233–7.

521. Simon MA. Right ventricular adaptation to pressure overload. Curr Opin Crit Care. 2010;16:237–43.
522. Dias CA, Assad RS, Caneo LF. Reversible pulmonary trunk banding. II. An experimental model for rapid pulmonary ventricular hypertrophy. J Thorac Cardiovasc Surg. 2002;14:999–1006.
523. Badano LP, Ginghina C, Easaw J. Right ventricle in pulmonary arterial hypertension: haemodynamics, structural changes, imaging, and proposal of a study protocol aimed to assess remodelling and treatment effects. Eur J Echocardiogr. 2010;11:27–37.
524. Jardin F, Gurdjian F, Fouilladieu JL. Pulmonary and systemic hemodynamic disorders in adult respiratory distress syndrome. Intensive Care Med. 1979;5:127–33.
525. Pfisterer M. Right ventricular involvement in myocardial infarction and cardiogenic shock. Lancet. 2003;362:392–4.
526. Neidhart PP, Suter PM. Changes of right ventricular function with positive end-expiratory pressure (PEEP) in man. Intensive Care Med. 1988;14 Suppl 2:471–3.
527. Huemer G, Kolev N, Kurz A. Influence of positive end-expiratory pressure on right and left ventricular performance assessed by Doppler two-dimensional echocardiography. Chest. 1994;106:63–7.
528. Terragni P, Rosbosh G, Tealdi A. Tidal hyperinflation during low tidal volume ventilation in acute respiratory distress syndrome. Am J Respir Crit Care Med. 2007;175:160–6.
529. Mekontso-Dessap A, Charron C, Devaquet J. Impact of acute hypercapnia and augmented positive end-expiratory pressure on right ventricle function in severe acute respiratory distress syndrome. Intensive Care Med. 2009;35:1850–8.
530. Meade MO, Cook DJ, Guyatt GH. Ventilation strategy using low tidal volumes, recruitment maneuvers, and high positive end expiratory pressure for acute lung injury and acute respiratory. JAMA. 2008;299:637–45.
531. Hickling KG, Walsh J, Henderson S. Low mortality rate in adult respiratory distress syndrome using low-volume, pressure-limited ventilation with permissive hypercapnia: a prospective study. Crit Care Med. 1994;22:1568–87.
532. Mekontso-Dessap A, Boissier F, Leon R. Prevanlence and prognosis of shunting across patent foramen ovale during acute respiratory distress syndrome. Crit Care Med. 2010;38:1786–92.
533. Hickling KG. The pressure–volume curve is greatly modified by recruitment. A mathematical model of ARDS lungs. Am J Respir Crit Care Med. 1998;154:194–202.
534. Jonson B, Svantesson C. Elastic pressure–volume curves: what information do they convey? Thorax. 1999;54:82–7.
535. Schultz MJ. Lung-protective mechanical ventilation with lower tidal volumes in patients not suffering from acute lung injury: a review of clinical studies. Med Sci Monit. 2008;14:RA22–6.
536. Tenney SM, Remmers JE. Comparative quantitative morphology of the mammalian lung: diffusing area. Nature. 1963;197:54–6.
537. Gajic O, Dara SI, Mendez JL. Ventilator-associated lung injury in patients without acute lung injury at the onset of mechanical ventilation. Crit Care Med. 2004;32:1817–24.
538. Reis M, Gommers D, Struijs A. Ventilation according to the open lung concept attenuates pulmonary inflammatory response in cardiac surgery. Eur J Cardiothorac Surg. 2005;28:889–95.
539. Gajic O, Frutos-Vivar F, Estaban A. Ventilator settings as a risk factor for acute respiratory distress syndrome in mechanically ventilated patients. Intensive Care Med. 2005;31:922–6.
540. Mascia L, Pasero D, Slutsky AS. Effect of a lung protective strategy for organ donors on eligibility and availability of lungs for transplantation: a randomised controlled trail. JAMA. 2010;304:2620–7.
541. Lee PG, Helsmoortel CN, Cohn SM. Are low tidal volumes safe? Chest. 1990;97:430–4.
542. Fernandez-Perez ER, Keegan MT, Brown DR. Intraoperative tidal volume as a risk factor for respiratory failure after pneumonectomy. Anesthesiology. 2006;105:14–8.
543. Weingarten TN, Whalen FX, Warner DO. Comparison of two ventilator strategies in elderly patients undergoing major abdominal surgery. Br J Anaesth. 2010;104:16–22.
544. Futire E, Constantin JM, Paugaen- Burtz C. A trail of intraoperative low – tidal volume ventilation in abdominal surgery. N Engl J Med. 2013;369:428.

545. Fernandez-Perez ER, Sprung J, Afessa B. Intraoperative ventilator settings and acute lung injury after elective surgery: a nested case control study. Thorax. 2009;64:121–7.
546. Hughes CG, Weavind L, Banerjee A. Intraoperative risk factors for acute respiratory distress syndrome in critically ill patients. Anesth Analg. 2010;111:464–7.
547. Stuber F, Wrigge H, Schroeder S. Kinetic and reversibility of mechanical ventilation-associated pulmonary and systemic inflammatory response in patients with acute lung injury. Intensive Care Med. 2002;28:834–41.
548. Nakamura T, Malloy J, McCaig L. Mechanical ventilation of isolated septic rat lungs: effects on surfactant and inflammatory cytokines. J Appl Physiol. 2001;91:811–20.
549. Frank JA, Gutierrez JA, Jones KD. Low tidal volume reduces epithelial and endothelial injury in acid-injured rat lungs. Am J Respir Crit Care Med. 2002;165:242–9.
550. Wrigge H, Zinserling J, Stuber F. Effects of mechanical ventilation on release of cytokines into systemic circulation in patients with normal pulmonary function. Anesthesiology. 2000;93:1413–7.
551. Wolthuis EK. Mechanical ventilation with lower tidal volumes and PEEP prevents pulmonary inflammation in patients without preexisting lung injury. Lung protective mechanical ventilation. Amsterdam: s.n., 2008. Vols. Dissertation, chapter 9. http://dare.uva.nl/record/274037. ISBN 9789071382338.
552. Pirttikangas CO, Salo M, Mansikka M. The influence of anaesthetic technique upon the immune response to hysterectomy. A comparison of propofol infusion and isoflurane. Anaesthesia. 1995;50:1056–61.
553. Crozier TA, Muller JE, Quittkat D. Effect of anaesthesia on the cytokine responses to abdominal surgery. Br J Anaesth. 1994;72:280–5.
554. Kor DJ, Warner DO, Alsara A. Derivation and diagnostic accuracy of the surgical lung injury prediction model. Anesthesiology. 2011;115:117–28.
555. Waaren OJ, Smith AJ, Alexiou A. The inflammatory response to cardiopulmonary bypass: part 1 – mechanisms of pathogenesis. J Cardiothorac Vasc Anesth. 2009;23:223–31.
556. Lellouche F, Dionne S, Simard S. High tidal volumes in mechanically ventilated patients increase organ dysfunction after cardiac surgery. Anesthesiology. 2012;116:1072–82.
557. Zupancich E, Paparella D, Turani F. Mechanical ventilation affects inflammatory mediators in patients undergoing cardiopulmonary bypass for cardiac surgery: a randomized clinical trial. J Thorac Cardiovasc Surg. 2005;130:378–83.
558. Reis Miranda D, Gommers D, Struijs A. Ventilation according to the open lung concept attenuates pulmonary inflammatory response in cardiac sugery. Eur J Cardiothorac Surg. 2005;28:889–95.
559. Koner O, Celebi S, Balci H. Effects of protective and conventional mechanical ventilation on pulmonary function and systemic cytokine release after cardiopulmonary bypass. Intensive Care Med. 2004;30:620–6.
560. Sundar S, Novack V, Jervis K. Influence of low tidal volume ventilation on time to extubation in cardiac surgical patients. Anesthesiology. 2011;114:1102–10.
561. Parsons PE, Matthay MA, Ware LB. Elevated plasma levels of soluble TNF receptors are associated with morbidity and mortality in patients with acute lung injury. Am J Physiol Lung Cell Mol Physiol. 2005;288:L426–31.
562. Parsons PE, Eisner MD, Thompson BT. Lower tidal volume ventilation and plasma cytokine markers of inflammation in patients with acute lung injury. Crit Care Med. 2005;33:230–2.
563. Determann RM, Royakkers A, Wolthuis EK. Ventilation with lower tidal volumes as compared with conventional tidal volumes for patients without acute lung injury: a preventive randomized controlled trail. Crit Care. 2010;14:R1.
564. Yilmaz M, Keegan MT, Iscimen R. Toward the prevention of acute lung injury: protocol-guided limitation of large tidal volume ventilation and inappropriate transfusion. Crit Care Med. 2007;35:1660–6; quiz 1667.
565. Fuller B, Mohr N, Drewry A. Lower tidal volume at initiation of mechanical ventilation may reduce progression to acute respiratory distress: a systematic review. Crit Care. 2013;17:R11.

566. Serpa NA, Cardoso SO, Manetta JA. Association between use of lung-protective ventilation with lower tidal volumes and clinical outcomes among patients without acute respiratory distress syndrome: a meta-analysis. JAMA. 2012;308:1651–9.
567. Kilickaya O, Gajic O. Initial ventilator settings for all critically ill patients. Crit Care. 2013;17:123.
568. Levy MM, Dellinger RP, Townsend SR. The surviving sepsis campaign: results of an international guideline-based performance improvement program targeting severe sepsis. Intensive Care Med. 2010;36:222–31.
569. Falke KJ. Zhe introduction of positive endexpiratory pressure into mechanical ventilation: a retrospective. In: Brochard L, Mancebo J, Pinsky MR, editors. Applied physiology in intensive care medicine. Berlin/Heidelberg: Springer; 2006. p. 291–4.
570. Dery R, Pelleter J, Jacques A. Alveolar collapse induced by denitrogenation. Can Anaesth Soc J. 1965;12:531–44.
571. Hedenstierna G, Edmark I. Mechanisms of atelectasis in the perioperative period. Best Pract Res Clin Anaesthesiol. 2010;24:157–69.
572. Briel M, Meade M, Mercat A. Higher vs lower positive end-expiratory pressure in patients with acute respiratory distress syndrome: systematic review and meta-analysis. JAMA. 2010;303:865–73.
573. Leatherman JW, McArthur C, Shapiro RS. To prolong expiration time is best accomplished of the by reduction respiratory rate. Crit Care Med. 2004;32:1542–5.
574. Unroe M, McIntyre N. Evolving approaches to assessing and monitoring patient-ventilator interactions. Curr Opin Crit Care. 2010;16:261–8.
575. Schmidt M, Demoule A, Polito A. Dyspnea in mechanically ventilated critically ill patients. Crit Care Med. 2011;39:2059–65.
576. Kallet RH, Campbell AR, Dicker RA. Effects of tidal volume on work of breathing during lung-protective ventilation in patients with acute lung injury and adult respiratory distress syndrome. Crit Care Med. 2006;34:8–14.
577. Hess DR, Thompson BT. Patient-ventilator dyssynchrony during lung protective ventilation: what's a clinician to do? Crit Care Med. 2006;34:231–3.
578. Stevens RD, Lazaridis C, Chalela JA. The role of mechanical ventilation in acute brain injury. Neurol Clin. 2008;26:543–63.
579. Pelosi P, Severgnini P, Chiaranda M. An integrated approach to prevent and treat respiratory failure in brain-injured patients. Curr Opin Crit Care. 2005;11:37–42.
580. Ngubane T. Mechanical ventilation and the injured brain. South Afr J Anaesth Analg. 2011;17:76–80.
581. Mascia L. Acute lung injury in patients with severe brain injury: a double – hit model. Neurocrit Care. 2009;11:417–26.
582. Bratton SL, Chestnut RM, Ghajar J. Hyperventilation. J Neurotrauma. 2007;24 Suppl 1:S87–90.
583. Lapinsky SE, Posadas-Colleja JG, McCullagh I. Clinical review: ventilatory strategies for obstetric, brain-injured and obese patients. Crit Care. 2009;13:206.
584. Huynh T, Messer M, Sing RF. Positive end-expiratory pressure alters intracranial and cerebral perfusion pressure. J Trauma. 2002;53:488–93.
585. Davis DP. Early ventilation in traumatic brain injury. Resuscitation. 2008;76:333–40.
586. Epstein SK. Complications associated with mechanical ventilation. In: Tobin JM, editor. Principles and practice of mechanical ventilation. 2nd ed. New York: McGraw-Hill; 2006. p. 877–902.
587. Hill NS. Noninvasive positive -pressure ventilation. In: Tobin MJ, editor. Principles and practice of mechanical ventilation. 2nd ed. New York: McGraw-Hill; 2006. p. 433–71.
588. Ambrosino N, Vagheggini G. Noninvasive positive pressure ventilation in acute care setting: where are we? Eur Respir J. 2008;31:874–86.
589. Ram FS, Picot J, Lightowler J. Non-invasive positive pressure ventilation for treatment of respiratory failure due to exacerbations of chronic obstructive pulmonary disease. Cochrane Database Syst Rev. 2004;3:CD004104.

590. Squadrone E, Frigerio P, Fogliati C. Noninvasive versus invasive ventilation in COPD patients with severe respiratory failure deemed to require ventilatory assistance. Intensive Care Med. 2004;30:1303–10.
591. Plant PK, Owen JL, Elliot MW. Noninvasive versus invasive ventilation for acute exacerbations of chronic obstructive pulmonary disease on general respiratory wards: a multicentre randomized controlled trail. Lancet. 2000;355:1931–5.
592. Hill NS, Brennan J, Garpestad E. Non-invasive ventilation in acute respiratory failure. Crit Care Med. 2007;35:2402–7.
593. Crane SD, Elliott MW, Gilligan P. Randomised controlled comparison of continuous positive pressure, bilevel non-invasive ventilation, and standard treatment in emergency department patients with acute cardiogenic pulmonary edema. Emerg Med J. 2004;21:155–61.
594. Masip J, Roque M, Sanchez B. Noninvasive ventilation in acute cardiogenic pulmonary edema. Systematic review and meta-analysis. JAMA. 2005;294:3124–30.
595. Collins SP, Mielniczuk LM, Whittingham HA. The use of noninvasive ventilation in emergency department patients with acute cardiogenic pulmonary edema: a systematic review. Ann Emerg Med. 2006;48:260–9.
596. Nava S, Ambrosino N, Clini E. Noninvasive mechanical ventilation in the weaning of patients with respiratory failure due to chronic obstructive pulmonary disease: a randomized, controlled trial. Ann Intern Med. 1998;128:721–8.
597. Burns KE, Adhikari NK, Meade MO. A meta-analysis of noninvasive weaning to facilitate liberation from mechanical ventilation. Can J Anaesth. 2006;53:305–15.
598. Ferrer M, Esquinas A, Arancibia F. Noninvasive ventilation during persistent weaning failure: a randomized controlled trail. Am J Respir Crit Care Med. 2003;168:70–6.
599. Confalonieri M, Calderini E, Terraciano S. Noninvasive ventilation for treating respiratory failure in AIDS patients with Pneumocystis carinii pneumonia. Intensive Care Med. 2002;28:1233–8.
600. Hilbert G, Gruson D, Vargas F. Noninvasive ventilation in immunosuppressed patients with pulmonary infiltrates, fever, and acute respiratory failure. N Engl J Med. 2001;344:481–7.
601. Squadrone V, Coha M, Cerutti E. Continuous positive airway pressure for treatment of postoperative hypoxemia. JAMA. 2005;293:589–95.
602. Jaber S, Delay JM, Chanques G. Outcome of patients with acute respiratory failure after abdominal surgery treated with noninvasive positive pressure ventilation. Chest. 2005;128:2688–95.
603. Kindgen-Milles D, Muller E, Buhl R. Nasal continuous positive airway pressure reduces pulmonary morbidity and length of stay following thoracoabdominal aortic surgery. Chest. 2005;128:821–8.
604. Meduri GU, Cook TR, Turner RE. Noninvasive positive pressure ventilation in status asthmaticus. Chest. 1996;110:767–74.
605. Fernandez MM, Villagra A, Blanch L. Noninvasive mechanical ventilation in status asthmaticus. Intensive Care Med. 2001;27:486–92.
606. Soroksky A, Stav D, Shpirer I. A pilot, prospective, randomized, placebo-controlled trail of bilevel positive airway pressure in acute asthmatic attack. Chest. 2003;123:1018–25.
607. Ram FS, Wellington S, Rowe BH. Noninvasive positive-pressure ventilation for the treatment of respiratory failure due to severe acute exacerbations of asthma. Cochrane Database Syst Rev. 2005;1:CD004360.
608. Nourdine K, Combes P, Carton MJ. Does noninvasive ventilation reduce the ICU nosocomial infection risk? A prospective clinical survey. Intensive Care Med. 1999;25:567–73.
609. Antonelli M, Conti G, Rocco M. A comparison of noninvasive positive-pressure ventilation and conventional mechanical ventilation in patients with acute respiratory failure. N Engl J Med. 1998;339:429–35.
610. Ferrer M, Esquinas A, Leon M. Noninvasive ventilation in severe hypoxemic respiratory failure: a randomized controlled trail. Am J Respir Crit Care Med. 2003;168:1438–44.
611. Puha J, Kong K, Lee KH. Noninvasive ventilation in hypercapnic acute respiratory failure due to chronic obstructive pulmonary disease versus other conditions: effectiveness and predictors of failure. Intensive Care Med. 2005;31:533–9.

612. Honrubia TG, Lopez FJ, Franco N. Noninvasive versus conventional mechanical ventilation in acute respiratory failure. A multicenter, randomized controlled trail. Chest. 2005;128:3916–24.
613. Fowler RA, Guest CB, Lapinsky SE. Transmission of severe scute respiratory syndrome during intubation and mechanical ventilation. Am J Respir Crit Care Med. 2004;169:1198–202.
614. Cheung TMT, Lau CWA, Poon E. Effectiveness of noninvasive positive pressure ventilation in the treatment of acute respiratory failure in severe acute respiratory syndrome. Chest. 2004;126:845–50.
615. Han F, Jiang YY, Zheng JH. Noninvasive positive pressure ventilation treatment for acute respiratory failure in SARS. Sleep Breath. 2004;8:97–106.
616. Vitacca M, Rubini F, Foglio K. Non-invasive modalities of positive pressure ventilation improve the outcome of acute exacerbations in COPD patients. Intensive Care Med. 1993;19:450–5.
617. Schettino G, Altobelli N, Kacmarek RM. Noninvasive positive-pressure ventilation in acute respiratory failure outside clinical trials. Experience at the Massachusetts General Hospital. Crit Care Med. 2008;36:441–7.
618. Yeow M-E, Santanilla JL. Noninvasive positive pressure ventilation in the emergency department. Emerg Med Clin North Am. 2008;26:835–47.
619. Bauman KA, Parsons PE, Wilson KC. Noninvasive positive pressure ventilation in acute respiratory failure. UpToDate. [Online] 20 June 2013. http://www.uptodate.com/contents/noninvasive-positive-pressure-ventilation-in-acute-respiratory-failure-in-adults. Cited: 17 Oct 2013.
620. Tang HM, Lau FL. Noninvasive ventilatory support in emergency department: initial experience in Hong Kong. Hong Kong J Emerg Med. 2001;8:140–5.
621. Kinnear W. Non-invasive treatment modalities. In: Black R, Hugher M, editors. Oxford specialist handbook in critical care: advanced respiratory critical care. New York: Oxford Univ Press; 2011. p. 73.
622. Roberts CM. Non-invasive ventilation in chronic obstructive pulmonary disease: management of acute type 2 respiratory failure. Clin Med. 2008;8:517–21.
623. Sethi S, Murphy TF. Infection in the pathogenesis and course of chronic obstructive pulmonary disease. N Engl J Med. 2008;359:2355–65.
624. Stoller JK. Management of acute exacerbations of chronic obstructive pulmonary disease. UpToDate. [Online] Wolter Kluwer, 14 June 2012. http://www.uptodate.com/contents/management-of-acute-exacerbations-ofchronic-obstructive-pulmonary-disease. Cited 30 Sept 2013.
625. Davidson AC. The pulmonary physician in critical care. 11: Critical care management of respiratory failure resulting from COPD. Thorax. 2002;57:1079–84.
626. Papi A, Belletatto CM, Braccioni F. Infections and airway inflammation in chronic obstructive pulmonary disease severe exacerbations. Am J Respir Crit Care Med. 2006;173:1114–21.
627. Seemungal TA, Harper-Owen R, Bhowmik A. Detection of rhinovirus in induced sputum at exacerbations of chronic obstructive pulmonary disease. Eur Respir J. 2000;16:677–83.
628. Wedzicha JA. Role of viruses in exacerbation of chronic obstructive pulmonary disease. Proc Am Thorac Soc. 2004;1:115–20.
629. Barnes PJ. Chronic obstructive pulmonary disease. N Engl J Med. 2000;343:269–80.
630. Sapey E, Stockley RA. COPD exacerbations.2: aetiology. Thorax. 2006;61:250–8.
631. Seneff MG, Wagner DP, Wagner RP. Hospital and one year survival of patients admitted to intensive care units with acute exacerbations of chronic obstructive disease. JAMA. 1995;274:1852–7.
632. Tillie-Leblond I, Marquette CH, Perez T. Pulmonary embolism in patients with unexplained exacerbation of chronic obstructive pulmonary disease: prevalence and risk factors. Ann Intern Med. 2006;144:390.
633. Miravitlles M, Guerrero T, Mayordomo C. Factors associated with increased risk of exacerbation and hospital admission in a cohort of ambulatory COPD patients: a multiple logistic regression analysis. Respiration. 2000;67:495–501.
634. Rascon-Aguilar IE, Pamer M, Wludyka P. Role of gastroesophageal reflux symptoms in exacerbations of COPD. Chest. 2006;130:1096–101.

635. Niewoehner DE, Lokhnygina Y, Rice K. Risk indexes for exacerbations and hospitalizations due to COPD. Chest. 2007;131:20–8.
636. Burgel P-R, Nesme-Meyer P, Chanez P. Cough and sputum production are associated with frequent exacerbations and hospitalizations in COPD subjects. Chest. 2009;135:975–82.
637. Pride NB, Macklem PT. Lung mechanics in disease. In: American Physiological Society, editor. Handbook of physiology, section 3, vol. III: the respiratory system. Mechanics of breathing, part2. Bethesda: American Physiological Society; 1986. p. 659–92.
638. O'Donoghue FJO, Catcheside PG, Jordan AS. Effect of CPAP on intrinsic PEEP, inspiratory effort, and lung volume in severe stable COPD. Thorax. 2002;57:533–9.
639. Kimball WR, Leith DG, Robins AG. Dynamic hyperinflation and ventilator dependence in chronic obstructive pulmonary disease. Am Rev Respir Dis. 1982;126:991–5.
640. Reddy RM, Guntupalli KK. Review of ventilatory techniques to optimize mechanical ventilation in acute exacerbation of chronic obstructive pulmonary disease. Int J Chron Obstruct Pulmon Dis. 2007;2:441–52.
641. Petrof BJ, Legare M, Goldberg P. Continuous positive airway pressure reduces work of breathing and dyspnea during weaning from mechanical ventilation in severe chronic pulmonary disease. Am Rev Respir Dis. 1990;141:281–9.
642. Broseqhini C, Brandolese R, Poggi R. Respiratory mechanics during the first day of mechanical ventilation in patients with pulmonary edema and chronic airway obstruction. Am Rev Respir Dis. 1988;138:355–61.
643. Rossi A, Gottfried SB, Higgs BD. Respiratory mechanics in mechanically ventilated patients with respiratory failure. J Appl Physiol. 1985;58:1849–58.
644. Derenne J, Fleury B, Pariente R. Acute respiratory failure of chronic obstructive pulmonary disease. Am Rev Respir Dis. 1988;138:1006–33.
645. Roussos C, Macklem PT. The respiratory muscles. N Engl J Med. 1982;307:786–97.
646. Tobin MJ. Respiratory muscles in disease. Clin Chest Med. 1988;9:263–86.
647. Orozco-Levi M. Structure and function of the respiratory muscles in patients with COPD: impairment or adoption? Eur Respir J. 2003;22 suppl 46:41s–51.
648. De Troyer A, Pride NB. The chest wall and respiratory muscles. In: Roussos C, editor. The thorax. New York: Dekker; 1995. p. 1975–2006.
649. Peress L, Sybrecht G, Macklem PT. The mechanism of increase in total lung capacity during acute asthma. Am J Med. 1976;61:165–9.
650. Dalley MT, Bui T. Severe asthma and COPD. In: Chui W, Flaxman A, Marshall J, Farcy D, editors. Critical care emergency medicine. New York: McGraw-Hill; 2011. Section III, Chapter 10, p. 98–108.
651. McFadden Jr ER, Ingram HR, Haynes RL. Predominant site of flow limitation and mechanisms of postexertional asthma. J Appl Physiol. 1977;42:746–52.
652. Corbridge TC, Hall JB. The assessment and management of adults with status asthmaticus. Am J Respir Crit Care Med. 1995;151:1296–316.
653. Collett PW, Brancatisano T, Engel LA. Changes in the glottic aperture during bronchial asthma. Am Rev Respir Dis. 1983;128:719–23.
654. Stanley D, Tunnicliffe W. Managemnt of life-threatening asthma in adults. Contin Edu Anaesth Crit Care Pain. 2008;8:95–9.
655. Mayo P, Radeos MS. The severe asthmatic: intubated and difficult to ventilate. In: Brenner B, editor. Emergency asthma. New York: Marcel Dekker; 1999. p. 469–87.
656. Colebatch HJ, Finucane KE, Smith MM. Pulmonary conductance and elastic recoil relationships in asthma and emphysema. J Appl Physiol. 1973;34:143–53.
657. Cormier Y, Lecours R, Legris C. Mechanisms of hyperinflation in asthma. Eur Respir J. 1990;3:619–24.
658. Carroll N, Elliot J, Morton A. The structure of large and small airways in nonfatal and fatal asthma. Am Rev Respir Dis. 1993;147:405–10.
659. Faul JL, Tormey VJ, Leonard C. Lung immunopathology in cases of sudden asthma death. Eur Respir J. 1997;10:301–7.
660. Saetta M, Di Stefano A, Rosina C. Quantitative structural analysis of peripheral airways and arteries in sudden fatal asthma. Am Rev Respir Dis. 1991;143:138–43.

661. Phipps P, Garrard CS. The pulmonary physician in critical care-12: acute severe asthma in the intensive care unit. Thorax. 2003;58:81–8.
662. Leatherman JW, Ravenscraft SA. Low measured auto-positive end-expiratory pressure during mechanical ventilation of patients with severe asthma: hidden auto-positive end-expiratory pressure. Crit Care Med. 1995;24:541–6.
663. McFadden ER, Kiser R, DeGroot WJ. Acute bronchial asthma — relations between clinical and physiologic manifestations. N Engl J Med. 1973;288:221–5.
664. Briscoe WA, McLEMORE Jr GA. Ventilatory function in bronchial asthma. Thorax. 1952;7:66–77.
665. Roca J. Distributions of alveolar ventilation and pulmonary blood flow. In: Marini AS, Slutsky JJ, editors. Physiological basis of ventilatory support. New York: Marcel-Dekker; 1998. p. 311.
666. Ranieri VM, Giuliani R, Cinnella G. Physiologic effects of positive end-expiratory pressure in patients with chronic obstructive mechanical ventilation. Am Rev Respir Dis. 1993;147:5–13.
667. Slutsky AS. Consensus conference on mechanical ventilation. Northbrook, 28–30 Jan, 1993. Part I. Intensive Care Med. 1994;20:64–79.
668. McIntyre N, Huang YC. Acute exacerbations and respiratory failure in chronic obstructive pulmonary disease. Proc Am Thorac Soc. 2008;5:530–5.
669. Darioli R, Perret C. Mechanical controlled hypoventilation in status asthmaticus. Am Rev Respir Dis. 1984;129:385–7.
670. Feihl F, Perret C. Permissive hypercapnia. How permissive should we be? Am J Respir Crit Care Med. 1994;150:1722–37.
671. Tuxen DV, Lane S. The effects of ventilatory pattern on hyperinflation airway pressures, circulation in mechanically ventilation of patients with severe airflow obstruction. Am Rev Respir Dis. 1987;136:872–9.
672. Brenner B, Corbridge T, Kazzi A. Intubation and mechanical ventilation of asthmatic patient in respiratory failure. Proc Am Thorac Soc. 2009;6:371–9.
673. Connors AF, McCaffree DR, Gray BA. Effect of inspiratory flow rate on gas exchange during mechanical ventilation. Am Rev Respir Dis. 1981;124:537.
674. Laghi F, Segal J, Choe WK. Effect of imposed inflation time on respiratory frequency and hyperinflation in patients with chronic obstructive pulmonary disease. Am J Respir Crit Care Med. 2001;163:1365.
675. Reddy VG. Auto-PEEP: how to detect and how to prevent – a review. Middle East J Anesthesiol. 2005;18:293–312.
676. Jurban A. Mechanical ventilation in acute respiratory failure complicating COPD. [Online] UpToDate, 30 Aug 2011. http://www.uptodate.com/contents/mechanical-ventilation-in-acute-respiratory-failure-complicating-copd?detectedLanguage=en&source=search_result&translation=mechanical+ventilation+in+acute+failure+respiratroy&search=mechanical+ventilation+in+acute+respirat. Cited 4 Oct 2013.
677. Tobin MJ, Alex CA, Fahey PJ. Fighting the ventilator. In: Tobin MJ, editor. Principles and practice of mechanical ventilation. New York: McGraw-Hill; 2006. p. 1121.
678. Marini JJ, Rodriguez RM, Lamb V. The inspiratory workload of patient-initiated mechanical ventilation. Am Rev Respir Dis. 1986;134:902–9.
679. Fernandez R, Benito S, Blanch L. Intrinsic PEEP: a cause of inspiratory muscle ineffectivity. Intensive Care Med. 1988;15:51–2.
680. Marini JJ. Should PEEP be used in airflow obstruction? Am Rev Respir Dis. 1989;140:1–3.
681. Tobin MJ, Perez W, Guenther SM. The pattern of breathing during successful and unsuccessful trials of weaning from mechanical ventilation. Am Rev Respir Dis. 1986;134:1111–8.
682. Georgopoulos D, Giannouli E, Patakas D. Effects of extrinsic positive end-expiratory pressure on mechanically ventilated patients with chronic obstructive pulmonary disease and dynamic hyperinflation. Intensive Care Med. 1993;19:197–203.
683. Dos Santos CC, Slutsky AS. Novel advancements in the management and diagnosis of acute respiratory failure. In: Anzuetro A, Cook DJ, Esteban A, editors. Evidence-based management of patients with respiratory failure. Berlin/Heidelberg: Springer; 2004. p. 149.

684. Aubier M, Murciano D, Fournier M. Central respiratory drive in acute respiratory failure of patients with chronic obstructive pulmonary disease. Am Rev Respir Dis. 1980;122:191–9.
685. Dick CR, Liu Z, Sassoon CS. O2-induced change in ventilation and ventilatory drive in COPD. Am J Respir Crit Care Med. 1997;155:609–14.
686. Hanson 3rd CW, Marshall BE, Frasch HF. Causes of hypercarbia with oxygen therapy in patients with chronic obstructive pulmonary disease. Crit Care Med. 1996;24:23–8.
687. Crossley DJ, McGuire GP, Barrow PM, Houston PL. Influence of inspired oxygen concentration on deadspace, respiratory drive, and PaCO2 in intubated patients with chronic obstructive pulmonary disease. Crit Care Med. 1996;25:1522–6.
688. Moloney ED, Kiely JL, McNicholas WT. Controlled oxygen therapy and carbon dioxide retention during exacerbations of chronic obstructive pulmonary disease. Lancet. 2001;357:526–8.
689. Brochard L, Mancebo J, Wysocki M. Noninvasive ventilation for acute exacerbations of chronic obstructive pulmonary disease. N Engl J Med. 1995;333:817–22.
690. Kramer N, Meyer TJ, Meharg J. Randomized, prospective trail of noninvasive positive pressure ventilation in acute respiratory failure. Am J Respir Crit Care Med. 1995;151:1799–806.
691. Lightowler JV, Wedzicha JA, Elliot MW. Noninvasive positive pressure ventilation to treat respiratory failure resulting from exacerbations of chronic obstructive pulmonary disease: Cochrane systematic review and meta-analysis. BMJ. 2003;326:185–9.
692. McIntyre NR, Cook DJ, Ely EW. Evidence-based guidelines for weaning and discontinuing ventilatory support: a collective task force facilitated by the American College of Chest Physicians, the American Association for Respiratory Care, and the American College of Critcal Care Medicine. Chest. 2001;120(Suppl):375S–95.
693. Elliott MW. Non-invasive ventilation in acute exacerbations of chronic pulmonary disease: a new gold-standard? Intensive Care Med. 2002;28:1691–4.
694. Keenan SP, Gregor J, Sibbald WJ. Noninvasive positive pressure ventilation in the setting of severe, acute exacerbations of chronic obstructive pulmonary disease: more effective and less expensive. Crit Care Med. 2000;28:2094–102.
695. Brochard L, Isabey D, Piquet J. Reversal of acute exacerbations of chronic obstructive lung disease by inspiratory assistance with a face mask. N Engl J Med. 1990;323:1523–30.
696. Angus RM, Ahmed AA, Fenwick LJ. Comparison of the acute effects on gas exchange of nasal ventilation and doxapram in exacerbations of chronic obstructive pulmonary disease. Thorax. 1996;51:1048–50.
697. Nava S, Navalesi P, Cont G. Time of non-invasive ventilation. Intensive Care Med. 2006;32:361–70.
698. Confalonieri M, Garuti G, Cattaruzza MS. A chart of failure risk for noninvasive ventilation in patients with COPD exacerbation. Eur Respir J. 2005;25:348–55.
699. Conti G, Antonelli M, Navalesi P. Noninvasive versus conventional mechanical ventilation in patients with chronic obstructive pulmonary disease after failure of medical treatment in the ward: a randomized trial. Intensive Care Med. 2002;28:1701–7.
700. Diaz GG, Alcaraz AC, Talavera JC. Noninvasive positive-pressure ventilation to treat hypercapnic coma secondary to respiratory failure. Chest. 2005;127:952–60.
701. Scala R, Naldi M, Archinucci I. Noninvasive positive pressure ventilation in patients with acute exacerbations of COPD and varying levels of consciousness. Chest. 2005;128:1657–66.
702. Scala R, Nava S, Conti G. Noninvasive versus conventional ventilation to treat hypercapnic encephalopathy in chronic obstructive pulmonary disease. Intensive Care Med. 2007;33:2101–8.
703. Celikel T, Sungur M, Ceyhan B. Comparison of noninvasive positive pressure ventilation with standard medical therapy in hypercapnic acute respiratory failure. Chest. 1998;114:1636–42.
704. Lieschning T, Kwok H, Hill NS. Acute applications of noninvasive positive pressure ventilation. Chest. 2003;124:699–713.
705. Vitacca M, Nava S, Confalonieri M. The appropriate setting of noninvasive pressure support ventilation in stable COPD patients. Chest. 2000;118:1286–93.

706. Ghoshal AG, Dhar R, Kundu S. Treatment of acute exacerbation of COPD. JAPI (J Assoc Physicians India). 2012;60(suppl):38–43.
707. Jubran A, Van de Graaff WB, Tobin MJ. Variability of patient-ventilator interaction with pressure support ventilation in patients with chronic obstructive pulmonary disease. Am J Respir Crit Care Med. 1995;152:129–36.
708. Leung P, Jubran A, Tobin MJ. Comparison of assisted ventilator modes on triggering, patient effort, and dyspnea. Am J Respir Crit Care Med. 1997;155:1940–8.
709. Cinnella G, Conti G, Lofaso F. Effects of assisted ventilation on the work of breathing: volume-controlled versus pressure-controlled ventilation. Am J Respir Crit Care Med. 1996;153:1025–33.
710. Turner-Stokes L, Higgins B. Non-invasive ventilation in chronic obstructive pulmonary disease: management of acute type 2 respiratory failure. National Guideline. In: Royal College of Physicians, editor. Concise guidance to good practice: a series of evidence-based guidelines for clinical management. London: RCP; 2008.
711. Wagner TO. Bronchialobstruktion in der Intensivmedizin. Internist. 2006;47:342–55.
712. Chiumello D, Pelosi P, Croci M. The effects of pressurization rate on breathing pattern, work of breathing, gas exchange and patient comfort in pressure support ventilation. Eur Respir J. 2001;18:107–14.
713. Goulet R, Hess D, Kacmarek RM. Pressure versus flow triggering during pressure support ventilation. Chest. 1997;111:1649–53.
714. Vargas F. Helmet with specific settings versus facemask for noninvasive ventilation. Crit Care Med. 2009;37:1921–7.
715. Schwartz AR, Kacmarek RM, Hess DR. Factors affecting oxygen delivery with bi-level positive airways pressure. Respir Care. 2004;49:270–5.
716. Evans TW, Albert RK, Angus DC. International Consensus Conferences in Intensive Care Medicine: noninvasive positive pressure ventilation in acute respiratory failure. Am J Respir Crit Care Med. 2001;163:283–91.
717. Esteban A, Alía I, Gordo F. Prospective randomized trial comparing pressure-controlled ventilation and volume-controlled ventilation in ARDS. Chest. 2000;117:1690–6.
718. Bach PB, Brown C, Gelfand SE. Management of acute exacerbations of chronic obstructive pulmonary disease: a summary and appraisal of published evidence. Ann Intern Med. 2001;134:600–20.
719. Turner MO, Patel A, Ginsberg S. Bronchodilator delivery in acute airflow obstruction: a meta-analysis. Arch Intern Med. 1997;157:1736–44.
720. Cooper CB, Tashkin DP. Recent developments in inhaled therapy in stable chronic obstructive pulmonary disease. BMJ. 2005;330:640–4.
721. Bone R, Boyars M, Braun S. In chronic obstructive pulmonary disease, a combination of ipratropium and albuterol is more effective than either agent alone. Chest. 1994;105:1411–9.
722. Shrestha M, T O'B, Haddox R. Decreased duration of emergency department treatment of chronic obstructive pulmonary disease exacerbations with the addition of ipratropium bromide to beta-agonist therapy. Ann Emerg Med. 1991;20:1206–9.
723. Cydulka RK, Emerman CL. Effects of combined treatment with glycopyrrolate and albuterol in acute exacerbation of chronic obstructive pulmonary disease. Ann Emerg Med. 1995;25:470–3.
724. Moayyedi P, Congleton J, Page RL. Comparison of nebulized salbutamol and ipratropium bromide with salbutamol alone in the treatment of chronic obstructive pulmonary disease. Thorax. 1995;50:834–7.
725. Patrick DM, Dales RE, Stark RM. Severe exacerbations of COPD and asthma. Incremental benefit of adding ipratropium to usual therapy. Chest. 1990;98:295–7.
726. Dhand R. Basic techniques for aerosol delivery during mechanical ventilation. Respir Care. 2004;49:611–22.
727. Anzueto A. Acute exacerbation of chronic obstructive pulmonary diseases: what is the impact of bronchodilators, corticosteroids and antibiotics? In: Anzueto A, Cook DJ, Esteban A, editors. Evidence – based management of patients with respiratory failure. Berlin/Heidelberg: Springer; 2004. p. 79.

728. Niewoehner DE, Erbland ML, Deupree RH. Effect of systemic glycocorticoids on exacerbations of chronic obstructive pulmonary disease. N Engl J Med. 1999;340:1941–7.
729. Davies R, Angus RM, Calverley PMA. Oral corticosteroids in patients admitted to hospital with exacerbations of chronic obstructive pulmonary disease: a prospective randomised controlled trail. Lancet. 1999;354:456–60.
730. Emerman CL, Connors AF, Lukens TW. A randomized controlled trail of methylprednisolone in the emergency treatment of acute exacerbation of COPD. Chest. 1989;95:563–7.
731. Bullard M, Liaw S, Tsai Y. Early corticosteroid use in acute exacerbations of chronic airflow obstruction. Am J Emerg Med. 1996;14:139–43.
732. Leuppi JD, Schuetz P, Bingisser R. Short-term vs conventional glucocorticoid therapy in acute exacerbations of chronic obstructive pulmonary disease: the REDUCE randomized trial. JAMA. 2013;309:2223–31.
733. Llor C, Moragas A, Hernandez S. Efficacy of antibiotic therapy for acute exacerbations of mild to moderate chronic obstructive pulmonary disease. Am J Respir Crit Care Med. 2012;186:716–23.
734. Saint S, Bent S, Vittinghoff E. Antibiotics in chronic obstructive pulmonary disease exacerbations: a meta-analysis. JAMA. 1995;253:957–60.
735. Sethi S, Evans N, Grant BJ. New strains of bacteria and exacerbations of chronic obstructive pulmonary disease. N Engl J Med. 2002;347:465–7.
736. Anthonisen NR, Manfreda J, Warren CP. Antibiotic therapy in exacerbations of chronic obstructive pulmonary disease. Ann Intern Med. 1987;106:196–204.
737. Stockley R, Pye A. Relationship of sputum color to nature and outpatient management of acute exacerbations of COPD. Chest. 2000;117:1638–45.
738. Hill NS. Indications for non-invasive ventilation. In: Brochard L, Slutsky AS, editors. Mechanical ventilation. Berlin/Heidelberg: Springer; 2004. p. 171.
739. Meduri GU, Turner RE, Abou-Shala N. Noninvasive positive pressure ventilation Via face mask: first-line intervention in patients with acute hypercapnic and hypoxemic respiratory failure. Chest. 1996;109:179–93.
740. Fernández MM, Villagrá A, Blanch L. Non-invasive mechanical ventilation in status asthmaticus. Intensive Care Med. 2001;27:486–92.
741. Environmental Protection Agency. Asthma facts. http://www.epa.gov/asthma/pdfs/asthma_fact_sheet_en.pdf. [Online].
742. McFadden Jr ER. Acute severe asthma. Am J Respir Crit Care Med. 2003;168:740–59.
743. Zimmermann JL, Dellinger RP, Shah AN. Endotracheal intubation and mechanical ventilation in severe asthma. Crit Care Med. 1993;21:1727–30.
744. Kohn MS. Intubation of the asthmatic patient. Clin Allergy Immunol. 1999;13:419–28.
745. Hall JB, Wood LDH. Management of the critically ill asthmatic patient. Med Clin North Am. 1990;74:779–96.
746. Ferguson N, Frutos-Vivar F, Esteban A. Airway pressures, tidal volumes, and mortality in patients with acute respiratory distress syndrome. Crit Care Med. 2005;33:21–30.
747. Williams TJ, Tuxen DV, Scheinkestel CD. Risk factors for morbidity in mechanically ventilated patients with acute severe asthma. Am Rev Respir Dis. 1992;146:1136–42.
748. Tuxen DV. Permissive hypercapnia. In: Tobin MJ, editor. Principles and practice of mechanical ventilation. New York: McGraw-Hill; 1994. p. 371.
749. Manser R, Reid D, Abramson M. Corticosteroids for acute severe asthma in hospitalised patients. Cochrane Dadabase Syst Rev. 2001;1:CD001740.
750. Cheuk DK, Vhau TC, Lee SL. A meta-analysis on intravenous magnesium sulphate for treating acute asthma. Arch Dis Child. 2005;90:74–7.
751. Rowe BH, Bretzlaff JA, Bourdon C. Intravenous magnesium sulphate treatment for acute asthma in the emergency department: a systematic review of literature. Ann Emerg Med. 2000;36:181–90.
752. Silverman RA, Osborn H, Runge J. IV magnesium sulfate in the treatment of acute severe asthma: a multicenter randomized controlled trail. Chest. 2002;122:489–97.
753. Sydow DM, Crozier TA, Zielmann S. High-dose intravenous magnesium sulfate in the management of life-threatening status asthmaticus. Intensive Care Med. 1993;19:467–71.

754. Rowe BH, Carmargo Jr CA. The use of magnesium sulfate in acute asthma: rapid uptake of evidence in North American emergency departments. J Allergy Clin Immunol. 2006;117:53–8.
755. Jolliet P, Tassaux D. Usefulness of helium-oxygen mixtures in the treatment of mechanically ventilated patients. Curr Opin Crit Care. 2003;9:45–50.
756. Kass JE, Castriotta RJ. Heliox therapy in acute severe asthma. Chest. 1995;107:757–60.
757. Rose JS, Panacek EA, Miller P. Prospective randomized trial of heliox-driven continuous nebulizers in the treatment of asthma in the emergency department. J Emerg Med. 2002;22:133–7.
758. Kress JP, Noth I, Gehlbach BK. The utility of albuterol nebulized with heliox during acute asthma exacerbations. Am J Respir Crit Care Med. 2002;165:1317–21.
759. Ho AMH, Lee A, Karmakar MK. Heliox vs air-oxygen mixtures for the treatment of patients with acute asthma: a systematic review. Chest. 2003;123:882–90.
760. Kass JE, Terregino CA. The effect of heliox in acute severe asthma: a randomized controlled trail. Chest. 1999;116:296–300.
761. Tassaux D, Jolliet P, Roeseler J. Effects of helium-oxygen on intrinsic positive end-expiratory pressure in intubated and mechanically ventilated patients with severe chronic obstructive disease. Crit Care Med. 2000;28:2721–8.
762. Henderson SO, Acharya P, Kilaghbian T. Use of heliox-driven nebulizer therapy in the treatment of acute asthma. Ann Emerg Med. 1999;33:141–6.
763. Coppadoro A, Bittner E, Berra L. Novel preventive strategies for ventilator-associated pneumonia. Crit Care. 2012;16:210.
764. American Thoracic Society and Infectious Diseases Society of America. Guidelines for the management of adults with hospital-acquired, ventilator-associated, and healthcare-associated pneumonia. Am J Respir Crit Care Med. 2005;171:388–416.
765. Torres A, el-Ebiary F, Gonzalez J. Gastric and pharyngeal flora in nosocomial pneumonia acquired during mechanical ventilation. Am Rev Respir Dis. 1993;148:352–7.
766. Inglis TJ, Millar MR, Jones JG. Tracheal tube biofilm as a source of bacterial colonization of the lung. J Clin Microbiol. 1989;27:2014–7.
767. Cairns S, Thomas JG, Hooper SJ. Molecular analysis of microbial communities in endotracheal tube bioflims. PLoS One. 2011;6:e14759.
768. Adair CG, Gorman SP, Feron BM. Implications of endotracheal tube biofilm for ventilator-associated pneumonia. Intensive Care Med. 1999;25:1072–6.
769. Torres A, Gatell JM, Aznar E. Re-intubation increases the risk of nosocomial pneumonia in patients needing mechanical ventilation. Am J Respir Crit Care Med. 1995;152:137–41.
770. Terrangi PP, Antonelli M, Fumagalli R. Early vs late tracheotomy for prevention of pneumonia in mechanically ventilated adult ICU patients: a randomised controlled trail. JAMA. 2010;303:1483–9.
771. Morris AC, Hay AW, Swann DG. Reducing ventilator-associated pneumonia in intensive care: impact of implementing a care bundle. Crit Care Med. 2011;39:2218–24.
772. Miller MA, Arndt JL, Konkle MA. A polyurethane cuffed endotracheal tube is associated with decreased rates of ventilator -associated pneumonia. J Crit Care. 2011;26:280–6.
773. Kollef MH, Afessa B, Anzueto A. Silver-coated endotracheal tubes and incidence of ventilator-associated pneumonia: the NASCENT randomized trial. JAMA. 2008;300:805–13.
774. Nseir S, Zerimech F, Fournier C. Continuous control of tracheal cuff pressure and microaspiration of gastric contents in critically ill patients. Am J Respir Crit Care Med. 2011;184:1041–7.
775. Delaney A, Gray H, Laupland KB. Kinetic bed therapy to prevent nosocomial pneumonia in mechanically ventilated patients: a systematic review and meta-analysis. Crit Care. 2006;10:R70.
776. Siempos II, Ntaidou TK, Falagas ME. Impact of the administration of probiotics on the incidence of ventilator-associated pneumonia: a meta-analysis of randomized controlled trails. Crit Care Med. 2010;38:954–62.

777. Koeman M, van der Ven AJ, Hak E. Oral decontamination with chlorhexidine reduces the incidence of ventilator-associated pneumonia. Am J Respir Crit Care Med. 2006;173:1348–55.
778. Flanagan ME, Welsh CA, Kiess C. A national collaborative for reducing health care -associated infections: current initiatives, challenges, and opportunities. Am J Infect Control. 2011; 39:685–9.
779. Muscedere J, Rewa O, McKechnie K. Subglottic secretion drainage for the prevention of ventilator-associated pneumonia: a systematic review and meta-analysis. Crit Care Med. 2011;39:1985–91.
780. Shorr AF, Zilberberg MD, Kollef MH. Cost-effectiveness of a silver-coated endotracheal tube to reduce the incidence of ventilator-associated pneumonia. Infect Control Hosp Epidemiol. 2009;30:759–63.
781. Nseir S, Brisson H, Marquette CH. Variations of endotracheal cuff pressure in intubated critically ill patients: prevalence and risk factors. Eur J Anaesthesiol. 2009;26:229–34.
782. Seegobin RD, van Hasselt GL. Endotracheal cuff pressure and tracheal mucosal blood flow: endoscopic study of effects of four large volume cuffs. Br Med J (Clin Res Ed). 1984;288:965–8.
783. Torres A, Serra-Batlles J, Ros E. Pulmonary aspiration of gastric contents in patients receiving mechanical ventilation: the effect of body position. Ann Intern Med. 1992;116:540–3.
784. Drakulovic MB, Torres A, Bauer TT. Supine body position as a risk factor for nosocomial pneumonia in mechanically ventilated patients: a randomized trial. Lancet. 1999;354:1851–8.
785. Mauri T, Berra L, Kumwilaisak K. Lateral-horizontal patient position and horizontal orientation of the endotracheal tube to prevent aspiration in adult surgical intensive care unit patients: a feasibility study. Respir Care. 2010;55:294–302.
786. Bassi GL, Zanella A, Cressoni M. Following tracheal intubation, mucus flow is reversed in the semirecumbent position: possible role in the pathogenesis of ventilator-associated pneumonia. Crit Care Med. 2008;36:518–25.
787. Gravity-VAP Network. Prospective, randomized, multi-center trial of lateral trendelenburg versus semi-recumbent body position in mechanically ventilated patients for prevention of ventilator-associated pneumonia. 2012. http://www.gravityvaptrail.org/joomla/documents/summaryProtocol.pdf.
788. Artinian V, Krayem H, DiGiovine B. Effects of early enteral feeding on the outcome of critically ill mechanically ventilated medical patients. Chest. 2006;129:960–7.
789. Ibrahim EH, Mehringer L, Prentice D. Early versus late enteral feeding of mechanically ventilated patients: results of a clinical trial. J Parenter Enteral Nutr. 2002;26:174–81.
790. Kompan L, Vidmar G, Spindler-Vesel A. Is early enteral nutrition a risk factor for gastric intolerance and pneumonia? Clin Nutr. 2004;23:527–32.
791. McClave SA, Martindale RG, Vanek VW. Guidelines for the Provision and Assessment of Nutrition Support Therapy in the Adult Critically Ill Patient: Society of Critical Care Medicine (SCCM) and American Society for Parenteral and Enteral Nutrition (A.S.P.E.N.). J Parenter Enteral Nutr. 2009;33:277–316.
792. Chen Y-C. Critical analysis of the factors associated with enteral feeding in preventing VAP: a systematic review. J Chin Med Assoc. 2009;72:171–8.
793. Craven DE. Preventing ventilator-associated pneumonia in adults: sowing seeds of change. Chest. 2006;130:251–60.
794. Alberda C, Gramlich L, Jones N. The relationship between nutritional intake and clinical outcomes in critically ill patients: results of an international multicenter observational study. Intensive Care Med. 2009;35:1728–37.
795. Villet S, Chiolero RL, Bollmann MD. Negative impact of hypocaloric feeding and energy balance on clinical outcome in ICU patients. Clin Nutr. 2005;24:502–9.
796. Casaer MP, Mesotten D, Hermans G. Early versus late parenteral nutrition in critically ill adults. N Engl J Med. 2011;365:506–17.
797. Vanderheyden S, Casaer M, Kesteloot S. Early versus late parenteral nutrition in ICU patients: cost analysis of the EPaNIC trail. Crit Care. 2012;16:R96.

798. Berger M, Pichard C. Best timing for energy provision during critical illness. Crit Care. 2012;16:215.
799. Thomason JWW, Ely EW. Liberating patients from mechanical ventilation: what have we learned about protocolizing care? In: Brochard L, Slutsky AS, editors. Mechnaical ventilation, Update in intensive care medicine. Berlin/Heidelberg: Springer; 2005. p. 135.
800. Epstein SK, Ciubotaru RL, Wong JB. Effect of failed extubation on the outcome of mechanical ventilation. Chest. 1997;112:186–92.
801. MacIntyre NR, Cook DJ, Ely EW. Evidence-based guidelines for weaning and discontinuing ventilatory support. Chest. 2001;120:375S–95.
802. Epstein SK. Decision to extubate. Intensive Care Med. 2002;28:535–46.
803. Manthous CA. The anarchy of weaning technique. Chest. 2002;121:1738–40.
804. Stroetz RW, Hubmayr RD. Tidal volume maintenance during weaning with pressure support. Am J Respir Crit Care Med. 1995;152:1034.
805. Boulain T. Unplanned extubations in the adult intensive care unit. A prospective multicenter study. Am J Respir Crit Care Med. 1998;157:1131–7.
806. Listello D, Sessler CN. Unplanned extubation. Clinical predictors for reintubation. Chest. 1994;105:1496–503.
807. Esteban A, Frutos-Vivar F, Ferguson ND. Noninvasive positive-pressure ventilation for respiratory failure after extubation. N Engl J Med. 2004;350:2452–60.
808. Meade M, Guyatt G, Cook D. Predicting success in weaning from mechanical ventilation. Chest. 2001;120:400S–20.
809. Esteban A, Alia I, Gordo F. Extubation outcome after spontaneous breathing trails with T-tube or pressure support ventilation. The Spanish Ling Failure Collaborative Group. Am J Respir Crit Care Med. 1997;156:459–65.
810. Esteban A, Frutos A, Tobin MJ. A comparison of four methods of weaning patients from mechanical ventilation: Spanish Lung Failure Collaborative Group. N Engl J Med. 1995;332:345–50.
811. Brochard L, Rauss A, Benito S. Comparison of three methods of gradual withdrawal from ventilatory support during weaning from mechanical ventilation. Am J Respir Crit Care Med. 1994;150:896–903.
812. Ely EW, Baker AM, Dunagan DP. Effect of the duration of mechanical ventilation of identifying patients capable of breathing spontaneously. N Engl J Med. 1996;335:1864–9.
813. Bendixin HH, Egbert LD, Hedley-Whyte J. Respiratory care. St Louis: Mosby; 1965. p. 149–50.
814. Tobin MJ, Jubran A. Weaning from mechanical ventilation. In: Tobin MJ, Jubran A, editors. Principles and practise of mechanical ventilation. New York: McGraw-Hill; 2006. p. 1185.
815. Lermitte J, Garfield MJ. Weaning from mechanical ventilation. Contin Edu Anaesth Crit Care Pain. 2005;5:113–7.
816. McConville JF, Kress JP. Weaning patients from the ventilator. N Engl J Med. 2012;367:2233–9.
817. Blackwood B, Alderdice F, Burns KEA. Protocolized versus non-protocolized weaning for reducing the duration of mechanical ventilation in critically ill adult patients. Cochrane Database Syst Rev. 2010;5:CD006904.
818. Dries DJ, McGonigal MD, Malian MS. Protocol-driven ventilator weaning reduces use of mechanical ventilation, rate of early reintubation, and ventilator-associated pneumonia. J Trauma. 2004;56:943–51.
819. Kress JP, Pohlman AS, O'Connor F. Daily interruption of sedative infusions in critically ill patients undergoing mechanical ventilation. N Engl J Med. 2000;342:1471–7.
820. Shafer A. Complications of sedation with midazolam in the intensive care unit and a comparison with other sedative regimens. Crit Care Med. 1998;26:947–56.
821. Jacobs JR, Reves JG, Glass PSA. Rationale and technique for continous infusions in anesthesia. Int Anesthesiol Clin. 1991;29:23–38.
822. Kollef MH, Levy NT, Ahrens TS. The use of continous i.v. sedation is associated with prolongation of mechanical ventilation. Chest. 1998;114:541–8.

823. Boles JM, Bion J, Connors A. Weaning from mechanical ventilation. Eur Respir J. 2007;29:1033–56.
824. Thomason JWW, Ely EW. Liberating patients from mechanical ventilation: what have we learned about protocolizing care? In: Brochard L, Slutsky AS, editors. Mechanical ventilation. Berlin/Heidelberg/New York: Springer; 2004. p. 135.
825. Ely EW, Meade MO, Haponik EF. Mechanical ventilator weaning protocols driven by nonphysician health-care professionals. Chest. 2001;120(6 Suppl):454S–63.
826. Krishnan JA, Moore D, Robeson C. A prospective, controlled trial of a protocol-based strategy to discontinue mechanical ventilation. Am J Respir Crit Care Med. 2004;169:673–8.
827. Marelich GP, Murin S, Battistella F. Protocol weaning of mechanical ventilation in medical and surgical patients by respiratory care practitioners and nurses. Chest. 2000;118:459–67.
828. Khamiees M, Raju P, DeGirolamo A. Predictors of extubation outcome in patients Who have successfully completed a spontaneous breathing trial. Chest. 2001;120:1262–70.
829. Yang KL, Tobin MJ. A prospective study of indexes predicting the outcome of trials of weaning from mechanical ventilation. N Engl J Med. 1991;324:1445–50.
830. Vassilakopoulos T, Routsi C, Sotiropoulou C. The combination of the load/force balance and the frequency/tidal volume can predict weaning outcome. Intensive Care Med. 2006;32:684–91.
831. Cook DJ. Weaning from mechanical ventilation: the McMaster evidence based medicine practice center. Washington, DC: Agency for Healthcare Research and Quality; 2000.
832. Alia I, Esteban A. Weaning from mechanical ventilation. Crit Care. 2000;4:72–80.
833. Dojat M, Harf A, Touchard D. Evaluation of a knowledge-based system providing ventilatory management and decision for extubation. Am J Respir Crit Care Med. 1996;153:997–1004.
834. Reyes A, Vega G, Blancas R. Early vs conventional extubation after cardiac surgery with cardiopulmonary bypass. Chest. 1997;112:193–201.
835. Esteban A, Alia I, Tobin MJ. Effect of spontaneous breathing trial duration on outcome of attempts to discontinue mechanical ventilation. Am J Respir Crit Care Med. 1999;159:512–8.
836. Vallverdu I, Calaf N, Subirana M. Clinical characteristics, respiratory functional parameters, and outcome of two hour t-piece trail in patients weaning from mechanical ventilation. Am J Respir Crit Care Med. 1998;158:1855–62.
837. Farias JA, Alia I, Esteban A. Weaning from mechanical ventilation in pediatric intensive care patients. Intensive Care Med. 1998;24:1070–5.
838. Richard C, Teboul J-L. Weaning failure from cardiovascular origin. Intensive Care Med. 2005;31:1605–7.
839. Eikermann M, Koch G, Gerwig M. Muscle force and fatigue in patients with sepsis and multiorgan failure. Intensive Care Med. 2006;32:251–9.
840. MacIntyre NR, Cheng KCG, McConnell R. Applied PEEP during pressure support reduces the inspiratory threshold load of intrinsic PEEP. Chest. 1997;111:188–93.
841. Sydow M, Golisch W, Buscher H. Effect of low-level PEEP on inspiratory work of breathing in intubated patients, both with healthy lungs and with COPD. Intensive Care Med. 1995;21:887–95.
842. Bonmarchand G, Chevron V, Jusserand C. Increased initial flow rate reduces inspiratory work of breathing during pressure support ventilation in patients with exacerbation of chronic obstructive pulmonary disease. Intensive Care Med. 1996;22:1147–57.
843. Goldstone J. The pulmonary physician in critical care 10: difficult weaning. Thorax. 2002;57:986–91.
844. Laghi F, D'Alfonso N, Tobin MJ. Pattern of recovery from diaphragmatic fatigue over 24 hours. J Appl Physiol. 1995;79:539–46.
845. Rochester DF, Arora NS. Respiratory muscle failure. Med Clin North Am. 1983;67:573–97.
846. Leith DE, Bradley M. Ventilatory muscle strength and endurance training. J Appl Physiol. 1976;41:508–16.
847. Williams AG, Rayson MP, Jubb M. Physiology: the ACE gene and muscle performance. Nature. 2000;403:614.

848. Woods DR, Humphries SE, Montgomery HE. The ACE I/D polymorphism and human physical performance. Trends Endocrinol Metab. 2000;11:416–20.
849. Chieveley-Williams S, Field D, Woods D. Genetic determinants of respiratory muscle training. Am J Respir Crit Care Med. 2001;163:A287.
850. Lemaire F, Teboul J-L, Cinotti L. Acute left ventricular dysfunction during unsuccessful weaning from mechanical ventilation. Anesthesiology. 1988;69:171–9.
851. Hurford WE, Lynch KE, Strauss HW. Myocardial perfusion as assessed by thallium-201 scintigraphy during the discontinuation of mechanical ventilation in ventilator-dependent patients. Anesthesiology. 1991;74:1007–16.
852. Srivastava S, Chatila W, Amoateng-Adjepong Y. Myocardial ischemia and weaning failure in patients with coronary artery disease: an update. Crit Care Med. 1999;27:2109–12.
853. Pinsky MR. Breathing as exercise: the cardiovascular response to weaning from mechanical ventilation. Intensive Care Med. 2000;26:1164–6.
854. Lamia B. Weaning -induced cardiac dysfunction. In: Vincent J-L, editor. Yearbook of intensive care and emergency medicine. Berlin/Heidelberg: Springer; 2005. p. 239.
855. Robotham JL, Lixfeld W, Holland L. Effects of respiration on cardiac performance. J Appl Physiol. 1978;44:703–9.
856. Mathru M, Rao T, El Etr A. Hemodynamic response to changes in ventilatory patterns in patients with normal and poor left ventricular reserve. Crit Care Med. 1982;10:423–6.
857. Gaasch SK, Weintraub WB, Skillman JJ. Cardiorespiratory and sympathoadrenal responses during weaning from controlled ventilation. Surgery. 1977;82:233–40.
858. Rasanen J, Nikki P, Heikkila J. Acute myocardial infarction complicated by respiratory failure: the effects of mechanical ventilation. Chest. 1984;85:21–8.
859. Scharf SM, Boanco JA, Tow DE. The effects of large negative intrathoracic pressure on left ventricular function in patients with coronary disease. Circulation. 1981;63:871–4.
860. Hurford WE, Favorito F. Association of myocardial ischemia with failure to wean from mechanical ventilation. Crit Care Med. 1995;23:1475–80.
861. Cabello B. Withdrawal from mechanical ventilation in patients with COPD: the issue of congestive heart failure. In: Vincent J-L, editor. Yearbook of intensive care and emergency medicine. Berlin/Heidelberg: Springer; 2003. p. 295.
862. Robotham JL, Mitzner W. A model of the effects of respiration on left ventricular performance. J Appl Physiol. 1979;46:411–8.
863. Buda AJ, Pinsky MR, Ingels NB. Effect of intrathoracic pressure on left ventricular performance. N Engl J Med. 1979;301:453–9.
864. Cotter G, Felker M, Adams KF. The pathophysiology of acute heart failure—is it all about fluid accumulation? Am Heart J. 2008;155:9–18.
865. Cotter G, Metra M, Milo-Cotter O. Fluid overload in acute heart failure — re-distribution and other mechanisms beyond fluid accumulation. Eur J Heart Fail. 2008;10:165–9.
866. Lamia B, Maizel J, Ochagavia A. Echocardiographic diagnosis of pulmonary artery occlusion pressure elevation during weaning from mechanical ventilation. Crit Care Med. 2009;37:1696–701.
867. Calle V, Amiel JB, Charron C. Echocardiography: a help in the weaning process? Crit Care. 2010;14:R120.
868. Papanikolaou J, Makris D, Saranteas D. New insights into weaning from mechanical ventilation: left ventricular diastolic dysfunction is a key player. Intensive Care Med. 2011;37:1976–85.
869. Voga G. Early and simple detection of diastolic dysfunction during weaning from mechanical ventilation. Crit Care. 2012;16:137.
870. Malbrain MLHG, Chiumello D, Pelosi P. Prevalence of intra-abdominal hypertension in critically ill patients: a multicentre epidemiological study. Intensive Care Med. 2004;30:822–9.
871. Mergoni M, Martinelli A, Volpi A. Impact of positive end-expiratory pressure on chest wall and lung pressure–volume curve in acute respiratory failure. Am J Respir Crit Care Med. 1997;156:846–54.

872. Girault C, Daudenthun I, Chevron V. Noninvasive ventilation as a systematic extubation and weaning technique in acute-on-chronic respiratory failure. Am J Respir Crit Care Med. 1999;160:86–92.
873. Jiang J-S, Kao S-J, Wang S-N. Effect of early application of biphasic positive airway pressure on the outcome of extubation in ventilator weaning. Respirology. 1999;4:161–5.

Chapter 2
Acute Respiratory Distress Syndrome (ARDS)

2.1 Definition

The sudden onset of acute, substantial hypoxic respiratory failure elicited by a serious, diffuse lung disease characterized by marked bilateral pulmonary infiltrates consistent with pulmonary oedema is termed acute respiratory distress syndrome (ARDS) or acute lung injury (ALI), if less serious [1–4]. Initially described as a distinct entity by Ashbaugh et al. [3] in 1967, the definition was modified and expanded by Murray et al. [5] in 1988, and finally in 1994, an American-European consensus conference (AECC) defined the criteria still used today to diagnose ALI/ARDS [1]:

ALI/ARDS is "a syndrome of inflammation and increasing permeability that is associated with a constellation of clinical, radiological and physiologic abnormalities that cannot be explained by, but may coexist with, left atrial or pulmonary capillary hypertension". It is most often associated with sepsis syndrome, aspiration, primary pneumonia or multiple trauma and less commonly with cardiopulmonary bypass, multiple transfusions, fat embolism, pancreatitis or others [1].

Criteria suggestive of ALI/ARDS and supporting the diagnosis are:

- Acute onset
- $P_aO_2/FiO_2 \leq 300$ mmHg (ALI); $P_aO_2/FiO_2 \leq 200$ mmHg (ARDS)
- Bilateral infiltrates seen on frontal chest radiograph
- PCWP $\leq$ 18 mmHg when measured or no clinical evidence of left atrial hypertension in the context of patients suffering from acute respiratory failure without a history of chronic lung disease and in the presence of risk factors/known aetiologies of ALI/ARDS [1]

The ratio between arterial oxygen tension and fraction of inspired oxygen (P_aO_2/FiO_2) distinguishes between ALI and ARDS; the required exclusion of left atrial hypertension rules out acute heart failure precipitating the pulmonary infiltrates [6, 7].

The definition has been reviewed and challenged [8–12] but does have advantages, including its simplicity to apply in daily practice [2, 13] and in facilitating

W. Krüger, A.J. Ludman, *Core Knowledge in Critical Care Medicine*,
DOI 10.1007/978-3-642-54971-7_2, © Springer-Verlag Berlin Heidelberg 2014

research and insights in an otherwise heterogeneous condition [13]. However, several shortcomings are apparent; arguably, the most significant are the determining levels of hypoxaemia. These are an essential feature in the definition of ALI/ARDS [14], are somewhat arbitrary [7] and do not correlate with the clinical course, disease severity or mortality [15–17]. As such, the P_aO_2/FiO_2 ratio may be influenced simply by the oxygen content administered to the patient or may be significantly affected by the positive end-expiratory pressure (PEEP) applied [10, 11, 13, 14, 18].

Critically ill patients often suffer from pulmonary oedema [19], and an elevated pulmonary wedge pressure is reported in approximately 30–50 % of all ALI/ARDS patients [13, 20]. However, this may just be the result of transmitted airway pressures [21], due to a concurrent illness, or comorbidity such as heart failure (systolic or diastolic) [10, 13]. Sepsis-related, inflammatory-mediated cardiomyopathies and heart failure are both frequent features in ARDS and present in up to 60 % of all septic cases [22, 23]. In these cases, pulmonary oedema may develop with or without a PCWP exceeding 18 mmHg [24] but displays a better prognosis and requires a different treatment to diffuse alveolar injury [23]. Thus, although left atrial hypertension may be present with a PCWP exceeding the determined limit of 18 mmHg, ALI/ARDS may also be present [10, 13]. On the other hand, occlusion pressures of less than 18 mmHg are not sufficient to exclude permeability-induced oedema (a hallmark of ALI/ARDS [2, 4, 25–27]) since a reduced colloid oncotic pressure may precipitate pulmonary oedema without elevated PCWP [28, 29]. Hence, in some really challenging cases, it may be impossible to distinguish between hydrostatic cardiogenic pulmonary oedema and that provoked by increased permeability of the microvasculature of the lungs, as typical for ALI/ARDS [23, 30]. Furthermore, both pathomechanisms may contribute to the oedema as an increase in intravascular volume and/or pressure is shown to amplify the quantity of extravascular lung water in the setting of raised lung vascular permeability [31].

Other concerns regarding the diagnostic criteria are the imprecise timelines – namely, over what time frame is meant by "acute" [32], the poor reliability of chest radiograph criteria, and the difficult interpretation of chest x-rays with low interobserver agreements and radiographic abnormalities being caused by high mean airway pressures and other features rather than by permeability-induced oedema [13, 23, 33, 34].

However, despite the limitations, a reasonable correlation has been demonstrated between the current clinical definition and ARDS autopsy findings – namely, diffuse alveolar damage [35–38]. A combination of clinical criteria and radiological (diffuse airspace disease) findings throughout the entire ICU stay gives a diagnostic sensitivity of 75 % and a specificity of 84 % [38]. Unfortunately, an assessment on a daily basis reduces the specificity to 50 %, although the sensitivity remains unchanged [39].

In 2011 a panel of experts convened in Berlin and developed a new definition (Berlin definition) subsequently published in 2012 [27]. The term acute lung injury was abandoned and three grades of ARDS were introduced. The chest x-ray criteria were clarified, the PaO_2/FiO_2 ratio was related to a minimum PEEP required for sufficient oxygenation, the PCWP criterion was replaced by other measures (i.e.

echocardiography) to exclude cardiac causes for the bilateral infiltrates and the timeline of "acute" is defined as within 7 days [32].

The Berlin definition in detail:

Acute respiratory distress syndrome	
Timing	Within 1 week of a known clinical insult or new or worsening respiratory symptoms
Chest imaging	Bilateral opacities – not fully explained by effusion, lobar/lung collapse or nodules
Origin of oedema	Respiratory failure not fully explained by cardiac failure or fluid overload Needs objective assessment (e.g. echocardiography) to exclude hydrostatic oedema if no risk factor (for ARDS) is present
Oxygenation[a]	
Mild	200 mmHg $<PaO_2/FiO_2 \leq$ 300 mmHg with PEEP or CPAP $\geq$ 5 cm H_2O
Moderate	100 mmHg $<PaO_2/FiO_2 \leq$ 200 mmHg with PEEP $\geq$ 5 cm H_2O
Severe	$PaO_2/FiO_2 \leq$ 100 mmHg with PEEP $\geq$ 5 cm H_2O

PaO_2 partial pressure of arterial oxygen, *FiO_2* fraction of inspired oxygen, *CPAP* continuous positive airway pressure, *PEEP* positive end-expiratory pressure

[a]If the altitude is higher than 1,000 m, the PaO_2/FiO_2 ratio needs to be corrected by $PaO_2/FiO_2 \times$ barometric pressure/760

Despite the refined definition, substantial criticisms remain. The research upon which it is based is largely empirical, and the conclusions reached are based on consensus views. In daily practice, it lacks predictive ability and demonstrates poor correlation between severity classification and post-mortem findings. Furthermore, the definition has not brought us forward to reliably identify and diagnose patients suffering from mild ARDS [23, 32, 40, 41].

To increase its utility, the definition of ARDS may actually need to include some of the central pathological features [40, 42, 43], for example, the increased microvascular permeability of the pulmonary capillary vessel network causing increased extravascular lung water and pulmonary oedema [2, 23, 27, 42, 44]. Chew [40] acknowledges a clear pathophysiological and prognostic relationship between extravascular lung water (EVLW) as a measure of permeability-induced pulmonary oedema and lung injury. This was based on the results of several studies assessing EVLW and its diagnostic, severity and prognostic properties in ARDS patients [2, 42, 44–46].

To overcome these shortcomings, the current scientific discussion and recent studies have focused on the central pathology, namely, the considerably enhanced permeability of the pulmonary microvasculature. This cardinal feature of an acute inflammatory reaction may be provoked by a multiplicity of miscellaneous triggers, in turn activating the innate immune system [47, 48], resulting in vascular leakiness and extravascular accumulation of protein-rich fluid [2, 4, 49, 50], called oedema. The aim is to sharpen diagnostic reliability, accuracy, sensitivity, specificity, prognostic validity and power [30, 40, 42, 44, 45]. Furthermore, the earliest pathological alteration displayed in ALI/ARDS is the increase in microvascular endothelial permeability with consecutive fluid extravasation [2, 51]. Extravascular lung water is recognized to reflect the fluid within the pulmonary tissue, in the interstitium and in the alveolar spaces but outside the vascular bed [30, 51, 52]. Indexing EVLW to pulmonary blood volume, called the pulmonary vascular permeability index (PVPI), resolves some of the uncertainties and imprecision. Measuring extravascular lung water (EVLW) and the pulmonary blood volume (PBV) by the semi-invasive method of transpulmonary thermodilution [30, 53–55] generates an indexed PVPI (the ratio of EVLW to PBV) which is able to distinguish between heart failure-/volume overload-induced hydrostatic pulmonary oedema and pulmonary oedema secondary to increased pulmonary microvascular permeability as seen in ALI/ARDS [42, 45, 56–58]. The role of EVLW and hence PVPI as a diagnostic parameter of ALI/ARDS is still under debate [57–60], although growing evidence is emerging that EVLW is useful and that it may be a decisive additional criteria for diagnosing ARDS [42, 45, 57]. This is substantiated by a clear pathophysiological relationship between lung injury and EVLW [2, 40], and by the substantial prognostic validity and plausibility in recognizing increasing amounts of EVLW (thus severity of pulmonary oedema) which correlate with increasing mortality rates [44, 46, 57, 61]. A PVPI $\geq$ 3.0 in the presence of elevated EVLWI $\geq$ 10 ml/kg body weight may allow the diagnosis of ALI/ARDS in the presence of a typical clinical setting with a 85 % sensitivity and a specificity well above 90 % [42, 45]. However, it is under debate whether to index to ideal body weight or to actual body weight [57, 60]. Indexing to predicted ideal body weight may be superior as lung size does not increase with body weight in adult persons [58], and this seems to correlate better to prognosis and with markers of the severity of lung injury such as lung compliance [30, 58, 62, 63].

The negative predictive value of PVPI may be even higher; a ratio $\leq$ 1.7 virtually excludes (specificity 95 %) the diagnosis of ALI/ARDS [64].

2.2 Epidemiology and Prognosis

The reported incidences of ALI/ARDS are quite variable. A study reported in 1999 from Sweden, Denmark and Iceland found 13.5 cases of ARDS per 100,000 and 17.5 cases of ALI per 100,000 population [65]. Approximately 75 % of all ALI patients will progress to ARDS [66]. However, a more recent study reported approximately 190,000 ALI/ARDS cases per year in the United States and evaluated the ALI incidence as between 17.9 and 78.9 per 100,000 person-years and that of ARDS as between 13.5 and 58.7 per 100,000 person-years [16].

In the ICU the overall incidence is 10–15 % and in ventilated patients up to 20 % may acquire ALI/ARDS [67–69]. Patients diagnosed with sepsis face the highest risk at about 40 % [16, 67–69], with sepsis being the most frequent reason for ALI/ARDS [15, 16, 70–72]. Other predisposing clinical settings with a significant risk for ALI/ARDS are pneumonia (community-acquired pneumonia being the most common aetiology of ARDS originating outside the hospital), aspiration pneumonia, severe trauma, near drowning, multiple blood transfusion, burns, drug toxicity and pancreatitis [16, 70–73].

Following the establishment of low tidal volume ventilation [74], the overall in-hospital mortality rates have fallen significantly from 40 to 60 % [15, 70, 75–77] to ranges as low as 25–30 % [78–81]. Notably, the majority of patients die as a result of sepsis or multiorgan failure (MOF) rather than from respiratory failure [15, 70, 75, 82–84]. Other conditions responsible for death are ALI/ARDS in the setting of liver dysfunction/chronic disease, advanced age and non-pulmonary organ dysfunction [15, 75, 82, 85, 86]. Furthermore, in ALI/ARDS survivors, persistent abnormal exercise endurance is frequently present with shortened 6-min walking distances (only 66–76 % of predicted value) even after 5 years [87–89].

2.3 Aetiology

Multifactorial and heterogeneous pathologies may elicit ALI/ARDS with the potentially incentive inflammation being central in its pathogenesis [4, 6, 10, 37, 49, 56, 73]. Most often this syndrome occurs as part of a systemic process [89, 90], and underlying aetiologies may be classified into inherent lung pathologies or as being extrapulmonary [2, 10, 17]. In the majority of cases (55–75 %), a direct insult to the lungs has occurred with extrapulmonary causes found in 25–45 % [67].

Previously held assumptions that extrapulmonary-induced ARDS was more severe with an even worse prognosis have been challenged, and in fact, if adjusted for age, gas exchange features, severity of non-pulmonary disease and underlying illness, no significant differences can be found [75, 92–94]. Overall, more than 60 causes and predisposing factors of ALI/ARDS are recognized [2, 10, 15, 76], although only a few account for the vast majority of cases of ALI/ARDS [71, 72, 95, 96] (see Table 2.1).

Predisposing factors identified:

- Alcoholism [98]
- Obesity [99, 100]
- Chronic lung disease [71]
- Cigarette smoking [101]
- Low serum pH [71]
- Some classify near drowning, cute pancreatitis, pneumonectomy and cardiopulmonary bypass as risk factors, not as independent aetiologies [102, 103, 104, 105].

Table 2.1 Common disorders potentially eliciting ALI/ARDS [2, 10, 37, 73]

Direct lung injury	Indirect lung injury
Pneumonia[a]	*Sepsis*[a]
Gastric aspiration[a]	*Severe trauma with shock and multiple transfusions*[a]
Pulmonary contusion	Drug toxicity
Fat and amniotic fluid embolism	Cardiopulmonary bypass
Smoke and toxic gas inhalation	Acute pancreatitis
Near drowning	Transfusion of blood products
Alveolar haemorrhage	Shock
Reperfusion injury, e.g. post embolectomy, pleural effusion, post lung transplantation	

[a]Most common [71, 72, 96, 97]

Especially remarkable is some recent data suggesting a genetic predisposition which may augment the individual risk of ALI/ARDS [106–109]. The more predisposing features present, the greater the likelihood for that person to develop ALI/ARDS [71, 72, 97, 98, 110].

Sepsis is the predominant aetiology of ALI/ARDS [1, 15, 16, 71, 72, 111] and is most frequently triggered by pneumonia with more than 40 % of patients with pulmonary sepsis precipitating ALI/ARDS [16]. Patients with known chronic alcohol abuse and septic shock have a particularly high risk of (approximately 70 %) to developing ALI/ARDS [112]. Aspiration pneumonia evolves to ALI/ARDS in one third of cases [72, 97, 113]. Severe trauma may be complicated by ALI/ARDS and is reported to induce this directly [114, 115], but fortunately, ALI/ARDS associated with trauma has a relatively good prognosis when compared to non-traumatic disorders. Although this may merely represent younger patients suffering from trauma [116]. ALI/ARDS is also relatively common in bilateral lung contusion [117] and following fat embolism attributed to long-bone fractures [118].

Blood transfusion-related ALI is characterized by noncardiogenic pulmonary oedema following transfusion of blood products [119], usually within 6 h [120]. The resolution of ALI is typically immediate and a low mortality is seen (less than 10 %) [121]. Transfusions of blood in patients afflicted with ALI/ARDS will increase their risk of mortality [122]. Interestingly, other rarer causes of ALI/ARDS include overdose of opioids, tricyclic antidepressants, aspirin, phenothiazines [123] and the use of radiocontrast solutions in susceptible patients [124].

2.4 Pathophysiology

Various pulmonary and extrapulmonary conditions may provoke ALI/ARDS [2, 10, 37]. The histology is typified by diffuse alveolar damage [2, 35–37, 125, 126] in any phase of the syndrome [37], with a complex underlying pathophysiology comprising cellular, molecular, and physiological issues and obviously insult-specific/threat-dependent, diverse pathways in its pathogenesis [127, 128].

Increased permeability of the pulmonary microvascular vessel bed provoking fluid extravasation of a protein-rich solution into the lung tissues will subsequently elicit pulmonary edema, a patho-physiological hallmark and central in the pathobiology of ALI/ARDS [3, 4, 10, 25–27, 129]. Insults by endotoxins, microbes or viral properties, nonmicrobial toxins, drugs, microemboli, reactive oxygen species, immune complexes, mechanical/physical vigour (i.e. alveolar overdistension and shear stress by mechanical ventilation) and chemicals (e.g. gastric acid) may all harm pulmonary tissues, in particular the pulmonary vasculature endothelium [4, 25, 130, 131]. The pulmonary endothelium with its large surface area is vulnerable from both the vascular and the air space interfaces [132]. As a result of the endothelial irritation, lesion or even damage, local and potentially widespread systemic endothelial activation and dysfunction may ensue [4, 25, 133–135]. The current pathological paradigm suggests that the endothelial affliction is paramount for precipitating inflammation and the activation of the innate immune response [136].

Endothelial activation and dysfunction are characterized by a variety of histological structural (including focal gaps between endothelial cells [137]) and functional alterations, overview by Ware and Price et al. [4, 25]. Inflammatroy response with the release of proinflammatory mediators [2, 6, 138–141], imbalanced vascular tone with a shift to vasoconstriction (due to uneven production and release of vasoactive mediators) [142, 143], activated coagulation [144–146], expression of endothelium-derived adhesion molecules which interact and activate neutrophils in particular [147–149] and increased microvascular permeability [2, 134, 147, 150, 151], all of which consecutively provoke pulmonary oedema [3, 4, 25, 129]. As mentioned, the increased pulmonary microvascular permeability is a hallmark of the syndrome and has been confirmed in several studies inclusive the application of radiolabeled tracer proteins [151–154]. The excessive inflammatory response is often referred to as "biotrauma" [155] and has also been widely accepted as a major and integral part of ALI/ARDS pathology, playing a decisive role in its pathogenesis [140]. The inflammatory injury to the alveolar–capillary membrane is considered as fundamental in the pathobiology of ALI/ARDS [4, 47, 135], moreover, the inflammation may propagate to the vascular beds of other remote organs promoting multiorgan dysfunction or failure (MODS/MOF) [91, 135]. Locally evolved proinflammatory mediators leak into the systemic circulation [156], and pulmonary inflammation, if associated with the loss of pulmonary compartmentalization in the setting of acute lung injury, may contribute further to multiorgan dysfunction [157]. This process is a balance between the injurious and protective innate and adaptive immune responses and pathways whether the alveolar damage continues and spreads or is repaired and resolves [47]. As patterns of mechanical ventilation can induce or amplify alveolar injury with consecutive inflammation [2, 158, 159], artificial ventilation may contribute via multiple inflammation-associated pathways [2, 91, 160, 161].

The inflammatory process encompasses a complex network of cytokines, other proinflammatory molecules and miscellaneous cells and their products, which are able to initiate, amplify, modulate and perpetuate the pathophysiology [2, 6, 70, 162]. The main players seem to be the neutrophils and the multiple mediator cascades [163, 164]. However, as denoted, the balance between the activated cascades (e.g. released pro-inflammatory mediators such as IL-1 and TNF) and their antagonistic physiological anti-inflammatory feedback mechanisms (e.g. IL-1 receptor antagonist and IL-10) is of considerable importance for the disease course [162, 165–167]. Notably, imbalances within these cascades are repeatedly observed: (I) The potency of the inflammatory response and the extent of cell and tissue injury depend on the balance between proinflammatory and anti-inflammatory components [2, 165]. (II) The effects induced by destructive products released by the activated neutrophils may be attenuated by a composition of endogenous antioxidants and anti-proteases [4, 168–170]. The balance between destructive and protective compounds may limit the extent of tissue damage [166]. (III) Reported elevations of vasoconstrictive mediators such as endothelin-1 and thromboxane A_2, coincidentally with significantly low levels of vasodilating agents and molecules like nitric oxide (due to a diminished endothelial NO synthase expression and/or activity), resulting in mismatched pulmonary vascular tone creating a vasoconstrictive milieu [25, 171–175]. (IV) Coagulation homeostasis is disrupted [10, 176–179] with a shift to a procoagulant status [180,

181], definitely precipitated by the inflammatory response, but overlaps with the inflammation and may influence and affect the pathology considerably [182–184].

The noncardiogenic pulmonary oedema in ALI/ARDS, particularly the degree of alveolar flooding, is not only caused by increased microvascular permeability and disrupted tissue integrity (cellular gaps) but by an altered fluid balance. This is attributed to the overcharged and/or reduced capacity of the alveolar epithelium to actively remove fluid from the airspaces [185, 186], altered epithelial cell layer integrity and dysfunction [47, 129, 187] and the overburdened capacity of the lymphatic system [150]. Thus, impaired alveolar fluid clearance contributes to oedema development and its extent [188]. Moreover, as Calfee and Matthay [31] demonstrated, increases in hydrostatic vascular pressure due to altered vascular tone or following increased intravascular volume in the presence of increased permeability will amplify the extent of EVLW, thus the amount of pulmonary oedema. Accordingly, a fluid-restricted management strategy in ARDS patients has been shown to be beneficial, as reductions in hydrostatic vascular pressures in conditions where the microvascular pulmonary permeability is enhanced will diminish pulmonary oedema [189, 190]. Jozwiak even reported of a better prognosis if the hydrostatic pressure predominantly contributes to oedema formation in relation to the enhanced permeability and the severity of alveolar injury [44]. Pulmonary oedema formation is generally precipitated either by an increase in pulmonary capillary hydrostatic pressure as predominantly found in cardiogenic pulmonary oedema, or increased pulmonary capillary permeability as typical and prevalent in ALI/ARDS or both [129].

A generally well-balanced systemic reaction and response is crucial for a positive outcome in ARDS [2, 167] as multiple pathways can excite or hinder lung injury [191, 192]. Furthermore, isolated endothelial disorders are not sufficient to provoke ALI [193], and generally, consecutive and/or coexisting abnormalities are also present in clinically manifest ARDS [4, 25], namely:

A. Microvascular cell sequestration may be jointly responsible for pulmonary vascular obstruction and even destruction of the vessel bed since micro-vascular infarcts, arterial thrombosis of larger vessels, thromboemboli and fibrinous conglomerates may all cause vascular occlusion [4, 25, 133, 194–196]. In pathophysiological consequence, the dead-space fraction accumulates and the development of pulmonary hypertension is facilitated [4, 25]. In addition, Nuckton et al. [197] suggests the degree of dead-space ventilation is an independent predictor of mortality. Pulmonary (micro)vascular pathology is another hallmark of ALI/ARDS [4].
B. Epithelial cell involvement with loss of the epithelial cell layer integrity, fluid transport dysfunction (both contributing to oedema development and extent) and altered surfactant production (causing alveolar collapse) have become important issues in the pathology of ALI/ARDS [50, 107, 191, 193]. Direct insults will injure primarily the alveolar epithelium while capillary endothelial structures initially stay more or less intact [198]. However, epithelial irritation or destruction will activate alveolar macrophages, neutrophils and the inflammatory cascade inducing intrapulmonary inflammation. This results in epithelial cell

dysfunction and disrupted continuity of epithelial cells (gaps) and hence oedema formation, which in turn will affect the endothelium [187]. Epithelial cell permeability seems to have a crucial contribution to alveolar flooding [129, 193].

Even if the endothelium is the primary site of insult, ultrastructural studies have demonstrated the involvement of the epithelium [2, 199]. In cases where the alveolar epithelium is substantially or recurrently injured or disorganized, its repair may end up in fibrosis [200]. The degree of the damaged epithelium is reported to be an important predictor of outcome [151, 188]. Furthermore, removal of the oedema depends to a considerable extent on the clearance function, which requires the integrity of the epithelial cell layers and restored epithelial cell function [4, 49].

C. Attracted by the endothelium-derived adhesion molecules, a marked number of neutrophils are pulled in and retained in the lung vessels and tissue [50, 201]. Once activated, they release potentially destructive products such as reactive oxygen and nitrogen species, cytokines and proteolytic enzymes which destroy the extracellular matrix or induce epithelial apoptosis [191, 202, 203].
D. Inappropriate generation of and the concomitant absence of an adequate concentration of antioxidants may result in accumulation of reactive oxygen species following endothelial and/or epithelial injury and is associated with alveolar macrophage cell activation which further contributes to cellular injury [4, 191, 204].

Fibrosing alveolitis histologically occurs 5–7 days after disease onset [191, 205] following the initial, exudative and proliferative stage of ARDS [2, 6, 10]. This is considered a maladaptive proliferative repair [4] promoted obviously already early in the course of the disease since proinflammatory mediators activate fibrogenesis [206–209]. Established fibrosis is a very serious condition as it precipitates a marked reduction in lung compliance with diminished tidal volume, increased alveolar dead space and persistent hypoxia resulting in high breathing work load, and, finally, CO_2 retention [6, 201, 205]. Unfortunately, neither the processes of recovery and repair nor the mechanisms regulating these are well understood [49, 91].

The clinical features of ARDS are dominated by respiratory failure which arises as a direct consequence of the extensive derangement of physiology detailed above [4, 6, 10, 25, 73].

The main appearances are:

I. Arterial hypoxaemia
II. Decreased lung and/or thoracic compliance
III. Increased dead space ventilation
IV. Pulmonary hypertension

I. *Arterial hypoxaemia*

This is attributed to:

(a) Ventilation–perfusion mismatch resulting from pulmonary oedema and cellular infiltration [4, 6, 10, 210–212], blunted hypoxic pulmonary vasoconstriction [213–215] and reduced cardiac output [216]
(b) Diffusion abnormalities due to alveolar flooding and cellular infiltrates [6, 10]

II. *Decreased lung and/or thoracic (chest wall) compliance*
Altered and diminished lung compliance is another hallmark of ALI/ARDS [217]. Alveolar oedema and surfactant dysfunction with consecutive alveolar collapse lead to diffuse atelectasis, increased dead space [218, 219], and cellular infiltrates which all impair lung (predominantly) but also thoracic compliance [4, 6, 10, 73, 217]. In the case of thoracic trauma, pleural effusion, anatomical thoracic abnormalities, sedation and paralysis [220–222] or extrapulmonary dysfunction such as abdominal hypertension or compartment syndrome, the chest wall compliance will be considerably affected (stiffened), provoking a marked increase in respiratory work, dyspnoea and respiratory failure [223, 224].
Extrapulmonary ALI/ARDS displays quite different mechanical properties, lung morphology and ventilatory requirements in comparison to primary pulmonary ARDS [99, 224–228]. Extrapulmonary ALI/ARDS may exhibit more diffuse oedema attributed to the extrapulmonary focus of inflammatory mediators [224], compression atelectasis in the dependent lung areas secondary to the increase in lung weight [229], diffuse collapse resulting from diffuse oedema, and lower transpulmonary pressures all discharging radiomorphologically into ground glass opacification. In contrast, in pulmonary ALI/ARDS, consolidation rather than collapse of lung tissue is typical [227, 228]. Furthermore, since the transpulmonary pressure applied is responsible for the strain on lung architecture [223], lower ventilatory pressures are required in mechanical ventilation of extrapulmonary ALI/ARDS [230–232]. In the prone position, patients with extrapulmonary ALI/ARDS will show a greater potential for improvement in oxygenation [233, 234] and benefits from recruitment manoeuvres [235, 236].

III. *Increased dead space ventilation*
Increased dead space ventilation is a result of alveolar collapse/diffuse atelectasis following oedema, surfactant dysfunction, compression [4, 6, 10, 212, 217, 218], damage or destruction of the pulmonary vascular bed [10, 197] and regional alveolar overdistension [10]. Arterial carbon dioxide content rises with increasing dead space [197, 211, 212], and hypercapnia reflects involvement of pulmonary microcirculation [25].

IV. *Pulmonary hypertension*
Pulmonary hypertension (PH) is common in patients with ALI/ARDS [237] with up to 25 % affected [143, 238, 239]. We are still missing a consensus definition of *acute* pulmonary hypertension [25]. It is uncertain what definitions may be transferred from the chronic setting, and assessing the clinical impact of PH is difficult [25, 73]. However, since lung protective ventilation has become standard, the incidence of acute right heart failure in that setting, with its associated higher mortality [240, 241], has fallen to 10–25 % [239]. Some authors even report incidence rates of less than 10 % [240, 242], whereas previously, up to 50 % of all patients with ALI/ARDS developed acute right heart failure [243, 244].

The pathological mechanisms which provoke an increase in pulmonary vascular resistance (PVR) and thus PH in ALI/ARDS are [25, 73, 245] detailed below [25]:

- Endothelial dysfunction [133, 134, 197, 246].
- Hypoxic vasoconstriction [143, 247].

- Pulmonary vascular occlusion (i.e. capillary thrombosis and microthrombi, injury and/or destruction of the pulmonary vascular bed) [14, 134, 184, 194, 195, 246, 248, 249].
- Increased vascular tone particularly attributed to the imbalance between vasoconstrictive and vasodilative agents, i.e. excess production/release of ET-1, TXA-1, PDE and 5-HT, while reduced levels of NO and prostanoids are found [25, 132, 143, 250]; additionally, endotoxins will affect the vasculature and its tone [143, 251].
- Extrinsic vessel compression/occlusion (e.g. positive airway pressure, airway collapse, parenchymal destruction) [159, 252].
- Vascular remodeling [134, 194].

In general, a significant rise in PVR strains the right ventricle (RV) [245] which poorly tolerates an acute increase in its afterload [253, 254]. Subsequently, acute RV dysfunction (dilated RV with reduced EF but preserved SV) or even RV failure (dilated RV with reduced EF and reduced SV) [255, 256] may be precipitated with the potential serious sequelae of systemic hypotension and shock [25, 245]. Acute cor pulmonale (ACP) is defined as acute RV failure in the setting of an acutely elevated PVR associated with a primary pulmonary disease [239, 257, 258]. The severity of RV outflow load is strongly related to RV dysfunction/ACP and mortality [243]. Patients suffering from ALI/ARDS are at particular risk of developing ACP, as both lung injury itself [4, 25] and mechanical ventilation (particularly PEEP) may cause further increase in PVR "afterloading" the right ventricle [259, 260].

As described in detail in Chap. 1, a previously normal RV is poorly adaptive to a sudden increase in afterload [253, 261] and is not able to acutely generate increased systolic PA pressures above 40–45 mmHg [262, 263]; hence, rapid hemodynamic compromise and an early onset of hypotension and circulatory shock may ensue [254]. Sudden, substantial afterloading of the RV attributed to an increase in PVR will overwhelm the RV contractile capability [264, 265] and will result in a mismatch between RV elastance (largely a load-independent index of RV contractility [266, 267]) and pulmonary arterial elastance (representing RV outflow impedance [82, 266–268]), causing a progressive disruption of normal ventriculopulmonary coupling [264, 269]. Ventriculo-pulmonary coupling is a major determinant of RV function [266, 267] and thus potentially leads to RV dysfunction or RV failure/ACP, if inappropriately altered [264, 268]. Typically, the RV (and right atrium) dilates immediately in response to a sudden increase in afterload [239, 253, 270, 271] as a compensatory mechanism, utilizing preload reserve [272], trying to preserve SV via the Frank-Starling mechanism [239]. In the acute setting, the non-distensible pericardium maintains an unchanged total cardiac volume [274, 276] as constraining biventricular filling [274, 275], and affecting interventricular volume proportioning through diastolic ventricular interaction (DVI) [273]. But since the interventricular septum is (mediated by DVI) displaced towards the left ventricle following RV dilatation, LV enddiastolic filling gets diminished [276]. RV enlargement reflecting progressive RV dysfunction [277, 278] is accompanied by an increase in RV end-diastolic pressure (RVEDP) [274, 279] directly impeding the LV [280, 281] and subsequently even further reducing LV filling [280–283]. Accordingly, LV stroke volume and CO will be substantially compromised potentially leading to hypotension and/or circulatory collapse [25, 263]. Additionally, myocardial ischaemia (particularly RV ischaemia) may occur through significant supply–demand mismatch which results from increased RVEDP impairing delivery while the RV myocytes require more oxygen [272, 284, 285]. Volume loading under these conditions may be deleterious and shows paradoxical effects, as further reducing LV filling [286] due to progressive RV enlargement and impaired pulmonary blood flow [261], typifying the failing right heart. Positive pressure mechanical ventilation and especially PEEP may worsen the situation as heart–lung and diastolic ventricular interaction [287–292] leads to RV enlargement and dysfunction, even more elevated afterload, reduced venous return and subsequently reduced cardiac output [261, 290, 293, 294]. Ongoing hypoxia due to hypoxic pulmonary vasoconstriction [247, 295] and hypercapnia [293, 294] may create and/or maintain an elevated PVR. Tachyarrhythmia, particularly atrial fibrillation which frequently occurs in this setting, will further impair RV function and hemodynamics [261, 296, 297]. Infections, anaemia, trauma, surgery and pulmonary embolism also may aggravate RV dysfunction further [261], with infections being the strongest predictor of death [298]. Moreover, patients exhibiting RV dysfunction/failure are at a significant risk of developing concomitant infection as the intestinal barrier is compromised in cases of low CO and elevated venous pressure [299, 300], both arises in RV failure.

2.5 Diagnosis and Clinical Features

Diagnosing ARDS may often be challenging, as this complex syndrome elicits a broad range of clinical phenotypes with unspecific symptoms and findings [47]. As discussed, ALI/ARDS is histologically characterized by diffuse alveolar damage [2, 10, 301], *clinically* hallmarked by severe respiratory failure indicated by marked hypoxemia [1, 27], and *radiomorphologically* by bilateral infiltrates/opacities [1, 27, 302], the latter representing and consistent with manifest non-hydrostatic pulmonary edema attributed to increased pulmonary capillary permeability [1, 3, 10, 126]. The diagnosis is made in the absence of significant left heart failure [1, 18, 304, 305]. More serious cases may be denoted by signs of systemic inflammation following massive release of proinflammatory mediators [4, 303].

Even the most recent definition has decisive shortcomings [23, 40] due to ambiguous and non-specific clinical findings [32], and even the histology is non-specific [13, 306] which leads to misdiagnosis and delayed recognition [13, 41].

In daily practice, **ARDS is principally clinically diagnosed** [1, 10, 27, 307–309], although a chest x-ray and blood gas analysis are required [308], in order to fulfill the criteria listed in the new Berlin definition. A clinical setting predisposing to ARDS such as sepsis, aspiration, trauma, etc. has to be present and the hypoxemic respiratory failure is classified as mild, moderate or severe, requiring at least non-invasive ventilatory support (e.g. CPAP) for adequate oxygenation while applying a suitable FiO_2 [27].

Of course, other differential diagnoses causing similar clinical and radiological pictures should be excluded [56], the most common being pneumonia [16, 307]. In one cohort, pneumonia was suspected in 20 % of patients, whereas 58 % had histological features at autopsy [310]. During the first 24 h, the chest x-ray may be quite normal due to only slight oedema [311]. However, over the following 2–3 days, widespread, generally homogeneously distributed, opacification should become obvious [312].

It needs to be stressed that a bedside chest x-ray has quite low diagnostic accuracy to detect alveolar consolidation and the alveolar interstitial syndrome [312]. Endo and co-workers reported that ARDS developed within 3 days of ICU admission in roughly 50 % of patients with a normal chest x-ray but clinically suffering from acute respiratory failure [313]. CT and lung ultrasound are clearly superior in detecting lung abnormalities consistent with ALI/ARDS [314, 315]. Patients with severely injured lungs may be identified by a FiO_2/PaO_2 ratio of less than 100 mmHg and with poor response to conventional ventilation [316, 317]. They may qualify early for rescue treatments.

Echocardiography may be a valuable first-line examination to exclude heart failure for the clinical finding of pulmonary oedema and the bilateral opacities on chest x-ray, in a typical setting suggestive for lung injury, and thus makes ALI/ARDS very likely [25, 27, 318]. Furthermore, recognizing right heart dysfunction/failure with its worsening impact on clinical conditions, on respiratory and hemodynamic features as well as on gas exchange may be diagnosed and recognized early [239, 277, 319, 320]. There is an association between ARDS mortality and RV dysfunction [240]. Moreover, RV function is a major determinant of the prognosis of pulmonary hypertension [321–323], and morbidity and mortality in PH are shown to be strongly related to RV failure [321, 324]. Tricuspid annular plane systolic excursion (TAPSE) is indicative of RV-EF, reflecting systolic RV function [325, 326]. This relationship between TAPSE and RV-EF has been verified in several different types (etiologies) of PH [327, 328]. TAPSE is easily obtained by echocardiography and a value < 15–18 mm suggests reduced RV function [326, 327]. As there is no clear cut-off, the lower the value of TAPSE, the more compromised the RV function [329]. Normal serum natriuretic peptide levels favour a diagnosis of ALI/ARDS and have a high negative predictive value in excluding heart failure [330]. Jardin and Vieillard-Baron provided proof of a strong relationship between the incidence of ACP and plateau pressure (P_{PLAT}) [244]; thus, the evaluation of RV function is helpful to determine the level of PEEP to be set and to avoid ACP development by setting a too high PEEP [331].

Due to the non-specific clinical and radiological findings in ALI/ARDS and the deficiencies of the definition(s) as discussed above, many authors ask for additional, more specific and sensitive features to identify and diagnose ALI/ARDS to be included or added to the definition [23, 40, 58, 332]. In particular, the extravascular lung water (EVLW) and the derived pulmonary vascular permeability index (PVPI) may be very valuable markers as quite a large body of literature concludes [42, 57, 58, 61, 332]. Even the contributors to the Berlin definition, suggest "measurement of EVLW or pulmonary vascular permeability will be an important advance and could be incorporated into a future definition of ARDS" [333].

The pulmonary oedema in patients with acute lung injury is largely due to increased permeability of the pulmonary microvasculature as the central pathophysiological feature [3, 4, 25, 27]. As pulmonary oedema is an accumulation of fluid in the interstitial and alveolar space, it can be termed extravascular lung water (EVLW) [30, 55]. A substantial number of publications have offered proof that EVLW can be accurately estimated at the bedside by double- and single-indicator [334–336] transpulmonary thermodilution techniques, even if compared to the gold standard, gravimetric assessments [53, 55, 337–339]. Indexing EVLW to predicted (ideal) body weight (PBW) turned out to be superior to actual body weight [30] as originally used. "Height and sex are the most important factors which determine lung volumes" as "lung size will not increase with body weight" [58, 340]. Indexing EVLW (EVLWI) to actual body weight underestimates lung water in obesity [62], while indexing to PBW has substantially better sensitivity to detect elevated extravascular fluid in the lungs [58, 341] and thus oedema formation and ARDS, providing very good prognostic predictions [57, 58, 62, 63, 338, 342]. PEEP may cause inaccuracy of measurements with underestimations most likely [343] but may directly increase EVLW by diminishing lymphatic flow [19]. Overestimations are reported in cases of major lung resection [344], the presence of large aortic aneurysms, significant intracardiac shunts and if the arterial catheter is placed too peripherally [335].

Elevated values for EVLWI and thus pulmonary oedema are EVLWI > 10 ml/kg [57, 58, 61, 345], and several authors advocate including an EVLWI above 10 ml/kg in the setting of clinically suggested ARDS as a diagnostic criterion [56, 342, 346, 347]. Furthermore, the determination of EVLW is currently the most sensitive bedside method (compared to clinical evaluation, x-ray and pulmonary artery catheter) to recognize the early stages of pulmonary oedema formation [335, 348, 349]. Normally, the extravascular fluid content of the lungs is ≤ 7 ml/kg PBW, and indeed, some authors have strictly defined EVLWI > 7 ml/kg as being elevated [338, 348, 350]. However, we should consider the range between > 7 and ≤ 10 ml/kg as a grey area signalling a potential risk of developing pulmonary oedema and that preventive strategies, particularly fluid balance strategies, will need to be considered to prevent development of pulmonary oedema.

The recently introduced pulmonary vascular permeability index (PVPI), defined as the ratio between EVLW and pulmonary blood volume (PBV), is considered to be able to distinguish between predominantly hydrostatic pulmonary oedema from that largely due to permeability-induced oedema and as such is a reflection of pulmonary microvascular permeability [45, 55, 56, 351, 352]. A PVPI ≥ 3.0 represents high permeability suggestive of disturbed capillary barrier, while a PVPI < 1.7 is typically found in a predominantly hydrostatic-mediated pathogenesis [45]. Astonishingly, this, as described by Monnet, demonstrated a specificity of 100 % with satisfying sensitivity of 85 % if the PVPI ratio was ≥ 3.0 [45]. Kushimoto largely confirmed these results, reporting a specificity of 90–95 % in the diagnosis of permeability-induced pulmonary oedema (PVPI ≥ 3.0), while a PVPI < 1.7 ruled out ALI/ARDS with a specificity of 95 % [42].

Interestingly, Chew recently reported from a trial assessing patients with shock that EVLW indexed to actual body weight rather than PBW performed better in diagnosing ALI/ARDS and in regard to prognostication [57]. Nevertheless, although the debate on which index is superior is currently ongoing [57, 60], the criterion, "elevated EVLW", needs to be included in the definition of ARDS [40, 332, 342, 346, 347]. Until the debate on indexing is solved and because lung volume does not increase with body weight [57], an EVLWI > 10 ml/kg (PBW) clearly indicates elevated extravascular lung water typical in ARDS, and the PVPI should be used to help distinguish oedema pathogenesis in daily practice to substantiate the diagnosis of ARDS, if a typical clinical setting is present [42, 45, 353]. If the diagnosis remains uncertain, lung biopsy can be performed safely and has been demonstrated to alter treatment in up to 80 % of patients in some studies [37, 354, 355].

2.6 Therapeutic Considerations

2.6.1 Respiratory Support/Mechanical Ventilation

The vast majority of patients suffering from ARDS require mechanical respiratory support as a result of exhaustion from elevated respiratory workload and are unable to further raise the ventilatory effort required to ensure suitable oxygenation and ventilation [356, 357].

In the acute phase, ARDS lungs on computed tomography show combined alveolar flooding (which may be attributed to oedema or infection), interstitial inflammation (related to the oedema) and tissue atelectasis due to compression [358]. However, the abnormal patterns are found to be highly heterogeneous throughout the lungs. Alongside areas of well-aerated normal tissue (usually in nondependent areas), poorly inflated, densely consolidated and fluid-filled/atelectatic areas without any identifiable vessel structures (primarily recognized in dependent areas) may be identified. There are also uninflated, collapsed but potentially recruitable regions with still identifiable vessels (ground-glass opacification) demonstrable [227, 312, 357]. This heterogeneous involvement of the lung tissue with consecutively variable regional compliance [359] and aeration [302, 358] predisposes to further injury by mechanical ventilation, termed ventilator-induced lung injury (VILI). This arises because the morphologically normal areas are most compliant [360] and will predominately be inflated during inspiration potentially overdistending the small airways and alveoli, particularly if large tidal volumes are applied [361–366]. As the amount of healthy lung tissue may be diminished to around one third of total tissue, the term "baby lung" was coined by Gattinoni and co-workers to describe ARDS lungs [227]. Later, taking into account the heterogeneity of lung tissue and aeration in ARDS patients and the act of "proning", which leads to density redistribution, led to the concept of "baby lung" being a functional rather than an anatomical phenomenon, implying that the decisive component of ventilation is the ratio of tidal volume (V_T) to anticipated "baby lung" size, not V_T to kg body weight [367]. Thus, it is literally not surprising that the application of low tidal volumes, being based on sound evidence, has been recommended and accepted as the approach of choice

[368–372]. Moreover, atelectrauma affects the areas recruited during inspiration but these collapse/de-recruit during expiration [360–363], and setting a suitable PEEP is critical to avoid cyclical collapsing and reopening [373, 374]. Although optimizing PEEP levels is another crucial component of mechanical ventilation, in contrast to the tidal volume limits now universally applied, the optimum PEEP is as yet undetermined.

Lung protective ventilation (LPV) aims to prevent artificially ventilated, diseased lungs from further, mechanical ventilation-inherent injury by applying low tidal volumes and limiting airway pressures [356, 375]. Basically, a volume-controlled ventilation (VCV) applying a tidal volume of ≤ 6 ml/kg PBW while concurrently limiting plateau pressure to ≤ 30 cm H_2O is the essential setting [368–370]. A reduced V_T will consecutively cause lower inflation and plateau pressures [376, 377]. Restricting V_T is the only feature with definitive evidence [368–370, 375, 378]; however, there is no disagreement that the application of a suitable PEEP is indispensable and improves outcomes in ALI/ARDS [370, 379–381]. This ensures lung recruitment and the prevention of lung de-recruitment likely following the application of low tidal volumes [382, 383], and avoids the expiratory collapsing and inspiratory re-openings (recruitment/de-recruitment) of alveolar units potentially accessorily insulting the lung parenchyma [360, 384, 385]. "The aim of lung protective ventilation is to prevent alveolar overdistension in open lung while avoiding repetitive opening and closing of the alveoli in collapsed lung" [386]. PEEP titration ideally prevents cyclic alveolar recruitment and collapse while minimizing overdistension in open lung [387], achieved by the lowest PEEP value possible [388]. As such, the original description of LPV needs to be expanded by this important issue. In those studies clearly demonstrating evidence of significant outcome benefit, PEEP was set empirically following a table based on FiO_2 required, aiming for an arterial oxygen saturation (SaO_2) of 88–95 % or a PaO_2 of 55–80 mmHg [369]. The results of the largest study, conducted by the ARDS network, were indeed impressive, with an absolute reduction in mortality of 9 % just by applying low tidal volumes (6 ml/kg PBW) [74].

LPV has been demonstrated to considerably attenuate pulmonary oedema formation which has been attributed to preserving alveolar endothelial and epithelial barrier function [389–392] and to down regulating mechanosensitive proinflammatory pathways [160, 393]. Furthermore, extrapulmonary organ function is affected less often and less severely, e.g. renal function [369, 394], as we become aware of the effects of mechanical ventilation on remote organ function [2, 160].

2.6.2 *Optimal PEEP*

The way to select and titrate an optimal PEEP has been an issue of research and discussion for several decades [155], but unfortunately, there is still no distinct answer [155, 302, 356, 375, 395, 396]. The key is to titrate PEEP by integrating the other ventilator settings of LPV to the patient's unique respiratory and haemodynamic requirements [397], hence an individualized approach. This is supported by

Table 2.2 Anticipated PEEP settings at various FiO_2 requirements, aiming for an oxygenation goal of PaO_2 55–80 mmHg, SpO_2 88–95 %, respectively, aiming to reduce FiO_2 to ≤0.60 within 24 h

FiO_2 [%]	0.3	0.4	0.4	0.5	0.5	0.6	0.7	0.7	0.7	0.8	0.9	0.9	0.9	1.0
PEEP [cm H_2O]	5	5	8	8	10	10	10	12	14	14	14	16	18	20–24

Since high fractions of inspired oxygen are considered harmful (i.e. resulting in reabsorption atelectasis [413]), are generally FiO_2 contents of ≤0.60, for long-term application ≤0.40 advocated, although no evidence-based thresholds are available [413–415]. However, prolonged high oxygen concentration in the inspired air may be injurious to the lung via oxygen toxicity [415]

several studies [379, 386, 398] and large meta-analyses [302, 396, 399], as a uniform PEEP selection in unselected patient groups yields no overall benefits at all, only in subgroups [302, 396].

There are many methods described to set PEEP including pressure–volume curve analysis [160, 368, 370], incremental PEEP adjustment to FiO_2 requirements using empirically derived tables based on oxygenation targets (FiO_2–PEEP tables) [369, 386, 400], oesophageal pressure measurements [401, 402], so-called stress index measurements [403] and incremental PEEP using plateau pressure [379]. The most popular has been the pressure–volume curve method, adjusting PEEP slightly higher (2–3 cm H_2O) than the lower inflection point (LIP) of the curve's inflation limb [160, 368] indicating ensuing recruitment [404, 405] and thus the point where PEEP should be set. However, even the acquisition of the curve is complicated; the patients need to be sedated or even paralyzed [397], and three different techniques to acquire the P/V curve are described [406–408]. Unfortunately, LIP is just the beginning of recruitment and the upper inflection point does not conclusively mark alveolar overdistension [382, 409, 410]. Thus, the upper inflection point of the deflating limb of the pressure–volume curve is, more recently, considered to be the most precise marker of lung mechanics indicating suitable PEEP selection [411, 412].

However, there is indeed no best method to determine proper PEEP as each approach has shortcomings [356]. Most of the methods may be technically challenging, and thus, the FiO_2–PEEP tables are frequently employed in daily practice [356]. The ARDS network recommends the following [369] (Table 2.2):

Ascertaining the most suitable/optimal PEEP level deserves further discussion. The ARDS network table explanation and the huge bulk of literature dealing with this issue offer more insights on how to proceed in daily practice and how to individualize the PEEP setting in each single patient [160, 302, 304, 370, 379, 386, 396, 399].

Traditionally, patients suffering from ARDS were treated with PEEP levels between 5 and 12 cm H_2O [69, 396, 416]. The rationale to try and apply higher PEEP levels in order to improve outcomes in patients with more severe ARDS was based on a number of theoretical considerations, experimental animal studies and human data [39, 373, 391, 417, 418]. Low tidal volumes may leave collapsed areas untouched [419] or may even lead to atelectasis and lung injury [69, 382, 383, 420, 421] and, in turn, lower tidal volumes may need compensatory higher PEEP levels to avoid collapse [34, 386, 422, 423]. The results of recent studies [259, 379, 386, 400, 402] comparing low PEEP with high PEEP in the setting of low tidal volume ventilation and larger meta-analyses [302, 396, 399, 424] did not reveal any significant difference in outcome in unselected patients with ARDS. However, meta-analyses identified subgroups suffering from severe ARDS allocated to the higher PEEP group who benefitted from higher PEEP levels revealing a trend towards lower mortality [302, 399, 424] or even a significant reduction in lethality [396]. Therefore, in unselected study cohorts, the application of high PEEP levels is not suitable and may be harmful, whereas in patients with severe ARDS, higher PEEP levels are advantageous and may even reduce mortality [302, 396]. To illustrate what high PEEP really means and how to act in daily practice, it is crucial to recognize that in the overwhelming number of studies, the average high PEEP was approximately 15 cm H_2O [375, 379, 386, 398, 425]. However, PEEP, particularly higher PEEP, may exert detrimental effects on right heart function and central haemodynamics potentially causing circulatory depression [259, 359, 388, 426]. Therefore, it is important to be aware of the effects of PEEP on the circulation in the context of heart and lung interactions and thus to allow further individualization in your patients [290, 291, 319, 331, 427, 428].

Positive pressure ventilation and especially PEEP application are, mediated by changing lung volume, related to PVR by a bifashioned function [260, 430], subsequently affecting right heart function and central hemodynamics [291, 428]. The net effect of the change in lung volume on PVR arises from the proportion of lung tissue being recruited by the air volume inflated with PEEP keeping recruited areas open, preventing de-recruitment [360, 373] and the alveolar overdistension provoked [290, 291]. It is quite obvious that the best PEEP level will vary considerably during the disease course [244, 428, 431]. In addition, since inappropriately high PEEP levels markedly affect RV outflow impedance and hence RV function [259, 260, 291, 319, 428], the PEEP setting should be tailored to haemodynamics to avoid RV dysfunction or failure [291, 294, 319]. Both, respiratory mechanical/gas exchange demand and cardiac (especially RV) function need to be considered and harmonized when determining ventilator and optimal PEEP setting [291, 331]. Recent publications, emphasizing the attractiveness of this integrative pulmonary and cardiovascular approach, use airway plateau pressure (P_{PLAT}) as a reasonable tool to ascertain optimal PEEP level. To avoid relevant and detrimental circulatory effects of PEEP [244, 319, 331, 432], a $P_{PLAT} \leq 26$ (<26–28) cm H_2O should be used as the cut-off point [239, 244, 365, 432, 433]. Thus, airway pressures applied should in total not exceed this margin, and a well-balanced combination of sufficient inflation/inspiration pressure and PEEP to secure gas exchange, while paying attention to cardiocirculatory function, is necessary in each phase of the disease.

We therefore recommend that multiple factors should be taken into account when applying ventilator settings, not only gas exchange and mechanical respiratory features but coequal cardiocirculatory aspects. Accordingly, the P_{PLAT}, which is easily assessable, is one corner post in our assessment and may be supported by echocardiography to integrate haemodynamics which may be the pivotal portion to successful treatment as substantiated by the result of the ARMA study [369] and some smaller trials. Dysfunction of the heart and circulation, particularly acute cor pulmonale (APC), in the setting of positive pressure mechanical ventilation may be a significant, independent factor of mortality [82, 240, 244, 359, 434] with a strong correlation between P_{PLAT} and APC established [244].

Alternatively, in the future, Cortes suggests that monitoring respiratory measures by FRC changes and FRC progression to determine ventilator settings will become the method of choice. This is because lung tissue recruitment improves functional residual capacity (FRC) [290, 291], and the change in lung volume is related to PVR [260, 435]. FRC correlates best with the functional lung size [435] which in case of ARDS has to be regarded as a functional baby lung as Gattinoni established [367].

The LPV recommendation of a P_{PLAT} limit of ≤ 30 cm H_2O has been challenged, particularly since the limit advocated is somewhat arbitrary [69, 436]. Of course, an individual approach is necessary and well established. Patients with a non-pulmonary disorder but stiff chest wall (indicating compromised chest wall compliance) as common in abdominal diseases triggering ALI/ARDS [437] or in the obese will tolerate higher peak and plateau pressures well above 30 cm H_2O [438], without any lung injury. Much of the pressure will dissipate against the stiff chest [439] leaving the actual distending pressure (transpulmonary pressure) reaching the airways and the alveoli markedly below the initial airway pressure applied [92, 223, 225].

2.6.3 *Permissive Hypercapnia*

Lung protective ventilation applying low tidal volumes only, may in turn, lead to carbon dioxide accumulation (hypercapnia) [440–442]. Allowing an increase in $PaCO_2$ concentration while protecting the lungs from ventilator-induced injury, notably from alveolar overinflation/overdistension by applying LPV [369, 440, 442], is termed permissive hypercapnia [440, 441] and is suggested to be beneficial. Hypercapnia improves arterial and tissue oxygenation as (a) it shifts the oxyhaemoglobin dissociation curve to the right; (b) it amplifies hypoxic pulmonary vasoconstriction, thus ameliorating the local alveolar ventilation/perfusion ratio; and (c) it increases CO and thus systemic O_2 delivery rises [294, 443–445]. Furthermore, contrary to the pulmonary vessel bed, the systemic peripheral microvasculature dilates under increasing carbon dioxide concentrations, promoting tissue perfusion and oxygen supply [446]. Moreover, CO_2 mitigates the inflammatory response via several mechanisms, including reduced generation of free radical and nitrogen species, as well as proinflammatory mediators, inhibition of enzymes activated in the inflammatory process [447, 448] and by blunting several metabolic and inflammatory pathways displaying protective properties against associated local and remote organ affliction [449–451].

Clinically, Hickling first recognized better outcomes in ventilated patients with tolerated modest hypercapnia [440, 441]. However, at that time, high tidal volumes had been routinely applied, and an increase in pCO_2 might have reflected reduced tidal volume ventilation with a consecutive reduction in lung tissue injury. Nevertheless, subsequent important clinical studies confirmed [368, 369] the beneficial results seen by Hickling and the protective effects of modestly increased carbon dioxide concentrations. Higher concentrations will induce pulmonary vasoconstriction affecting PVR and thus RV function with the risk of developing ACP and negating any positive effects of hypercapnia [452]. Unfortunately, it is impossible to dissect the important trials on lung protective ventilation in order to differentiate between the effects produced by low tidal volumes applied and by permissive hypercapnia [368]. A post hoc analysis of the ARMA study designed and undertaken by the ARDS network [369] revealed in the 12 ml/kg group a lower mortality at day 28 if moderate hypercapnia (pH 7.15–7.35; $PaCO_2$ 45–65 mmHg) on day 1 was present, consistent with a protective effect of elevated $PaCO_2$ [453]. Thus, although not conclusive, the data available suggest a direct (attenuation of inflammatory response and consecutively diminished disease progression) and indirect (mitigation of the intensity of mechanical ventilation) benefit from modest hypercapnia which contributes to the significantly reduced mortality of ARDS patients treated by LPV [454].

The ARDS network guideline explicitly allows for modest hypercapnia [369]. Recommended treatment of acidosis is to increase the respiratory rate (RR). In cases of pH$<$7.15–7.3, the RR should be increased up to 35/min aiming for a pH$>$7.3, $PaCO_2 < 25$ mmHg. If pH$<$7.15, increase RR to 35 pm; tidal volume can be increased by 1 ml/kg steps until pH$>$7.15 and even the P_{PLAT} may exceed 30 cm H_2O; further small amounts of buffer are acceptable. However, the effect of an increase in RR is uncertain, as it may provoke auto-PEEP, and thus, the RR should be chosen below an auto-PEEP-inducing threshold [455, 456]. The increase in additional cyclic opening and closing (due to the elevation of RR) of the small airways and alveoli will be roughly around 25,000–30,000 per day which may represent an additional source of lung injury [457]. Data is lacking as to whether buffering is really a beneficial therapeutic option. It is clear however, that bicarbonate should be abandoned as it worsens intracellular acidosis. Tris-hydroxymethyl-aminomethane (THAM) buffer should be used as it is effective, reduces intracellular acidosis and does not convert into CO_2, with renal dysfunction being a relative contraindication [458]. Permissive hypercapnia may be contraindicated in patients with cerebral injury and particularly if the intracranial pressure is elevated, as higher $PaCO_2$ concentration causes vasodilation in the systemic and thus cerebral vasculature. Caution is also recommended in heart diseases because hypercapnia is depressing myocardial contractility, promotes the release of endogenous catecholamines and diminishes renal blood flow [70, 452].

2.6.4 Treating Triggers

Conditions underlying and triggering the lung injury need to be addressed as their treatment may calm down the pathways activated. As such, early and appropriate management of sepsis is crucial and may even avoid lung injury development [459]. Blood transfusions (II. hit) may potentially amplify an already activated inflammatory- and vascular endothelium-stimulating process [460]. Antibodies to leukocytes transfused by administering blood may result in neutrophil-mediated vascular endothelial injury and subsequent increased permeability and thus ALI/ARDS [460, 461]. Hence, a restrictive regimen is advisable [459]. The use of statins has shown promise in murine models, reducing leukocyte adherence [462], mitigating release of proinflammatory cytokines [463, 464] and attenuating lung injury and pulmonary vascular permeability [463, 464], and has been associated with an improved outcome [462, 463, 465]. Observational studies in humans seem to confirm a beneficial effect of statins with modulation or even impediment of ALI evolution [466–469].

2.6.5 Respiratory Modes

Although not explored in such big studies as volume-controlled protective mechanical ventilation, pressure-controlled ventilation (PCV) may be equally applied. Recommended are to set V_T between 4 and 8 ml/kg PBW and to limit plateau pressure (P_{PLAT}) to<30–35 cm H_2O [357, 470]. Moreover, detrimental transalveolar forces occurring with monotonous, fixed-flow air delivery during VCV are not seen in PCV [471]. Further advantages are reported, such as reduced respiratory work for the intermittent spontaneously breathing patient, higher patient comfort and synchrony with the ventilator [472, 473], lower peak inspiratory pressures and improved oxygenation resulting from higher mean airway pressures related to breaths of variable flows in the pressure-targeted mode compared to the VCV mode with its fixed flows [474–476]. However, if applying a decelerating flow wave pattern in contrast to square wave patterns as common in VCV, the assumed disadvantages of VCV may be overcome and counterbalanced [476, 477]. As already depicted in the chapter on mechanical ventilation, no significant and relevant differences in outcomes have been ever described [386, 478], although the comparison may be difficult sometimes since different strategies have been used [479].

BiPAP (biphasic positive pressure ventilation) is basically a pressure-controlled mode and may be applied instead of VCV or PCV as this form of ventilation allows for spontaneous breaths at any time during the respiratory cycle, making patient–ventilator interaction easier and much more comfortable while securing sufficient ventilation in a pressure-limited matter [480, 481]. BiPAP has become popular in the treatment of patients with hypoxaemic respiratory failure, not at least due to spontaneous breathing activity recruiting dependent lung tissues [482] and improving ventilation/perfusion ratio and thus oxygenation [483].

Airway pressure release ventilation (APRV) is, in principle, identical to BiPAP except that the high-pressure inspiration period may be considerably extended while the low-pressure expiration period is set very short [484]. This provides a non-physiological inspiration to expiration ratio [356, 485], also known as inversed or inverted ratio ventilation [486]. APRV may be particularly helpful if the patient has poor oxygenation, as the prolongation of the inspiration period with a high pressure level will potentially recruit collapsed, atelectatic parenchyma [487]. As typical for pressure-targeted modes, the V_T generated is a function of respiratory compliance, of airway resistance and in this case of the periodicity and duration of the pressure release phase [488]. The big advantage of APRV (and BiPAP) is the preservation of spontaneous breathing activity with consecutive preservation of respiratory muscle contraction, providing full ventilation [489]. Furthermore, spontaneous and particularly diaphragmatic breathing will contribute to recruitment of the dependent lung regions [482, 490]. As studies evaluating APRV depict, the application of this mode may lead to improved lung compliance and oxygenation, less sedative and vasopressor demand and shorter duration of mechanical ventilation and ICU stay [491, 492]. APRV may therefore be an alternative mode, with which to ventilate patients with ARDS [356].

2.6.6 Rescue Measures

Approximately 10–15 % of ARDS patients suffer from refractory, life-threatening hypoxaemia, and although this may be fatal, the vast majority of deaths are caused by multiorgan failure [70, 78]. Patients with refractory hypoxaemia may require rescue measures, including prone positioning, high-frequency oscillatory ventilation, extraordinary high PEEP or recruitment manoeuvres, heavy sedation and neuromuscular blockade or extracorporeal circuits [47, 70, 375, 493]. All these measures lack comprehensive evaluation and sufficient and resilient data and neither have provided proof of proper efficacy nor have been ever verified empirically [47, 493, 494].

Placing a patient in the prone position (proning) has, unfortunately, not been convincingly shown to significantly improve outcome [378, 419, 495–497]. Significant and sustained improvements in oxygenation have persistently been stated [316, 498, 499], but oxygenation has been a poor surrogate marker of survival. The absolute level of hypoxaemia obviously is an individual specific issue [10, 67, 493, 500].

Hypoxaemia persisting longer than 24 h bears a higher risk of mortality [18, 501]. A most recent meta-analysis only demonstrated benefits in subgroups, including more severely ill patients with severe hypoxaemia who have lower mortality rates if proned early on [316, 317]. Benefit was seen in patients with severe hypoxaemia and a FiO_2/PaO_2 ratio < 100 mmHg [316, 317]. Taking into account some potential adverse effects of proning such as an increased risk of tracheal pressure ulcers and endotracheal tube as well as intravenous line dislodgements, Sud prefers to use a cut-off of 100 mmHg as well, although post hoc analysis has shown a benefit in subgroups up to 140 mmHg [316]. This positive effect of proning was confirmed in a study-level meta-analysis by Abroug and co-workers [502] including roughly 1,600 patients, indicating that those with severe ARDS (assessed by FiO_2/PaO_2 ratio) benefitted from proning if applied for more than 17 h (up to 24 h) when compared to shorter periods of 6–8 h [496, 498]. This is in line with the strong physiological rationale to recruit dependent, collapsed supine areas and thus homogenize pleural pressure gradients and improve respiratory and chest wall compliance [233, 316]. Further, this also fits with the concept proposed by Gattinoni that severely diseased lungs have a higher amount of potentially recruitable lung tissues than less severely injured ones [304]. Interestingly, Abourg did not report significant adverse effects associated with proning [502]. A very recently (2013) published multicentre, randomized controlled clinical trial (PROSEVA study) [503] obviously adds more evidence that *early* (within 24 h since diagnosed and without any improvement on conventional ventilation) and *complete proning for at least 16 h* can be very valuable in ARDS patients suffering from severe hypoxaemia ($Pa:FiO_2 < 150$ mmHg on $FiO_2 \geq 0.6$ and a PEEP ≥ 5 cm H_2O) reducing significantly the mortality (28-day mortality rate in proning group 16.0 % versus 32.8 % in the conventionally ventilated group, 90-day mortality rate 23.6 and 41 %, respectively) if the centre is experienced (>5 years) with prone positioning of patients – thus confirming the findings and results reported above.

High-frequency ventilation (HFV) is widely used in the treatment of paediatric ALI/ARDS [504–507] despite lacking evidence of better outcomes. Usually, the high-frequency oscillatory ventilation (HFOV) method is applied [356, 504].

LPV aims to limit alveolar overdistension, to recruit collapsed, non-aerated tissues and to hinder cyclic recruitment/de-recruitment [504, 508]. HFV, utilizing respiratory rates up to 900 pm combined with the generation of only small tidal volumes between 1 and 4 (5) ml/kg PBW and constant lung recruitment fulfills these requirements, and may be a practical and promising alternative for adults in the future [508–510]. The small tidal volumes in ranges less than the anatomical dead space minimize swings in alveolar pressure [511, 512]; thus, a constant distension pressure [513] can be applied, recruiting lung tissue while concurrently reducing the risk of artificial lung injury (volu- and atelectrauma) [493, 511, 512]. Although in HFOV the mean pressure applied will usually be initially set 5 cm H_2O above that used in conventional ventilation prior to commencement [356], the often quite significant pressure swings in the proximal airways are markedly attenuated until they reach the periphery allowing for just small V_T's [514, 515]. Inspiration and expiration cannot be differentiated any longer [513], and ventilation happens in this technique due to oscillations, at 4–5 Hz in adults, of the whole gas column within the airways [513] and is directly related to the pressure amplitude of oscillation and inversely to the oscillatory frequency set [516]. Gas exchange occurs by unusual, non-convective flow mechanisms, such as the pendelluft effect or longitudinal dispersion [511, 516]. Remarkably for practical purposes, if a higher carbon dioxide release is necessary, a lower frequency has to be set whereas conventionally the respiratory rate would be increased. To improve oxygenation, the airway pressure applied needs to be increased. Other alternative ventilatory methods include high-frequency percussive ventilation (HFPV), high-frequency positive pressure ventilation and high-frequency jet ventilation [356].

HFOV is generally well tolerated, safe and effective in ARDS patients unsuccessfully treated by conventional ventilation. It improves oxygenation [517, 518] but does not compromise haemodynamics [356, 517, 519–521]. An early trial of HFOV in cases where conventional ventilation has failed is more likely to be advantageous [518, 519], and a failure to improve oxygenation is reported to be associated with poor outcome [356]. A noteworthy disadvantage of HFOV is that quite a substantial number of patients require sedation and may need to be paralyzed [522, 523].

Albeit beneficial effects are anecdotally seen when applying this technique, unfortunately, HFV in general has not displayed a convincing survival benefit as yet [356, 524–527]. A recent systematic review and meta-analysis by Sud and co-workers concludes that HFOV might improve survival and is unlikely to cause harm

[508]. However, two recently finished or even stopped (OSCILLATE) larger clinical trials, OSCAR [528] and OSCILLATE [529], did not find an improved outcome at 30 days (OSCAR), unfortunately even an increased in-hospital mortality (OSCILLATE) in populations of patients with moderate to severe ARDS being oscillated compared to those ventilated lung protective (PEEP was set by the ARDS network recommended FiO_2–PEEP requirement tableau). Several explanations for these disappointing results concomitantly contradicting and challenging the clear theoretical advantages of HFOV over conventional ventilator settings (HFOV potentially minimizes the risk for volu- and atelectrauma by applying very low tidal volumes while nonetheless allowing for higher mean airway pressures [530]) are under consideration [531]: Primarily suggested, the right heart function might have been meaningfully affected and compromised by the oscillator settings provoking RV dysfunction or even APC [531–533]. It is supposed that high levels of airway pressures might have occurred in the oscillated patient groups in both studies [531]. High airway pressures are known to potentially recruit lung tissue in the morphological and functional highly heterogeneous ARDS baby lungs as long as there is "recruitable" tissue [304, 534]. This lung recruitment will result in reduction in alveolar stress and strain, improvement of lung function (FRC) and concomitantly attenuation of RV afterload [535], the latter convenient for maintenance of a sufficient RV function [531]. However, if lung tissue cannot be recruited, the high airway pressure will not only enhance alveolar stress and strain but, due to an increase in PVR (and further impairment of venous return), may substantially compromise right heart function [294, 536]. Again, the artificial ventilation needs to be (right) heart protective, while coincidentally, RV dysfunction in the setting of mechanical ventilation may indicate poor respirator settings and will blunt the patient's prognosis [319].

Other possible factors contributing to the bad news of the studies are the amount of fluids administered (over-replenished), the necessary neuromuscular blockade and the probably inadequate timing of HFOV initiation [531, 537].

In the light of the new results, HFOV might not even be applied as a rescue measure in ARDS patients failing to properly respond to conventional ventilation as so far suggested by several authors [493, 508, 538].

So-called recruitment manoeuvres were quite popular in the 1990s following Lachmann's open lung concept [539, 540]. This technique aims to open and recruit collapsed areas of the lung by temporarily applying high pressure and/or by sustained inflation and then keeping them open by a suitable PEEP [493]. Various procedures have been reported, most commonly used have been sustained inflation, high pressure in a pressure-controlled mode, high incremental PEEP and high periodic hyperventilation, sigh volumes [541]. Meanwhile, sustained inflation should be abandoned [542] as it has been recognized as being ineffective [543, 544], associated with considerable risk of volu- and barotrauma [386], of transitory effect [545], and serious haemodynamic compromise may occur [546]. Preferred recruitment procedures include (1) incremental increase in PEEP but with an inspiratory pressure limit [547], (2) slow increase in inspiratory pressure to a maximum of 40 cm H_2O [548], (3) slow but continuous PEEP titration up to 15 cm H_2O

combined with end-inspiratory pauses lasting for 7 s twice a minute during a 15-min period [549] or (4) the application of intermittent sighs targeted to a specific P_{PLAT} [550]. However, no one method has demonstrated superiority over another, and recruitment manoeuvres show no convincing evidence of a favourable effect on outcome [356, 493, 551]. Even the effect of improved oxygenation is fleeting [551–553]. However, patients in life-threatening situations, as indicated by a poor FiO_2/PaO_2 ratio not responding to conventional measures and with a potential volume of non-aerated parenchyma, may benefit from an early recruitment manoeuvre [493, 541].

2.6.7 *Initial Ventilator Settings [160, 302, 368, 369, 375, 554, 555]*

LPV determined by ventilator settings comprising a V_T 6 ml/kg PBW, restriction of P_{PLAT} to <30 cm H_2O and suitable PEEP setting allowing for modest permissive hypercapnia is a grade 1 A recommendation [384].

- Mode: VCV, PCV, BiPAP and APRV.
- V_T *6 ml/kg PBW*, a grade 1 A recommendation [369, 556].
- $P_{PLAT} \leq$ *26 cm* H_2O, adopted due to substantial evidence [239, 244, 432, 433] (≤30 is a grade1 B recommendation [369]), if evidence of poor chest wall compliance, higher levels (up to 35 cm H_2O [438]) are acceptable. As outlined, we strongly recommend a $P_{PLAT} \leq 26$ cm H_2O (heart protective).
- PEEP *8–15 cm* H_2O [375, 379, 386] (5–12 cm H_2O [69, 416]), adopted from the FiO_2–PEEP Table 2.2.

  PEEP set may be even higher if individual cardiocirculatory function is considered, avoiding the development of RV dysfunction/ACP [319, 331, 370]. To set a PEEP at all is a grade B recommendation [369].

- Respiratory rate *12–20/min* (5–35/min [375]), avoidance of auto-PEEP development.
- I:E ratio *1:1*, range may be 1:2 or up to 9:1 [356].
- Decelerating flow pattern, at least if VCV mode is applied [476].
- $FiO_2 \leq 0.6$ after 24 h, initially 1.0 is acceptable.
- Rescue measures should be considered if no improvement occurs, multiorgan dysfunction develops or persistent FiO_2/PaO_2 ratio < 100 (140) mmHg [47, 316, 317, 493].

  Predicted (ideal) body weight (PBW) calculation [557, 558]:

Male: PBW = 50 kg + 2.3 [height (inches) – 60] kg or 50 kg + 0.91 [height (cm) – 152.4]kg

Female: PBW = 45.5 kg + 2.3 [height (inches) – 60] kg or 45.5 kg + 0.91 [height (cm) – 152.4]kg

Furthermore, implementation and application of standardized ventilation protocols following the recommendations are predicted to contribute to a better outcome [559, 560].

2.6.8 Conservative (Restrictive) Fluid Management

In recent years restrictive fluid management has been advocated [189] by some authors as being a valuable tool in ALI/ARDS treatment [47, 561, 562]. Pathophysiological data indicate that reducing pulmonary microvascular hydrostatic pressure by limited volume loading will reduce pulmonary oedema formation in patients with acute lung injury [190, 561–563]. In the setting of increased pulmonary vascular permeability, the main underlying pathophysiological feature in ALI/ARDS oedema generation, increasing pulmonary capillary hydrostatic pressure amplifies fluid transcapillary filtration and thus fluid losses into the interstitial and alveolar space [47, 359, 561, 564]. Hence, with decreasing hydrostatic microvascular pressures, the hydrostatic pressure conditional portion of the oedema in ARDS patients will be mitigated [47, 561, 562]. Of note, recent research work revealed that lower mircovascular pulmonary pressures resulting from lesser fluid volumes in the microvascular pulmonary network attenuate the generation and release of inflammatory mediators or display even an anti-inflammatory effect [159, 565].

Wiedemann [189] showed in a clinical trial of 1,000 patients that restriction of fluid administration in ARDS patients may yield beneficial effects in the conservative fluid group such as significantly less ventilator days (2.5 days), improved oxygenation and shorter ICU stay (2 days). The 7-day cumulative fluid balance in the conservative group was 135 ± 500 ml, while in the liberal fluid group, a highly positive balance of nearly 7,000 ± 500 ml was administered [189]. Unfortunately, there was only a trend towards a lower mortality rate in the group with restricted fluids [189]. The results of this basic study, however, confirmed the results by Mitchell from 1992 guiding the fluid management in critically ill patients by using pulmonary artery catheter [348].

In the study by Wiedemann [189], only haemodynamically stable patients were included [561]; however, in daily practice, quite a substantial number of patients are initially haemodynamically unstable with fluid loading being a necessary and valuable tool for stabilization [566], whereas fluid restriction may result in haemodynamic deterioration, multiorgan failure and worse outcome [78]. As will be stressed in the chapters on shock (Chap. 3) and sepsis (Chap. 4) and consistent with the results by Wiedemann, the cumulative fluid balance is important in the treatment of critically ill patients and ARDS patients. After approximately 4–7 days, a neutral to slightly negative cumulative balance is desirable but fluid resuscitation strategies may need to vary over time [566, 567]. Early in the clinical course, fluids may be life-saving, and frequently, ARDS is associated with severe sepsis or septic shock which is usually sensitive to fluids [561]. Potentially, a biphasic fluid strategy may be optimal, with a positive balance during the first two consecutive days [568] and then later a restrictive regimen achieving a cumulative neutral to mildly negative balance between days 4 to 7 [561]. However, more clinical trials are warranted to substantiate this approach. Liberal and conservative fluid strategies are complementary to each other, with each being correct at the right time. In haemodynamically stable ARDS patients, a well-managed fluid balance right from the beginning is advisable [189, 561, 562], and a recently published post hoc analysis of the large ARDS network trial published in 2000 [369] confirmed benefit from both a lower tidal volume ventilation and a conservative fluid therapy in all patients and severity groups (FiO_2/PaO_2) [569].

2.6.9 *Treatment of Acute Right Heart Dysfunction (RV-D)/ Acute RV Failure (RV-F) (With the Focus on ARDS)*

Pulmonary hypertension resulting from an increase in PVR, potentially causing right heart dysfunction or failure [569], may be a result of pulmonary vascular dysfunction attributed to several pathological features. These include pulmonary endothelial dysfunction, imbalanced vasoactive mediators, exaggerated hypoxic pulmonary vasoconstriction, altered microvascular permeability, microvascular thrombosis, disturbed pulmonary metabolism and vascular remodeling in progressive disease states [194, 245, 261, 570, 571].

Accordingly, the therapeutic approach comprises measures that affect the underlying pathology and aims to aid the cardiovascular system [245, 261]:

I. Treat underlying triggering factors like infections, anaemia, arrhythmias, comorbidities, pulmonary embolism and myocardial infarction. Avoid intubation and apply sufficient oxygen.
II. Optimize RV preload.
III. Reduce RV afterload.
IV. Strengthen RV systolic function/contractility.
V. Maintain perfusion pressure, guided by mean arterial pressure (MAP) [572].
VI. If therapy resistant, consider extracorporeal life support or lung transplantation.
VII. Miscellaneous.

2.6.9.1 Treat Underlying Triggering Factors

Measures to address the underlying condition or causative factors, thus preventing an increase in PVR or attenuating an elevated PVR affecting the RV function, such as LPV [244, 319, 331] should be applied. Examples are administration of airway dilating agents (β-agonists, ipratropium bromide) in obstructive airway diseases (COPD, asthma) [573–575], thrombolysis in haemodynamically relevant pulmonary embolism [286], antibiotics in patients suffering from pneumonia or sepsis [576], treatment of myocardial ischaemia (particularly if the right heart is affected) [577–579], control of tachyarrhythmias which may significantly affect atrial contraction and limit ventricular filling (and function) [288, 580] and, finally, identification of abdominal illnesses associated with abdominal hypertension affecting respiratory compliance with consecutively elevated intrathoracic and intrapulmonary pressures [581, 582].

2.6.9.2 Optimize RV Preload

Both, hypo- and hypervolaemia are associated with poor outcomes [583] as they may reduce cardiac output [584–586]. Physiologically, RV-EF is basically dependent on RV preload [587]. In contrast, with RV dilation, as typical in RV-D and notably in RV-F, following an acute rise in PVR with concurrent PH, volume loading above the physiological limit will just lead to more pronounced RV dilation with subsequent reduction in LV filling and LV-SV. This is mediated by interventricular interdependence and may seriously worsen systemic haemodynamics potentially resulting in cardiogenic shock [286, 588–590]. Furthermore, pulmonary hypertension puts the RV at risk of ischaemia [285, 591] resulting from increased oxygen demand attributed to PH, elevated filling pressure (RVEDP) related to RV dilation and (increasing) RV filling volume; this may be accompanied by compromised coronary perfusion due to low MAP further contributing to RV malfunction [272, 273, 279, 592].

In general, no volume loading and a negative fluid balance are the treatments of choice [261, 593–595], although in acute massive pulmonary embolism [253, 596] and in acute myocardial infarction with marked RV involvement [273], initial cautious fluid loading may be beneficial. However, if a fluid challenge is considered in the setting of acutely increased PVR as may occur in mechanical ventilation to stabilize the circulation and to optimize RV filling, volume loading under these conditions of potential RV-D/RV-F should never be performed without close monitoring [597, 598].

2.6.9.3 Reduce RV Afterload

RV unloading by reduction of PVR and thereby improving the disturbed/mismatched coupling between RV and pulmonary artery ameliorating cardiac function may be achieved by several measures. Firstly, a lung protective ventilation in ARDS patients significantly reduces the incidence of ACP, largely by avoidance of the development of pulmonary hypertension associated with high tidal volume mechanical ventilation and PEEP setting [260, 277, 319, 427]. LPV implies modest hypercapnia being allowed; however, substantial hypercapnia will cause pulmonary vasoconstriction and should be avoided [294]. Secondly, sufficient oxygen is applied to attenuate or even abolish hypoxic pulmonary vasoconstriction [261, 599]. Finally, selective pulmonary vasodilators affecting RV afterload and hypoxic vasoconstriction are used [245]. Inhaled formulas exert less effect on systemic circulation with less hypotension, and they provoke less ventilation–perfusion mismatch than intravenous preparations [245, 600]. In ARDS patients, inhaled prostacyclin [601, 602] and prostaglandin E1 [603] have shown to reduce PVR and to improve oxygenation; the systemic effect was minimal. Intravenous (IV) prostacyclin augmented RV function and diminished PVR but may facilitate intrapulmonary shunting [600]. A promising combination is IV prostaglandin I_2 and inhaled NO, substantially reducing PVR without relevant adverse effects [604]. Nitric oxide improves cardiac output and oxygenation and reduces PVR substantially in ARDS patients [149, 605, 606]. The inhalation needs to be continuous as a rebound effect may be seen and thus NO cessation must be gradual [607]. Despite a number of studies, no outcome benefit has yet been documented [608–610]. Phosphodiesterase inhibitors (PDEs) like sildenafil and vardenafil are able to lower PVR and increase CO [611, 612]; however, in ARDS patients so far, only a small cohort has been studied and with 50-mg oral sildenafil causing a better RV performance but inducing a drop in mean arterial pressure and oxygenation and promoted ventilation–perfusion mismatch [613].

2.6.9.4 Improve RV Systolic Function/Contractility

A compromised RV pump function is part of the definition of both RV-D and RV-F [245, 255, 256]. Positive inotropic agents such as dobutamine, PDE3 inhibitors (amrinone, milrinone) and levosimendan all strengthen the RV performance by improving the RV contractility and by reducing RV afterload [574, 614–617].

Dobutamine is still the agent of choice [556, 618–620], displaying positive inotropic and chronotropic effects [621]. It decreases sympathetic tone resulting in lower peripheral resistance [622], without a significant drop in MAP attributed to the compensatory increase in stroke volume (SV) [255]. In higher, greater than 10 μg/kg/min, doses (dose range 2.5–20 μg/kg/min), tachycardia, increases in PVR and systemic hypotension may occur [623]. Dobutamine is associated with an increased risk of arrhythmia [584], and tachyphylaxis has to be expected after a 72-h continuous infusion with doses above 10–15 μg/kg/min [624].

Phosphodiesterase (PDE) 3 inhibitors show positive inotropic, lusitropic and vasodilatory effects with reduction of the vascular resistance of the systemic and pulmonary vessel bed [625, 626]. Due to their site of action (via intracellular inhibition of type III phosphodiesterase thus increasing cardiac cyclic AMP concentration [627]), they may be administered even if the patient is on beta-blockers [628]. Unfortunately, there is growing evidence that phosphodiesterase inhibitors increase mortality and complications when compared with other treatment regimes [629–632].

Levosimendan has recently been shown to improve haemodynamics and RV efficiency [633], significantly improves systolic and diastolic RV function and RV–PA coupling [634–636] and reduces PVR particularly in ischaemic RV failure [637–640] and in ARDS [246]. The main impact probably comes from unloading of the RV by pulmonary vasodilation [246, 641]. As there is evidence that the beneficial effects exerted by levosimendan are of particular value for the right ventricle [637], further display anti-inflammatory properties [638, 642, 643] and, compared to dobutamine, turned out to be superior in some studies [634, 640], it may appear to be the preferable drug in predominantly RV failure. This is especially true if inotropic support is required in the setting of acute lung injury [634, 637, 643]. It is important for its safe use to be aware of its vasodilative properties on the systemic circulation; normovolaemia and sufficient blood pressure to guarantee suitable right coronary artery (RCA) perfusion are prerequisites before commencing levosimendan [644]. Avoidance of the loading dose and commencing with the continuous infusion of 0.05–0.2 μg/kg/min [645] may prevent a systemic pressure drop [644]; otherwise, combination with norepinephrine may be required [644, 646]. Levosimendan's action lasts several days, theoretically up to 7–10 days after completion of the advocated 24-h infusion [647], further a favourable effect on renal function with diuresis facilitation has been noted [648].

2.6.9.5 Maintain Perfusion Pressure

Preserving adequate MAP, organ perfusion and notably RV myocardial perfusion in this setting is crucial, as the combination of high PVR and RV ischaemia is the worst case scenario which potentially leads to circulatory shock and death [279, 649, 650]. Sufficient coronary perfusion is a prerequisite for good cardiac function [651]. In the case of compromised right coronary artery (RCA) perfusion as found in systemic hypotension and in the presence of increased demand [273, 279, 650], acute RV ischaemia will inevitably ensue, heralding the final step in the pathophysiological cascade of RV-F [255, 279, 649]. RV afterload and myocardial ischaemia are the decisive factors determining the pathophysiology of RV-D/RV-F [255, 273, 279, 285, 591].

Since vasopressors directly increase systemic blood pressure and thus improve the perfusion pressure of the RCA, timely and suitable administration of vasopressors is pivotal in the treatment of RV-D/RV-F [652–654]. Vlahakes [652] demonstrated that an increase in RCA perfusion pressure will directly increase the net perfusion of the myocardium, certainly for the right ventricle and probably for the LV as well.

Norepinephrine has become the vasopressor of choice [655]; as in hypotensive, life-threatening situations, it not only restores arterial pressure [653, 654, 656] but improves RV myocardial contractility as well [273]. Remarkably, RV-D/RV-F develops in up to 40 % of septic patients [657], and they are unable to increase SV and perfusion pressure by fluid loading alone [657, 658]. Accordingly, to prevent progression of right heart dysfunction, early and often simultaneous application of norepinephrine is required [657, 658].

Norepinephrine principally causes pulmonary vasoconstriction via α-1 receptors [659, 660]; however, this effect probably occurs only at higher doses, since from animal studies it is known that pulmonary vascular dysfunction ensues at norepinephrine doses above 0.5 μg/kg/min [641]. If higher doses are necessary to achieve an adequate MAP (>70–80 mmHg), dobutamine due to its vasodilative effects might be added to counteract the pulmonary vasoconstrictive actions of norepinephrine [614, 641, 661].

2.6.10 *Extracorporeal Techniques*

Extracorporeal techniques, especially membrane oxygenation, is one of the rescue options applied if conservative treatment regimes are not successful. It is employed in patients with severe respiratory failure (ARDS) in order to minimize the injury caused by mechanical ventilation (mostly accompanied by right heart failure) and for bridging in those conditions prior to lung transplantation [662, 663]. However, until recently, the results for this indication (therapy-resistant respiratory failure with and without consecutive ACP) have not been encouraging with a high and unaltered mortality rate still and no demonstrated benefit at all in adults [664–666]. New technologies using veno-venous access are promising as the CESAR study revealed [667], but the indications remain controversial [668–670]. Indeed, this and other procedures are concentrated to only a few centres, and further explanations are beyond the scope of this book.

2.6.11 *Miscellaneous*

(a) AV sequential stimulation. In order to optimize RV filling, maintaining and even improving RV function, AV synchronous stimulation is essential [671, 672]. Therefore, sustaining or restoring sinus rhythm by chemical or electrical cardioversion or temporary pacing is pivotal as a normal atrial function is essential to optimize RV and LV filling [671].

(b) In pulmonary hypertension, hypercoagulation in the pulmonary vascular tree will always be present [673, 674] and the development of microthrombi is highly likely [675]. Accordingly, prophylactic anticoagulation is necessary [674, 676] as these critically ill patients are already at increased risk for development of thromboembolic complications [677–679].

(c) Diuretics are clearly indicated in volume-overloaded patients who have a dilated RV with leftward septal shift [618]. Diuretics may induce metabolic alkalosis potentially resulting in aggravation of hypoventilation and hypercapnia [680]. Moderate peripheral oedema should be tolerated [681].

(d) Oxygen therapy, regardless of the underlying illness, is indicated in PH [599] as it will reduce hypoxaemic pulmonary vasoconstriction unloading the RV [682, 683]. Continuous application of oxygen has been shown to reduce mortality [683].

References

1. Bernard GR, Artigas A, Brigham KL. The American-European consensus conference on ARDS. Definitions, mechanisms, relevant outcomes, and clinical trial coordination. Am J Respir Crit Care Med. 1994;149:818–24.

2. Ware LB, Matthay MA. Medical progress: the acute respiratory distress syndrome. N Engl J Med. 2000;342:1334–49.
3. Asbaugh DG, Bigelow DB, Petty TL. Acute respiratory distress in adults. Lancet. 1967;2: 319–23.
4. Ware LB. Pathophysiology of acute lung injury and the acute respiratory distress syndrome. Semin Respir Crit Care Med. 2006;27:337–49.
5. Murray JF, Matthay MA, Luce JM. An expanded definition of the adult respiratory distress syndrome. Am Rev Respir Dis. 1988;138:720–3.
6. Bellingan GJ. The pulmonary physician in critical care. 6: the pathogenesis of ALI/ARDS. Thorax. 2002;57:540–6.
7. Hansen-Flaschen J, Siegel MD, Parsons PE. Acute respiratory distress syndrome: definition; clinical features; and diagnosis. UpToDate. 2012, http://www.uptodate.com/contents/acute-respiratory-distress-syndrome-clinical-features-and-diagnosis?source=search_result&search=ards&selectedTitle=1~150
8. Schuster DP. What is acute lung injury? What is ARDS? Chest. 1995;107:1721–6.
9. Flori HR. Towards a more accurate definition of acute lung injury and the acute respiratory distress syndrome in pediatric patients. Pediatr Crit Care Med. 2006;7:393–4.
10. Wheeler AP, Bernard GR. Acute lung injury and acute respiratory distress syndrome: a clinical review. Lancet. 2007;369:1553–64.
11. Estenssoro E, Dubin A, Laffaire E. Impact of positive end-expiratory pressure on the definition of acute respiratory distress syndrome. Intensive Care Med. 2003;29:1936–42.
12. Ferguson ND, Davis AM, Slutsky AS. Development of a clinical definition for acute respiratory distress syndrome using the Delphi technique. J Crit Care. 2005;20:147–54.
13. Phua J, Stewart TE, Ferguson ND. Acute respiratory distress syndrome 40 years later: time to revisit definition. Crit Care Med. 2008;36:2912–21.
14. Aboab J, Louis B, Jonson J. Relation between PaO2/FiO2 ration and FiO2: a mathematical description. Intensive Care Med. 2006;32:1494–7.
15. Doyle RL, Szaflarsky N, Modin GW. Identification of patients with acute lung injury. Predictors of mortality. Am J Respir Crit Care Med. 1995;152:1818.
16. Rubenfeld GD, Caldwell E, Peabody E. Incidence and outcomes of acute lung injury. N Engl J Med. 2005;353:1685.
17. Atabai K, Matthay MA. The pulmonary physician in critical care. 5: acute lung injury and the acute respiratory distress syndrome: definitions and epidemiology. Thorax. 2002;57:452–8.
18. Villar J, Mendez-Perez L, Lopez J. An early PEEP/FiO2 trial identifies different degrees of lung injury in ARDS patients. Am J Respir Crit Care Med. 2007;176:795–804.
19. Michard F. Bedside assessment of extravascular lung water by dilution methods: temptations and pitfalls. Crit Care Med. 2007;35:1186–92.
20. Wheeler AP, Bernard GR, Thompson BT. Pulmonary-artery versus central venous catheter to guide treatment of acute lung injury. N Engl J Med. 2006;354:2213–24.
21. Ferguson ND, Meade MO, Hallet DC. High values of the pulmonary artery wedge pressure in patients with acute lung injury and acute respiratory distress syndrome. Int Care Med. 2002;28:1073–7.
22. Furian T, Aguiar C, Prado K. Ventricular dysfunction and dilation in severe sepsis and septic shock: relation to endothelial function and mortality. J Crit Care. 2012;27:319.e9–e15.
23. Phillips CR. The Berlin definition: real change or the emperor's new clothes? Crit Care. 2013;17:174.
24. Ferguson ND, Meade MO, Hallett DC. High values of the pulmonary artery wedge pressure in patients with acute lung injury and acute respiratory distress syndrome. Intensive Care Med. 2002;28:1073–7.
25. Price LC, McAuley DF, Marino PS. Pathophysiology of pulmonary hypertension in acute lung injury. Am J Physiol Lung Cell Mol Physiol. 2012;302:L803–15.
26. Anderson WR, Thielen K. Correlative study of the adult respiratory distress syndrome by light scanning, and transmission electron microscopy. Ultrastruct Pathol. 1992;16: 615–28.

27. Force, ARDS Definition Task, Ranieri VM, Rubenfeld GD, Thompson BT. Acute respiratory distress syndrome: the Berlin definition. JAMA. 2012;307:2526–33.
28. Mangialardi RJ, Martin GS, Bernard GR. Hypoproteinemia predicts acute respiratory distress syndrome development, weight gain, and death in patients with sepsis. Ibuprofen in Sepsis. Crit Care Med. 2000;28:3137–45.
29. Martin GS, Mangialardi RJ, Wheeler AP. Albumin and furosemide therapy in hypoproteinemic patients with acute lung injury. Crit Care Med. 2002;30:275–2182.
30. Brown LM, Liu KD, Matthay MA. Measurement of extravascular lung water using the single indicator method in patients: research and potential clinical value. Am J Physiol Lung Cell Mol Physiol. 2009;297:L547–58.
31. Calfee CS, Matthay MA. Nonventilatory treatments for acute lung injury and ARDS. Chest. 2007;131:913–20.
32. Angus DC. The respiratory distress syndrome: what's in a name? JAMA. 2012;307:2542–4.
33. Rubenfeld GD, Caldwell E, Granton J. Interobserver variability in applying a radiographic definition for ARDS. Chest. 1999;116:1347–53.
34. Meade MO, Cook RJ, Guyatt GH. Interobserver variation in interpreting chest radiographs for the diagnosis of acute respiratory distress syndrome. Am J Respir Crit Care Med. 2000;161:85–90.
35. Katzenstein A, Askin F. Surgical pathology of non-neoplastic lung disease. Philadelphia: WB Saunders; 1990.
36. Crouch E. Pathobiology of pulmonary fibrosis. Am J Physiol. 1990;3:L159–84.
37. Kao K-C, Tsai Y-H, Wu Y-K. Open lung biopsy in early-stage acute respiratory distress syndrome. Crit Care. 2006;10:R106.
38. Esteban A, Fernandez-Segoviano P, Frutos-Vivar F. Comparison of clinical criteria for the acute respiratory distress syndrome with autopsy findings. Ann Intern Med. 2004; 141:440–5.
39. Ferguson ND, Frutos-Vivar F, Esteban A. Acute respiratory distress syndrome: underrecognition by clinicians and diagnostic accuracy of three clinical definitions. Crit Care Med. 2005;33:2228–34.
40. Chew MS. Extravascular lung water in acute respiratory distress syndrome and the Berlin definition: time for a real change. Crit Care. 2013;17:463.
41. Thompson BT, Matthay MA. The Berlin definition of ARDS versus pathological evidence of diffuse alveolar damage. Am J Respir Crit Care Med. 2013;187:675–7.
42. Kushimoto S, Taira Y, Kitzazawa Y. The clinical usefulness of extravascular lung water and pulmonary vascular permeability index to diagnose and characterize pulmonary edema: a prospective multicenter study on the quantitative differential diagnostic definition for acute lung injury/acute. Crit Care. 2012;16:R232.
43. Abraham E, Matthay MA, Dinarello CA. Consensus conference definitions for sepsis, septic shock, acute lung injury, and acute respiratory distress syndrome: time for reevaluation. Crit Care Med. 2000;28:232–5.
44. Jozwiak M, Silva S, Persichini R. Extravascular lung water is an independent prognostic factor in patients with acute respiratory distress syndrome. Crit Care Med. 2013;41:472–80.
45. Monnet X, Anguel N, Osman D. Assessing pulmonary permeability by transpulmonary thermodilution allows differentiation of hydrostatic pulmonary oedema from ALI/ARDS. Intensive Care Med. 2007;33:448–53.
46. Chung FT, Lin HC, Kuo CH. Extravascular lung water correlates with multiorgan dysfunction and mortality in sepsis. PLoS One. 2010;5:e15265.
47. Matthay MA, Ware LB, Zimmerman GA. The acute respiratory distress syndrome. J Clin Invest. 2012;122:2731–40.
48. Opitz B, van Laak V, Eitel J. Innate immune recognition in infectious and noninfectious diseases of the lung. Am J Respir Crit Care Med. 2010;181:1294–309.
49. Matthay MA, Zimmerman GA. Acute lung injury and the acute respiratory distress syndrome. Four decades of inquiry into pathogenesis and rational management. Am J Respir Cell Mol Biol. 2005;33:319–27.

50. Bachofen A, Weibel ER. Alterations of the gas exchange apparatus in the adult respiratory insufficiency associated with septicemia. Am Rev Respir Dis. 1977;116:589–615.
51. Baudendistel LJ, Kaminski DL, Dahms TE. Evaluation of extravascular lung water by single thermal indicator. Crit Care Med. 1986;14:52–6.
52. Isakow W, Schuster DP. Extravascular lung water measurements and hemodynamic monitoring in the critically ill: bedside alternatives to pulmonary artery catheter. Am J Physiol Lung Cell Mol Physiol. 2006;291:L1118–31.
53. Fernandez-Mondejar E, Rivera-Fernandez R, Garcia-Delgado M. Small increases in extravascular lung water are accurately detected by transpulmonary thermodilution. J Trauma. 2005;59:1420–3; discussion 1424.
54. Fernandez-Mondejar E, Castano-Perez J, Rivera-Fernandez R. Quantification of lung water by transpulmonary thermodilution in normal and edematous lungs. J Crit Care. 2003;18:253–8.
55. Katzenelson R, Perel A, Berkenstadt H. Accuracy of transpulmonary thermodilution versus gravimetric measurement of extravascular lung water. Crit Care Med. 2004;32:1550–4.
56. Schuster D. The search for "objective" criteria of ARDS. Intensive Care Med. 2007;33:400–2.
57. Chew MS, Ihtman L, During J. Extravascular lung water index improves the diagnostic accuracy of lung injury in patients with shock. Crit Care. 2012;16:R1.
58. Berkowitz DM, Danai PA, Eaton S. Accurate characterization of extravascular lung water in acute respiratory distress syndrome. Crit Care Med. 2008;36:1803–9.
59. Mallat M. Is extravascular lung water index useful for the diagnostic accuracy of lung injury in patients with shock? We need more evidence. Crit Care. 2012;16:420–2.
60. Camporota L, De Neef M, Beale R. Extrapulmonary lung water in acute respiratory distress syndrome: potential clinical value, assumptions and limitations. Crit Care. 2012;16:114–5.
61. Sakka G. Prognostic value of extrapulmonary water in critically ill patients. Chest. 2002;122:2080–6.
62. Phillips CR, Chesnutt MS, Smith SM. Extravascular lung water in sepsis-associated acute respiratory distress syndrome: indexing with predicted body weight improves correlation with severity of illness and survival. Crit Care Med. 2008;36:69–73.
63. Craig TR, Duffy MJ, Shyamsundar M. Extravascular lung water indexed to predicted body weight is a novel predictor of intensive care unit mortality in patients with acute lung injury. Crit Care Med. 2010;38:114–20.
64. Deeks JJ, Altman DG. Diagnostic tests 4: likelihood ratios. BMJ. 2004;329:168–9.
65. Luhr OR, Antonson K, Karlsson M. Incidence and mortality after acute respiratory failure and acute respiratory distress syndrome in Sweden, Denmark, and Iceland. The ARF Study Group. Am J Respir Crit Care Med. 1999;159:1849–61.
66. Pilbeam S, Cairo JM. Mechanical ventilation. St. Louis: Mosby/Elsevier; 2006.
67. Frutos-Vivar F, Nin N, Esteban A. Epidemiology of acute lung injury and acute respiratory distress. Curr Opin Crit Care. 2004;10:1.
68. Estenssoro E, Dubin A, Laffaire E. Incidence, clinical course, and outcome in 217 patients with acute respiratory distress syndrome. Crit Care Med. 2002;30:2450.
69. Esteban A, Anzueto A, Frutos-Vivar F. Characteristics and outcomes in adult patients receiving mechanical ventilation: a 28-day international study. JAMA. 2002;287:345.
70. Tsushima K, King LS, Aggarwal NR. Acute lung injury review. Intern Med. 2009;48:621–30.
71. Hudson LD, Milberg JA, Anardi D. Clinical risks for development of the acute respiratory distress syndrome. Am J Respir Crit Care Med. 1995;151:293.
72. Pepe PE, Potkin TR, Reus DH. Clinical predictors of the adult respiratory distress syndrome. Am J Surg. 1982;144:124–30.
73. Siegel MD, Parsons PE. Acute respiratory distress syndrome: epidemiology; pathophysiology; pathology; and epidemiology. UpToDate. 2012. http://www.uptodate.com/contents/acute-respiratory-distress-syndrome-epidemiology-pathophysiology-pathology-and-etiology?source=search_result&search=acute+respiratory+distress+syndrome&selectedTitle=2~150
74. Network, ARDS. Ventilation with lower tidal volumes as compared to traditional tidal volumes for acute lung injury and the acute respiratory distress syndrome. N Engl J Med. 2000;342:1301–8.

75. Zilberberg MD, Epstein SK. Acute lung injury in the medical ICU: comorbid conditions, age etiology, and hospital outcome. Am J Respir Crit Care Med. 1998;157:1159–64.
76. Sloane PJ, Gee MH, Gottlieb JE. A multicenter registry of patients with acute respiratory distress syndrome: physiology and outcome. Am Rev Respir Dis. 1992;146:419–26.
77. Milberg JA, Davis DR, Steinberg KP. Improved survival of patients with acute respiratory distress syndrome (ARDS): 1983–1993. JAMA. 1995;273:306–9.
78. Stapleton RD, Wang BM, Hudson LD. Causes and timing of death in patients with ARDS. Chest. 2005;128:525–32.
79. Erikson SE, Martin GS, Davis JL. Recent trends in acute lung injury mortality: 1996–2005. Crit Care Med. 2009;37:1574.
80. Zambon M, Vincent JL. Mortality rates for patients with acute lung injury/ARDS have decreased over time. Chest. 2008;133:1120.
81. National Heart, Lung, and Blood Institute Acute Respiratory Distress Syndrome (ARDS) Clinical Trail Network. Efficacy and safety of corticosteroids for persistent acute respiratory distress syndrome. N Engl J Med. 2006;354:1671–84.
82. Monchi M, Bellenfant F, Cariou A. Early predictive factors of survival in the acute respiratory distress syndrome. A multivariate analysis. Am J Respir Crit Care Med. 1998;158:1076–81.
83. Ferring M, Vincent JL. Is outcome from ARDS related to the severity of respiratory failure? Eur Respir J. 1997;10:1297–300.
84. Bersten AD, Edibam C, Hunt T. Incidence and mortality of acute lung injury and the acute respiratory distress syndrome in three Australian states. Am J Respir Crit Care Med. 2002;165:443.
85. Roupie E, Lepage E, Wyocki M. Prevalence, etiologies and outcome of the acute respiratory distress syndrome among hypoxemic ventilated patients. SRLF Collaborative Group on Mechanical Ventilation. Intensive Care Med. 1999;25:920–9.
86. Ely WE, Wheeler AP, Thompson BT. Recovery rate and prognosis in older persons who develop lung injury and the acute respiratory distress syndrome. Ann Intern Med. 2002;136:25–36.
87. Neff TA, Stocker R, Frey HR. Long-term assessment of lung function in survivors of severe ARDS. Chest. 2003;123:845.
88. Cheung AM, Tansey CM, Tomlinson G. Two-year outcomes, health care use, and costs of survivors of acute respiratory distress syndrome. Am J Respir Crit Care Med. 2006;174:538.
89. Herridge MS, Tansey CM, Matte A. Functional disability 5 years after acute respiratory distress syndrome. N Engl J Med. 2011;364:1293.
90. Bion J. Pathophysiology of acute respiratory distress syndrome and acute lung injury. In: Webb A, Shapiro MJ, Singer M, Suter PM, editors. Oxford textbook of critical care. Oxford: Oxford University Press; 1999. p. 923–6.
91. Matthay MA, Zimmermann GA, Esmon C. Future research directions in acute lung injury: summary of a National Heart, Lung and Blood Institute working group. Am J Respir Crit Care Med. 2003;167:1027–35.
92. Gattinoni L, Pelosi P, Sutter PM. Acute respiratory distress syndrome caused by pulmonary and extrapulmonary disease: different syndromes? Am J Respir Crit Care Med. 1998;158:3–11.
93. Tugrul S, Akinci O, Ozcan PE. Effects of sustained inflation and postinflation positive expiratory pressure in acute respiratory distress syndrome: focusing on pulmonary and extrapulmonary forms. Crit Care Med. 2003;31:738.
94. Rocco PR, Zin WA. Pulmonary and extrapulmonary acute respiratory distress syndrome: are they different? Curr Opin Crit Care. 2005;11:10.
95. Villar J, Blanco J, Anon JM. The ALIEN study: incidence and outcome of acute respiratory distress syndrome in the era of lung protective ventilation. Intensive Care Med. 2011;37:1932.
96. Zaccardelli DS, Pattishall EN. Clinical diagnostic criteria of the adult respiratory distress syndrome in the intensive care unit. Crit Care Med. 1996;24:247.
97. Moss M, Bucher B, Moore FA. The role of chronic alcohol abuse in the development of acute respiratory distress syndrome in adults. JAMA. 1996;275:50–4.
98. Anzueto A, Frutos-Vivar F, Esteban A. Influence of body mass index on outcome of the mechanically ventilated patients. Thorax. 2011;66:66.

99. Gong MN, Bajwa EK, Thompson BT. Body mass index is associated with the development of acute respiratory distress syndrome. Thorax. 2010;65:44.
100. Irbarren C, Jacobs Jr DR, Sidney S. Cigarette smoking, alcohol consumption, and risk of ARDS: a 15-year cohort in a managed care setting. Chest. 2000;117:163.
101. Messent M, Sullivan K, Keogh B. Adult respiratory distress syndrome following cardiopulmonary bypass: incidence and prediction. Anaesthesia. 1992;47:267.
102. Dulu A, Pastores SM, Park B. Prevalence and mortality of acute lung injury and ARDS after lung resection. Chest. 2006;130:73.
103. De Campos T, Deree J, Coimbra R. From acute pancreatitis to end-organ injury: mechanisms of acute lung injury. Surg Infect (Larchmt). 2007;8:107.
104. Cohen D, Matthay M, Cogan M. Pulmonary edema associated with salt water near-drowning: new insights. Am Rev Respir Dis. 1992;146:794.
105. Marshall RP, Webb S, Hill MR. Genetic polymorphisms associated with susceptibility and outcome in ARDS. Chest. 2002;121:68–9S.
106. Greene KE, Wright JR, Steinberg KP. Serial changes in surfactant associated proteins in lung and serum before and after onset of ARDS. Am J Respir Crit Care Med. 1999;160:1843–50.
107. Gong MN, Wei Z, Xu LL. Polymorphism in the surfactant protein-B gene, gender, and the risk of direct pulmonary injury and ARDS. Chest. 2004;125:203.
108. Lin Z, Pearson C, Chinchilli V. Polymorphisms of human SP-A, SP-B, SP-D genes: association of SP-B Thr131Ile with ARDS. Clin Genet. 2000;58:181.
109. Fowler A, Hamman R, Good J. Adult respiratory distress syndrome: risk with common predisposition. Ann Intern Med. 1983;98:593.
110. Hyers TM. Prediction of survival and mortality in patients with the adult respiratory distress syndrome. New Horiz. 1993;1:466–70.
111. Fein A, Lippman M, Holtzman H. The risk factors, incidence and prognosis of the adult respiratory distress syndrome following septicemia. Chest. 1983;83:40.
112. Moss M, Pearsons PE, Steinberg KP. Chronic alcohol abuse is associated with an increased incidence of acute respiratory distress syndrome and severity of multiple organ dysfunction in patients with septic shock. Crit Care Med. 2003;31:869–77.
113. Tietjen PA, Kaner RJ, Quinn CE. Aspiration emergencies. Clin Chest Med. 1994;15:117.
114. Demlinh R. Current concepts on the adult respiratory distress syndrome. Circ Shock. 1990;30:297.
115. Moore FA, Moore EE, Read RA. Postinjury multiple organ failure: role of extrathoracic injury and sepsis in adult respiratory distress syndrome. New Horiz. 1993;1:538.
116. Calfee CS, Eisner MD, Ware LB. Trauma-associated lung injury differs clinically and biologically from acute lung injury due to other clinical disorders. Crit Care Med. 2007;35:2243.
117. Sutyak JP, Wohltmann CD, Larson J. Pulmonary contusions and critical care management in thoracic trauma. Thorac Surg Clin. 2007;17:11.
118. Schonfeld SA, Ploysongsang Y, DiLisio R. Fat embolism prophylaxis with corticosteroids. Ann Intern Med. 1983;99:438.
119. Looney MR, Gropper MA, Matthay MA. Transfusion-related acute lung injury: a review. Chest. 2004;126:249–58.
120. Kopko PM, Marshall CS, MacKenzie MR. Transfusion-related acute lung injury: report of clinical look-back investigation. JAMA. 2002;287:1968–71.
121. Popovsky MA, Moore SB. Diagnostic and pathogenetic considerations in transfusion-related acute lung injury. Transfusion. 1985;25:573–7.
122. Gong MN, Thompson BT, Williams P. Clinical predictors of and mortality in acute respiratory distress syndrome: potential role of red blood cell transfusion. Crit Care Med. 2005;33:1191.
123. Parsons P. Respiratory failure as a result of drugs, overdoses, and poisonings. Clin Chest Med. 1994;15:93.
124. Borish L, Matloff S, Findlay S. Radiologic contrast media-induced non-cardiogenic pulmonary edema: case report and review of the literature. J Allergy Clin Immunol. 1984;74:104.
125. Tomashefski Jr JF. Pulmonary pathology of the adult respiratory distress syndrome. Clin Chest Med. 1990;11:593–619.

126. Piantadosi CA, Schwartz DA. The acute respiratory distress syndrome. Ann Intern Med. 2004;141:460.
127. Liu M. Searching for acute respiratory distress syndrome genes: aren't we there yet? Am J Respir Crit Care Med. 2005;171:298–9.
128. dos Santos CC, Okutani DD, Hu P. Differential gene profiling in acute lung injury identifies injury specific gene expression. Crit Care Med. 2008;36:855–65.
129. Ware LB, Matthay MA. Clinical practice: acute pulmonary edema. N Engl J Med. 2005;353:2788–96.
130. Nathan CF. Secretory products of macrophages. J Clin Invest. 1987;79:319–26.
131. Orfanos SE, Mavrommati I, Korovesi I. Pulmonary endothelium in acute lung injury: from basic science to the critically ill. Intensive Care Med. 2004;30:1702–14.
132. Wort SJ, Evans TW. The role of the endothelium in modulating vascular control in sepsis and related conditions. Br Med Bull. 1999;55:30–48.
133. Tomashefski Jr JF. Pulmonary pathology of acute respiratory distress syndrome. Clin Chest Med. 2000;21:435–66.
134. Tomashefski Jr JR, Davies P, Boggis C. The pulmonary vascular lesions of the adult respiratory distress syndrome. Am J Pathol. 1983;112:112–26.
135. Zimmerman GA, Albertine KH, McIntyre TM. Pathogenesis of sepsis and septic-induced lung injury. In: Matthay MA, Lefant C, editors. Acute respiratory distress syndrome. New York: Marcel Dekker; 2003. p. 245–87.
136. Ward PA, Hunninghake GW. Lung inflammation and fibrosis. Am J Respir Crit Care Med. 1998;157:S123–9.
137. Hurley JV. Types of pulmonary microvascular injury. Ann N Y Acad Sci. 1982;384:269–86.
138. Martin TR. Lung cytokines and ARDS: Roger S. Mitchell Lecture. Chest. 1999;116:2S.
139. Donnelly SC, Strieter RM, Reid PT. The association between mortality rates and decreased concentrations of interleukin – 10 and interleukin-1 receptor antagonist in the lung fluids of patients with the adult respiratory distress syndrome. Ann Intern Med. 1996;125:191.
140. Andrews P, Azoulay E, Antonelli M. Year in review in Intensive Care Medicine, 2006. II. Infections and sepsis, hemodynamics, elderly, invasive and noninvasive mechanical ventilation, weaning, ARDS. Intensive Care Med. 2007;33:214–29.
141. Fagan KA, McMurtry IF, Rodman DM. Role of endothelin-1 in lung disease. Respir Res. 2001;2:90–101.
142. Myers PR, Parker JL, Tanner MA. Effects of cytokines, tumor necrosis factor alpha and interleukin-1- beta on endotoxin-mediated inhibition of endothelium-derived relaxing factor bioactivity and nitric oxide production in vascular endothelium. Shock. 1994;1:73–8.
143. Rossi P, Persson B, Boels PJ. Endotoxemic pulmonary hypertension is largely mediated by endothelin-induced venous constriction. Intensive Care Med. 2008;34:873–80.
144. Bone RC, Francis PB, Pierce AK. Intravascular coagulation associated with the adult respiratory distress syndrome. Am J Med. 1976;61:585–91.
145. Idell S, Koenig KB, Fair DS. Serial abnormalities of fibrin turnover in evolving adult respiratory distress syndrome. Am J Physiol. 1991;261:L240–8.
146. Saldeen T. Blood coagulation and shock. Pathol Res Pract. 1979;165:221–52.
147. Zimmerman GA, Albertine KH, Carveth HJ. Endothelial activation in ARDS. Chest. 1999;116:18S–24.
148. Ley K, Laudanna C, Cybulsky MI. Getting to the site of inflammation: the leukocyte adhesion cascade updated. Nat Rev Immunol. 2007;7:678–89.
149. Hsu CW, Lee DL, Lin SL. The initial response to inhaled nitric oxide treatment for intensive care unit patients with acute respiratory distress syndrome. Respiration. 2008;75:288–95.
150. Calandrino Jr FS, Anderson DJ, Mintun MA. Pulmonary vascular permeability during the adult respiratory distress syndrome: a positron emission tomographic study. Am Rev Respir Dis. 1988;138:421.
151. Matthay MA, Wiener-Kronish JP. Intact epithelial barrier function is critical for the resolution of alveolar edema in humans. Am Rev Respir Dis. 1990;142:1250–7.
152. Sprung C, Rachow E, Fein I. The spectrum of pulmonary edema: differentiation of cardiogenic intermediate and non-cardiogenic forms of pulmonary edema. Am Rev Respir Dis. 1981;124:718–22.

153. Fein A, Grossman RF, Jones JG. The value of edema protein measurements in patients with pulmonary edema. Am J Med. 1979;67:32–9.
154. Raijmakers PGHM, Groeneveld ABJ, Teule GJJ. The diagnostic value of the 67-gallium pulmonary leak index in pulmonary edema. J Nucl Med. 1996;37:1316–22.
155. Goligher EC, Fan E, Slutsky AS. Year in review 2012: critical care – respirology. Crit Care. 2013;17:249.
156. Gullo A, Berlot G, Vivani M. The role of adult respiratory distress syndrome in the multiple organ dysfunction syndrome. Acta Anaesthesiol Scand. 1996;109(Suppl):70–3.
157. Heuer JF, Sauter P, Pelosi P. Effects of pulmonary acid aspiration on the lungs and extra-pulmonary organs: a randomized study in pigs. Crit Care. 2012;16:R35.
158. Frank JA, Gutierrez JA, Jones K. Low tidal volume reduces epithelial and endothelial injury in acid-induced rat lungs. Am J Respir Crit Care Med. 2002;165:2492–9.
159. Kuebler WM, Ying X, Singh B. Pressure is proinflammatory in lung venular capillaries. J Clin Invest. 1999;104:495–502.
160. Ranieri VM, Suter PM, Tortadella C. Effect of mechanical ventilation on inflammatory mediators in patients with acute respiratory distress syndrome: a randomized controlled trial. JAMA. 1999;282:54–61.
161. Uhlig S, Ranieri VM, Slutsky AS. Biotrauma hypothesis of ventilator-induced lung injury. Am J Respir Crit Care Med. 2004;69:314–5.
162. Goodman R, Pugin J, Lee JJS. Cytokine mediated inflammation in acute lung injury. Cytokine Growth Factor Rev. 2003;14:523–35.
163. Bellingan GGJ. Leukocyte: friend or foe? Intensive Care Med. 2000;26:111–8.
164. Chollet-Martin S, Jourdain B, Gibert C. Interactions between neutrophils and cytokines in blood and alveolar spaces during ARDS. Am J Respir Crit Care Med. 1996;154:594–601.
165. Martin TR. Cytokines and the acute respiratory distress syndrome (ARDS): a question of balance. Nat Med. 1997;3:272–3.
166. Gadek JE, Pacht ER. The interdependence of lung antioxidants and antiprotease defense in ARDS. Chest. 1996;110:273S–7.
167. Ward NS, Casserly B, Ayala A. The compensatory anti-inflammatory response syndrome (CARS) in critically ill patients. Clin Chest Med. 2008;29:617–viii.
168. Weiland JE, Davis B, Holter JF. Lung neutrophils in the adult respiratory distress syndrome: clinical and pathophysiologic significance. Am Rev Respir Dis. 1986;133:218–25.
169. Fowler AA, Walchak S, Giclas PC. Characterization of antiprotease activity in the adult respiratory distress syndrome. Chest. 1982;81:50S–1.
170. Idell S, Kucich U, Fein A. Neutrophil elastase releasing factors in bronchoalveolar lavage from patients with adult respiratory distress syndrome. Am Rev Respir Dis. 1985;132:1098–105.
171. Bernard GR, Reines HD, Halushka PV. Prostacyclin and thromboxane A2 formation is increased in human sepsis syndrome. Effects of cyclooxygenase inhibition. Am Rev Respir Dis. 1991;144:1095–101.
172. Brett SJ, Evans TW. Measurement of endogenous nitric oxide in the lungs of patients with the acute respiratory distress syndrome. Am J Respir Crit Care Med. 1998;157:993–7.
173. Druml W, Steltzer H, Waldhausl W. Endothelin-1 in the adult respiratory distress syndrome. Am Rev Respir Dis. 1993;148:1169–73.
174. Giaid A, Saleh D. Reduced expression of endothelial nitric oxide synthase in the lungs of patients with pulmonary hypertension. N Engl J Med. 1995;333:214–21.
175. Huttemeier PC, Watkins WD, Peterson MB. Acute pulmonary hypertension and lung thromboxane release after endotoxin infusion in normal and leukopenic sheep. Circ Res. 1982;50:688–94.
176. Gando S, Kameue T, Matsuda N. Imbalances between the levels of tissue factor and tissue factor pathway inhibitor in ARDS patients. Thromb Res. 2003;109:119–24.
177. Ribes JA, Francis CW, Wagner DD. Fibrin induces release of von Willebrand factor from endothelial cells. J Clin Invest. 1987;79:117–23.
178. Bastatache JA, Wang L, Geiser T. The alveolar endothelium can initiate the extrinsic coagulation cascade through expression of tissue factor. Thorax. 2007;62:608–16.

179. Ware LB, Bastarache JA, Wang L. Coagulation and fibrinolysis in human acute lung injury: new therapeutic targets? Keio J Med. 2005;54:142–9.
180. Idell S. Coagulation, fibrinolysis, and fibrin deposition in acute lung injury. Crit Care Med. 2003;31:S213–20.
181. Gunther A, Mosavi P, Heinemann S. Alveolar fibrin formation caused by enhanced procoagulant and depressed fibrinolytic capacities in severe pneumonia: comparison with acute respiratory distress syndrome. Am J Respir Crit Care Med. 2000;161:454–62.
182. Ware LB, Fang X, Matthay MA. Protein C and thrombomodulin in human acute lung injury. AM J Physiol Lung Cell Mol Physiol. 2003;285:L514–21.
183. Bertozzi P, Astedt B, Zenzius L. Depressed bronchoalveolar urokinase activity in patients with adult respiratory distress syndrome. N Engl J Med. 1990;322:890–7.
184. Matthay MA, Ware LB. Plasma protein C levels in patients with acute lung injury: prognostic significance. Crit Care Med. 2004;32:S229–32.
185. Matthay MA, Folkesson HG, Clerici C. Lung epithelial fluid transport and the resolution of pulmonary edema. Physiol Rev. 2002;82:569–600.
186. Sznajder JL. Strategies to increase alveolar epithelial fluid removal in the injured lung. Am J Respir Crit Care Med. 1999;160:1441–2.
187. Meduri GU, Carratu P, FFreire AX. Evidence of biological efficacy for prolonged glucocorticoid treatment in patients with unresolving ARDS. Eur Respir J. 2003;22 Suppl 42:57s–64.
188. Ware LB, Matthay MA. Alveolar fluid clearance is impaired in the majority of patients with acute lung injury and the acute respiratory distress syndrome. Am J Respir Crit Care Med. 2001;163:1376–83.
189. Wiedemann HP, Wheeler AP, Bernard GR. Comparison of two fluid-management strategies in acute lung injury. N Engl J Med. 2006;354:2564–75.
190. Prewitt RM, McCarthy J, Wood LD. Treatment of acute low pressure pulmonary edema in dogs: relative effects of hydrostatic and oncotic pressure, nitroprusside, and positive end-expiratory pressure. J Clin Invest. 1981;67:409–18.
191. Pittet JE, MacKersie RC, Martin TR. Biological markers of acute lung injury: prognostic and pathogenetic significance. Am J Respir Crit Care Med. 1997;155:1187–205.
192. Matthay MA. Conference summary: acute lung injury. Chest. 1999;116(Suppl):119S–26.
193. Wiener-Kronish JP, Albertine KH, Matthay MA. Differential responses of endothelial and epithelial barriers of the lung in sheep to Escherichia coli endotoxin. J Clin Invest. 1991;88:864–75.
194. Snow RL, Davies P, Pontoppidan H. Pulmonary vascular remodeling in adult respiratory distress syndrome. Am Rev Respir Dis. 1982;126:887–92.
195. Greene R, Lind S, Jantsch H. Pulmonary vascular obstruction in severe ARDS: angiographic alterations after i.v. fibrinolytic therapy. Am J Roentgenol. 1987;148:501–8.
196. Powe JE, Short A, Sibbald WJ. Pulmonary accumulation of polymorphonuclear leukocytes in the adult respiratory distress syndrome. Crit Care Med. 1982;10:712–8.
197. Nuckton TJ, Alonso JA, Kallet RH. Pulmonary dead space fraction as a risk factor for death in the acute respiratory distress syndrome. N Engl J Med. 2002;346:1281–6.
198. Capelozzi VL, Negri EM, Menezes SLS. Pulmonary and extrapulmonary acute respiratory distress syndrome: inflammatory and ultrastructural morphometric analysis. Eur Respir J. 2002;20:339.
199. Albertine KH. Ultrastructural abnormalities in increased-permeability pulmonary edema. Clin Chest Med. 1985;6:345–69.
200. Bitterman PB. Pathogenesis of fibrosis in acute lung injury. Am J Med. 1992;92: 39S–43.
201. Bachofen M, Weibel ER. Structural alterations of lung parenchyma in the adult respiratory distress syndrome. Clin Chest Med. 1982;3:35–56.
202. Worthen GS, Downey GP. Mechanisms of neutrophil mediated injury. In: Haslett C, Evans TW, editors. ARDS, acute respiratory distress in adults. London: Chapman and Hall; 1996. p. 99–114.

203. Suzuki T, Moraes TJ, Vachon E. Proteinase-activated receptor-1 mediates elastase-induced apoptosis of human lung epithelial cells. Am J Respir Cell Mol Biol. 2005;33:231–47.
204. Chabot F, Mitchell JA, Gutteridge JMC. Reactive oxygen species in acute lung injury. Eur Respir J. 1998;11:745–57.
205. Pratt PPC, Vollmer RT, Shelburne JD. Pulmonary morphology in a multihospital collaborative extracorporeal membrane oxygenation project. I. Light microscopy. Am J Pathol. 1979;95:191–214.
206. Marshall R, Bellingan G, Laurent G. The acute respiratory distress syndrome: fibrosis in the last lane. Thorax. 1998;53:815–7.
207. Lindroos PM, Coin PG, Osornio-Vargas AR. Interleukin-1-beta (IL-1beta) and the IL-1 beta alpha 2-macroglobulin complex upregulate the platelet -derived growth factor alpha on rat pulmonary fibroblasts. Am J Respir Crit Care Med. 1995;13:455–65.
208. Martinet Y, Menard O, Vaillant P. Cytokines in human lung fibrosis. Arch Toxicol Suppl. 1996;18:127–39.
209. Olman MA, White KE, Ware L. Microarray analysis indicates that pulmonary edema fluid from patients with acute lung injury mediates inflammation, mitogen gene expression, and fibroblast proliferation through bioactive interleukin-1. Chest. 2002;121:69S–70.
210. Cressoni M, Caironi P, Polli F. Anatomical and functional intrapulmonary shunt in acute respiratory distress syndrome. Crit Care Med. 2008;36:669–75.
211. Dantzker DR, Brook CJ, Dehart P. Ventilation-perfusion distributions in the adult respiratory distress syndrome. Am Rev Respir Dis. 1979;120:1039.
212. Kiiski R, Takala J, Kari A. Effect of tidal volume on gas exchange and oxygen transport in the adult respiratory distress syndrome. Am Rev Respir Dis. 1992;146:1131.
213. Schuster DP, Anderson C, Kozlowski J. Regional pulmonary perfusion in patients with acute pulmonary edema. J Nucl Med. 2002;43:863–70.
214. Baraka AS, Taha SK, Yaacoub CI. Alarming hypoxemia during one-lung ventilation in a patient with respiratory bronchiolitis – associated interstitial lung disease. Can J Anaesth. 2003;50:411–4.
215. Gust R, McCarthy TJ, Kozlowski J. Response to inhaled nitric oxide in acute lung injury depends on distribution of pulmonary blood flow prior to its administration. Am J Respir Crit Care Med. 1999;159:563–70.
216. Smith G, Cheney Jr FW, Winter PM. The effect of change in cardiac output on intrapulmonary shunting. Br J Anaesth. 1974;46:337–42.
217. Roupie E, Dambrosio M, Servillo G. Titration of tidal volume and induced hypercapnia in acute respiratory distress syndrome. Am J Respir Crit Care Med. 1995;152:121.
218. Lewis JF, Jobe AH. Surfactant and the adult respiratory distress syndrome. Am Rev Respir Dis. 1993;147:218–33.
219. Gregory TJ, Longmore WJ, Moxley MA. Surfactant chemical composition and biophysical activity in acute respiratory distress syndrome. J Clin Invest. 1991;88:1976–81.
220. Pelosi P, Ravagnan I, Giurati G, Panigada M. Positive end-expiratory pressure improves respiratory function in obese patients but not in normal subjects during anesthesia and paralysis. Anesthesiology. 1999;91:1221–31.
221. Sousa AS, Moll RJ, Pontes CF. Mechanical and morphometrical changes in progressive bilateral pneumothorax and pleural effusion in normal rats. Eur Respir J. 1995;8:99–104.
222. Krell WS, Rodarte JR. Effects of acute pleural effusion on respiratory system mechanics in dogs. J Appl Physiol. 1985;59:1458–63.
223. Gattinoni L, Chiumello D, Carlesso E. Bench-to-bedside review: chest wall elastance in acute lung injury/acute respiratory distress syndrome patients. Crit Care. 2004;8:350–5.
224. Pelosi P, D'Onofrio D, Chiumello D. Pulmonary and extrapulmonary acute respiratory distress syndrome are different. Eur Respir J Suppl. 2003;42:48S–56.
225. Ranieri VM, Brienza N, Santostasi S. Impairment of lung and chest wall mechanics in patients with acute respiratory distress syndrome. Am J Respir Crit Care Med. 1997;156:1082–91.
226. Pelosi P, Cereda M, Foti G. Alterations of lung and chest wall mechanics in patients with acute lung injury: effects of positive end-expiratory pressure. Am J Respir Crit Care Med. 1995;152:531–7.

227. Gattinoni L, Caironi P, Pelosi P. What has computed tomography taught us about the acute respiratory distress syndrome? Am J Respir Crit Care Med. 2001;164:1701–11.
228. Goodman LR, Fumigalli R, Tagliabue M. Adult respiratory distress syndrome due to pulmonary and extrapulmonary causes: CT, clinical and functional correlations. Radiology. 1999;213:545–52.
229. Brismar B, Hedenstierna G, Lundquist H. Pulmonary densities during anaesthesia with muscular relaxation. Anesthesiology. 1985;62:422–8.
230. Gattinoni L, Vagginelli F, Chiumello D. Physiologic rationale for ventilator setting in acute lung injury-acute respiratory distress syndrome. Crit Care Med. 2003;31:300s–4.
231. Gattinoni L, Carlesso E, Cadringher P. Physical and biological triggers of ventilator induced lung injury and its prevention. Eur Respir J. 2003;47:15s–25.
232. Brower RG, Shanholtz CB, Fessler HE. Prospective, randomised, controlled clinical trial comparing traditional versus reduced tidal volume in acute respiratory distress syndrome patients. Crit Care Med. 1999;27:1492–8.
233. Pelosi P, Tubiolo D, Mascheroni D. Effects of prone position on respiratory mechanics and gas exchange during acute lung injury. Am J Respir Crit Care Med. 1998;157:387–93.
234. Lim CM, Kim EK, Lee JS. Comparison of the response to the prone position between pulmonary and extrapulmonary acute respiratory distress syndrome. Intensive Care Med. 2001;27:477–85.
235. Grasso S, Mascia L, Del Turco M. Effects of recruiting maneuvers in patients with acute respiratory distress syndrome ventilated with protective ventilatory strategy. Anesthesiology. 2002;96:795–802.
236. Pelosi P, Cadringher P, Bottino N. Sigh in acute respiratory distress syndrome. Am J Respir Crit Care Med. 1999;159:872–80.
237. Zapol WM, Snider MT. Pulmonary hypertension in severe acute respiratory failure. N Engl J Med. 1977;296:476–80.
238. Villar J, Blazquez MA, Lubillo S. Pulmonary hypertension in acute respiratory failure. Crit Care Med. 1989;17:523.
239. Viellard-Baron AA, Schmitt JM, Augarde R. Acute cor pulmonale in acute respiratory distress syndrome submitted to protective ventilation: incidence, clinical implications, and prognosis. Crit Care Med. 2001;29:1551.
240. Osman D, Monnet X, Castelain V. Incidence and prognostic value of right ventricular failure in acute respiratory distress syndrome. Intensive Care Med. 2009;35:69–76.
241. Ribeiro A, Lindmarker P, Juhlin-Dannfelt A. Echocardiography Doppler in pulmonary embolism: Right ventricular dysfunction as a predictor of mortality rate. Am Heart J. 1997; 134:479–87.
242. Cepkova M, Kapur V, Ren X. Pulmonary dead space fraction and pulmonary artery systolic pressure as early predictor of clinical outcome in acute lung injury. Chest. 2007; 132:836–42.
243. Jardin F, Gueret P, Dubourg O. Two-dimensional echocardiographic evaluation of right ventricular size and contractility in acute respiratory failure. Crit Care Med. 1985;13:952–6.
244. Jardin F, Vielleiard-Baron A. Is here a safe plateau pressure in ARDS? The right heart only knows. Intensive Care Med. 2007;33:444–7.
245. Price L. Pulmonary vascular and right ventricular dysfunction in adult critical care: current and emerging options for management, a systematic literature review. Crit Care. 2010;14:R169.
246. Morelli A, Teboul JL, Maggiore SM. Effects of levosimendan on the right ventricular afterload in patients with acute respiratory distress syndrome: a pilot study. Crit Care Med. 2006;34:2287–93.
247. Marshall BE, Marshall C, Frasch F. Role of hypoxic pulmonary vasoconstriction in pulmonary gas exchange and blood flow distribution. Intensive Care Med. 1994;20:291–7.
248. Zapol WM, Kobayashi K, Snider MT. Vascular obstruction causes pulmonary hypertension in severe respiratory failure. Chest. 1977;71 Suppl 2:306–7.
249. Pistolesi M, Miniati M, Di Ricco C. Perfusion lung imaging in the adult respiratory distress syndrome. J Thorac Imaging. 1986;1:11–24.

250. Druml W, Steltzer H, Waldhausl W. Endothelin-1 in epithelial lining fluid of patients with acute respiratory distress syndrome. Am Rev Respir Dis. 1993;148:1169–73.
251. Lambermont B, Ghuysen A, Kolh P. Effects of endotoxic shock on right ventricular systolic function and mechanical efficiency. Cardiovasc Res. 2003;59:412–8.
252. West J. Respiratory physiology. The essentials. 7th ed. Philadelphia: Lippincott, Williams and Wilkins; 2005.
253. Piazza G, Goldhaber SZ. The acutely decompensated right ventricle: pathways for diagnosis and management. Chest. 2005;128:1836–52.
254. Lee FA. Hemodynamics of the right ventricle in normal and disease states. Cardiol Clin. 1992;10:59–67.
255. Wiedemann HP, Matthay RA. Acute right heart failure. Crit Care Clin. 1985;1:631.
256. Raabe Jr DS, Chester AC. Right ventricular infarction. Chest. 1978;73:96–9.
257. Vieillard-Baron A, Prien S, Chergui K. Echo–Doppler demonstration of acute cor pulmonale at the bedside in the medical intensive care unit. Am J Respir Crit Care Med. 2002; 166:1310–9.
258. Jardin F, Dubourg O, Bourdarias J-P. Echocardiographic pattern of acute cor pulmonale. Chest. 1997;111:209–17.
259. Jardin F, Vieillard-Baron A. Right ventricular function and positive pressure ventilation in clinical practice: from hemodynamic subsets to respirator settings. Intensive Care Med. 2003;29:1426–34.
260. Whittenberger JL, McGregor M, Berglund E. Influence of state of inflation of the lung on pulmonary vascular resistance. J Appl Physiol. 1960;15:878–82.
261. Hoeper MM, Granton J. Intensive care unit management of patients with severe pulmonary hypertension and right heart failure. Am J Respir Crit Care Med. 2011;184:1114–24.
262. Chin KM, Kim NHS, Rubin LJ. The right ventricle in pulmonary hypertension. Coron Artery Dis. 2005;16:13–8.
263. Via G, Braschi A. Pathophysiology of severe pulmonary hypertension in the critically ill patient. Minerva Anestesiol. 2004;70:233–7.
264. Bronicki RA, Baden HP. Pathophysiology of right ventricular failure in pulmonary hypertension. Pediatr Crit Care Med. 2010;11:S15–22.
265. Watts JA, Marchick MR, Kline JA. Right ventricular heart failure from pulmonary embolism: key distinctions from chronic pulmonary hypertension. J Card Fail. 2010;16:250–9.
266. Champion HC, Michelakis ED, Hassoun PM. Comprehensive invasive and noninvasive approach to the right ventricle–pulmonary circulation unit: state of the art and clinical and research implications. Circulation. 2009;120:992–1007.
267. Fourie PR, Coetzee AR, Bolliger CT. Pulmonary artery compliance: its role in right ventricular–arterial coupling. Cardiovasc Res. 1992;26:839–44.
268. Kasper W, Konstantinides S, Geibel A. Prognostic significance of right ventricular afterload stress detected by echocardiography in patients with clinically suspected pulmonary embolism. Heart. 1997;77:346–9.
269. Brimioulle S, Wauthy P, Ewalenko P. Single-beat estimation of right ventricular end-systolic pressure–volume relationship. Am J Physiol Heart Circ Physiol. 2003;284:H 1625–30.
270. McIntyre KM, Sasahara AA. Determinants of right ventricular function and hemodynamics after pulmonary embolism. Chest. 1974;65:534–43.
271. Pinsky MR. Pulmonary artery occlusion pressure. Intensive Care Med. 2003;29:19–22.
272. Wood KE. Major pulmonary embolism: review of a pathophysiologic approach to the golden hour of hemodynamically significant pulmonary embolism. Chest. 2002;121:877–905.
273. Jacobs AK, Leopold JA, Bates E. Cardiogenic shock caused by right ventricular infarction. A report from the SHOCK registry. J Am Coll Cardiol. 2003;41:1273–9.
274. Jardin F, Dubourg O, Guéret P. Quantitative two-dimensional echocardiography in massive pulmonary embolism: emphasis on ventricular interdependence and leftward septal displacement. J Am Coll Cardiol. 1987;10:1201–6.
275. Lee JM, Boughner DR. Tissue mechanics of canine pericardium in different test environments. Evidence for time-dependent accommodation, absence of plasticity, and new roles for collagen and elastin. Circ Res. 1981;49:533–44.

276. Hoffman EA, Ritman EL. Invariant total heart volume in the intact thorax. Am J Physiol Heart Circ Physiol. 1985;249:H883–90.
277. Vieillard-Baron A, Page B, Augarde R. Acute cor pulmonale in massive pulmonary embolism: incidence, echocardiographic pattern, clinical implications and recovery rate. Intensive Care Med. 2001;27:1481–6.
278. Vincent J-L. What is right ventricular (Dys)function? Crit Care Med. 1994;22:2014.
279. Prewitt RM, Ghignone M. Treatment of right ventricular dysfunction in acute respiratory failure. Crit Care Med. 1983;11:346–52.
280. Gan CT, Lankhaar JW, Marcus JT. Impaired left ventricular filling due to right-to-left ventricular interaction in patients with pulmonary arterial hypertension. Am J Physiol Heart Circ Physiol. 2006;290:H1528–33.
281. Kingma I, Tyberg JV, Smith ER. Effects of diastolic transseptal pressure gradient on ventricular septal position and motion. Circulation. 1983;68:1304–14.
282. Janicki JS, Weber KT. The pericardium and ventricular interaction, distensibility, and function. Am J Physiol Heart Circ Physiol. 1980;238:H494–503.
283. Olsen CO, Tyson GS, Maier GW. Dynamic ventricular interaction in the conscious dog. Circ Res. 1983;52:85–104.
284. de Groote P, Millaire A, Foucher-Hossein C. Right ventricular ejection fraction is an independent predictor of survival in patients with moderate heart failure. J Am Coll Cardiol. 1998;32:948–54.
285. Ghio S, Gavazzi A, Campana C. Independent and additive prognostic value of right ventricular systolic function and pulmonary artery pressure in patients with chronic heart failure. J Am Coll Cardiol. 2001;37:183–8.
286. Belenkie I, Dani R, Smith ER. Effects of volume loading during experimental acute pulmonary embolism. Circulation. 1989;80:178–88.
287. Pinsky MR. Recent advances in the clinical application of heart-lung interactions. Curr Opin Crit Care. 2002;8:26–31.
288. Stojnic BB, Brecker SJ, Xiao HB. Left ventricular filling characteristics in pulmonary hypertension: a new mode of ventricular interaction. Br Heart J. 1992;68:16–20.
289. Taylor RR, Covell JW, Sonnenblick EH. Dependence of ventricular distensibility on filling of the opposite ventricle. Am J Physiol. 1967;213:711–8.
290. Luecke T, Pelosi P. Clinical review: positive end-expiratory pressure and cardiac output. Crit Care. 2005;9:607–21.
291. Cherpanath TGV, Lagrand WK, Schultz MJ. Cardiopulmonary interactions during mechanical ventilation in critically ill patients. Neth Heart J. 2013;21:166–72.
292. Fessler HF. Heart lung interactions: application in the critically ill. Eur Respir J. 1997;10:26–37.
293. Balanos GM, Talbot NP, Dorrington KL. Human pulmonary vascular response to 4 h hypercapnia and hypocapnia measured using Doppler echocardiography. J Appl Physiol. 2003;94:1543–51.
294. Mekontso Dessap A, Charron C, Devaquet J. Impact of acute hypercapnia and augmented positive end-expiratory pressure on right ventricular function in severe acute respiratory distress syndrome. Intensive Care Med. 2009;35:1850–8.
295. Moudgil R, Michelakis ED, Archer SL. Hypoxic pulmonary vasoconstriction. J Appl Physiol. 2005;98:390–403.
296. Tongers J, Schwerdtfeger B, Klein G. Incidence and clinical relevance of supraventricular tachyarrhythmias in pulmonary hypertension. Am Heart J. 2007;153:127–32.
297. Goldstein JA, Harada A, Yagi Y. Hemodynamic importance of systolic ventricular interaction, augmented right atrial contractility and atrioventricular synchrony in acute right ventricular. J Am Coll Cardiol. 1990;16:181–9.
298. Sztrymf B, Souza R, Bertoletti L. Prognostic factors of acute heart failure in patients with pulmonary arterial hypertension. Eur Respir J. 2010;35:1286–93.
299. Krack A, Sharma R, Figulla HR. The importance of the gastrointestinal system in the pathogenesis of heart failure. Eur Heart J. 2005;26:2368–74.
300. Niebauer J, Volk HD, Kemp M. Endotoxin and immune activation in chronic heart failure: a prospective cohort study. Lancet. 1999;353:1838–42.

301. Katzenstein AL, Bloor CM, Leibow AA. Diffuse alveolar damage – the role of oxygen, shock, and related factors. A review. Am J Pathol. 1976;85:209–28.
302. Putensen C, Theuerkauf N, Zinserling J. Meta-analysis: ventilation strategies and outcome of acute respiratory distress syndrome and acute lung injury. Ann Intern Med. 2009;151:566–76.
303. Pontes-Arruda A, DeMichele SJ. Enteral nutrition with anti-inflammatory lipids in ALI/ARDS. In: Vincent J-L, editor. Yearbook of intensive care and emergency medicine. Berlin/Heidelberg/New York: Springer; 2009. p. 695.
304. Gattinoni L, Caironi P, Cressoni M. Lung recruitment in patients with the acute respiratory distress syndrome. N Engl J Med. 2006;354:1775–86.
305. Lennon M, Ashbaugh DG, Bigelow DB. Acute respiratory distress in adults. The Lancet, Saturday 12 August 1967. Crit Care Resusc. 2005;7:60–1.
306. American Thoracic Society/European Respiratory Society. ATS/European Respiratory Society International Multidisciplinary Consensus Classification of the Idiopathic Interstitial Pneumonias. Am J Respir Crit Care Med. 2002;165:277–304.
307. Saguil A, Fargo M. Acute respiratory distress syndrome: diagnosis and management. Am Fam Physician. 2012;85:352–8.
308. Mackay A, Al-Haddad M. Acute lung injury and acute respiratory distress syndrome. Contin Educ Anaesth Crit Care Pain. 2009;9:152–6.
309. Laycock H, Rajah A. Acute lung injury and acute respiratory distress syndrome: a review article. BJMP. 2010;3:324–30.
310. Andrews CP, Coalson JJ, Smith JD. Diagnosis of nosocomial bacterial pneumonia in acute, diffuse lung injury. Chest. 1981;80:254.
311. Dushianthan A, Grocott MP, Postle AD. Acute respiratory distress syndrome and acute lung injury. Postgrad Med J. 2011;87:612–22.
312. Chiumello D, Froio S, Bouhemad B. Clinical review: lung imaging in acute respiratory distress syndrome patients – an update. Crit Care. 2013;17:243.
313. Endo S, Shibata S, Sato N. A prospective cohort study of ALI/ARDS in the Tohoku district of Japan (second report). J Anesth. 2010;24:351–8.
314. Lichtenstein D, Goldstein I, Mourgeon E. Comparative diagnostic performances of auscultation, chest x-ray, and lung ultrasonography in acute respiratory distress syndrome. Anesthesiology. 2004;100:9–15.
315. Xirouchaki N, Magkanas E, Vaporidi K. Lung ultrasound in critically ill patients: comparison with bedside chest x-ray. Intensive Care Med. 2011;37:1488–93.
316. Sud S, Friedrich JO, Taccone P. Prone ventilation reduces mortality in patients with respiratory failure and severe hypoxemia: systematic review and meta-analysis. Intensive Care Med. 2010;36:585–99.
317. Gattinoni L, Carlesso E, Taccone P. Prone positioning improves survival in severe ARDS: a pathophysiologic review and individual patient meta-analysis. Minerva Anestesiol. 2010;76:448–54.
318. Vignon P, Colreavy F, Slama M. Pulmonary edema: which role for echocardiography in the diagnostic workup? In: Cholley BP, Slama M, De Backer D, editors. Hemodynamic monitoring using echocardiography in the critically ill. Berlin/Heidelberg: Springer; 2011. p. 177–94.
319. Bouferrache K, Vieillard-Baron A. Acute respiratory distress syndrome, mechanical ventilation, and right ventricular function. Curr Opin Crit Care. 2011;17:30–5.
320. Caille V, Vieillard-Baron A. The right ventricle during the acute respiratory distress syndrome. Open Nucl Med J. 2010;2:119–24.
321. Blaesdale RA. Prognostic importance of right ventricular dysfunction. Heart. 2002;88:323–4.
322. Kind T, Mauritz G-J, Marcus JT. Right ventricular ejection fraction is better reflected by transverse rather than longitudinal wall motion in pulmonary hypertension. J Cardiovasc Magn Reson. 2010;12:35.
323. McLaughlin VV, Presberg KW, Doyle RL. Prognosis of pulmonary arterial hypertension. Chest. 2004;125(suppl):S78–92S.
324. D'Alonzo GE, Barst RJ, Ayres SM. Survival in patients with primary pulmonary hypertension: results from a national prospective registry. Ann Intern Med. 1991;115:343–9.

325. Miller D, Farah MG, Liner A. The relation between quantitative right ventricular ejection fraction and indices of tricuspid annular motion and myocardial performance. J Am Soc Echocardiogr. 2004;14:443–7.
326. Kjaergaard J, Akkan D, Iversen KK. Right ventricular dysfunction as an independent predictor of short- and long-term mortality in patients with heart failure. Eur J Heart Fail. 2007;9:610–6.
327. Forfia PR, Fisher MR, Mathai SC. Tricuspid annular displacement predicts survival in pulmonary hypertension. Am J Respir Crit Care Med. 2006;174:1034–41.
328. Ghio S, Klersy C, Magrini C. Prognostic relevance of the echocardiographic assessment of right ventricular function in patients with idiopathic pulmonary arterial hypertension. Int J Cardiol. 2010;140:272–8.
329. Tamborini G, Pepi M, Galli CA. Feasibility and accuracy of a routine echocardiographic assessment of right ventricular function. Int J Cardiol. 2007;115:86–9.
330. Levitt JE, Vinayak AG, Gehlbach BK. Diagnostic utility of B-type natriuretic peptide in critically ill patients with pulmonary edema: a prospective cohort study. Crit Care. 2008;12:R3.
331. Gordo-Vidal F, Enciso-calderon V. Acute respiratory distress syndrome, mechanical ventilation, and right heart function. Med Intensiva. 2012;36:138–42.
332. Perel A. Extravascular lung water and the pulmonary vascular permeability index may improve the definition of ARDS. Crit Care. 2013;17:108.
333. Ferguson ND, Fan E, Camporota L. The Berlin definition of ARDS: an expanded rational, justification, and supplementary material. Intensive Care Med. 2012;38:1573–82.
334. Mihm FG, Feeley TW, Jamieson SW. Thermal dye double indicator dilution measurement of lung water in man: comparison with gravimetric measurements. Thorax. 1987;42:72–6.
335. Sakka SG, Rühl CC, Pfeiffer UJ. Assessment of cardiac preload and extravascular lung water by single transpulmonary thermodilution. Intensive Care Med. 2000;26:180–7.
336. Neumann P. Extravascular lung water and intrathoracic blood volume: double versus single indicator dilution technique. Intensive Care Med. 1999;25:216–9.
337. Sakka SG. Extravascular lung water in ARDS patients. Minerva Anestesiol. 2013;79:274–84.
338. Kuzkov VV, Kirov MY, Sovershaev MA. Extravascular lung water determined with single transpulmonary thermodilution correlates with the severity of sepsis-induced acute lung injury. Crit Care Med. 2006;34:1647–53.
339. Cordemans C, De Laet I, Van Regenmortel N. Fluid management in critically patients: the role of extravascular lung water, abdominal hypertension, capillary leak and fluid balance. Ann Intensive Care. 2012;2(Suppl):S1.
340. Capro RO, Morris AH, Clayton PD. Lung volumes in healthy nonsmoking adults. Bull Eur Physiopathol Respir. 1982;18:419–25.
341. Phillips CR, Smith SM. Predicted body weight-indexed extravascular lung water is elevated in acute respiratory distress syndrome. Crit Care Med. 2009;37:377–8.
342. LeTourneau JL, Pinney J, Phillips CR. Extravascular lung water predicts progression to acute lung injury in patients with increased risk. Crit Care Med. 2012;40:847–54.
343. Ruiz-Bailen M, Fernandez-Mondejar F, Hurtado-Ruiz B. Immediate application of positive end-expiratory pressure is more effective than delayed positive end-expiratory pressure to reduce extravascular lung water. Crit Care Med. 1999;27:380–4.
344. Naidu BV, Dronavalli VB, Rajesh PB. Measuring lung water following major lung resection. Interact Cardiovasc Thorac Surg. 2009;8:503–6.
345. Martin GS, Eaton S, Mealer M. Extravascular lung water in patients with severe sepsis: a prospective cohort study. Crit Care. 2005;9:R74–82.
346. Schuster DP. Identifying patients with ARDS: time for a different approach. Intensive Care Med. 1997;23:1197–203.
347. Michard F, Fernandez-Mondejar E, Kirov MY. A new and simple definition for acute lung injury. Crit Care Med. 2012;40:1004–6.
348. Mitchell JP, Schuller D, Calandrino FS. Improved outcome based on fluid management in critically ill patients requiring pulmonary artery catheterization. Am Rev Respir Dis. 1992;145:990–8.
349. Combes A, Berneau J-B, Luyt C-E. Estimation of left ventricular systolic function by single transpulmonary thermodilution. Intensive Care Med. 2004;30:1377–80.

350. Eisenberg PR, Hansbrough JR, Anderson D. A prospective study of lung water measurements during patient management in an intensive care unit. Am Rev Respir Dis. 1987;136:662–8.
351. Van der Heijden M, Groeneveld A. Extravascular lung water to blood volume ratios as measure of pulmonary capillary permeability in nonseptic critically ill patients. J Crit Care. 2010;25:16–22.
352. Groeneveld AB, Verheij J. Extravascular lung water to blood volume ratios as a measure of permeability in sepsis-induced ALI/ARDS. Intensive Care Med. 2006;32:1315–21.
353. Sakka SG. Measurement of extravascular lung water in critically ill patients. Curr Respir Med Rev. 2007;3:206–13.
354. de Hemptinne Q, Remmelink M, Brimioulle S. ARDS: a clinicopathological confrontation. Chest. 2009;135:944–9.
355. Patel SR, Krampaliotis D, Ayas NT. The role of open-lung biopsy in ARDS. Chest. 2004;125:197–202.
356. Esan A, Hess D, Sessler C. Ventilator strategies in severe hypoxaemia respiratory failure. Eur Respir Mon. 2012;55:19–39.
357. Haas CF. Mechanical ventilation with lung protective strategies: what works? Crit Care Clin. 2011;27:469–86.
358. Rouby J-J, Puybasset L, Nieszkowska A. Acute respiratory distress syndrome: Lessons from computed tomography of the whole lung. Crit Care Med. 2003;31(Suppl):S285–95.
359. Soni N, Williams P. Positive pressure ventilation: what is the real cost? Br J Anaesth. 2008;101:446–57.
360. Slutsky AS. Lung injury caused by mechanical ventilation. Chest. 1999;116(1 Suppl):9S–16.
361. Lionetti V, Recchia FA, Ranieri VM. Overview of ventilator-induced lung injury mechanisms. Curr Opin Crit Care. 2005;11:82–6.
362. Carny D, Di Rocco J, Nieman G. Dynamic alveolar mechanics and ventilator-induced lung injury. Crit Care Med. 2005;33(suppl):S122–8.
363. Tremblay LN, Slutsky AS. Ventilator-induced lung injury: from bench to the bedside. Intensive Care Med. 2006;32:24–33.
364. Gattinoni L, Pelosi P, Crotti S. Positive end-expiratory pressure on regional distribution of tidal volume and recruitment in adult respiratory distress syndrome. Am J Respir Crit Care Med. 1995;151:1807–14.
365. Terragni P, Rosbosh G, Tealdi A. Tidal hyperinflation during low tidal volume ventilation in acute respiratory distress syndrome. Am J Respir Crit Care Med. 2007;175:160–6.
366. Chiumello D, Carlesso E, Cadringher P. Lung stress and strain during mechanical ventilation for acute respiratory distress syndrome. Am J Respir Crit Care Med. 2008;178:346–55.
367. Gattinoni L, Pesenti A. The concept of "baby lung". Intensive Care Med. 2005;31:776–84.
368. Amato MB, Barbas CS, Medeiros DM. Effect of a protective-ventilation strategy on mortality in the acute respiratory distress syndrome. N Engl J Med. 1998;338:347–54.
369. Network, ARDS. Ventilation with lower tidal volumes as compared with traditional tidal volumes for acute lung injury and the acute respiratory distress syndrome. The Acute Respiratory Distress Syndrome Network. N Engl J Med. 2000;342:1301–8.
370. Villar J, Kacmarek RM, Perez-Mendez L. A high positive end-expiratory pressure, low tidal volume ventilatory strategy improves outcome in persistent acute respiratory distress syndrome: a randomized, controlled trial. Crit Care Med. 2006;34:1311–8.
371. Esteban A, Ferguson ND, Meade MO. Evolution of mechanical ventilation in response to clinical research. Am J Respir Crit Care Med. 2008;177:170–7.
372. Checkley W, Brower R, Korpak A. Effect of a clinical trial on mechanical ventilation practices in patients with acute lung injury. Am J Respir Crit Care Med. 2008;177:1215–22.
373. Dreyfuss D, Sauman G. Ventilator-induced lung injury: lessons from experimental studies. Am J Respir Crit Care Med. 1998;157:294–323.
374. Grasso G, Terrangi P, Mascia L. Airway pressure – time profile (stress index) detects tidal recruitment/hyperinflation in experimental acute lung injury. Crit Care Med. 2004;32:1018–27.

375. Girard TD, Bernard GR. Mechanical ventilation in ARDS. A state of the art review. Chest. 2007;131:921–9.
376. Esteban A, Immaculada A, Gordo F. Prospective randomized trial comparing pressure-controlled ventilation and volume-controlled ventilation in ARDS. Chest. 2000;117:1690–6.
377. Retamal J, Libuy J, Jimenez M. Preliminary study of ventilation with 4 ml/kg tidal volume in acute respiratory distress syndrome: feasibility and effects on cyclic recruitment – derecruitment and hyperinflation. Crit Care. 2013;17:R16.
378. Mancebo J, Fernandez R, Blanch L. A multicenter trial of prolonged prone ventilation in severe acute respiratory distress syndrome. Am J Respir Crit Care Med. 2006;173:1233–9.
379. Mercat A, Richard JC, Vielle B. Positive end-expiratory pressure setting in adults with acute lung injury and acute respiratory distress syndrome: a randomized controlled trial. JAMA. 2008;299:646–55.
380. Chelucci GL, Dall'Ava-Santucci J, Dhainaut JF. Association of PEEP with two different inflation volumes in ARDS patients: effects on passive lung deflation and alveolar recruitment. Intensive Care Med. 2000;26:870–7.
381. Burns D, West TA, Hawkins K. Immediate effects of positive end-expiratory pressure and low and high tidal volume ventilation upon gas exchange and compliance in patients with acute lung injury. J Trauma. 2001;51:1177–81.
382. Crotti S, Mascheroni D, Caironi P. Recruitment and derecruitment during acute respiratory failure: a clinical study. Am J Respir Crit Care Med. 2001;164:131–40.
383. Richard JC, Maggiore SM, Jonson B. Influence of tidal volume on alveolar recruitment. Respective role of PEEP and a recruitment maneuver. Am J Respir Crit Care Med. 2001;163:1609–13.
384. Pinhu L, Whitehead T, Evans T. Ventilator-associated lung injury. Lancet. 2003;361:332–40.
385. Caironi P, Cressoni M, Chiumello D. Lung opening and closing during ventilation of acute respiratory distress syndrome. Am J Respir Crit Care Med. 2010;181:578–86.
386. Meade MO, Cook DJ, Guyatt GH. Ventilation strategy using low tidal volumes, recruitment maneuvers, and high positive end-expiratory pressure for acute lung injury and acute respiratory. JAMA. 2008;299:637–45.
387. Brochard L. New goals for positive end-expiratory pressure in acute respiratory distress syndrome: a paradigm shift or the end of an area of uncertainty? Am J Respir Crit Care Med. 2010;181:528–30.
388. Schmitt JM, Vieillard-Baron A, Roch A. Positive end-expiratory pressure titration in acute respiratory distress syndrome patients: impact on right ventricular outflow impedance evaluated by pulmonary artery Doppler flow velocity measurements. Crit Care Med. 2001;29:1154–8.
389. Dreyfuss D, Soler P, Basset G. High inflation pressure pulmonary edema. Respective effects of high airway pressure, high tidal volume, and positive end-expiratory pressure. Am J Respir Crit Care Med. 1988;137:1159–64.
390. Parker JC, Townsley MI, Rippe B. Increased microvascular permeability in dog lungs due to high peak airway pressure. J Appl Physiol. 1984;57:1809–16.
391. Webb HH, Tierney DF. Experimental pulmonary edema due to intermittent positive pressure ventilation with high inflation pressures. Protection by positive end-expiratory pressure. Am Rev Respir Dis. 1974;110:556–65.
392. Calfee C. Plasma receptor for advanced glycation end products and clinical outcomes in acute lung injury. Thorax. 2008;63:1083–9.
393. Parsons PE. Lower tidal volume ventilation and plasma cytokine markers of inflammation in patients with acute lung injury. Crit Care Med. 2005;33:1–6.
394. Imai Y. Injurious mechanical ventilation and end-organ epithelial cell apoptosis and organ dysfunction in an experimental model of acute respiratory distress syndrome. JAMA. 2003;289:2104–12.
395. Rubenfeld GD. How much PEEP in acute lung injury. JAMA. 2010;303:883–4.

396. Briel M, Meade MO, Mercat A. Higher vs lower positive end-expiratory pressure in patients with acute lung injury and acute respiratory distress syndrome. Systematic review and meta-analysis. JAMA. 2010;303:865–73.
397. LaFollette R, Hojnowski K, Norton J. Using pressure-volume curves to set proper PEEP in acute lung injury. BACCN J. 2007;12:231–41.
398. ARDS Clinical Trail Network, ALVEOLI-Trail. Higher versus lower positive end-expiratory pressures in patients with the acute respiratory distress syndrome. N Engl J Med. 2005;351:327–36.
399. Oba Y, Thameem DM, Zaza T. High levels of PEEP may improve survival in acute respiratory distress syndrome: a meta-analysis. Respir Med. 2009;103:1174–81.
400. Brower RG, Lanken PN, MacIntyre NR. Higher versus lower positive end-expiratory pressures in patients with the acute respiratory distress syndrome. N Engl J Med. 2004;351:327–36.
401. Talmor D, Sarge T, O'Donnell CR. Esophageal and transpulmonary pressures in acute respiratory failure. Crit Care Med. 2006;34:1389–94.
402. Talmor D, Sarge T, Malhotra A. Mechanical ventilation guided by esophageal pressure in acute lung injury. N Engl J Med. 2008;359:2095–104.
403. Grasso S, Stripoli T, De Michele M. ARDSnet ventilatory protocol and alveolar hyperinflation: role of positive end-expiratory pressure. Am J Respir Crit Care Med. 2007;176:761–7.
404. Schiller HJ, Steinberg J, Halter J. Alveolar inflation during generation of a quasi-static pressure/volume curve in the acutely injured lung. Crit Care Med. 2003;31:1126–33.
405. Jonson B, Svantesson C. Elastic pressure-volume curves: what information do they convey? Thorax. 1999;54:82–7.
406. Harris RS. Pressure-volume curves of the respiratory system. Respir Care. 2005;50:78–98.
407. Neve V, de la Roque ED, Leclerc F. Ventilator -induced overdistension in children: dynamic versus low-flow inflation volume-pressure curves. Am J Respir Crit Care Med. 2000;162:139–47.
408. Stengvist O. Practical assessment of respiratory mechanics. Br J Anaesth. 2003;91:92–105.
409. Downie JM, Nam AJ, Simon BA. Pressure-volume curve does not predict steady state lung volume in canine lavage lung injury. Am J Respir Crit Care Med. 2004;169:957–62.
410. Pelosi P, Goldner M, McKibben A. Recruitment and derecruitment during acute respiratory failure. an experimental study. Am J Respir Crit Care Med. 2001;164:122–30.
411. Albaiceta GM, Taboada F, Parra D. Tomographic study of the inflection point of the pressure-volume curve in acute lung injury. Am J Respir Crit Care Med. 2004;170:1066–72.
412. Gattinoni L, Carlesso E, Valenza F. Acute respiratory distress syndrome, the critical care paradigm: what we learned and what we forgot. Curr Opin Crit Care. 2004;10:272–8.
413. Carvalho CR, de Paula Pinto Schettino G, Maranhao B. Hyperoxia and lung disease. Curr Opin Pulm Med. 1998;4:300–4.
414. Durbin CG, Wallace KK. Oxygen toxicity in the critically ill patient. Respir Care. 1993;93:739–50.
415. Capellier G, Maupoil V, Boussat S. Oxygen toxicity and tolerance. Minerva Anestesiol. 1999;65:388–92.
416. Thompson BT, Hayden D, Matthay MA. Clinicians' approaches to mechanical ventilation in acute lung injury and ARDS. Chest. 2001;120:1622–7.
417. de Durante G, del Turco M, Rustichini L. ARDSNet lower tidal volume ventilatory strategy may generate intrinsic positive end-expiratory pressure in patients with acute respiratory distress syndrome. Am J Respir Crit Care Med. 2002;165:1271–5.
418. Gattinoni L, Caironi P. Refining ventilatory treatment for acute lung injury and acute respiratory distress syndrome. JAMA. 2008;299:691–3.
419. Taccone P, Pesenti A, Latini R, Prone-Supine II Study Group. Prone positioning in patients with moderate to severe acute respiratory distress syndrome: a randomized controlled trial. JAMA. 2009;302:1977–84.
420. Muscedere JG, Mullen JB, Gan K. Tidal ventilation at low airway pressures can augment lung injury. Am J Respir Crit Care Med. 1994;149:1327–34.
421. Kallet RH, Siobal MS, Alonso JA. Lung collapse during low tidal volume ventilation in acute respiratory distress syndrome. Respir Care. 2001;46:49–52.

422. Abdualnour R-EE, Levy BD. Section 4: pulmonary, chapter 14: the acute respiratory distress syndrome. In: Dodson TF, Winawer NH, Lubin MF, editors. Medical management of the surgical patient. Cambridge: Cambridge University Press; 2013. p. 152–69.
423. Allen G, Lundblad LK, Parsons P. Transient mechanical benefits of a deep inflation in the injured mouse lung. J Appl Physiol. 2002;93:1709–15.
424. Dasenbrook EC, Needham DM, Brower RG. Higher PEEP in patients with acute lung injury: a systematic review and meta-analysis. Respir Care. 2011;56:568–75.
425. Ferguson ND, Frutos-Vivar F, Esteban A. Airway pressures, tidal volumes, and mortality in patients with acute respiratory distress syndrome. Crit Care Med. 2005;33:21–30.
426. Pinsky MR. The hemodynamic consequences of mechanical ventilation: an evolving story. Intensive Care Med. 1997;23:493–503.
427. Jardin F, Vieillard-Baron A. Acute cor pulmonale. Curr Opin Crit Care. 2009;15:67–70.
428. Smeding L, Lust E, Plötz FB. Clinical implications of heart-lung interactions during mechanical ventilation: an update. Neth J Med. 2010;68:56–61.
429. Vieillard-Baron A, Loubieres Y, Schmitt J-M. Cyclic changes in right ventricular output impedance during mechanical ventilation. J Appl Physiol. 1999;87:1644–50.
430. Shekerdemian L, Bohn D. Cardiovascular effects of mechanical ventilation. Arch Dis Child. 1999;80:475–80.
431. Canada E, Benumof IL, Tousdale FR. Pulmonary vascular resistance correlates in intact normal and abnormal canine lungs. Crit Care Med. 1982;10:719–21.
432. Hager D, Krishnan J, Hayden D. Tidal volume reduction in patients with acute lung injury when plateau pressure are not high. Am J Respir Crit Care Med. 2005;172:1241–5.
433. Jardin F. Tidal volume reduction in patients with acute lung injury when plateau pressures are not high. Am J Respir Crit Care Med. 2006;173:685–6.
434. Pinsky MR. Hemodynamic effects of artificial ventilation. In: Ayres SN, Grenvik A, Shoemaker WC, editors. Textbook of critical care. Philadelphia: WB Saunders; 1995.
435. Cortes GA, Marini JJ. Two steps forward in bedside monitoring of lung mechanics: transpulmonary pressure and lung volume. Crit Care. 2013;17:219.
436. Slutsky A. Mechanical ventilation. Am College of Chest Physicians' Consensus Conference. Chest. 1993;104:1833–59.
437. Malbrain M. Abdominal pressure in the critically ill. Curr Opin Crit Care. 2000;6:1–21.
438. Tobin M. Advances in mechanical ventilation. N Engl J Med. 2001;344:1986–96.
439. Richard J-C, Marini JJ. Transpulmonary pressure as a surrogate of plateau pressure for lung protective strategy: not perfect but more physiologic. Intensive Care Med. 2012;38:339–41.
440. Hickling KG, Walsh J, Henderson S. Low mortality rate in adult respiratory distress syndrome using low-volume, pressure-limited ventilation with permissive hypercapnia: a prospective study. Crit Care Med. 1994;22:1568–78.
441. Hickling KG, Henderson SJ, Jackson R. Low mortality associated with low volume pressure limited ventilation with permissive hypercapnia in severe adult respiratory distress syndrome. Intensive Care Med. 1990;16:372–7.
442. Nichol AD. Permissive hypercapnia in acute respiratory distress syndrome — is now the time to get strict? Crit Care Resusc. 2011;13:142–3.
443. Swenson ER, Robertson HT, Hlastala MP. Effects of inspired carbon dioxide on ventilation-perfusion matching in normoxia, hypoxia, and hyperoxia. Am J Respir Crit Care Med. 1994;149:1563–9.
444. Domino KB, Emery MJ, Swenson ER. Ventilation heterogeneity is increased in hypocapnic dogs but not pigs. Respir Physiol. 1998;111:89–100.
445. Turek Z, Kreuzer F. Effect of shifts of the O2 dissociation curve upon alveolar-arterial O2 gradients in computer models of the lung with ventilation-perfusion mismatching. Respir Physiol. 1981;45:133–9.
446. Komori M, Takada K, Tomizawa Y. Permissive range of hypercapnia for improved peripheral microcirculation and cardiac output in rabbits. Crit Care Med. 2007;35:2171–5.
447. Somero GN. Protons, osmolytes, and fitness of internal milieu for protein function. Am J Physiol Regul Integr Comp Physiol. 1986;251:R197–213.

448. Nishio K, Suzuki Y, Takeshita K. Effects of hypercapnia and hypocapnia on [Ca2+]i mobilization in human pulmonary artery endothelial cells. J Appl Physiol. 2001;90:2094–100.
449. Kitakaze M, Weisfeldt ML, Marban E. Acidosis during early reperfusion prevents myocardial stunning in perfused ferret hearts. J Clin Invest. 1988;82:920–7.
450. Currin RT, Gores GJ, Thurman RG. Protection by acidotic pH against anoxic cell killing in perfused rat liver: evidence for a pH paradox. FASEB J. 1991;5:207–10.
451. Laffey JG, Tanaka M, Engelberts D. Therapeutic hypercapnia reduces pulmonary and systemic injury following in vivo lung reperfusion. Am J Respir Crit Care Med. 2000;162:2287–94.
452. Feihl F, Perret C. Permissive hypercapnia. How permissive should we be? Am J Respir Crit Care Med. 1994;150:1722–37.
453. Kregenow DA, Rubenfeld GD, Hudson LD. Hypercapnic acidosis and mortality in acute lung injury. Crit Care Med. 2006;34:1–7.
454. Curley GF, Laffey JG, Kavanagh BP. CrossTalk proposal: there is added benefit to providing permissive hypercapnia in the treatment of ARDS. J Physiol. 2013;591:2763–5.
455. Richecoeur J, Lu Q, Vieira SR. Expiratory washout versus optimization of mechanical ventilation during permissive hypercapnia in patients with severe acute respiratory distress syndrome. Am J Respir Crit Care Med. 1999;160:77.
456. Siegel MD, Parsons PE, Wilson KC. Mechanical ventilation in acute respiratory distress syndrome. http://wwww.uptodate.com/contents/mechanical-ventilation-in-acute-respiratory-distress-syndrome. [Online] 16 Aug 2013. [Cited: 05 Dec 2013.] http://www.uptodate.com/contents/mechanical-ventilation-in-acute-respiratory-distress-syndrome
457. Hotchkiss Jr JR, Blanch L, Murias G. Effects of decreased respiratory frequency on ventilator-induced lung injury. Am J Respir Crit Care Med. 2000;161:463–8.
458. Kallet JH, Jasmer RM, Luve JM. The treatment of acidosis in acute lung injury with tris-hydroxymethyl aminomethane (THAM). Am J Respir Crit Care Med. 2000;161:1149–53.
459. Levitt JE, Matthay MA. Clinical review: early treatment of acute lung injury – paradigm shift toward prevention and treatment prior to respiratory failure. Crit Care. 2012;16:223.
460. Sillman CC. The two-event model of transfusion-related acute lung injury. Crit Care Med. 2006;34(5 Suppl):S124–31.
461. Bux J, Sachs UJ. The pathogenesis of transfusion-related acute lung injury (TRALI). Br J Haematol. 2007;136:788–99.
462. Merx MW, Liehn EA, Janssens U. HMC-CoA reductase inhibitor simvastatin profoundly improves survival in murine model of sepsis. Circulation. 2004;109:2560–5.
463. Ando H, Takamura T, Ota T. Cerivastatin improves survival of mice with lipopolysaccharide-induced sepsis. J Pharmacol Exp Ther. 2000;294:1043–6.
464. Jacobson JR, Barnard JW, Grigoryev DN. Simvastatin attenuates vascular leak and inflammation in murine inflammatory lung injury. Am J Physiol. 2005;288:L1026–32.
465. Merx MW, Liehn EA, Graf J. Statin treatment after onset of sepsis in a murine model improves survival. Circulation. 2005;112:117–24.
466. Donnino MW, Cocchi MN, Howell M. Statin therapy is associated with decreased mortality in patients with infection. Acad Emerg Med. 2009;16:230–4.
467. Liappis AP, Kan VL, Rochester CG. The effect of statins on mortality in patients with bacteremia. Clin Infect Dis. 2001;33:1352–7.
468. Chalmers JD, Singanayagam A, Murray MP. Prior statin use is associated with improved outcomes in community-acquired pneumonia. Am J Med. 2008;121:1002–7.e1.
469. Thomsen RW, Riis A, Kornum JB. Preadmission use of statins and outcomes after hospitalization with pneumonia: population – based cohort study of 29,900 patients. Arch Intern Med. 2009;168:2081–7.
470. MacIntyre NR, Sessler CN. Are there benefits or harm from pressure targeting during lung-protective ventilation? Respir Care. 2010;55:175–83.
471. Marini JJ. Point: is pressure assist-control preferred over volume assist- control mode for lung protective ventilation in patients with ARDS? Yes. Chest. 2011;140:286–90.
472. Yang LY, Huang YC, MacIntyre NR. Patient-ventilator synchrony during pressure- targeted versus flow-targeted small tidal volume assisted ventilation. J Crit Care. 2007;22:252–7.

473. MacIntyre NR, McConnel R, Cheng KC. Patient-ventilator flow dyssynchrony: flow – limited versus pressure-limited breaths. Crit Care Med. 1997;25:1671–7.
474. Rappaport SH, Shpiner R, Yoshihara G. Randomized, prospective trial of pressure-limited versus volume-controlled ventilation in severe respiratory failure. Crit Care Med. 1994;22:22–32.
475. Cinnella G, Conti G, Lofaso F. Effects of assisted ventilation on the work of breathing: volume-controlled versus pressure-controlled ventilation. Am J Respir Crit Care Med. 1996;153:1025–33.
476. Davies Jr K, Branson RD, Campbell RS. Comparison of volume control and pressure control ventilation: is flow wave form the difference? J Trauma. 1996;41:808–14.
477. Munoz J, Guerrero JE, Escalante JL. Pressure-controlled ventilation versus controlled mechanical ventilation with decelerating inspiratory flow. Crit Care Med. 1993;21:1143–8.
478. Kallet RH, Campbell AR, Dicker RA. Work of breathing during lung- protective ventilation in patients with acute lung injury and acute respiratory distress syndrome: a comparison between volume and pressure-regulated breathing modes. Respir Care. 2005;50:1623–31.
479. MacIntyre NR. Counterpoint: is pressure-assist control preferred over volume-assist control mode for lung protective ventilation in patients with ARDS? No. Chest. 2011;140:290–2.
480. Hormann C, Baum M, Putensen C. Biphasic positive airway pressure (BIPAP)–a new mode of ventilatory support. Eur J Anaesthesiol. 1994;11:37–42.
481. Bahn E. BIPAP. Two steps forward in ventilation. Lübeck: Dräger Medical AG; 2004.
482. Putensen C, Muders T, Varelmann D. The impact of spontaneous breathing during mechanical ventilation. Curr Opin Crit Care. 2006;12:13–8.
483. Santanilla JI, Daniel B, Yeow M-E. Mechanical ventilation. Emerg Med Clin North Am. 2008;26:849–62.
484. Rose L, Hawkins M. Airway pressure release ventilation and biphasic positive airway pressure: a systematic review of definitional criteria. Intensive Care Med. 2008;34:1766–73.
485. Downs JB, Stock MC. Airway pressure release ventilation: a new concept in ventilatory support. Crit Care Med. 1987;15:459–61.
486. Mercat A, Graïni L, Teboul JL. Cardiorespiratory effects of pressure-controlled ventilation with and without inverse ratio in the adult respiratory distress syndrome. Chest. 1993;104:871–5.
487. Neumann P, Golisch W, Strohmeyer A. Influence of different release times on spontaneous breathing pattern during airway pressure release ventilation. Intensive Care Med. 2002;28:1742–9.
488. Siau C, Stewart TE. Current role of high frequency oscillatory ventilation and airway pressure release ventilation in acute lung injury and acute respiratory distress syndrome. Clin Chest Med. 2008;29:265–75.
489. Putensen C, Zech S, Wrigge H. Long-term effects of spontaneous breathing during ventilatory support in patients with acute lung injury. Am J Respir Crit Care Med. 2001;164:43–9.
490. Yoshida T, Rinka H, Kaji A. The impact of spontaneous ventilation on distribution of lung aeration in patients with acute respiratory distress syndrome: airway pressure release ventilation versus pressure support ventilation. Anesth Analg. 2009;109:1892–900.
491. Maxwell RA, Green JM, Waldrop J. A randomized prospective trial of airway pressure release ventilation and low tidal volume ventilation in adult trauma patients with acute respiratory failure. J Trauma. 2010;69:501–10.
492. Varpula T, Valta P, Niemi R. Airway pressure release ventilation as a primary ventilatory mode in acute respiratory distress syndrome. Acta Anaesthesiol Scand. 2004;48:722–31.
493. Pipeling MR, Fan E. Therapies for refractory hypoxemia in acute respiratory distress syndrome. JAMA. 2010;304:251–2527.
494. Diaz JV, Brower R, Calfee CS. Therapeutic strategies for severe acute lung injury. Crit Care Med. 2010;38:1644–50.
495. Guerin C, Gaillard S, Lemasson S. Effects of systematic prone positioning in hypoxemic acute respiratory failure: a randomized controlled trial. JAMA. 2004;292:2379.
496. Gattinoni L, Tognoni G, Pesanti A. Effect of prone positioning on the survival of patients with acute respiratory failure. N Engl J Med. 2001;345:568.

497. Alsaghir AH, Martin CM. Effect of prone positioning in patients with acute respiratory distress syndrome: a meta-analysis. Crit Care Med. 2008;36:603.
498. Guerrin C, Gaillard S, Lemasson S. Effects of systematic prone positioning in hypoxemic acute respiratory failure: a randomized controlled trial. JAMA. 2004;292:2379–87.
499. Voggenreiter G, Aufmkolk M, Stiletto RJ. Prone positioning improves oxygenation in post-traumatic acute lung injury – a prospective randomized trial. J Trauma. 2005;59:333–41; discussion 341–3.
500. Ware LB. Prognostic determinants of acute respiratory distress syndrome in adults: impact on clinical design. Crit Care Med. 2005;33(suppl):S217–22.
501. Villar J, Perez-Mendes L, Kacmarek RM. Current definitions of acute lung injury and the acute respiratory distress syndrome do not reflect their true severity and outcome. Intensive Care Med. 1999;25:930–5.
502. Abroug F, Ouanes-Besbes L, Dachraoui F. An updated study-level meta-analysis of randomized controlled trials on proning in ARDS and acute lung injury. Crit Care. 2011;15:R6.
503. Guerin C, Reignier J, Richard J-C. Prone positioning in severe acute respiratory distress syndrome. N Engl J Med. 2013;368:2159–68.
504. Khemani RG, Newth JL. The design of future pediatric mechanical ventilation trials for acute lung injury. Am J Respir Crit Care Med. 2010;182:1465–74.
505. Ten IS, Anderson MR. Is high-frequency ventilation more beneficial than low tidal volume conventional ventilation. Respir Care Clin N Am. 2006;12:437–51.
506. Ventre KM, Arnold JH. High frequency oscillatory ventilation in acute respiratory failure. Paediatr Respir Rev. 2004;5:323–32.
507. Arnold JH, Hanson JH, Toro-Figuero LO. Prospective, randomized comparison of high-frequency oscillatory ventilation and conventional mechanical ventilation in pediatric respiratory failure. Crit Care Med. 1994;22:1530–9.
508. Sud S, Sud M, Friedrich JO. High frequency oscillation in patients with acute lung injury and acute respiratory distress syndrome (ARDS): a systematic review and meta-analysis. BMJ. 2010;340:c2327.
509. Krishnan JA, Brower RG. High-frequency ventilation for acute lung injury and ARDS. Chest. 2000;118:795–807.
510. Fessler HE, Hess DR. Respiratory controversies in the critical care setting. Does high-frequency ventilation offer benefits over conventional ventilation in adult patients with acute respiratory distress syndrome? Respir Care. 2007;52:595–605.
511. Hager DN, Fessler HE, Kaczka DW. Tidal volume delivery during high-frequency oscillatory ventilation in adults with acute respiratory distress syndrome. Crit Care Med. 2007;35:1522–9.
512. Drazen JM, Kamm RD, Slutsky AS. High – frequency ventilation. Physiol Rev. 1984;64:505–43.
513. Müller T, Budweiser S, Muders F. Hochfrequenzoszillationsventilation beim akuten Lungenversagen des Erwachsenen. Dt Ärzteblatt. 2004;101:A928–34.
514. Fessler HE, Hager DN, Brower RG. Feasibility of very high-frequency ventilation in adults with acute respiratory distress syndrome. Crit Care Med. 2008;36:1043–8.
515. Ritacca FV, Stewart TE. Clinical review: high-frequency oscillatory ventilation in adults – a review of the literature and practical applications. Crit Care. 2003;7:385–90.
516. Chan KP, Stewart TE, Mehta S. High-frequency oscillatory ventilation for adult patients with ARDS. Chest. 2007;131:1907–16.
517. Ferguson ND, Chiche JD, Kacmarek RM. Combining high-frequency oscillatory ventilation and recruitment maneuvers in adults with early acute respiratory distress syndrome: treatment with oscillation and an open lung strategy (TOOLS) trial pilot study. Crit Care Med. 2005;33:479–86.
518. Fort P, Farmer C, Westerman J. High – frequency oscillatory ventilation for adult respiratory distress syndrome – a pilot study. Crit Care Med. 1997;25:937–47.
519. Mehta S, Lapinsky SE, Hallett DC. Prospective trial of high-frequency oscillation in adults with acute respiratory distress syndrome. Crit Care Med. 2001;29:1360–9.

520. Mehta S, Granton J, MacDonald RJ. High-frequency oscillatory ventilation in adults: the Toronto experience. Chest. 2004;126:518–27.
521. Claridge JA, Hostetter RG, Lowson SM. High-frequency oscillatory ventilation can be effective as rescue therapy for refractory acute lung dysfunction. Ann Surg. 1999;65:1092–6.
522. Finkielman JD, Gajic O, Farmer JC. The initial Mayo Clinic experience using high-frequency oscillatory ventilation for adult patients: a retrospective study. BMC Emerg Med. 2006;6:2.
523. Andersen FA, Guttormsen AB, Flaatten HK. High frequency oscillatory ventilation in adult patients with acute respiratory distress syndrome–a retrospective study. Acta Anaesthesiol Scand. 2002;46:1082–8.
524. Bollen CW, van Well GT, Sherry T. High-frequency oscillatory ventilation compared with conventional mechanical ventilation in adult respiratory distress syndrome: a randomized, controlled trail. Crit Care. 2005;9:R430–9.
525. Derdak S, Mehta S, Stewart TE. High – frequency oscillatory ventilation for respiratory distress syndrome in adults: a randomized, controlled trial. Am J Respir Crit Care Med. 2002;166:801–8.
526. Kacmarek ND. Counterpoint: high-frequency ventilation is not the optimal physiological approach to ventilate ARDS patients. J Appl Physiol. 2008;104:1232–5.
527. Ferguson ND, Slutsky AS. Point: high -frequency ventilation is the optimal physiological approach to ventilate ARDS patients. J Appl Physiol. 2008;104:1230–1.
528. Young D, Lamb S, Shah S. High-frequency oscillation for acute respiratory distress syndrome. N Engl J Med. 2013;368:806–13.
529. Ferguson ND, Cook DJ, Guyatt GH. High-frequency oscillation in early acute respiratory distress syndrome. N Engl J Med. 2013;368:795–805.
530. Ferguson ND, Slutsky AAS. Point: counterpoint: high-frequency ventilation is/is not the optimal physiologic approach to ventilate ARDS patients. J Appl Physiol. 2008;104:1230–1.
531. Goligher EC, Ferguson ND. Re-evaluating high-frequency oscillation for ARDS: would a targeted approach be successful? Crit Care. 2013;17:133.
532. Liaudet L. High-frequency oscillation for ARDS. N Engl J Med. 2013;368:2231 (letter to the editor).
533. Guervilly C, Roch A, Papazian L. High-frequency oscillation for ARDS. N Engl J Med. 2013;368:2233 (letter to the editor).
534. Puybasset L, Gusman P, Muller JC. Regional distribution of gas and tissue in acute respiratory distress syndrome. III. Consequences for the effects of positive end-expiratory pressure. CT-Scan ARDS Study Group. Adult Respiratory Distress Syndrome. Intensive Care Med. 2000;26:1215–27.
535. Protti A, Cressoni M, Santini A. Lung stress and strain during mechanical ventilation: any safe threshold? Am J Respir Crit Care Med. 2011;183:1354–62.
536. Guervilly C, Forel JM, Hraiech S. Right ventricular function during high-frequency oscillatory ventilation in adults with acute respiratory distress syndrome. Crit Care Med. 2012;40:1539–45.
537. Ferguson ND, Slutsky AS, et al. High – frequency oscillation for ARDS. N Engl J Med. 2013;368:2231–4, Correspondence.
538. Matsuda N, Hattori Y. Vascular biology in sepsis: pathophysiological and therapeutic significance of vascular dysfunction. Smooth Muscle Res. 2007;43:117–37.
539. Lachmann B. Open up the lung and keep the lung open. Intensive Care Med. 1992;18: 319–21.
540. Lachmann B. Open lung in ARDS. Minerva Anestesiol. 2002;68:637–43.
541. Fan E, Wilcox ME, Brower RG. Recruitment maneuvers for acute lung injury: a systematic review. Am J Respir Crit Care Med. 2008;178:1156–63.
542. Marini JJ. Recruitment by sustained inflation: time for a change. Intensive Care Med. 2011;37:1572–4.
543. Pelosi P, de Abreu MG, Rocco PR. New and conventional strategies for lung recruitment in acute respiratory distress syndrome. Crit Care. 2010;14:210.

544. Villagra A, Ochagavia A, Vatua S. Recruitment maneuvers during lung protective ventilation in acute respiratory distress syndrome. Am J Respir Crit Care Med. 2002;165:165–70.
545. Brower RG, Morris A, MacIntyre N. Effects of recruitment maneuvers in patients with acute lung injury and acute respiratory distress syndrome ventilated with high positive end-expiratory pressure. Crit Care Med. 2003;31:2592–7.
546. Odenstedt H, Aneman A, Karason S. Acute hemodynamic changes during lung recruitment in lavage and endotoxin-induced ALI. Intensive Care Med. 2005;31:112–20.
547. Rzezinski AF, Oliveira GP, Santiago VR. Prolonged recruitment manoeuvre improves lung function with less ultrastructural damage in experimental mild acute lung injury. Respir Physiol Neurobiol. 2009;169:271–81.
548. Riva DR, Oliveira MB, Rzezinski AF. Recruitment maneuver in pulmonary and extrapulmonary experimental acute lung injury. Crit Care Med. 2008;36:1900–8.
549. Odenstedt H, Lindgren S, Olegard C. Slow moderate pressure recruitment maneuver minimizes negative circulatory and lung mechanic side effects: evaluation of recruitment maneuvers using electric impedance tomography. Intensive Care Med. 2005;31:1706–14.
550. Steimback PW, Oliveira GP, Rzezinski A. Effects of frequency and inspiratory plateau pressure during recruitment manoeuvres on lung and distal organs in acute lung injury. Intensive Care Med. 2009;35:1120–8.
551. Meade MO, Cook DJ, Griffith LE. A study of the physiologic responses to a lung recruitment maneuver in acute lung injury jury. Respir Care. 2008;53:1441–9.
552. Oczenski W, Hörmann C, Keller C. Recruitment maneuvers after a positive End-expiratory pressure trial do not induce sustained effects in early adult respiratory distress syndrome. Anesthesiology. 2004;101:620–5.
553. ARDS Clinical Trials Network, National Heart, Lung, and Blood Institute, National Institutes of Health. Effects of recruitment maneuvers in patients with acute lung injury and acute respiratory distress syndrome ventilated with high positive end-expiratory pressure. Crit Care Med. 2003;31:2592–7.
554. Burns KE, Adhikari NK, Slutsky AS. Pressure and volume limited ventilation for the ventilatory management of patients with acute lung injury: a systematic review and meta-analysis. PLoS One. 2011;6:e14623.
555. Brochard L, Roudot-Thoraval F, Roupie E. Tidal volume reduction for prevention of ventilator-induced lung injury in acute respiratory distress syndrome. Am J Respir Crit Care Med. 1998;158:1831–8.
556. Dellinger RP, Levy MM, Rhodes A. Surviving sepsis campaign: international guidelines for management of severe sepsis and septic shock: 2012. Crit Care Med. 2013;41:580–637.
557. Russel F. Management of sepsis. N Engl J Med. 2006;355:1699–713.
558. Brower RG, Matthay MA, Morris A. Ventilation with lower tidal volumes as compared with traditional tidal volumes for acute lung injury and respiratory distress syndrome. N Engl J Med. 2000;342:1301–8.
559. Morris AH. Developing and implementing computerized protocols for standardization of clinical decisions. Ann Intern Med. 2000;132:373–83.
560. Blaser R, Schnabel M, Biber C. Improving pathway compliance and clinician performance by using information technology. Int J Med Inform. 2007;76:151–6.
561. Roch A, Guervilly C, Papazian L. Fluid management in acute lung injury and ARDS. Ann Intensive Care. 2011;1:16.
562. Wiedemann HP. Fluid restriction is superior in acute lung injury and ARDS. Cleve Clin J Med. 2008;75:42–8.
563. Staub NC. Pulmonary edema: physiologic approaches to management. Chest. 1978;74:559–64.
564. Guyton AC. Interstitial fluid pressure. II. Pressure-volume curves of interstitial space. Circ Res. 1965;16:452–60.
565. Calfee CS. Plasma angiopoietin-2 in clinical acute lung injury: prognostic and pathogenetic significance. Crit Care Med. 2012;40:1731–7.
566. Rivers E, Nguyen B, Havstad S. Early goal-directed therapy in the treatment of severe sepsis and septic shock. N Engl J Med. 2001;345:1368–77.

567. Paton-Gay JD, Brindley PG, McDermid RC. Best evidence in critical care medicine: fluid management in acute lung injury: friend or foe? Can J Anaesth. 2007;54:73–5.
568. Murphy CV, Schramm GE, Doherty JA. The importance of fluid management in acute lung injury secondary to septic shock. Chest. 2009;136:102–9.
569. Britos M, Smoot E, Liu KD. The value of positive end-expiratory pressure and FiO2 criteria in the definition of acute respiratory distress syndrome. Crit Care Med. 2011;39: 2025–30.
570. Gillis CN, Pitt BR, Wiedemann HP. Depressed prostaglandin E1 and 5-hydroxytryptamine removal in patients with adult respiratory distress syndrome. Am Rev Respir Dis. 1986;134:739–44.
571. Greene R, Zapol WM, Snider MT. Early bedside detection of pulmonary vascular occlusion during acute respiratory failure. Am Rev Respir Dis. 1981;124:593–601.
572. Mirsky MR, Payen D. Functional hemodynamic monitoring. Crit Care. 2005;9:566.
573. Jardin F, Gueret P, Prost J-F. Two dimensional echocardiographic assessment of left ventricular function in chronic obstructive pulmonary disease. Am Rev Respir Dis. 1984;129:135–44.
574. Mebazaa A, Karpati P, Renaud E. Acute right heart failure- from pathophysiology to new treatment. Intensive Care Med. 2004;30:185–96.
575. Zwissler B. Das akute rechtsherzversagen ätiologie – pathophysiologie – diagnostik – therapie. Der Anaesthesist. 2000;49:788–809.
576. Vieillard-Baron A, Prin S, Chergui K. Hemodynamic instability in sepsis: bedside assessment by Doppler echocardiography. Am J Respir Crit Care Med. 2003;168: 1270–6.
577. Goldstein JA. Pathophysiology and management of right heart ischemia. JACC. 2002;40:841–53.
578. Bowers TR, O'Neill WW, Grines C. Effect of reperfusion on biventricular function and survival after right ventricular infarction. N Engl J Med. 1998;338:933–40.
579. Hochman JS, Sleeper LA, White HD. One-year survival following early revascularization for cardiogenic shock. JAMA. 2001;285:190–2.
580. Louie EK, Rich S, Brundage BH. Doppler echocardiographic assessment of impaired left ventricular filling in patients with right ventricular pressure overload due to primary pulmonary hypertension. JACC. 1986;8:1298–306.
581. Stassen NA, Lukan JK, Dixon MS. Abdominal compartment syndrome. Scand J Surg. 2002;91:104–8.
582. Cheatham ML. Abdominal compartment syndrome: pathophysiology and definitions. Scand J Trauma Resusc Emerg Med. 2009;17:10.
583. James KB, Stelmach K, Armstrong R. Plasma volume and outcome in pulmonary hypertension. Tex Heart Inst J. 2003;30:305–7.
584. Lualdi JC, Goldhaber SZ. Right ventricular dysfunction after acute pulmonary embolism: pathophysiologic factors, detection, and therapeutic implications. Am Heart J. 1995;130:1276–82.
585. Naeije R, Vachiery JL. Medical therapy of pulmonary hypertension conventional therapies. Clin Chest Med. 2001;22:517–22.
586. Layish DT, Tapson VF. Pharmacologic hemodynamic support in massive pulmonary embolism. Chest. 1997;111:218–24.
587. Reuse C, Vincent JL, Pinsky MR. Measurements of right ventricular volumes during fluid challenge. Chest. 1990;98:1450–4.
588. Mathru M, Venus B, Smith RA. Treatment of low cardiac output complicating acute pulmonary hypertension in normovolemic goats. Crit Care Med. 1986;14:120–4.
589. Belenkie I, Sas R, Mitchell J. Opening the pericardium during pulmonary artery constriction improves cardiac function. J Appl Physiol. 2004;96:917–22.
590. Ghignone M, Girling L, Prewitt RM. Volume expansion versus norepinephrine in treatment of low cardiac output complicating an acute increase in right ventricular afterload in dogs. Anesthesiology. 1984;60:132–5.

591. Bhorade S, Christenson J, O'Connor M. Response to inhaled nitric oxide in patients with acute right heart syndrome. Am J Respir Crit Care Med. 1999;159:571–9.
592. Matthay RA, Arroliga AC, Wiedemann HP. Right ventricular function at rest and during exercise in chronic obstructive pulmonary disease. Chest. 1992;101:255S–62.
593. Madger S. The cardiovascular management of the critically ill patient. In: Pinsky MR, editor. Applied cardiovascular physiology. Berlin: Springer; 1997. p. 28–35.
594. Kimichi A, Ellrodt AG, Berman DS. Right ventricular performance in septic shock: a combined radionuclide and hemodynamic study. JACC. 1984;4:945–51.
595. Parker MM, McCarthy KE, Ognibene FP. Right ventricular dysfunction and dilatation, similar to left ventricular changes, characterize the cardiac depression of septic shock in humans. Chest. 1990;97:126–31.
596. Mercat A, Diehl JL, Meyer G. Hemodynamic effects of fluid loading in acute massive pulmonary embolism. Crit Care Med. 1999;27:540–4.
597. Forrest P. Anaesthesia and right ventricular failure. Anaesth Intensive Care. 2009;37:370–85.
598. Goldhaber SZ. The approach to massive pulmonary embolism. Semin Respir Crit Care Med. 2000;21:555–61.
599. Roberts DH, Lepore JJ, Maroo A. Oxygen therapy improves cardiac index and pulmonary vascular resistance in patients with pulmonary hypertension. Chest. 2001;120: 1547–55.
600. Radermacher P, Santak B, Wust HJ. Prostacyclin and right ventricular function in patients with pulmonary hypertension associated with ARDS. Intensive Care Med. 1990;16:227–32.
601. Zwissler B, Kemming G, Habler O. Inhaled prostacyclin (PGI2) versus inhaled nitric oxide in adult respiratory distress syndrome. Am J Respir Crit Care Med. 1996;154:1671–7.
602. van Heerden PV, Barden A, Michalopoulos N. Dose response to inhaled aerosolized prostacyclin for hypoxemia due to ARDS. Chest. 2000;117:819–27.
603. Meyer J, Theilmeier G, Van Aken H. Inhaled prostaglandin E1 for treatment of acute lung injury in severe multiple organ failure. Anesth Analg. 1998;86:753–8.
604. Kuhlen R, Walbert E, Frankel P. Combination of inhaled nitric oxide and intravenous prostaglandin for successful treatment of severe pulmonary hypertension in a patient with acute respiratory distress syndrome. Intensive Care Med. 1999;25:752–4.
605. Rossaint R, Falke KJ, Lopez F. Inhaled nitric oxide for the adult respiratory distress syndrome. N Engl J Med. 1993;328:399–405.
606. Bigatello LM, Hurford WE, Kacmarek RM. Prolonged inhalation of low concentrations of nitric oxide in patients with severe adult respiratory distress syndrome: effects on pulmonary hemodynamics and oxygenation. Anesthesiology. 1994;80:761–70.
607. Christenson J, Lavoie A, O'Connor M. The incidence and pathogenesis of cardiopulmonary deterioration after abrupt withdrawal of inhaled nitric oxide. Am J Respir Crit Care Med. 2000;161:1443–9.
608. Taylor RW, Zimmerman JL, Dellinger RP. Low-dose inhaled nitric oxide in patients with acute lung injury: a randomized controlled trial. JAMA. 2004;291:1603–9.
609. Adhikari NK, Burns KE, Freidrich JO. Effect of nitric oxide on oxygenation and mortality in acute lung injury: systematic review and meta-analysis. BMJ. 2007;334:779.
610. Sokol J, Jacobs SE, Bohn D. Inhaled nitric oxide for acute hypoxic respiratory failure in children and adults: a meta-analysis. Anesth Analg. 2003;97:989–98.
611. Prasad S, Wilkinson J, Garzoulis MA. Sildenafil in primary pulmonary hypertension. N Engl J Med. 2000;343:1342.
612. Preston IR, Klinger JR, Houtches J. Acute and chronic effects of sildenafil in patients with pulmonary arterial hypertension. Respir Med. 2005;99:1501–10.
613. Cornet AD, Hofstra JJ, Swart EL. Sildenafil attenuates pulmonary arterial pressure but does not improve oxygenation during ARDS. Intensive Care Med. 2010;36:758–64.
614. Leier CV, Binkley PE. Parenteral inotropic support for advanced congestive heart failure. Prog Cardiovasc Dis. 1998;41:207–24.
615. Honerjager P. Pharmacology of the bipyridine phosphodiesterase III inhibitors. Am Heart J. 1991;121:1939–43.

616. Follath F, Cleland JG, Just H. Efficacy and safety of intravenous levosimendan compared to dobutamine in severe low output heart failure (the LIDO study): a randomized double-blind trial. Lancet. 2002;360:196–202.
617. Slawsky MT, Colucci WS, Gottlieb SS. Acute hemodynamic and clinical effects of levosimendan in patients with severe heart failure: study investigators. Circulation. 2002;102:2222–7.
618. Zamanian RT, Haddad F, Doyle RL. Management strategies for patients with pulmonary hypertension in the intensive care unit. Crit Care Med. 2007;35:2037–50.
619. Stepinska J. Heart failure. Munich: ESC Congress, session 135. Munich, 25–29 Aug 2012.
620. Geppert A, Dzavik V. General management: pharmacotherapy and mechanical ventilation. In: Ohman EM, Hochman JS, editors. Cardiogenic shock. Dallas: American Heart Association; 2009. p. 45. ISBN 978-1-4051-7926-2.
621. Goldhaber SZ, Elliott CG. Acute pulmonary embolism: part II: risk stratification, treatment, and prevention. Circulation. 2003;108:2834–8.
622. Ruffolo Jr RR, Messick K. Systemic hemodynamic effects of dopamine, (+/–)-dobutamine and the (+)-and (–)-enantiomers of dobutamine in anesthetized normotensive rats. Eur J Pharmacol. 1985;109:173–81.
623. Bradford KK, Deb B, Pearl RG. Combination therapy with inhaled nitric oxide and intravenous dobutamine during pulmonary hypertension in the rabbit. J Cardiovasc Pharmacol. 2000;36:146–51.
624. Leier CV, Webel J, Bush CA. The cardiovascular effects of the continuous infusion of dobutamine in patients with severe cardiac failure. Circulation. 1977;56:468–72.
625. Ryan TJ, Antman EM, Brooks NH. 1999 update: ACC/AHA guidelines for the management of patients with acute myocardial infarction: executive summary and recommendations. Circulation. 1999;100:1016–30.
626. Colucci WS, Wright RF, Braunwald E. New positive inotropic agents in the treatment of congestive heart failure. N Engl J Med. 1986;314:349–58.
627. Shipley J, Tolman D, Hastillo A. Milrinone: basic and clinical pharmacology and acute and chronic management. Am J Med Sci. 1996;311:286–91.
628. Böhm M, Deutsch HJ, Hartmann D. Improvement of postreceptor events by metoprolol treatment in patients with chronic heart failure. JACC. 1997;30:992–6.
629. Abraham WT, Adams KF, Fonarow GC. In-hospital mortality in patients with acute decompensated heart failure requiring intravenous vasoactive medications. JACC. 2005;46:57–64.
630. Cuffe MS, Califf RM, Adams Jr KF. Short-term intravenous milrinone for acute exacerbation of chronic heart failure: a randomized controlled trial. JAMA. 2002;287:1541–7.
631. Felker GM, Benza RL, Chandler AB. Heart failure etiology and response to milrinone in decompensated heart failure. JACC. 2003;41:997–1003.
632. Thackray S, Witte K, Clark AL. Clinical trials update: OPTIME-CHF, PRAISE-2, ALL-HAT. Eur J Heart Fail. 2000;2:209–12.
633. Ukkonen H, Saraste M, Akkila J. Myocardial efficiency during levosimendan in congestive heart failure. Clin Pharmacol Ther. 2000;68:522–31.
634. Kerbaul F, Rondelet B, Demester JP. Effects of levosimendan versus dobutamine on pressure load-induced right ventricular failure. Crit Care Med. 2006;34:2814–9.
635. Missant C, Rex S, Segers P. Levosimendan improves right ventriculovascular coupling in a porcine model of right ventricular dysfunction. Crit Care Med. 2007;35:707–15.
636. Leather HA, Ver Eycken K, Segers P. Effects of levosimendan on right ventricular function and ventriculovascular coupling in open chest pigs. Crit Care Med. 2003;31:2339–43.
637. Poelzl G, Zwick RH, Grander W. Safety and effectiveness of levosimendan in patients with predominant right heart failure. Herz. 2008;33:368–73.
638. Parissis JT, Paraskevaidis I, Bistola V. Effects of levosimendan on right ventricular function in patients with advanced heart failure. Am J Cardiol. 2006;98:1489–92.
639. Duygu H, Ozerkan F, Zoghi M. Effects of levosimendan on right ventricular systolic and diastolic functions in patients with ischemic heart failure. Int J Clin Pract. 2008;62:228–33.
640. Yilmaz MB, Yontar C, Erdem A. Comparative effects of levosimendan and dobutamine on right ventricular function with biventricular heart failure. Heart Vessels. 2009;24:16–21.

641. Kerbaul F, Rondelet B, Motte S. Effects of norepinephrine and dobutamine on pressure load- induced right ventricular failure. Crit Care Med. 2004;32:1035–40.
642. Parissis J, Adamopoulos S, Antoniodes C. Effects of levosimendan on circulating pro-inflammatory cytokines and soluble apoptosis mediators in patients with decompensated advanced. Am J Cardiol. 2004;93:1309–12.
643. Alibaz-Oner F, Gurbuz OZ, Oner E. Impact of levosimendan on right ventricular functions by using novel tissue Doppler derived indices in patients with ischaemic left ventricular failure. Kardiol Pol. 2013;71:1036–41.
644. Delle-Karth G, Heinz G. Levosimendan in Kardiologie und Intensivmedizin. Wien Klin Wochenschr. 2004;116:6–14.
645. Kasikcioglu HA, Cam N. A review of levosimendan in the treatment of heart failure. Vasc Health Risk Manag. 2006;2:389–4000.
646. Rabuel C, Mebazaa A. Septic shock: a heart story since the 1960s. Intensive Care Med. 2006;32:799–807.
647. Kivikko M, Antila S, Eha J. Pharmacokinetics of levosimendan and its metabolites during and after a 24-hour continuous infusion in patients with severe heart failure. Int J Clin Pharmacol Ther. 2002;40:465–71.
648. Yilmaz MB, Yalta K, Yontar C. Levosimendan improves renal function in patients with acute decompensated heart failure: comparison with dobutamine. Cardiovasc Drugs Ther. 2007;21:431–5.
649. Gold FL, Bache RJ. Transmural right ventricular blood flow during acute pulmonary artery hypertension in the sedated dog. Evidence for subendocardial ischemia despite residual vasodilator reserve. Circ Res. 1982;51:196–204.
650. Brooks H, Kirk ES, Vokonas PS. Performance of the right ventricle under stress: relation to right coronary flow. J Clin Invest. 1971;51:2176–83.
651. Angle MR, Molloy DW, Penner B. The cardiopulmonary and renal hemodynamic effects of norepinephrine in canine pulmonary embolism. Chest. 1989;95:1333–7.
652. Vlahakes GJ, Turley K, Hoffman JI. The pathophysiology of failure in acute right ventricular hypertension: hemodynamic and biochemical correlations. Circulation. 1981;63:87–95.
653. Levy B, Dusang B, Annane D. Cardiovascular response to dopamine and early prediction of outcome in septic shock: a prospective multiple-center study. Crit Care Med. 2005;33:2172–7.
654. Sakr Y, Reinhart K, Vincent JL. Does dopamine administration in shock influence outcome? Results of the sepsis occurrence in acutely Ill patients (SOAP) study. Crit Care Med. 2006;34:589–97.
655. De Backer D, Biston P, Devriendt J. Comparison of dopamine and norepinephrine in the treatment of shock. N Engl J Med. 2010;362:779–89.
656. Menon V, Hochman JS. Management of cardiogenic shock complicating acute myocardial infarction. Heart. 2002;88:531–7.
657. Redl G, Germann P, Plattner H. Right ventricular function in early septic shock states. Intensive Care Med. 1993;19:3–7.
658. Schneider AJ, Teule GJ, Groeneveld AB. Biventricular performance during volume loading in patients with early septic shock, with emphasis on the right ventricle: a combined hemodynamic and radionuclide study. Am Heart J. 1988;116:103–12.
659. Bergofsky EH. Humoral control of the pulmonary circulation. Annu Rev Physiol. 1980;42:221–3.
660. Hanson EL, O'Connor NE, Trinker PA. Hemodynamic response to controlled ventilation during hypoxia in man and animals. Surg Forum. 1972;23:207–9.
661. Pagnamenta A, Fesler P, Vaninvit A. Pulmonary vascular effects of dobutamine in experimental pulmonary hypertension. Crit Care Med. 2003;31:1140–6.
662. Schuerer DJE, Kolovos NS, Boyd KV. Extracorporeal membrane oxygenation. Chest. 2008;134:179–84.
663. Christie JD, Edwards LB, Kucheryavaya AY. The registry of the International Society for Heart and Lung Transplantation: twenty-seventh official adult lung and lung-heart transplant report – 2010. J Heart Lung Transplant. 2010;29:745–55.

664. Zapol WM, Snider MT, Hill JD. Extracorporeal membrane oxygenation in severe acute respiratory failure. A randomized prospective study. JAMA. 1979;242:2193–6.
665. Morris AH, Wallace CJ, Menlove RL, et al. Randomized clinical trial of pressure-controlled inverse ratio ventilation and extracorporeal CO2 removal or adult respiratory distress syndrome. Am J Respir Crit Care Med. 1994;149:295–305.
666. Combes A, Bacchetta M, Brodie D. Extracorporeal membrane oxygenation for respiratory failure in adults. Curr Opin Crit Care. 2012;18:99–104.
667. Peek M, Mugford M, Tiruvoipati R. Efficacy and economic assessment of conventional ventilatory support versus extracorporeal membrane oxygenation for severe adult respiratory failure (CESAR): a multicentre randomised controlled trial. Lancet. 2009;374:1351–63.
668. Hubmayr RD, Farmer JC. Should we 'rescue' patients with 2009 influenza A(H1N1) and lung injury from conventional mechanical ventilation? Chest. 2010;137:745–7.
669. Morris AH, Hirshberg E, Miller RR. Counterpoint: efficacy of extracorporeal membrane oxygenation in 2009 influenza A(H1N1): sufficient evidence? Chest. 2010;138:778–81; discussion 782–4.
670. Park PK, Dalton HJ, Bartlett RH. Point: efficacy of extracorporeal membrane oxygenation in 2009 influenza A(H1N1): sufficient evidence? Chest 2010. Chest. 2010;138:776–8.
671. Topol EJ, Goldschlager N, Ports TA. Hemodynamic benefit of atrial pacing in right ventricular myocardial infarction. Ann Intern Med. 1982;96:594–7.
672. Isner JM, Fisher GP, Del Negro AA. Right ventricular infarction with hemodynamic decompensation due to transient loss of active atrial augmentation: successful treatment with atrial pacing. Am Heart J. 1981;102:792–4.
673. Fuster V, Steele PM, Edwards WD. Primary pulmonary hypertension: natural history and the importance of thrombosis. Circulation. 1984;70:580–7.
674. Welsh CH, Hassell KL, Badesch DB. Coagulation and fibrinolytic profiles in patients with severe pulmonary hypertension. Chest. 1996;110:710–7.
675. Ghofrani HA, Wiedemann R, Rose F. Sildenafil for treatment of lung fibrosis and pulmonary hypertension: a randomised controlled trial. Lancet. 2002;360:895–900.
676. Thompson BT, Spence CR, Janssens SP. Inhibition of hypoxic pulmonary hypertension by heparins of differing in vitro antiproliferative potency. Am J Respir Crit Care Med. 1994;149:1512–7.
677. Rajaskhar A, Beyth R, Crowther MA. Newer anticoagulants in critically ill patients. Crit Care Clin. 2012;28:427–51.
678. Fries D. Thrombosis prophylaxis in critically ill patients. Wien Med Wochenschr. 2011;161:68–72.
679. Adriance SM, Murphy CV. Prophylaxis and treatment of venous thromboembolism in the critically ill. Int J Crit Illn Inj Sci. 2013;3:143–51.
680. Brijker F, Heijdra YF, van den Elshout FJJ. Discontinuation of furosemide decreases PaCO(2) in patients with COPD. Chest. 2002;121:377–82.
681. Heinemann HO. Right-sided heart failure and the use of diuretics. Am J Med. 1978;64:367–70.
682. Timms RM, Khaja FU, Williams GW. Hemodynamic response to oxygen therapy in chronic obstructive pulmonary disease. Ann Intern Med. 1985;102:29–36.
683. Beuckelmann DJ. Pulmonale hypertonie. Internist. 1997;38:1020–33.

Chapter 3
Shock

3.1 Definition

Circulatory shock specifies a serious circulatory malfunction at which systemic hypoperfusion and/or disproportionately distributed blood flows precipitate inadequate tissue supply of oxygen and other vital nutrients – or, rarely, inappropriate use of metabolic substrates by the tissues [1, 2] – resulting in generalized cellular hypoxia [1, 3–5].

The emphasis of the definition aims to describe the imbalance between tissue perfusion, notably the supply of oxygen, and the cellular metabolic requirements [6–8]. Oxygen debt following substantially impaired effective systemic perfusion causes general cellular hypoxia/ischaemia and affects essential biochemical processes [9, 10].

A distinctive, infrequent occurrence of shock, occasionally called "cytopathic shock", indicates the circumstance of an altered oxygen pathway on the cellular level (the mitochondria may not be able to utilize O_2 properly to produce the energy necessary to maintain cellular function – inappropriate use of metabolic substrates) as found in septic shock and MOF [2].

3.2 Epidemiology

Patients suffering from shock account for about 1 % of all emergency department presentations [9].

In a recently published multicenter trial, septic shock (62.2 %) was the most frequently reported type of shock, followed by cardiogenic (16.7 %) and hypovolaemic (15.7 %) shock with nearly equal fraction [11]. The incidence of septic shock seems to be increasing within recent years [12, 13].

The overall mortality rates depend on the underlying aetiology as well as the duration of dysfunction and vary between 23 and 75 % [9, 14–18]. Although

W. Krüger, A.J. Ludman, *Core Knowledge in Critical Care Medicine*,
DOI 10.1007/978-3-642-54971-7_3, © Springer-Verlag Berlin Heidelberg 2014

cardiogenic shock associated with acute myocardial infarction still bears the highest risk [19], in-hospital mortality rates seem to have fallen (from over 60 % in 1995 to roughly 48 % in 2004) as the (US) National Registry of Myocardial Infarction (NRMI) suggests [20].

Fifty percent of all deaths caused by shock occur within the first 24 h after admission [21].

3.3 Aetiology

Inaugurated [22] and reclassified by Weil and Shubin [23], shock states are traditionally assorted into four types based on their aetiology and pathophysiological features [24–27]: *hypovolaemic*, *cardiogenic*, *obstructive and vasodilatory or distributive* shock.

However, this allocation is somewhat random and incoherent [6, 28] and has been strongly criticized for being wrong, too simple for a very complex condition, not even distinguishing symptoms which are grossly conform in all categories [28]. Recently, Bonanno calls for a holistic reflection of this serious systemic derangement causing generalized hypoxia and organ dysfunction and suggests to allocate three types, *haemorrhagic*, cardiogenic and *inflammatory* shock [29]. The latter should pose as an "umbrella term" as whatever the kind of circulatory fault; the innate, initially localized, displayed inflammatory response is verifiable in any case and uniform but may extend and overwhelm the compensatory efforts in the haemorrhagic and cardiogenic sort, if not treated and controlled in time, running into the typical picture of *inflammatory* (distributive) shock with its substantially altered microcirculation [29].

However, the traditional classification is in principal still widely accepted [5, 25, 27, 30–32], although increasingly only three shock types, hypovolaemic, cardiogenic and distributive, are in use [5, 27] – not at least because this allocation provides some rough clues for the initial therapeutic approach [26, 30, 31].

Hypovolaemic shock is determined by a substantial loss of intravascular circulating volume (true hypovolaemia) – mostly caused by trauma and haemorrhagic [33], but capillary leakage as well [25] – or by relative hypovolaemia, e.g. spinal cord injury, and vasodilating drugs [25, 34], resulting in decreased venous return, impaired cardiac filling and hence diminished cardiac output (CO); consecutively peripheral vascular resistance increases [6, 34–36]. It may be depicted as failure of the peripheral macrocirculation [29].

Cardiogenic shock literally describes cardiac failure attributed to a considerably impaired pump function causing circulatory disruption [6, 19, 25, 34, 37]. As such, *myocardial contractility* may be *impaired*, e.g. loss of myocardium due to myocardial ischaemia/infarction, perimyocarditis, myocardial depression during sepsis and acute exacerbated severe chronic heart failure, *or pump function* may be *disturbed*, e.g. valvular heart disease and right heart failure; furthermore *altered diastolic properties* are present, e.g. hypertensive cardiomyopathy, infiltrative cardiomyopathies and right heart failure [19, 25, 38]. Haemodynamic hallmark is a markedly declined CO in the presence of elevated filling pressures while the peripheral resistance shows only a mild to moderate, in the light of the severity of the affected function, inappropriate compensatory rise [39–42].

Obstructive shock (of the heart or major arteries and veins [37]) indicates any condition of impeded central forward blood flow. Common examples include pulmonary embolism and acute pulmonary hypertension affecting RV (RV afterload climbs) or hypertrophic obstructive cardiomyopathy and aortic dissection (elevated LV afterload) as well as left atrial myxoma hindering LV filling, and blunted biventricular filling as in pericardial tamponade [6, 25, 31]. Typically, CO is reduced and SVR increased [6, 25].

Distributive shock, often termed *vasodilative/vasogenic* shock as well, is characterized by both marked peripheral vasodilation and *maldistribution* of blood flow [6, 30]. Whereas formerly the profound vasodilation was the subject of experimental and clinical focus [43], more recent study results give evidence of a substantial microcirculatory dysfunction generating a heterogeneous distribution of a normal or even increased CO as being decisively involved in the pathogenesis [37, 44–46]. The peripheral vasodilation per se leads to some disordered allocation [47], "the shunting of oxygen transport to the tissues is the main pathogenic feature of distributive shock" [44]. Septic shock as a special exponent of distributive shock with a low peripheral resistance in presence of preserved or even elevated CO being pathognomonic [6, 22, 25, 48].

Furthermore, distributive shock is the common final pathway of prolonged or severe shock of any cause [43].

Thus, in contrast to the other shock forms, in distributive shocks it is the *microcirculation* which is *initially* altered/affected displaying maldistribution, subsequently causing dysoxia/hypoxia of the tissue areas [44, 49, 50] and/or an impaired use of substrates following defects in cellular oxygen utilization [2].

Accordingly, in hypovolaemic, cardiogenic and obstructive shock, tissue hypoperfusion is largely attributed to a decrease in CO with compensatory increase in peripheral vascular resistance [30, 37, 47] leading to anaerobic tissue metabolism [37]. The distributive form typically displays an oxygen tissue debt due to profound maldistribution at the microcirculatory level accompanied by reduced peripheral vascular resistance in the presence of normal or even raised CO [25, 37, 47, 48].

Some more detailed schedule of aetiologies [5, 6, 23, 27, 32, 51]:

Hypovolaemic Shock [35]

- *Intravascular (true) hypovolaemia*:
  - *Haemorrhagic*: gastrointestinal, *trauma*, aortic dissection, other intestinal sources
  - Renal losses: diabetes insipidus, diabetes mellitus, diuretics, polyuria
  - Gastrointestinal losses: diarrhoea, vomiting, gastric suctioning, surgical stoma
  - Extravascular loses: *capillary leakage*, e.g. *sepsis*, *burn*, *trauma*, *postoperative*
- Increased resistance to venous return (*relative hypovolaemia*):
  - High intrathoracic pressure: mechanical ventilation and PEEP application, tension pneumothorax, massive pleural effusion
  - *High abdominal pressure*: ascites, obesity, *abdominal trauma/infection (peritonitis)/surgery and abdominal compartment syndrome*

- Decreased venous tone (decreased venous compliance, relative hypovolaemia): drugs, e.g. sedatives, narcotics

Cardiogenic Shock [19, 42, 52]

- *Myopathic*:
  - *Ischaemic*, *acute myocardial infarction* but reversible ischaemic conditions like stunning and hibernating myocardium as well as *myocarditis, cardiomyopathies, septic myocardial depression*
    Takotsubo syndrome, myocardial contusion, acute exacerbated severe chronic heart failure, cardiodepressive drugs, e.g. anthracyclines, *anaesthetics*
- Mechanical:

  Valvular heart disease, particularly *acute* aortic and mitral insufficiency attributed to *endocarditis*, complications of acute myocardial infarction, i.e. papillary muscle/chordae rupture, ventricular septal defect
- Arrhythmic:

  Severe bradycardic and tachycardic disorders
- Obstructive issues as follows

Obstructive Shock [51]

- Constricted diastolic filling (reduced preload):
- Increased thoracic pressure, e.g. *mechanical ventilation*, particularly if PEEP is applied or even in case of intrinsic PEEP, *tension pneumothorax*
  Asthma, impaired cardiac compliance as in *cardiac tamponade* and constrictive pericarditis
- Confined systolic function (elevated RV or LV afterload):
- Pulmonary embolism
- Acute pulmonary hypertension
- Aortic dissection
- Left atrial myxoma
- Malignant hypertension
- Aortic coarctation

Distributive (Vasodilative/Inflammatory) Shock (DS/IS) [43]

Sepsis
Systemic inflammatory response (SIRS), e.g. pancreatitis, trauma, major tissue injury
final common pathway of prolonged or severe shock of any *cause*, e.g. cardiogenic, hypovolaemic, prolonged cardiopulmonary bypass and
neurogenic such as spinal cord injury and high spinal anaesthetic carbon monoxide intoxication (inadequate tissue oxygenation)

Shock with probable vasodilation: *metformin intoxication*, severe lactic acidosis, cyanide poisoning, *PEA arrest*, some mitochondrial diseases

Most authors allocate *anaphylaxis*, *liver failure* and *endocrinologic* maladies such as adrenal insufficiency, thyrotoxicosis and pheochromocytoma to the distributive shock form, but the data are inconclusive [43].

As obvious while reading the various aetiologies and taking the pathophysiology into consideration, quite a number of diseases may be classified in more than one category as different types of shock may coexist, and the pathologies are overlapping and mixing – with sepsis being an exceptional condition [6, 31, 53]. However, the foremost disorder will lead the pathology [37, 43, 44].

Bonanno [29] suggests a revised classification *with* a repatriation of terms like toxic shock, meanwhile lost to the gynaecologists describing just a variant of septic shock (Gram-positive exotoxic shock) [54], to the terminology of critical illness *and* the application of the new knowledge, particularly the mechanisms of cellular hypoxia *and* the central role of inflammatory response applicable in and linking all shock states and forms, ensuring more concise disease terms and depiction [29, 55]. As such, postspinal and posttraumatic neurogenic hypotension (neurologic shock), severe dehydration (hypovolaemic shock) and acute adrenal insufficiency (distributive shock) should no longer be considered and named as a shock entity: Microcirculatory haemodynamics and the cellular ecosystem are not upset during their disease course [29]. Moreover, acute adrenal insufficiency, often associated with septic shock, has to be interpreted as an epiphenomenon attributed to the fatigue of the hypothalamic–pituitary–adrenal axis leading to reduced production of ACTH and cortisone in case of protracted stress response [56].

In anaphylactic shock, release of histamine causes bronchoconstriction, vasodilation, increased capillary permeability and pathologically raised mucous secretion – not at all typical shock signs [29]. Due to the fact that patients with anaphylaxis may die from acute CS, cardiac arrest or respiratory failure [57], it may be termed "pseudo shock" [29].

Single shock features are found in circumstances like cardiopulmonary bypass and extracorporeal membrane oxygenation, acute haemolytic anaemia, severe malaria and intraoperatively caused insults [29, 37]. However, a full-blown shock picture is missing; therefore, these conditions may be seen as partial or incomplete shocks [37, 58–62]. Furthermore, septic shock, although dedicated to DS, is a very special, unique shock form and meets, during its initial "hyperdynamic", vasodilative phase, only partially (incomplete) the shock criteria [29, 63].

3.4 Pathophysiology

3.4.1 General Pathophysiological Aspects and Remarks

The pathophysiology of shock states is very complex, multifactorial and cross-linked, involving several features and setups [4–6, 45, 64]. "Systemic consequences of tissue injury such as shock involve an integrated and complex response of various systems including neural, immune, hormonal, coagulation and metabolic pathways" [65]. Furthermore, "shock" is a continuous condition, passing through distinct stages and finally ending up irreversible if not intervened upon in time [66, 67]. However, *inappropriate perfusion*, literally *hypoperfusion and maldistribution* of the blood flow finally resulting in *cellular and tissue dysoxia/hypoxia* as the *fundamental and central hallmarks of shock states* [1, 3–6], will decisively affect the cellular metabolism [68–71]: A mismatch between cellular oxygen supply and demand ensues, proceeding in diminished ATP production [69, 72]. Compensatory, ineffective anaerobic metabolism – only 2 mol ATP per mol glucose compared to 38 mol ATP via aerobic metabolism will be produced [72] – with lactate as the result of metabolic way of ATP generation arises [35, 72, 73]. The appearance of pathological lactate concentrations in blood serum in the clinical setting of shock is an *early sign* of insufficient mitochondrial oxygenation [74] and a measure of severity and duration of oxygen debt [75, 76]. However, under these conditions, cellular function may be maintained in the first instance [73]. Reperfusion on time can improve the situation, but unfortunately may simultaneously display an aggravation, the so-called reperfusion injury [77, 78].

If oxygen debt persists, marked cellular ATP depletion becomes evident [69, 70, 73], and several consequences emerge: cell swelling with loss of membrane integrity and cell dysfunction attributed to altered enzyme and protein production, blunted DNA repair, mitochondrial dysfunction [79] and disturbed intracellular signal transduction, generation of oxygen free radicals, cell deaths, etc. [80–83]. Severe and/or prolonged oxygen deprivation causing hypoxic cellular injury, initially reversible, can progress from the cellular level to systemic effects [66, 84]. Untreated, an irreversible stage with cell death, MOF and finally death arise [35, 64, 85]. Shoemaker demonstrated that cellular dysfunction, organ injury and death depend on the degree and the duration of oxygen debt associated with tissue hypoperfusion [86].

Beyond the common, certainly evidence-based understanding that *hypoperfusion*, particularly if prolonged and severe as in shock states – further depending on the organ condition prior to the hypoxic insult [87] – definitely may cause hypoxia [29, 87–89], we had to recognize during recent years that *malperfusion* (that means disturbed blood distribution with normo-, hypo-, and even hyperperfusion of neighbouring areas as well as arteriolar–venular/venous (AV) shunting) will *expose* cells and tissues to *hypoxia/ischemia* [30, 50, 90, 91]. The *consecutive hypoxic/ischaemic lesions* are obviously the *initial trigger* for a complex pathophysiology with potentially irreversible organ damage and/or MOF [87, 92–94] – thus a very complex, multifactorial, cross-linked regional reaction and, in case of shock, systemic reaction [66, 84] induced and provided by multiple contributors with hypoperfusion and cellular hypoxia being the source of origin [87, 92–94].

Notably, not only hypoxia/ischaemia itself rather than *reperfusion* may contribute or aggravate local (LIR) and promote systemic inflammatory response (SIR) [29, 77, 78, 95–97] – these coupled threats are called I–R (I, ischaemia; R, reperfusion) phenomenon [29, 63, 77]. Reperfusion injury is classically found in disturbed limb blood flow (i.e. gangrene), burns or posttrauma, crush syndrome and other entities, but may be even most important in regard to the gut's attitude in the setting and treatment of hypoxia/ischaemia, particularly if the remedies fail or are delayed [29, 98–100]. Substantial experimental and clinical evidence underlines the considerable potential involvement of the gut in the course of circulatory shock states [101–107]: Shock states themselves with their characteristic hypoperfusion but notably the redistribution following the adopted compensatory mechanisms [108–111] create cellular hypoxia. Cellular hypoxia is reported to predispose tissues to reperfusion injury attributed to regional malperfusion including local vasoconstriction and further shunting, the vicinity of hypo- and hyperperfusion, thrombosis, release of proinflammatory agents and direct cell damage [87]. Moreover, even proinflammatory mediators originating from extraintestinal areas may exert jeopardizing effects on the gut (gut as distant remote organ being affected by the inflammatory response of other tissues and organs), as proinflammatory substances are key modulators of the gut tight junction expression and localization within the gut epithelium engendering increased permeability and thus dilute gut mucosa barrier function [112]. Further, Hassoun et al. [113] found that shock-induced gut hypoperfusion may, in turn, be followed by intestinal reperfusion injury with the production and release of proinflammatory mediators potentially amplifying the systemic inflammatory response syndrome, SIRS, by an I–R phenomenon. Alverdy et al. [114] demonstrated that interactions between host and bacterial pathogens may lead to gut-associated sepsis and this independent of intestinal permeability changes (remote organ involvement). Hence, the intestine, sometimes termed as the "motor" of critical illness [112], may initiate (more infrequent) but most frequently propagate SIRS and thus MODS and critical illness as the result of (remote) second organ involvement preferentially related to *I–R injuries*, also known as *second hit* [102, 104, 107]. In haemorrhagic (hypovolaemic) and distributive shock, if not treated causally and on time, even if macrohaemodynamics are reversed to normal or near normal, before a no-return point is accomplished during the shock process, the patients are jeopardized to a second hit, and most of the deaths in critically ill patients are related to this [115, 116].

Meanwhile, there is quite substantial evidence that "some elements of the gut contribute to critical illness" [112]: Unfortunately, no unifying thesis embedding the possible pathways the gut may contribute in propagation and its impact on the disease course exists yet [112]. The "gut elements" (the *epithelium*, mucosal barrier function; the *mucosal immune system*; and the *gut content* such as bacteria/bacterial toxins and products and even faeces) involved are leading once more (a second time) to the release of inflammatory, vasoactive, proteolytic and several other substances and mediators. Together with the local and at least latent systemic immune response and bacteria- and faeces-associated toxic damage, all these may provoke continuation and boost the mediator cascades already activated and running [101, 103, 105, 117, 118]. Furthermore, strong evidence is present confirming loss of gut barrier function early in the course of critical illness, albeit bacterial translocation has not been definitely disposed [119]. Nevertheless, there is no doubt and persuading evidence that intestinal material and products are released into the body kindling critical illness [112]. Most recently, a contribution of the crosstalk between the enteric and central nervous system has joined the debate; see overview by Clark [112]. Insofar, the intestine undoubtedly is of crucial relevance in pathobiology of shock states and the evolution of critical illness [112]. The main target of this (mostly) systemic response of the intestine is predominantly the lung mucosa potentially causing acute lung injury (ALI) or even acute respiratory distress syndrome (ARDS) [101, 102, 120, 121].

Basically, generalized inflammatory response (SIR) – traditionally termed systemic inflammatory response syndrome (SIRS) – commonly and conventionally attributed to *sepsis and septic shock* may be similarly displayed in case of *non-septic, toxic* cellular and tissue insults *such as hypoxia/ischaemia and reperfusion injury* or *trauma*, able to *elicit a shock setting* [122, 123]. Hence, both septic and non-septic triggered responses are allied clinical entities but evoked by *different underlying aetiologies* [29, 63, 123, 124]: "Tissue injury induces systemic inflammatory response and immune dysfunction (even) in the absence of infection" [122].

Inflammation associated with shock is a complex process [63, 123, 124] and part of the reaction of vascular tissues to deleterious insults [125]. It is considered as being the organism's first-line stereotype, ontogenetic and phylogenetic answer in terms of innate immune response to cellular and tissue injury [29, 126, 127]: Tissue damage triggers a haemodynamic, cellular inflammatory and coagulatory and humoral–metabolic response [63, 124]. The magnitude of this reaction correlates with the extent of tissue injury [128] or, even more precisely, is related to the degree of oxygen debt at the cellular level [129] – as does the shock state. It seems that types, amount and the composition of the inflammatory mediators and toxic substances released rely upon the initial stimulus, hence may initially differ in septic- and hypoxic-induced shock states [63]. This is in line with the fact that the *early septic shock is characterized by vasodilation* [43, 63], whereas *CS and hypovolaemic, particularly haemorrhagic, shock initially and during the early disease stages show peripheral vasoconstriction* [35, 52, 63]. The recognized difference in composition is further substantiated by the finding, that in the *early* critical illness, submerged proinflammatory cytokines and stress hormones, promoting capillary leakage and vasodilation, at least in distributive, especially in septic shock, are less coined in most cases of CS and clinically irrelevant in haemorrhage [130, 131].

Both tissue hypoxia/ischaemia and bacterial substrates and products like LPS elicit a large number of diverse, mainly vasoactive and inflammatory mediators primarily provoking local inflammation (LIR) [29] which may progress to a systemic inflammatory syndrome (SIRS) ending up in MOD/MOF presenting as an identical clinical picture [132–135]. The intensity of the systemic impact of the initial local damage considerably determines the disease course and may consecutively cause metabolic derangements and MODS/MOF [136]. Duration, quantity and quality of the insulting stimulus essentially modify the progress [29]. Ongoing systemic hypoperfusion causing cellular and tissue hypoxia/ischaemia has been accused of leading to SIRS and irreversible MODS [137]. Nathan [138] and

Jarrar et al. [139] demonstrated that the inflammatory response may even affect healthy tissues and thus further aggravate inflammation probably ceasing in organ failure or even death [140]. The gut undoubtedly plays a pivotal role in disease propagation and systemic manifestation, being mostly affected, perturbed and activated via a second hit under the outlined conditions [101, 102, 106, 107, 112–114]. Most recently, the impact of crosstalk within the affected organ(s), particularly the intestine, but also between different organs, has been stressed [112, 141], e.g. acute lung injury/ARDS or acute kidney injury complicating sepsis or mechanical ventilation [142, 143].

Not at least, it has to be noted discerning that the condition shock, by definition, encompasses MODS/MOF as MODS/MOF describes a progression of the disease state, not a *complication* of shock [63].

Among the above-mentioned contributors to the shock pathobiology, the *endothelium* and the *microcirculation* have crystallized to be of considerable impact and crucial importance. Already suggested by Kovach and Lefer in 1993, "endothelial cells seem to be extremely sensitive to shock-induced dysfunction" [92], and the most recent evidence has revealed a central role of the endothelium in the pathology of septic and other shock forms [144, 145]: *Endothelial cell activation and dysfunction* are widely recognized and an accepted appearance in circulatory shock [92, 145–149]. Hypoxic/ischemic injury, inevitably following a marked depletion of ATP [69, 70, 150], affiliated to hypo- and malperfusion will insult endothelial cells causing endothelial activation and dysfunction [92, 146, 147, 151, 152]. It has been demonstrated that hypoxia provokes via modified gene expression changes in endothelial phenotype resulting in endothelial activation and dysfunction [148, 153, 154].

Endothelial cells are lined up inwardly along the vessels [155], are pervasive within the organism, cover a surface area of roughly 4,000 m^2 [156] and may be allowed organ status [155, 157]. This 'organ' provides an "interface" between circulating blood and the vessel wall, the endothelium is involved in many physiological sequences [157], such as the regulation of the vascular tone and permeability, balance between coagulation and fibrinolysis as well as between proinflammatory and anti-inflammatory activity. Endothelial cells are recognized to considerably participate in the initiation and amplification of the inflammatory response [152, 155, 158–160]: Together with other cells like monocytes, tissue macrophages and circulating neutrophiles (due to activated endothelial cells adhered) are the endothelial cells *substantially involved in the innate immune response* to injury [144, 152, 161, 162].

Thus, the endothelium plays a crucial role in vascular homeostasis [163–165]; it responds to various hormones, neurotransmitters and vasoactive factors by producing and releasing vasodilative mediators such as nitric oxide (NO) and prostacyclin (PGI_2) and vasoconstrictive agents like endothelin-1 (ET-1) and thromboxane (TXA_2), trying to keep a well-balanced vascular tone meeting the cellular and tissue metabolic demands [152, 163, 164]. Endothelial cells show crosstalk communicating upstream information about haemodynamic constitution downstream further modulating the local blood flow [166].

In case of endothelial affliction, e.g. secondary to hypoxia and hypoperfusion, endothelial cells may be activated and dysfunction can occur [148, 152, 153] – represented as blunted vasomotor response with diminished endothelium-dependent vascular relaxation and augmented endothelium-dependent vascular constriction [145, 157], raised vascular permeability, disturbed balances preferring (pro)inflammatory and prothrombotic activity [167–170], platelet activation and aggregation, and, in cooperation with other cells like circulating neutrophiles, launching innate immune response to injury [157, 162, 170–172]. Endothelial activation represents a spectrum – not an all-or-nothing phenomenon – of cellular structural and functional changes ranging from physiological adoptions to pathological features [144, 152]. Ten and Pinsky discuss in their paper the differences between an appropriate adaptive reaction and what becomes dysfunction [152]. Vasodilation and increased permeability following laceration are necessary for wound healing and thus are physiological [152, 173]. On the other hand, the activated and potentially dysfunctional endothelium may exhibit a considerable disruption of its important abilities to regulate the microcirculation with loss of smooth muscle control and electrophysiological communication blunting signal transduction pathways [148, 166]. Thus, the valuable crosstalk between the cells is impaired [174, 175] and information about distal flow haemodynamics of the small vessel system does not reach proximal endothelial cells [175]. The nitric oxide system, within the mediator network contributing to a suitable tissue perfusion [163] may be substantially perturbed due to a heterogeneous expression of the NO synthase in different areas [176, 177] and thus uneven NO concentrations in nearby tissue areas [178]. This results in a mismatch between vasoconstricting and vasodilative mediators provoking shunting and inhomogeneous blood flows occur [174, 179]. Furthermore, smooth muscle cells as situated alongside the endothelial cells lining the arterioles and regulating perfusion as effector cells [145, 180, 181] lose their sensitivity to respond to catecholamines [180, 182]. Capillary leak (syndrome) is an expression of disturbed endothelial barrier function in response to several stimuli including hypoxia [183]. Insofar, "inflammation can be seen as a vascular response" [184], and the "endothelial response to hypoxia primes vasculature for inflammatory response" [152]. Anyhow, if the threat and its response are not controlled on-site and especially if prolonged or severe, amplification of the inflammation and direct irreversible cellular injuries ensue; the process generalizes and shock state manifests [144, 152].

In conclusion, the endothelium takes a decisive part of the innate immunological response to insults, it 'orchestrates' the local answer by promoting the adhesion and transmigration of leukocytes, altering local vasomotor tone, increasing the permeability, changing the balance within the vasoactive mediator composition affecting perfusion, activating coagulation disturbances and the complement cascade, all may contribute and trigger cell death [29, 185].

Altered *microcirculation* has emerged as being an essential element in the pathobiology of shock states and other critical illness maladies [29, 90, 91, 186–188]. A well-functioning microcirculation is a crucial assumption and inevitable for adequate tissue oxygenation, since microcirculation is the place where the exchange of substrates and oxygen delivery to the tissues and cells happens [44]. Initially, studies in septic shock, the most common shock syndrome were conducted, because altered microcirculation is found to be frequent in sepsis [91] helping to understand its previously not completely explainable pathophysiology [44]. In sepsis, disturbed microcirculation plays a key role in the pathology [29, 189]. However, meanwhile altered haemodynamics of the small vessel networks have been demonstrated aside from septic shock in several other critical illness states including cardiogenic shock and severe acute heart failure as well as haemorrhagic shock [73, 90, 186, 190–192].

Microcirculatory alterations encompass decreased capillary density and, of most impact, a *heterogeneity in blood flows* as capillaries may not be perfused or only intermittently, others are underperfused, while some show normal perfusion or are even clearly more than normal perfused (hyperperfusion) [44, 90, 164, 189, 193]. This generates hypoxic and non-hypoxic areas in close vicinity – dysoxic areas occur [174, 194], even if total organ blood supply and flow are preserved [195]. Further, the coexistence of differently perfused areas is suggestive for shunting [37, 50, 90], in sepsis the predominant pathomechanism [23, 44]. Spronk et al. [196] and De Backer et al. [91] found capillary obstruction along with normal blood flow in larger vessels in septic shock patients, thus a direct illustration of *functional shunting* particularly in distributive shock, but, as reported, in other shock types as well [193, 197, 198]. Moreover, both, autoregulation (an important mechanism to provide an appropriate blow flow distribution ensuring that oxygen and nutrient supply meets the actual metabolic demands of each organ region and tissue area) is substantially blunted attributed to too low perfusion/blood pressure, a macrocirculatory failure, and a markedly attenuated or even no regulation of vasomotor tone ("vasoplegia"), a microcirculatory failure, contribute to disproportionate local blood flows [199–201]. *The heterogeneity of microcirculatory blood flows* promoting dysoxia is *characteristic and a hallmark* in sepsis and septic shock, but may apply as a decisive pathophysiological feature in various conditions, particularly in all *shock state diseases* [90, 91, 202, 203].

Heterogeneous perfusion is associated with disturbed oxygen extraction capabilities [204–207] and causes more severely perturbed tissue oxygenation than homogeneously diminished perfusion [205, 207]. Those haemodynamic abnormalities of the peripheral vessel networks may lead to cellular injury [208], and the severity of microcirculatory disturbance is associated with the development of organ dysfunction and death [90, 91, 195, 209–211].

The depicted circulatory aberrations are basically referred to as a *loss of regulation of the peripheral vasomotor tone associated with endothelial cell dysfunction and activation* [145, 148, 179].

Noteworthy and important for understanding, monitoring and therapy, at least in more advanced stages of critical illness, a *dissociation between micro- and macrocirculation* is evident: Despite improvements in systemic haemodynamics, microcirculatory abnormalities may persist [212–215]. This may be time and stage dependent as microvascular abnormalities early in the course of sepsis are obviously not uncoupled from macrocirculation [199, 216, 217]. However, the link between micro- and macrocirculation is loose! [174]

Failure of the microcirculatory system "is a conditio sine qua non for irreversibility" in all shock types [29]. Thus, while the macrocirculation may simply be considered as a reservoir for the microcirculation, it is the malfunction of microcirculation which determines the course of the malady [29], and it is the *arteriolar–capillary* and/or the *cellular*, accurately the *mitochondrial*, site which *will finally collapse* [2, 29, 145], with either:

I. The destruction of the arteriolar and capillary reaction to signals, hormones, mediators and the local milieu, etc., indicated by the loss of any vasomotor tone (e.g. resistance to catecholamines [218, 219], attenuated sympathetic tone secondary to altered perfusion of the central nervous system [220], and inadequate NO release [221, 222]).
II. The breakdown of one or more of the four main enzymatic pathways at the mitochondrial level resulting in the impossibility to produce energy. The points of no return and hence shock irreversibility will be attained in case of acute severe and/or persisting hypoxia [2, 6, 29, 66]. The two points of no return are independent of and show no correlation to macrocirculation [29].

3.4.2 Compensatory Mechanisms and Shock Stages

The response to any insult is complex and will launch a couple of setups in order to restore normal physiology, e.g. haemodynamic, humoral–metabolic and cellular inflammatory compensatory reactions [6, 35, 41, 63, 68, 73, 223, 224]. The activation of the sympatheticoadrenal and hypothalamic–adrenal system – as involved in any shock type [63, 223, 224] – is a more nonspecific answer and a general alarm to numerous threats and disorders such as hypovolaemia (particularly haemorrhage), hypo- and hyperthermia, hypoxia/ischaemia, pain, infection, pH imbalance and necrosis [29, 63, 88, 224]. Sympathetic discharge, release of ACTH, corticosteroids, catecholamines, ADH, insulin and glucagon, and the activation of the renin–angiotensin–aldosterone system (RAAS) are the main issues predominantly mobilized and set up [6, 29, 63, 224] supporting the macrocirculation by increasing the peripheral vascular resistance (vasoconstriction) [6, 32, 63, 224]. This aims to reinforce an appropriate mean arterial pressure essential for a sufficient autoregulatory pressure level and hence decisive for adequate organ perfusion [108, 110, 224], maintaining or even increasing intravascular fluids [6, 224–226] as well as causing blood redistribution to crucial organs like the heart and the brain, whereas the skin and skeletal muscles, the kidneys and the splanchic circulation become less perfused [68, 108–111]. Variable degrees of kidney dysfunction are obligatory in shock [227, 228].

Simultaneously, parasympathetic tone is attenuated, and particularly in more severe hypovolaemia, a substantial vasovagal response may lead to a transient and uncharacteristic bradycardia [229–231]. Details can be gathered from the diagram by Klabunde [224].

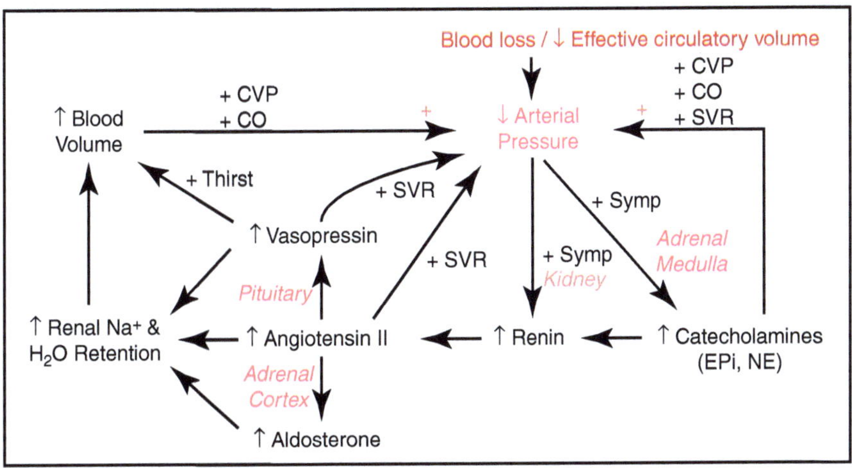

Although vasoconstriction defines any shock in its established form [43, 63, 224], a *hallmark and characteristic for distributive shocks*, particularly septic shocks, *is peripheral vasodilation* [43, 232] and low systemic vascular resistance [43], usually indicated by a low SVR in daily practice. Vascular resistance mainly applies at the level of the small arteries and arterioles [233, 234], an integral part of the microcirculation [224, 235, 236], where the compensatory mechanisms launch control of vasomotor tone to regulate perfusion pressure and perfusion within the physiological range [237, 238]. Astonishing, microscopic animal studies revealed that the compensatory sympathetic discharge following SNS (sympathetic nervous system) activation caused by relative or absolute hypovolaemia indeed elicits constriction of the arteries and the medium-sized arterioles, but does not affect the terminal arterioles, the intrinsic resistance vessels and as such substantial contributors to blood pressure magnitude; hence, concurrently dilation of the terminal vascular bed may occur in an otherwise sympathicotonic, vasoconstrictive milieu [239–241]. Moreover, distributive shock is a condition basically originating within and altering the microcirculation [29, 44, 49, 50]: The considerable systemic inflammatory response predominantly exerts influence on the microcirculation (provoking maldistribution with unevenly perfused areas, and cellular derangements with capillary fluid and inflammatory cell extravasation as well as dysoxia) but indirectly affects macrocirculation as well by causing vasodilation with relative hypovolaemia, most often accompanied by hypotension, following a substantial diminution of the total peripheral resistance attributed to arteriolar and venule dilation as a specific feature of this shock type [204, 242–245].

Nevertheless, sympathetic and neurohumoral compensation ensues in severe sepsis and septic shock (as low normal or mild to moderate hypotension, and an alleviated effective circulating volume attributed to fluid losses following increased permeability will be recognized by baro- and cardiopulmonary receptors, and other control mechanisms of the body) causing sympathetic stimulation, discharge and humoral responses, e.g. marked increases in catecholamines and an activated renin–angiotensin system, are verified [43, 63, 224, 246–248]. However, the systemic inflammatory reaction following local, septic injury by LPS toxins and other bacterial materials in the course of septic shock primarily affects the microcirculation *overwhelming the vasoconstrictive compensatory setups* [63].

Based on the currently available knowledge, the ischemic–reperfusion injury as the foremost event within toxic–inflammatory derangements seems to elicit a different composition of vasoactive mediators compared to the exposure to bacterial toxins or material [63], although the mainstream of inflammatory response is similar in both, hypoxic–toxic and septic conditions [29, 63]. Moreover, (I.) with the exception of distributive basically septic shock, all shock forms are primarily compromising the macrocirculation [29, 44, 49]; (II.) as described in detail by Landry and Oliver [43], the specific role of vascular smooth muscle failure to contract in distributive shock needs to be highlighted. Notably, the small vessels in septic shock are not plegic – vasoplegia is often stated [249] – rather than the frequency and the amplitude of their rhythmic vasomotions are altered [250]. Both may be responsible for the different course in the early phase of this special shock entity [43, 47]. Additionally, (III.) the modification of the venous system in septic shock syndrome with disproportionate impairment of peripheral vasoregulation affecting both, the veins and the arterioles may have an impact [251].

As adequate intravascular volume will maintain microcirculatory integrity and adequate tissue perfusion in the presence of low-resistance circulation [252–255], *timely* fluid supply in the early, hyperdynamic, fluid-responsive phase of septic shock, may stabilize haemodynamics, and thus, this septic stage may be seen in comparison to the other shock types as a somehow compensated, partially *incomplete* shock condition [29, 63]. Sufficient fluids applied well on time may prevent a progression to a vasoconstrictive, decompensated disease phase [63, 243].

Anyhow, the general clinical results of compensatory reactions are tachycardia, vasoconstriction, preserved mean circulatory blood pressure and intermittent control of ventilation, all trying to limit the damage by increasing the cardiovascular function, preserving fluids and enhancing the supply of the tissues with oxygen and nutrients in order to cover the requirements [6, 34, 224]. Synchronously, cellular immunological/inflammatory features ensue, interacting and interfering with the other mechanisms [256].

If the compensatory mechanisms are insufficient to reverse or at least to attenuate the mismatch of oxygen and nutrient supply of the tissues and to mitigate the inflammation and the concomitantly activated cascades, decompensation ensues [4, 84, 257]. Although vasoconstriction even in advanced (hypodynamic) septic shock is the predominant feature [29, 63, 258], in decompensated stages *clinical hypotension* results despite adequate volume resuscitation, expressing failure of compensation, in particular of the sympathetic nervous system [4, 29, 63, 84]. Septic shocks then migrates to the hypodynamic phase which, among others, is characterized by its unresponsiveness to fluids (no increase in BP and tissue perfusion following fluid administration) [259, 260]. Thus, BP drop in general and the haemodynamically prevailing hypotension clinically mark the transition from compensated to decompensated shock [261]. At this stage catecholamine plasma levels have been demonstrated to be considerably raised with the RAAS highly activated, but despite that, *vasodilation* ensues and may become established and therapy resistant with ongoing disease in all shock categories [43, 262, 263]. Several elements such as hyperpolarization of vascular smooth muscle cells [264], plugging of the capillaries [265], increased nitric oxide [266, 267], generation of free radicals [268] and vasopressin deficiency [269] are suggested contributing to persistent shock development.

One proposed contributor requires specific notation: As with therapy-resistant hypotension, organs like the heart and the brain will become hypoperfused and cerebral ischaemia will impair *central* vasomotor regulation causing substantial reduction or even complete loss of peripheral sympathetic tone and may abolish the initial (compensatory) vasoconstriction [220, 224, 261]. The so-called sympathetic escape may not only cause progressive hypotension and hypoperfusion, but the consecutively raised capillary pressure leads to fluid filtration into the organ tissues further precipitating hypovolaemia [261]. Otherwise, fluids administered in the irreversible shock phase will be sequestered within the capillary circulation, all provoking fluid overload with (organ) oedema and anasarca [239, 265].

Even at this advanced shock stage characterized by pronounced vasoplegia, the condition is potentially reversible [29] unless one of the two points of no return (NRP) are accomplished: significant (a) arteriole (small vessel) and/or (b) mitochondrial collapse [29]. The vasculature is unable to respond to any stimulation in order to restore peripheral vasoconstriction (vasomotor paralysis) and thus the tissue circulation and the mitochondria are unable to produce energy (cytopathic hypoxia) whatever enzymatic pathway may be activated [2, 29]. Adrenal failure may contribute to the vasomotor paralysis [270]. Of note, the NRPs are both independent of and do not correlate with macrocirculation; thus, even if macrocirculation could be (intermittently) re-established, irreversible shock will arise [29].

3.5 Special Pathophysiology of

3.5.1 Cardiogenic Shock

Cardiogenic shock (CS) describes a severely impaired, primarily heart associated, cardiac performance with evidence of systemic hypoperfusion in the presence of adequate intravascular filling (euvolaemia) [19, 52, 271, 272].

CS is clinically characterized by *hypotension* with a systolic blood pressure (sBP) <90 mmHg, lasting ≥30 minutes or

the necessity for catecholamines and/or IABP support in order to maintain sufficient macrocirculation with sBP ≥90 mmHg *and*

evidence for hypoperfusion as identified by cyanosis, cold peripheries (extremities), altered mental status and (persistent) oliguria [19, 52, 271–273].

Haemodynamically, most authors determine a CI ≤2.2 l/min/m^2 and a PCWP >15 mmHg [52, 274–276] as being typical features for CS.

In cardiogenic shock, the overwhelming majority of cases are caused by an abrupt depression and/or loss of myocardial contractility irrespective of loading conditions with subsequent significant fall in SV/CO and, frequently, blood pressure drop [52, 274, 277]. The predominant aetiology is LV failure associated with acute myocardial infarction [19]. The acute and for the affected patient critical loss of contractile properties [52] results in a significant decrease of total pump force [278], thus in systolic and due to diminish LV relaxation and compliance, in diastolic dysfunction as well [52, 276, 279]. The diastolic properties are further affected by the ischaemic process and cascades [276]: Ischaemia may be amplified, as a considerable fall in CO, and concomitant BP drop following CS further compromises coronary perfusion. Additionally, the compensatory response foremost activates sympathetic and renin–angiotensin (aldosterone) systems causing – among others – vasoconstriction and tachycardia; both may aggravate myocardial ischaemia by increasing afterload and myocardial oxygen demand as well shortening the duration of diastole and hence myocardial perfusion [42, 276, 280]: "'Ischemia begets ischemia', leading to a progressive downward spiral of worsening ischemia, progressive deterioration of myocardial function, and worsening shock" [52]. Ischaemia and tachycardia – the latter caused by compensatory mechanisms – promote fluid retention and impaired diastolic filling potentially provoking pulmonary congestion and hypoxia [52]. Even in non-infarcted areas hypotension and consecutive ischaemia and metabolic derangements may compromise contractile properties [281]. Hence, in CS, both diastolic and systolic dysfunctions are present [280, 282].

However, the development of CS is complex [41, 42] and obviously not solitary and exclusively dependent on the degree of impairment of cardiac contractility/performance: In many cases LVEF (representing systolic function [283, 284]) is found to be "only" moderately diminished [285]. The population investigated in the SHOCK trial showed an average LVEF of 30 % [274], and some patients even present with CS although their LVEF is fairly preserved and no severe mitral regurgitation is noticed [286]. Further, the EF is demonstrated to be absolutely the same, in the acute phase of CS and 2 weeks later in the presence of a quite different functional situation [287]; hence, severely impaired systolic function does not inevitably result in shock development [42]. The reason may be in part due to a deficient or incomplete acute ventricular adaptation because "acute" LV dilation may preserve or restore SV in the early phase, but roughly half of all CS patients show a small or normal LV size [287, 288]. In addition, diastolic properties may contribute to CS development [42]. Moreover, reperfusion within 6 h will not immediately elicit recovery of the regional myocardial function; as due to myocardial stunning, it may take days and even weeks until the salvaged myocardium resumes its contractile function [289, 290]. Further, persistently ischemic myocardium may enter a hibernating state with severe contractile dysfunction and this may contribute to the development of CS [52, 291–294].

CS is a central shock form, causing scarce peripheral and organ perfusion attributed to substantial pump failure, hence is a "totalitarian organ derangement right from the beginning" [29], macrohaemodynamically indicated by compromised CO and MAP [29, 63, 295]. Traditionally, CS is seen as a mechanical problem [52] with corresponding central nervous and humoral activation and response [280, 296]. The classical paradigm of CS was depicted by Antman [296] – represented by the "right side", the mechanical aspects of CS within the current concept as issued by Hochman and Reynolds [42], see below – who confirmed the work by Califf [280]: In CS, myocardial dysfunction directly leads to both, decreased SV and an increase in left ventricular filling pressure (LVEDP) [38, 52, 282]. In response, several compensatory features ensue causing systemic vasoconstriction secondary to neuroendocrine, particularly sympathetic activation and fluid retention [38, 52]. Unfortunately, systemic vasoconstriction and fluid load exert adverse loading conditions onto the already limited cardiac function which are not balanced by the potential increase in contractility attributed to the sympathetic discharge; hence, in particular the relation between cardiac performance and afterload becomes disproportionate, the so-called afterload mismatch [297] emerges, further diminishing the heart's performance [42, 297].

However, surprisingly, several recent studies on CS revealed a fundamentally different haemodynamic profile than expected in number of patients suffering from CS [39, 41, 274, 297, 298]: Although their contractility is severely impaired with a marked fall in SV accompanied by relevant diastolic dysfunction and (in the vast majority of cases) an elevated LVEDP, the patients exhibit a variable, only marginally to moderately elevated peripheral resistance, in some patients as low as typical for patients with severe sepsis and septic shock [40]; see figures by Cotter [39].

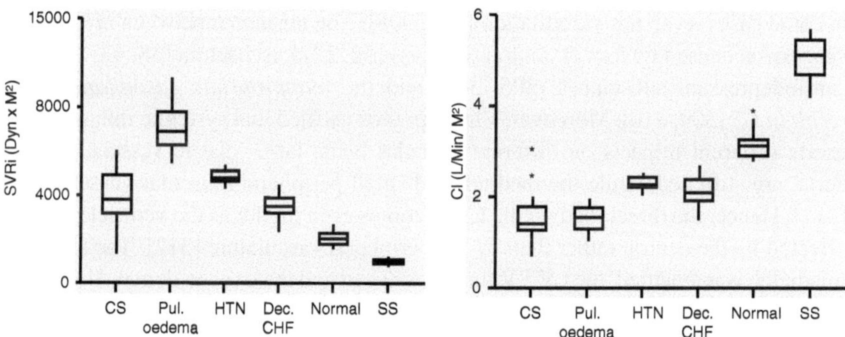

Diagram by Cotter et al. [39] The level of peripheral resistance in CS swings in a wide range and may be in single patients be as low as found in sepsis. On average it is comparable with that of acutely decompensated chronic heart failure but is clearly less than in pulmonary oedema or decompensation following hypertensive dysregulation. Likewise, CI in CS is, in general, not that bad and ranges, besides single cases, on average at the same level found in patients with pulmonary oedema; furthermore, it is not substantially lower than in acutely decompensated chronic heart failure.

CS cardiogenic shock, *Pul. oedema* pulmonary edema, *HTN* hypertensive crisis, *Dec. CHF* decompensated chronic heart failure, *SS* septic shock

This inappropriate (and functionally insufficient) vasoconstriction is consistent with the observation that acute myocardial infarction (AMI) may deploy a systemic inflammatory response syndrome (SIRS) and reflects that SIRS may be present in patients suffering from CS [19, 40, 41, 297–301]. Indeed, in CS elevated levels of interleukin-6 (IL-6) are reported, while the serum concentrations of procalcitonin are low compared to sepsis, hence giving evidence of an *inflammatory process* launching proinflammatory mediators rather than the shock situation itself [298, 302, 303]. Theroux demonstrated raised IL-6 and TNFα concentrations already on admission in patients with AMI and found a dramatic increase of these cytokines within 72 h after the onset of MI [304]. Both IL-6 and TNFα exhibit myocardial depressant effects; furthermore, TNFα may provoke coronary endothelial dysfunction further compromising coronary perfusion [305]. In the SHOCK trial 18 % of patients were as suspected being septic, of which 74 % had positive blood cultures [40], suggestive for bacterial transmigration secondary to dysfunctional intestinal wall barrier function associated with and induced by hypoperfusion of the intestinal tract [42]. Low vascular resistance early in the course of CS development turned out to be associated with positive blood cultures further complicating the disease [40]. Moreover, reperfusion may directly induce the release of vasodilatory-acting mediators, and Kohsaka detected high levels of inducible NO synthase (iNOS) subsequent to the release of inflammatory mediators in patients with acute myocardial infarction [40]. These features are consistent with a second hit phenomenon reported as being part of the general pathogenesis of shock syndromes [29, 42, 95, 97, 102]. High levels of iNOS are associated with LV dysfunction [300, 306]. Concomitant, high concentrations of NO and peroxyacyl nitrates (mediators with vasodilative and cardiodepressive effects) contribute to SIRS, counterbalancing the initially induced compensatory vasoconstriction and, in fact, may lead to an inappropriate circulatory response with, in some cases, even net vasodilation [307, 308]. The *mismatch* between *myocardial depression* caused by loss of contractile mass [52, 277], ischaemia [38, 42, 296] and cardiodepressant substances [305, 306] and the *inappropriate vasodilation* may *result in CS* [309, 310]. Moreover, Vlachopoulos verified that systemic inflammation exerts different impacts on different vascular beds: large, elastic vessels, e.g. the aorta, are stiffened, while the medium and small peripheral muscular vessels dilate [311]. Hence, the direct load on the LV becomes even higher as the ventricle is more affected by the central rather than by the peripheral vasculature [312]. The afterload mismatch is intensified, the LV-SV further reduced and the oxygen demand increased [311], all without proper increase in MAP since the peripheral vasomotor tone, which decisively determines the MAP, is compromised [313–315].

Along, low peripheral resistance predisposes patients with CS to endothelial damage [40].

So, the systemic inflammatory reaction contributes substantially to the pathogenesis and the course of CS [300, 316–318] with further reduction of myocardial contractility secondary to aggravated afterload mismatch (potentially causing further myocardial ischaemia), the generation of myocardial depressive mediators and the release of peripheral vasodilative mediators [282, 301, 316, 319–321]. This may lead to profound, persistent and refractory vasodilation and hypotension [41, 136, 316, 322] and to the development of MODS/MOF [41, 136, 298] with its deleterious outcome if not treated adequately and in time [41, 43, 136]. If the inflammatory markers are substantially elevated at presentation in stable AMI – patients without any evidence of shock, the risk to develop CS later in the course is found to be high [323]. Furthermore, concentrations of inflammatory agents are found to be more and considerably elevated in case of substantial myocardial necrosis compared to unstable angina [40], while in manifest CS the highest levels are found reaching amounts as high as in septic shock [298].

Thus, in our current understanding, *substantial (cardiac) myocardial damage/dysfunction* and the *associated systemic inflammatory response* (SIRS) *with subsequent vasodilative impact* on the peripheral vasculature are the main components in the pathogenesis of CS [41, 42, 300, 316–318]. The "only" marginal to moderate, *disproportionate peripheral vasoconstriction* – in some cases even net vasodilation – *indicated by an inappropriate low SVRI* due to vasodilative mediators (particularly iNOS) generated in the context of an inflammatory response in the setting of CS complicating AMI [41, 319] and counterbalancing the reflex and humoral compensatory mechanisms plays a significant and pivotal role in the pathophysiology of CS and, in turn, is pathognomonic for CS [39–41, 297, 298].

In 2003 Hochman [41] suggested a new cardiogenic shock paradigm integrating the newer pathophysiological aspects within the older concepts from Antman [296] and Califf [38] but was subsequently modified by Reynolds and Hochman [42]:

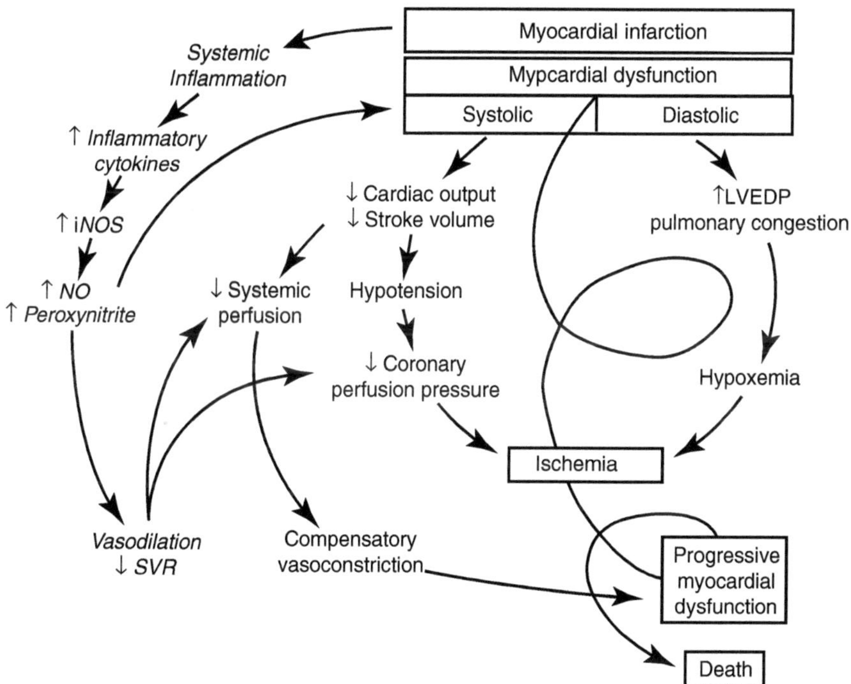

The traditional, mechanical and neurohumoral cardiogenic shock model depicted on the right side combined with the newly verified inflammatory response features (left side, cursive) demonstrated to be present in cardiogenic shock situations constituting the new cardiogenic shock paradigm. Modified with permission from Reynolds and Hochman [42]

Hypotension, better a critically low blood pressure (BP), usually accompanies cardiogenic shock as secondary to the decrease in MAP following the considerably reduced cardiac performance with impaired SV and inappropriate vasoconstriction; the BP is low [272, 274, 280, 298]. Autoregulated blood flow distribution occurring in the presence of regional vasodilation of arterial resistance vessels will be blunted in hypotensive conditions [200, 201]. As such, hypotension may engender even more ischaemia, even less myocardial contractility and shock [319]. Furthermore, hypotension reflects failure of the sympathetic nervous system and other compensatory mechanisms; however, normotension does not ensure haemodynamic stability [257]. In the SHOCK registry and trial, a small number of patients were clinically normotensive or only mildly hypotensive but showed systemic hypoperfusion – a hallmark of shock states [42] with low cardiac output and elevated left ventricular filling pressures (LVEDP) [16, 274, 324]. As the SVR was reported to be reasonably elevated, a fair BP could be maintained with the systemic inflammatory response obviously not overwhelming the compensatory features [273]. Menon therefore suggests to refer those situations as being a pre-shock [273].

Both systolic and diastolic dysfunctions may cause CS [38, 282]. Ischaemia impairs myocardial compliance, a diastolic property; subsequently LVEDP will rise [325, 326]. An elevated LVEDP puts the patient at risk of developing pulmonary oedema [327]. Furthermore, severely impaired systolic function/pump failure will precede an increase in LVEDP in order to maintain SV [38, 52, 328]. Thus both systolic and diastolic disorders are contributing to an augmented LVEDP potentially promoting pulmonary oedema [38, 329] and further ischaemia [52, 280].

LVEDP reflects well the compliance of the left ventricle [328], and an abnormally high LVEDP indicates abnormal stiffness of the LV [330]. It is well known that, particularly in critically ill patients, the compliance of the ventricles continuously varies, contributing to the heterogeneous response and changes of the LVEDP value [331, 332]. Even in healthy persons there is absolutely no correlation between changes in ventricular filling and a shift in LVEDP [333]. Hence, even severe acute heart failure as CS is not necessarily accompanied by high filling pressures (LVEDPs); some patients with CS will definitely show normal or even low LVEDPs [334–336]. Insofar, LVEDP may not adequately indicate LV preload or intravascular volume status and changes and may not correctly reflect different fluid loading [332, 333, 337].

The role and the involvement of the right ventricle (RV) need to be noticed: Sharing the interventricular septum and being enclosed by the "same" pericardium, interactions between the right and the left ventricle occur [338–340]. Thus, it is not surprising that RV function may be affected, altered and even contribute to CS [42], whereupon predominant RV – infarction and shock mainly attributed to CS – is rare (about 5 %) [341]. Significant RV acute myocardial infarction complicates 50 % of all inferior infarctions [342]. The haemodynamic alterations and the severity of circulatory compromise are determined by the damage of the RV itself, the ventricular interaction and the involvement of the LV in the ischemic injury [341, 343]. Foremost, RV function will be affected due to acute LV failure with increase in LVEDP provoking pulmonary hypertension and hence an acute increase in RV afterload [344, 345]: A sufficient RV pump function ensuring the delivery of an appropriate preload required to guarantee LV output [346, 347] is necessary to prevent (further) LV pump failure – series effect as the two ventricles are arranged in a row [346, 348]. However, in contrary to former recommendations vigorous volume loading intending to providing adequate RV filling and preload may be detrimental, as a consequence of ventricular interdependence increasing RV filling may shift the interventricular septum toward the LV, reducing LV cavity size, LV filling capacity and thus compromises LV output and LV systolic function even further [42, 349, 350].

In recent years, microcirculatory disturbances are also reported to participate and contribute to CS [59, 90, 351]. Microcirculatory flow is shown to be impaired [352–354], an increase in flow heterogeneity has been demonstrated [352–354] and the coexistence of well-perfused and non-perfused areas indicates microvasculatory shunting [50, 204, 355]. In patients with resuscitated cardiogenic shock, heterogeneous microvascular perfusion could be demonstrated [90]. Diminished leukocyte [356] and erythrocyte [352] deformity, interstitial oedema and microvascular vasoconstriction [357] attributed to sympathetic–vagal imbalance leading to microvascular constriction promoted by low CO [90] are suggested reasons for the perturbed microvascular circulation. Some authors discuss that the alterations may be due to inflammatory mediators [358, 359]; reactive oxygen species may be involved as well [360].

Notably, some of our treatment efforts may display adverse effects as β-blockers, angiotensin-converting enzyme inhibitors, morphine and diuretics are all have been associated with potentially undesired contributions to CS [52, 361–363].

3.5.2 Hypovolaemic Shock

Hypovolaemic shock with its core aetiologies haemorrhage and trauma-associated blood losses is precipitated by vascular volume (blood) deficit and consecutive failure of the *peripheral macrocirculation* [6, 29, 32, 35, 188]. This results in reduced venous return, impaired CO and, in turn if losses are severe, a fall in MAP [6, 29, 35]. Organs will be affected by hypoxia following hypoperfusion [29] – unfortunately, hypoperfusion is demonstrated to occur well before hypotension ensues [35, 364, 365]. MODS emerges as the consequence of a second hit inflammatory response attributed to reperfusion (I–R injury) [134, 135], with the gut being the origin of SIR(S) [29, 68]. The initial hit – e.g. bleeding – is not inducing systemic inflammation as the proinflammatory factors and other mediators released get lost in extravascular spaces or by bleeding [29, 55]. Haemorrhagic shock is a typical example of "cryptic" (subclinical) shock [31, 63] as hypotension – always signalizing decompensation of the situation [261] – occurs in the vast majority late in the course, often not until more than 20–25 % (15–30 % [366]) of the total blood volume is lost [63, 357, 365]; some authors report even if the losses exceed 40 % [35]. Hypovolaemic shock therefore has been classified into four stages, based upon the amount of the blood volume losses but with integration of relevant macrohaemodynamic and main clinical implications:

	Class 1	Class 2	Class 3	Class 4
Blood loss, %	<15	15–30	30–40	>40
Heart rate, bpm	<100	>100	>120	>140
Blood pressure, mmHg	Normal	Normal	Decreased	Decreased
Pulse pressure	Normal or increased	Decreased	Decreased	Decreased
Respiratory rate, breaths per min	14–20	20–30	30–40	>35
Mental status	Slightly anxious	Mildly anxious	Anxious, confused	Confused, lethargic

Modified from Cherkas [367], Bonanno[368], based on ATLS® guidelines of the American College of Surgeons, last update 2008

Hypotension consistently is present if TBV loss is >30 %. Hypotension signalizes decompensation. Persistent tachycardia with normalized sBP after fluid load signalizes partial compensation (stable unstable). Persistent tachycardia and hypotension after fluid load signalize severe life-threatening physiology (unstable unstable)

sBP systolic blood pressure, *TBV* total blood volume, *bpm* beats per minute

However, whatever the aetiology of blood volume loss, whether the loss is obvious (e.g. bleeding wound) or concealed (e.g. intra-abdominal) [32], absolute loss (e.g. open trauma bleeding) or losses due to fluid shift from the vascular space to extravascular compartments (i.e. ascites, capillary leakage) [87], hypovolaemia will result in reduced preload, filling pressures and consecutively diminished CO [36, 239, 369, 370]. In consequence, the low-pressure receptors in the atria, ventricles, pulmonary arteries and big veins will be excited; with further reduction in venous return and blood losses, the baroreceptors (high-pressure stretch receptors) in the aortic arch and carotid sinus will be unloaded and the sympathetic nervous system activated [32, 35, 68, 87]. Thereby first tachycardia and peripheral vasoconstriction (increased SVR) ensue [35, 224]. Thus, despite that CO may decrease, MAP will be maintained over a broad range of blood volume loss [32, 35, 224]. Nevertheless, a moderate drop in CO will entail a decline in pulse pressure and temporary orthostatic hypotension may occur [36, 371, 372]. Notably, up to 7 % of all patients may respond with transient relative bradycardia and attenuated rise in peripheral resistance instead of tachycardia and marked increase in peripheral vasoconstriction following a transitory imbalance between sympathetic activation and vagal/parasympathetic inhibition – probably caused and influenced by the release of opioids [229–231, 373, 374].

In contrast to CS and those shock types initially causing vasoconstriction affecting the macrocirculation [29, 35], hypovolaemic shock is a “blood tank” disorder, primarily without failure of the circulatory pump as in CS and SS and hence may be considered right at the beginning “just” as a “pure” peripheral macrocirculatory condition [29]. Furthermore, compared with the other shock syndromes, microcirculation in this shock type is only affected in the terminal stages and inflammation as in CS, but particularly in distributive shock states, is not seen until progression to end-stage shock – as in MOF/MODS [29, 63, 68]. This is related to loss of inflammatory compounds due to blood volume deprivation (e.g. bleeding). The volume loss, particularly in case of haemorrhage, at least in the initial stages remains the preponderant disorder with systemic inflammatory response occurring later and not synchronous with the initial disruption [29, 63, 68]. May be that the amplification is missing and the inflammatory reaction initially displayed is the unspecific central response as found to pain, etc. [29, 63, 88, 224].

In reply to the disrupted macrocirculation, redistribution of peripheral blood flow has to be noticed [35, 102, 375, 376]. Redistribution is partly the result of adoption of autoregulation provoked by the balance and interactions between vasodilative components such as NO – released by endothelial cells, local metabolic products, adenosine and the vasoconstrictive elements including sympathetic discharge, catecholamines and local endothelin [35, 377, 378].

Hypovolaemic shock is primarily not a microcirculatory disorder and the terminal arterioles may even, despite the constriction of the arteries and medium-sized arterioles, dilate in order to maintain tissue and cellular perfusion as shown in microscopical studies in animals [239–241]. However, at least at some stage – well before the terminal stages are commencing – and if hypovolaemia persists, diminished capillary blood flow and perfusion heterogeneities in neighbouring tissues independent of CO are demonstrated [241, 379, 380]. Decompensation occurs if the perfusion of the central nervous system becomes compromised and alterations of the patient's consciousness signalize ensuing disruption of cerebral perfusion [32] ending up in sympathetic escape and progressive, therapy-resistant hypotension, organ hypoperfusion, and MOF [32, 224].

3.5.3 Septic (Distributive–Vasodilative) Shock

Septic shock defines a condition of acute circulatory failure characterized by *sepsis-related*, fluid-unresponsive *hypotension* with a systolic blood pressure of ≤90 mmHg (or MAP ≤65 mmHg) despite adequate vascular filling persisting for at least 60 min or the requirement of catecholamines to ensure a reasonable perfusion pressure (the 90 mmHg/65 mmHg pressure level) [260, 381–383] *along with* the presence of *perfusion abnormalities* [384–386] or organ dysfunction [382, 387–389].

To meet the definition of *sepsis-related* circulatory abnormalities, the subsequent criteria need to be around:

I. Evidence of a systemic inflammatory response syndrome (SIRS) as characterized by at least two of the following so-called SIRS criteria [381, 382, 385]:

   (a) Body temperature of either ≥38 °C or ≤36 °C
   (b) Leukocyte count (WBC) of either ≥12,000 cells/µl or ≤4,000 cells/µl or >10 % immature cells
   (c) Tachycardia with a heart rate of ≥90 bpm

(d) Tachypnoea with a respiratory rate of ≥20 bpm or proof of hyperventilation by $PaCO_2$ ≤4.3 kPa (<32.33 mmHg)

and

II. *Evidence or suspicion* of an (bacterial) infection either by microbiological proof or by clinical criteria [260, 385]

Thus, sepsis depicts the systemic response to a presumed or verified, not necessarily, bacterial infection [388, 390], and septic shock characterizes the most severe form with predominant cardiovascular collapse [388, 391]. Sepsis in general is associated with a range of cardiovascular abnormalities [391] potentially displaying tissue hypoperfusion [243], a key factor in the development of MOF [391]. Systemic cardiovascular failure in the setting of septic shock [392] may result in global tissue hypoxia [393–395] where the diffuse endothelial activation, injury and disruption of the united cell structure following systemic inflammatory response to in general severe infections [148, 396] are responsible for and provoke global tissue hypoxia/ischaemia and various organ dysfunctions [144, 145].

The haemodynamic profile of severe sepsis and septic shock is predominantly and virtually eponymously distributive but includes elements of hypovolaemic and cardiogenic shock [45, 46, 397, 398] with absolute and relative hypovolaemia (secondary to capillary leakage and raised venous capacitance precipitating decreased preload [391]) and myocardial suppression (see below). The distributive issues encompass vasodilation, vasoregulatory failure, heterogeneous blood distribution, microcirculatory and mitochondrial dysfunction (cytopathic hypoxia) occurring via innate immune response to tissue injury [45, 388, 398, 399]. Of special significance and pivotal impact on the pathology hallmarking *the predominant haemodynamic disorder of septic shock is the altered microcirculation* [29, 44, 91, 189, 400]. Septic shock is a peripheral shock type with the microcirculation being affected right from the beginning as the terminal vascular bed is the initial target of SIRS triggered by bacterial endotoxins/LPS, etc., independent of the macrocirculatory features which may be quite variable [29]. The *microcirculatory alterations* are *largely related to endothelial cell injury and functional abnormalities* [148, 396]; furthermore, cohesive and in relation to the endothelial alteration, the vascular smooth muscle cells will lose their adrenergic sensitivity and tone and thus their ability to adequately regulate vascular tone and hence peripheral (organ) tissue perfusion [180, 182]. As a result of this predominant endothelial affliction, *malperfusion* (due to *maldistribution* of blood flow even in the presence of normal or augmented CO) arises, and, due to shunting (generating hypoxia/dysoxia in close vicinity to normoxic areas) [44, 49, 50, 401] *and impaired extraction of substrate, defects in cellular oxygen utilization and aerobic energy generation occur* [2, 402].

Peripheral vasodilation instead of the "physiologically" anticipated vasoconstriction characterizes and hallmarks (early) septic shock, the typical exponent of the distributive shock form [6, 30, 43, 63, 204, 224, 243, 245] – and this happens despite markedly elevated catecholamine serum concentrations [242, 243] and the activation of the RAAS-system [403].

Haemodynamic features typically found in septic shock include [43, 48, 178, 189]:

(a) Increased heart rate (baroreceptor mediated) [404, 405]
(b) Normal or even raised cardiac output [43, 48, 388, 406]
(c) Markedly impaired systemic vascular resistance (SVR) [48, 388, 407]
(d) Depleted vascular sensitivity to catecholamines [47, 408]
(e) Tissue hypoperfusion [243]
(f) Maldistribution [204]
(g) Loss of microvascular autoregulatory mechanisms [388, 409, 410]
(h) Tissue hyperpermeability secondary to microcirculatory/capillary leakage is an invariable hallmark of severe sepsis and septic shock [411]
(i) Hypovolaemia, absolute (capillary leakage) and/or relative (e.g. venous pooling) [45, 392]
(j) Reduced venous tone provoking fluid pooling [392]
(k) Vascular obstruction due to diminished deformability of erythrocytes and neutrophils [412]
(l) Myocardial dysfunction (to be discussed below)
(m) Elevated pulmonary vascular resistance potentially evoking RV dilation and dysfunction attributed to the increase in RV afterload [413]

The vasodilation typically indicated by a low SVR with, in the majority of patients, consecutive hypotension in the setting of severe sepsis and septic shock has meanwhile been referred to as being precipitated by:

I. *A failure of the vascular smooth muscles to constrict* [43, 47]
II. *The poor reactivity of the vascular smooth muscle cells to catecholamines* and other vasoconstrictive agents [43, 47]
III. *A relative vasopressin deficiency* attributed to inappropriate release with relatively low serum concentration [414–416] contributing to the attenuated peripheral vascular tone [43]

Peripheral and organ blood flow is largely regulated by the small arteries and arterioles [417]. Arterioles have substantial musculature – layers of vascular smooth muscle cells – and thus can actively relax and constrict modulating blood flow in the organs and the most remote areas [417]. They represent the so-called resistance vessels of the vascular system as the major pressure drop that must occur between large conduit arteries and capillaries to let the blood flow/stream be observed in this vessel section [233, 417, 418]. The small arteries and the arterioles account for up to 70 % of the total or overall systemic resistance to flow through the entire vascular system, the capillaries for another roughly 20 % [234, 419]. However, the exact anatomical location of the resistance vessels is difficult to assign [420]; commonly vessels with diameters between 300 and 15 μm are included [421], while some authors appoint vessel diameters of <500 μm and discriminate another functional/anatomical structure, precapillary sphincters, situated between arterioles and capillaries [422]. The resistance vessels are characterized by – in comparison to other vascular segments – as being able to exert substantial resistance to blood flow and by their ability to actively constrict and relax if appropriate and necessary and to modify myogenic tone, i.e. their intrinsic competence to contract in case of sudden rise in transmural pressure [417, 423]. This local vasomotor tone provided by the small arteries and particularly arterioles determines the distribution of the bloodstream into the organ areas according to the actual and special metabolic requirements [224, 418] – proportional increases or reductions by selective local constriction or dilation, respectively, are usually driven by local metabolic demands [424], principally autoregulated as long as the MAP is sufficient [25, 200, 201, 425]. To emphasize, the small arteries, but particularly arterioles and the capillary system, are part of and assigned to the microcirculation [224, 235, 236], and it is the microcirculatory failure which basically and pivotally designates the pathology of septic shock [29, 44, 189].

Systemic vascular resistance (SVR) is the most common parameter used to describe and assess resistance to blood flow [314, 426]. SVR largely indicates the vasomotor tone of the small arteries and arterioles [427–429] by reflecting the *non-pulsatile* component of the peripheral load [427, 428]. Since arterial pressure and blood flow are of oscillatory nature, SVR unfortunately does not include the other (*pulsatile*) components of arterial impedance such as arterial compliance, characteristic impedance and wave reflection [237, 429–431]. Arterial input impedance would be the more comprehensive specification encompassing all relevant elements determining the "complete" arterial load the ejecting ventricle is facing [430–432]. Furthermore, SVR is not homogeneously distributed along the arterial network [426]. However, despite all shortcomings, SVR reflects the major component of arterial resistance [233, 433] and displays the vasomotor tone of the small arteries and the arterioles [233, 427, 428] which is modulated and regulated by a complex arrangement of neurohormonal and local factors to provide and adjust organ and tissue blood flow in order to meet the metabolic requirements [418, 424, 429].

Meanwhile, *nitric oxide* (NO) has been identified [43, 47, 178] and widely recognized [47, 434] as being probably the *most important mediator of blood pressure and organ blood flow regulation* [178, 400, 435], NO is able to induce physiological vasodilation [435] and is responsible for the maintenance of microvascular homeostasis [47, 436], but also plays a crucial role in the pathomechanism of vasodilation in sepsis and particularly septic shock. *NO interferes with and impairs the ability of the vascular smooth muscle cells to constrict and hence to generate a basic vascular tone and to modify this tension in order to properly meet the metabolic demands of the cells and tissues* [43, 47, 266, 434, 437].

Proinflammatory agents like endotoxins, cytokines such as TNF-α, IL-2 and IL-6, and other substrates trigger the expression of the inducible form (iNOS) of the NO-synthesizing enzymes (nitric oxide synthases, NOS) [43, 249, 438–440]. The attitude of the other isoforms of NOS in the pathobiology of sepsis, endothelial NOS (eNOS), responsible for the basic production of NO [47], and the neuronal NOS, located in sympathetic and parasympathetic nerve endings and in cardiomyocytes probably substantially engaged in the regulation of the circulatory system and may contribute to vasodilatory environment during septic shock [441], is not yet defined [47, 441]. However, particularly in severe sepsis and septic shock, iNOS produces excessive amounts of NO, which, mediated via cGMP, provokes substantial relaxation of the vascular smooth muscle cells impeding vascular smooth muscles constriction (the latter being necessary to preserve vascular tone) and hence generates profound vasodilation – by some authors expressed as "vasoplegia" [43, 437, 442–444]. This increased expression of iNOS triggered by various proinflammatory and other substances with consecutive exorbitant NO production is found in many cell types of which endothelial and vascular smooth muscle cells may be of special interest [221, 438]. NO synthesized by endothelial cells will diffuse across them into the adjunct smooth muscle cells [163], bind to soluble guanyl cyclase [445], the enzyme which converts guanosine triphosphate to cGMP [446]. cGMP directly reduces the tension of the smooth muscle cells [446], reduces the release of Ca from the sarcoplasmic reticulum (SR) [447] and at the same time facilitates Ca re-storage [448]; hence, smooth muscle contraction is markedly impaired [163].

Moreover, sepsis-associated *changes of the membrane potential of vascular smooth muscle cells attenuate their ability to modulate vascular tone* [43, 449]. Several ion transporters and channels are contributing to the membrane potential, in this context the so-called potassium ATP channel (K_{ATP}) is of special interest [43, 47]. K_{ATP} is in general inactive under physiological conditions provided that appropriate intracellular ATP concentrations are present and hence keeping some degree of cell and vascular tone (some rate of vascular constriction). The channel will open if there is a drop in ATP concentration as found, e.g. in case of hypoperfusion with resulting hypoxia [47, 450, 451]. Increases in intracellular hydrogen ions and lactate as well as substances such as kinases, phospholipids, intracellular calcium and vasoactive hormones (e.g. prostacyclin, atrial natriuretic peptide, vasopressin, angiotensin II, endothelin) may provoke channel opening as well [450, 452, 453]. Opening the K_{ATP} channel allows potassium to leave the cell generating hyperpolarization of the cell surface. Consecutively, the voltage-gated Ca membrane channels close, inhibiting extracellular Ca (which is, mediating norepinephrine and angiotensin II-induced vasoconstriction but needs low normal membrane potential for influx) to enter the cytoplasm decreasing the cytosolic Ca concentration and by dephosphorization of myosin engenders vasodilation [451, 452, 454, 455]. Hence, here the *link becomes evident between intracellular metabolism and vascular tone/blood flow* which may be modulated by vasoactive hormones [451]. Further, the fact that even vasoconstrictors like angiotensin II may open the K_{ATP} channels indicates the physiologically well-balanced vascular tension with inherent feedback mechanism as sustained vasoconstriction needs extracellular Ca to enter the cells [449]. However, e.g. LPS as a bacterial component found in sepsis [264, 456, 457] and tissue hypoxia/dysoxia [264, 452] are all able to activate (open) K_{ATP} channels and hence elicit vasodilation by inhibiting smooth muscles to constrict [43, 47].

As hyperpolarization precludes Ca influx, catecholamine – or angiotensin II – effects inducing vasoconstriction via Ca influx will be impeded by LPS (keeping the channel open even in case of balanced metabolism) or other K_{ATP} channel-activating agents even in the presence of high plasma concentrations of these vasoconstrictive ligands [449, 451]. Notably, NO and other molecules may indirectly via cGMP activate the K_{ATP} channels [458], and, as a consequence, may attenuate or even abolish the vascular smooth muscle cells' sensitivity to catecholamines and other vasoconstrictors [458–460].

Recent research results have challenged the role of the K_{ATP} channels in sepsis-associated vasodilation [460] however, clinical studies in humans still verified an effect of low dose sterioids inhibiting the K_{ATP} channel activity and thus attenuating the LPS-induced vascular hyporeactivity fascilitating shock reversal [461, 462].

Furthermore, a relative vasopressin deficiency observed in prolonged and advanced shock states contributes to the diminished vascular tone promoting vasodilation [43, 47, 463–465]. Not only hypovolaemia and hypotension but also pain, nausea, hypoxia, hormones and mediators such as catecholamines and angiotensin II may all stimulate vasopressin release in the posterior pituitary of the neurohypophysis [466]. In the early phases of shock septic (and other shock types), vasopressin release by the neurohypophysis increases by 20–200 fold, mediating via activation of the V1 receptors of the vascular smooth muscles vasoconstriction [467–470]. In late-phase septic, cardiogenic and haemorrhagic shock, inappropriate low vasopressin plasma levels unresponsive to fluids and catecholamines are reported [416, 469, 471, 472]. This dynamic may be very fast as Morales demonstrated that during hypotension resulting from haemorrhage, the vasopressin concentrations from initially 300 pg/ml (physiologically <5 pg/ml) decreased to 30 pg/ml within an hour of sustained hypotension [469]. The underlying mechanisms still remain unclear; under discussion are autonomic dysfunction, depletion of hypophyseal vasopressin stores, high noradrenaline concentrations impeding the release of vasopressin from the supraoptic neurons and an increased release of NO within the pituitary gland attenuating the vasopressin production [472–475].

To conclude, "microcirculatory failure is determined as being the hallmark of septic shock state and central to septic pathophysiology" [400], thus may even qualify as being "the motor of sepsis" as Ince [44] points out characterizing the key role of microcirculatory dysfunction in the development and course of sepsis [44, 91, 189, 196, 476]. Septic pathobiology with the endothelium being concomitantly the source and target of the innate cellular response to injury [477] is multifactorial, ending up in a multifaceted disorder of microhaemodynamics [37, 44, 400, 478]. Shunting has been recognized as a main pathomechanism and may even be displayed in primary healthy regions [37, 44], markedly contributing to heterogeneous blood flows and uneven oxygen distribution in neighbouring areas inducing hypoxia and dysoxia nearby to normoxic layers [37, 44, 204, 400, 478]. Alongside, features like affected crosstalk influencing upstream microvessel recruitment [37, 175, 400] and disproportionately displayed endothelial response in different areas, but even of the same vessel territory have attracted our interest [479–481]. Especially noteworthy and with therapeutic implications, altered microcirculation may arise without relevant global macrocirculatory abnormalities [91, 193, 206, 212, 482], substantiating the characteristic and pathophysiologically seminal underlying microcirculatory pathology of sepsis, induced by endotoxins, bacterial substances and other threats [175], thus being a process basically emerging *within* the microcirculatory flow path [91, 400]. However, although the link between macro- and microcirculation is somewhat weak [174], of course, both yield some impact on each other [388, 399]. As such, the disrupted autoregulation – important for appropriate blood distribution within the organ and organ areas [200, 201] – in case of hypotension [199, 409, 410, 483, 484] may be re-established by macrocirculatory therapeutic measures (BP rise) and hence improve microcirculation [252, 254, 255]. Not titrated to the tissues requirements [355, 485], discordantly oxygenated neighbouring regions may provoke in general a diminished oxygen extraction by the tissues and cells as could be demonstrated in mathematical and animal models [245, 355, 485–487]. Moreover, a globally impaired oxygen extraction in remote and regional tissue in sepsis has been reported [44, 49, 91, 204, 486]: Oxygen is either not reaching the remote tissue (e.g. shunting) [50] or, as indicated by a high oxygen concentration in central venous blood return and verified by the above-mentioned studies, not utilized by the mitochondria for phosphorylation of energy-rich ATP [488]. This mitochondrial failure/dysfunction has been termed cytopathic hypoxia [2]; the combined abnormality as found in sepsis consisting of mitochondrial failure and altered microcirculation has been coined as *microcirculatory and mitochondrial distress syndrome* (MMDS) [44]. Altered microcirculatory haemodynamics are suggested as being an early trigger for the ensuing multiorgan failure in sepsis [212, 401, 489, 490] and, if not corrected in time, for death [144].

As the *specific macrocirculatory features* of this distributive, peripheral microcirculatory shock type, a *normal or even elevated CO in the presence of reduced and low SVR* is recognized [48, 392, 491]. This is a finding of the *early* so-called "warm "or "hyperdynamic" shock phase, where the cardiocirculatory system is responsive to fluids improving its function and establishing the hyperdynamic constellation [254, 258, 259, 492]. Since the administration of adequate amounts of fluids at this stage where peripheral vasodilation but not manifest microcirculatory disorders preludes established septic shock stages – characterized by microcirculatory dysfunction, vasoconstriction and finally "vasoplegia" – may prevent shock progression to *later* shock phases indicative by the fluid-unresponsive "cold" or hypodynamic conditions [258–260], Bonanno refers to this particularity of the early step of septic-associated circulatory abnormalities as to a pre- or uncompleted shock [29]. This interpretation and determination by Bonanno [29] is notably confirmed by the current definition of septic shock published by the Surviving Sepsis Campaign guidelines [259, 260]. Adequate volume resuscitation in time is demonstrated to restore macrocirculation but may prevent and/or resuscitate (subclinical) microcirculatory dysfunction before a complete shock stage ensues [216, 217, 493]. Moreover, it has to be emphasized that a "cold, hypodynamic situation" early in the septic disease course is suggestive for insufficient volume resuscitation and relative hypovolaemia rather than for an advanced shock stage [491, 492, 494].

Myocardial depression in sepsis, particularly in severe sepsis and septic shock, is a common feature contributing to circulatory failure [392, 491, 495–497]. The numbers of septic patients with affected systolic function reported in studies varies significantly from 18 % [498] to 94 % [499], obviously depending on the point in time of the septic process, the assessment method and the index used [500–502]. In animal studies, 100 % of the cohort is found to exhibit impaired systolic performance during the course [495]; hence, compromised myocardial contractility applies to the majority [495, 500, 502–504], if not to all patients [495]. Systolic dysfunction develops during the early (hyperdynamic) phase of septic shock when combined liberal volume replacement and adaptive LV dilation may preserve CO [406, 505] and thus may obscure myocardial restriction [491, 497]. Moreover, both compromised systolic and diastolic dysfunction have been depicted [502–504, 506–509]. There is clear evidence that *depressed intrinsic contractility* (a systolic property [510]) and *impaired myocardial compliance* (a diastolic property [511]) are the major cardiac features displayed by the septic process [500, 507, 508, 512, 513].

No difference is described in magnitude of impairment of myocardial performance whether it is a Gram-positive, Gram-negative or fungal sepsis [514]. Various factors and mechanisms may provoke an altered intrinsic contractility including myocardial depressant factors such as cytokines with myocardial depressive properties, e.g. TNFα [321, 515–518] released by endotoxins, or interleukin-1β playing a pivotal role in sepsis-induced myocardial dysfunction [519]. NO has direct adverse effects on cardiac contractility in case of NO overproduction [520–522]; NO may even cause apoptosis [523] or may display detrimental, cytotoxic effects via peroxynitrate, a powerful cytotoxic oxidant resulting from the reaction between NO and superoxide, a free radical [524, 525]. Furthermore, downregulated β-adrenergic receptors show an attenuated response to catecholamines [526] and myocardial hibernation, principally a protective mechanism allowing the myocytes to reduce oxygen consumption and energy demand, but diluting cardiac performance in the acute situation [527–529]. Recent studies seem to support previous results that depressed adrenergic responsiveness, most probably largely nitric oxide induced, may be a relevant causative issue for the impaired cardiac performance [408, 530, 531].

Definitely excluded is a contribution of global myocardial hypoxia/ischaemia to myocardial dysfunction in sepsis [532–534], although, of course, in patients with known coronary artery disease, regional ischaemia may have some minor degree of input [491, 495, 535].

Both the left and right ventricle exhibit myocardial dysfunction simultaneously [502, 536–538]. The attenuated RV performance is attributed to acute pulmonary hypertension – commonly associated with the septic conditions affecting the lungs (ALI) directly or as remote organ – and/or depressed RV contractility [536, 538].

As not all authors could demonstrate an association between reduction in RV-EF and acute increase in pulmonary pressure [538], compromised myocardial contractile function obviously is the crucial component [491]. Furthermore, RV diastolic properties are substantially influenced contributing to an altered right ventricular function in septic shock patients [538, 539].

Septic cardiomyopathy is characterized by:

I. Its acute and principally reversible condition [503, 540, 541].

II. The association of impaired systolic function and normal or even low filling pressures [502–504], the latter attributed to RV dysfunction and especially to the blunted LV compliance which allows "only" an adaptive, slight LV dilation and increase in SV [542]. Although a more non-compliant ventricle potentially may promote or aggravate pulmonary hypertension and lung congestion, attributed to substantial increases in filling pressures [496, 543], however marked increases of filling pressures in case of well-balanced fluid administration in the setting of severe sepsis and septic shock are not reported [501]. A relevant increase in LVEDP stays away due to the confined systolic function as the decrease in the PCWP/LVEDV index indicates [495, 501–503].

III. A substantially narrowed heart rate variability [544] reflecting a loss of balance between sympathetic and vagal tone [545].

IV. Tachycardic patients with small, "hypercontractile" ventricles (indicated by a supra-normal LV-EF and a high cardiac index) are shown as having the worst prognosis – 100 % mortality [498]. This quality, called "hyperkinetic profile" by Vieillard-Baron and co-workers, reflects a persistent and profound abnormal vasodilation and thus reduced vascular resistance/afterload allowing the ventricle to generate a high CO although the LV intrinsic contractility is perturbed [495, 498].

It seems that patients with adaptive, slightly dilated ventricles [406, 498, 504] and normal LV-EF/CI despite and in the presence of normalized afterload although the intrinsic LV contractility is impaired, the so-called normokinetic profile, show the best prognosis (24 % mortality rate), while patients with a hypokinetic profile indicated by clearly reduced LV-EF/CI (and consecutively tachycardia) under a normalized afterload, and, of course reduced intrinsic contractility, a finding in all profiles, show an intermediate mortality rate (43 %) [498].

Notably, in case of low LV-EF, hypovolaemia needs to be excluded as in such a situation fluid administration may be beneficial leading to an increase in CI/LV-EF [495]. Based on data surveyed in 2003 [546] and 2001 [498], Vieillard-Baron has summarized these constellations and relationships in his article about "septic cardiomyopathy" [495]; see figure below. It has to be notified that LV-EF and CI/CO, among other factors such as preload, heart volume, contractility and heart rate, are in particular dependent on afterload (afterload ↑→EF ↓ and vice versa [312], as well as flow, indicated by SV/CO, depends on afterload since afterload ↑→flow (CO/CI) ↓ [547]); thus, both parameters, insofar, largely reflect the actual afterload and not the intrinsic contractility at all [548, 549].

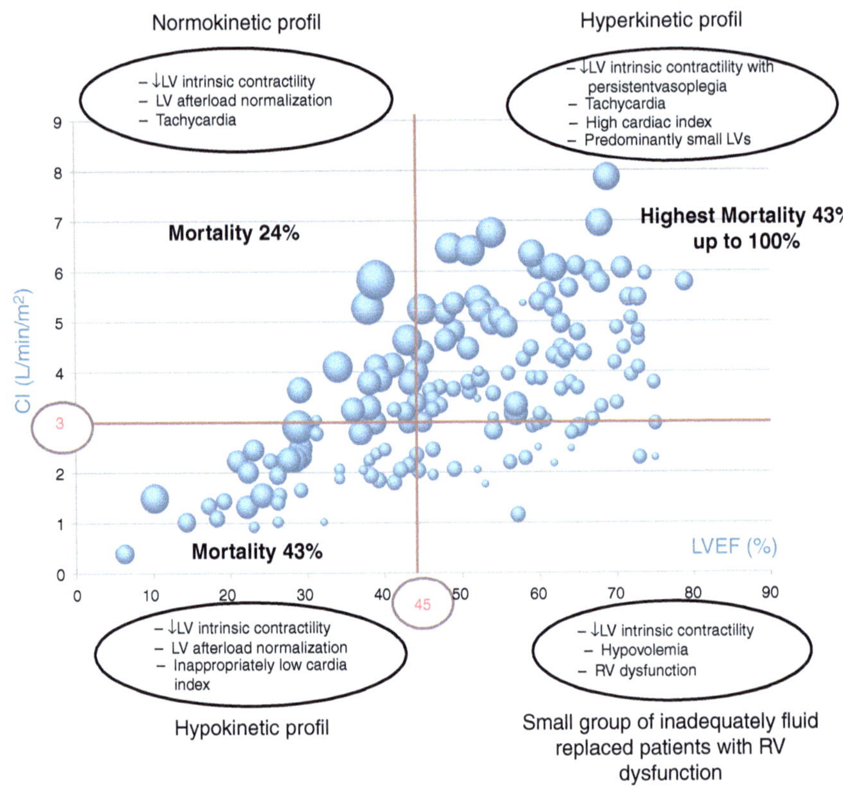

The above presented haemodynamic and associated prognostic features are consistent with results Parker already reported in 1987 where he found a favourable outcome if patients achieved a systemic vascular resistance index above 1529 $dyn \times s^{-1} \times cm^{-5} \times m^{-2}$, a heart rate of <95 bpm or a reduction by >18 bpm and a cardiac index of >0.5 $l \times min^{-1} \times m^{-2}$ within 24 h of their presentation [550]. On admission, a heart rate of less than 106 bpm was suggestive for a better survival [550].

Hence, LV-EF and CI/CO may be used to assess the haemodynamic situation. A low EF while SVR has returned to normal as well as in particular a supra-normal EF/CI in the presence of a persistently low SVR [551] are suggestive as being serious and worrying conditions associated with worse outcome [495, 546].

An overview of the essential characteristics of the different shock types is depicted in the following table.

Shock Type	Aetiologies	Macrohemodynamic profile /tissue perfusion								Pathophysiology and typical characteristics
		CO	SVR	CVP	PCWP	MAP	EDV	SvO2	VO2	
Cardiogenic Shock (CS)	Acute myocardial Septic cardiomyopathy Valvular heart disease	↓-↓↓↓	↑-↑↑	↑	↑-↑↑	↓-↓↓	↑-↑↑	↓	↓	• Basically failure of the central macro-circulation due to abrupt depression and / or loss of myocardial contractility irrespective of loading conditions causing scare peripheral perfusion • Significant fall in SV (CO) in the vast majority in the presence of ↑ filling pressures (LVEDP ↑), and stiffened central arteries • Inappropriate / overwhelmed compensation with only mild /moderate, disproportional vasoconstriction attributed to an associated systemic inflammatory response diluting sympathetic discharge and activated humoral systems by release of vasodilative operating mediators and by launching microcirculatory disorders • I-R phenomenon, II. hit • It is the **combination** of **compromised contractility** and the **vascular alterations** related to vasoactive mediators that coine the maladie as ↓↓ EF is found in acute heart failure as well. SVR only mildly/moderately elevated due to peripheral vasodilative effects
		Bonanno: hemodynamics of shock states ↓ CO → ↓ MAP ⊦ VC : ↑ TPR-MAP intervention: ↑cardiac performance → ↑ CO → ↑ MAP (by ↑ preload, myocardial contractility, ↓ afterload - esp. SVR) VC: Venus constriction, TPR: total peripheral restistance (equals SVR)								
Hypovolemic Shock (HS)	Hemorrhage Renal and Capillary leakage Elevated abdominal pressure	↓-↓↓	↑-↑↑↑	↓-↓↓	↓	n-↓	↓-↓↓	↓-↓↓	↓-↓↓	• Basically failure of the peripheral macro-circulation following blood loss / loss of circulating • Typically II. hit with I-R contribution to shock development • MAP largely maintained for a long time even in case of severe blood / volume losses despite drop in SV (CO) due to sustained peripheral vasoconstriction • Ensuing hypotension indicates decompensation / compensatory failure • Microcirculatory alterations noticed only in advanced stage • Central shock type, up to 7% may initially respond with brady- hypotension
		Bonanno: ↓ VR → ↓ CO → ↓MAP ⊦ VC : ↑ TPR - MAP intervention source control and blood / volume replacement → ↑VR → ↑CO → MAP↑ VR: Venus return								
distributive Shock, here Septic Shock specified (SS)	Sepsis (SS) SIRS (SIR) Final common pathway of all shock types	very early disease stage, pre fluid resuscitation								• Basically a peripheral micro-circulatory failure right from the beginning, characterized by peripheral malperfusion (maldistribution of blood flow and hypoperfusion/dysoxia) launching: • Inhomogenous blood flows and blood distribution causing areas of hypo/dysoxia in close vicinity to normally perfused regions due to shunting, disturbed cross-talk of vascular endothelial cell layers, and blunted autoregulation disregarding the cells' and tissues' metabolic requirements in the presence of general peripheral vasodilation and despite normal or elevated CO (the latter to be observed at least after sufficient volume resuscitation during the early, fluid - responsive disease stage) • Initially "incomplete" shock as fluid responsive early on creating hyperdynamic conditions • Altered microcirculation / gut motor of shock (disturbed mucosal barrier function with bacterial translocation) • Peripheral shock type • In full blown out phase is "vasoplegia" following a vasoconstrictive phase with vascular smooth muscle cells unresponsive to catecholamines → common, terminal pathway found in all shock types • Worst prognosis: hyperkinetic profile
		↓/n↑	↓-↓↓↓	↓/n(↑)	n/(↑)	↓-↓↓	n/↓	↑	n-↑	
		early post inital fluid resuscitation, hyperdynamic phase								
		↑↑	↓-↓↓	↑-↑↑	↑	↓-n	(↑)	↑-↑↑	↑-↑↑	
		Bonanno: hyperdynamic phase: normal microhemodyn; full blown out: ≠ microhemodyn, maldistribution SIR›VD: ↓TPR/MAP→ ↑ VR → ↑ CO - MAP → Dysoxia /hypoxia Intervention: source elimination, fluids and vasoconstriction → ≠ SIR → N (TPR/MAP/CO) VD: Vascular dilatiation, SIR : Systemic inflammatory response								
Obstruktive Shock	Cardiac tamponade Tension pneumothorax PE Aortic dissection LA myxoma	preload obstruction, CVP >PCWP								• Central shock type with either impeded forward flow (afterload affected) or hindered filling causing low preload and thus scare peripheral perfusion • Typically SV /CO diminished and SVR ↑
		↓-↓↓	↑	↑	n-↑	↓	↓	↓	↓	
		afterload obstruction								
		↓-↓↓	↑↑	↑	↑-↑↑	↓	↑	↓	↓	

3.6 Diagnostic and Clinical Issues

With the exception of obvious and clear pictures like stab wounds and/or profuse bleeding accompanied by immeasurable blood pressure, it is really challenging in the majority of cases to diagnose a “shock” timely and properly: There are no specific and quintessential symptoms, signs or markers explicitly diagnostic for shock; the assessment is subjective looking for imprecise and insensitive indicators in often subtle presentations [28, 552, 553]. Furthermore, findings are dependent on the type of shock and the stage of disease course at presentation [6, 53]. Moreover, different types of shock may coexist [53, 554], However, therapeutic measures should ensue as soon as possible [555, 556] since early treatment has shown to be beneficial for the outcome [556, 557].

As such, features like

- *pallor*, *ashen grey* or cyanotic *skin*;
- *cold peripheries* (forearms and/or lower legs [558]), *moist and clammy extremities* and *mottled skin* of the extremities;

- *altered (clouded) mental status*: quiet, apathetic, sometimes restless, apprehensive or confused;
- *diminished urine production/oliguria* (<30 ml/h or < 0.5 ml/kg/h for ≥2 h) [559];
- thready pulse of poor quality;
- (too) low skin temperature, especially reduced temperature of the great toe [560];
- prolonged capillary refill time (less than 2 s, but wide variation up to 4.5 s being normal in the elderly) [561, 562];
- *arterial hypotension*;
- tachycardia

are all *suggestive* for altered tissue perfusion and cellular oxygen shortage/inappropriate utilization in a *respective clinical setting*[4, 28, 52, 63, 188, 273, 552–564].

Arterial hypotension (sBP <90 mmHg or a pressure drop >40 mmHg of the habitual patient's sBP [53, 564]) is considered as being a typical and classical finding in shock maladies and one of the central symptoms of shock states [31, 42, 53, 63, 188, 552]. In hypotension, autoregulated blood flow distribution is blunted [199–201], promoting hypoperfusion [110, 224, 319]. Hypotension often indicates insufficient sympathetic compensation or overwhelmed compensatory mechanisms [4, 257, 552], as continual hypotension accompanied by tachycardia describes decompensated shock [63] and arterial hypotension unresponsive to volume expansion defines septic shock [259, 383]. However, hypotension, although seen in the vast majority of cases, is not a universal and diagnostic symptom since normotensive patients may well suffer from shock [31, 53, 317, 552, 565]. Shock may even be present although the patient displays completely "normal" vital signs [565–567]: As the link between macro- and microcirculation is weak and a dissociation between both is common, macrohaemodynamic variables may not indicate even substantial microcirculatory disturbances [63, 198, 566]. SvO_2 as an indicator of global tissue hypoxia does not notify regional hypoxia or even dysoxia as inherently found in early distributive shock forms [50, 552, 568] – SvO_2 seems to have some monitoring relevance in cardiogenic and haemorrhagic shock [568]. It is even difficult to define "normal" ranges, as, due to widely varying metabolic demands, there is *no normal cardiac output* [25]. Furthermore, compensatory mechanisms may early on keep macrocirculation in physiological ranges [63, 366, 565]. However, even by resuscitation normalized macrohaemodynamics do not guarantee adequate microcirculation and appropriate oxygen supply, the shock situation may rather persist and proceed [63, 552]. Moreover, normotension is far away from ensuring haemodynamic stability in the setting of shock [257, 317].

Clinical conditions featuring clinical signs of peripheral tissue hypoperfusion [63, 569] including mottled skin, cold extremities and/or diminished uo accompanied by elevated lactate levels >4 mmol/l [31, 145] in the presence of normal (or even high) blood pressure are referred to as *cryptic, pre-shock or normotensive/non-hypotensive shock* [31, 63, 273, 317, 569, 570]. Cryptic/pre-shock circumstances, often representing early disease stages [4, 565, 571], with their normal macrohaemodynamics [63, 569] require subtle examination and diagnostic suspicion [31, 552, 569, 571] as they are insidious [63, 134], may quite often cause MOF, e.g.

following I–R phenomenon in case of haemorrhage [134, 364], and thus carry a high mortality rate [569]. Particularly young, insofar healthy patients may maintain BP in normal ranges attributed to the compensatory mechanisms causing vasoconstriction and tachycardia [565] as notably seen in haemorrhage. In haemorrhagic shock, normotension is widely maintained even up to blood losses of 25 % (30 %) due to compensatory mechanisms [68, 565]. Particularly in haemorrhagic injuries, hypotension is a late sign [572]. Hence, ideally haemorrhagic shock is diagnosed well before hypotension applies [573] as tissue hypoperfusion ensues long before hypotension becomes apparent [63, 68, 364]. Clinical hints are tachycardia (100–120/min), tachypnoea (respiratory rates 20–24/min), reduced pulse pressure, cool and clammy skin and delayed capillary refill [565]. Unfortunately, paradoxical bradycardic response has been described in trauma patients [4, 574], and vagally mediated bradycardia was demonstrated in penetrating intraperitoneal injuries [575].

Non-hypotensive cardiogenic shock/cardiogenic pre-shock has been well depicted and characterized by Menon [19, 317]: clinical evidence of peripheral tissue hypoperfusion is attributed to markedly impaired myocardial function causing poor CO not meeting the tissue and organ metabolic requirements in the presence of an sBP >90 mmHg. Cardiogenic pre-shock has to be determined by clinical evaluation, thus is likely to be underdiagnosed [317]. As clinical signs of inadequately diminished peripheral perfusion are associated with enhanced mortality [576], Menon strongly recommends diagnosing cardiogenic pre-shock in all patients exhibiting signs of poor tissue perfusion in the setting of severe cardiac dysfunction irrespective of BP [273, 317]. Furthermore, in elderly patients known to be hypertensive or in patients suffering from atherosclerosis, even an sBP of 120 mmHg may be insufficient [577].

Alterations of mental status attributed to hypoperfusion may initially be subtle [565], but are common in the elderly [578]. Diaphoresis suggestive for physiological stress is reported to occur well before vital variables become abnormal [565], and tachypnoea is an often overlooked early sign [571] reflecting compensation for metabolic acidosis [565].

Only a few shock conditions present with jugular venous distension such as cardiac tamponade, tension pneumothorax, massive pulmonary embolism, acute myocardial infarction (in particular in right ventricle involvement), intra-cardiac and pericardial thrombi, and end-stage heart failure [579]. Pericardial tamponade classically exhibits a triage of hypotension, distended neck veins and muffled heart sounds – regrettably late findings [565].

Neurogenic deficits and hypotension suggest a neurologic aetiology [554].

Febrile response as a hallmark of infection [388] is blunted in roughly 50 % of the elderly [580], frequently in sepsis in that patient group are symptoms such as delirium, weakness, malaise, falls, urinary incontinence and mental modifications [578] – regrettably unspecific signs found in non-infectious diseases as well [578]. Under these circumstances, procalcitonin (PCT) serum concentration may be helpful to differentiate septic from other shock entities [581, 582].

Peripheral mottled cyanosis is clearly attributed to failure of microcirculation and a sign of irreversibility [29].

The so-called shock index, defined as heart rate (HR) divided by systolic blood pressure (sBP), should give rise to serious concern in the setting of circulatory compromise, if its ratio is above 1 (HR/sBP >1) [583, 584].

Elevated *lactic acid blood concentration* following anaerobic energy generation *causing metabolic acidosis* still remains one of the best/most useful biochemical markers for *global* tissue hypoperfusion [188, 585] and is a universal feature in shock [63]. Lactate may accumulate due to enhanced production or mitigated clearance [586]; in shock situations, both factors are most likely involved in the majority of patients [587]. Lactate is nonspecific and increases may occur due to co-morbidities of patients in shock states such as impaired clearance in the context of liver failure, applies during shivering or seizures, and may be associated with extensive neoplastic disease [188, 552, 585]. Furthermore, in sepsis it may not be attributed to anaerobic metabolism, and thus, tissue hypoxia as raised concentrations have been reported in the presence of normal intracellular oxygen supply [586, 588] – lactate may be generated by the epinephrine effect on Na/K–ATPase activity [589, 590]. However, in the setting of altered tissue perfusion, lactate levels proportionally increase with the severity of the perfusion disturbance and oxygen debt and are directly related to outcome [591–594]. They will decline with successful therapy and thus may be used to monitor disease management [556, 592, 594]. Persisting high levels of lactate are suggestive for ongoing hypoxaemia and indicative for a poor prognosis [7]. Unfortunately, increases may not occur in early shock stages (only locally) [552], and therefore, but notably in general, serial measurements are demonstrated to provide more valuable information [595].

Central venous ($ScvO_2$) and mixed venous (SvO_2) oxygen saturation are frequently used parameters to assess and to monitor critically ill patients [556, 583, 596–599]. These markers reflect the balance between tissue oxygen demand (consumption) and *systemic* oxygen supply (delivery) [600–603] and as such are considered to be indirectly indicative for tissue oxygenation [596, 604, 605]. Adequate cellular oxygenation is dependent on macrocirculatory flow (supply), microcirculation (distribution) and mitochondrial function (utilization), in which each element and the interaction of all three levels need to perform well [606]. Low values notify a mismatch between oxygen demand and supply due to a diminished delivery, an enhanced extraction or a combination of both; hence, in that case the net supply does not meet the cellular and tissue requirements, often attributed to low CO – a macrocirculatory failure [599, 606, 607]. However, several factors and circumstances are affecting the oxygen delivery and consumption (stress, pain, shivering, anaemia, hypoxia, CO), and low values are not specific for hypoxia following hypoperfusion at all [553, 599, 608]. High levels may either signalize microcirculatory (oxygen does not reach the remote tissues and cells due to functional shunting, maldistribution, augmented microvascular permeability or thrombosis of the small vessels [606]) disturbances or scare utilization reflecting mitochondrial dysfunction being unable to process oxygen [2, 193, 198, 400, 553, 608–611]. Unphysiologically high $SvcO_2/SvO_2$ levels attributed to microcirculatory failure and/or mitochondrial dysfunction are referred to as tissue *dysoxia* [606].

Both too low levels (of $ScvO_2$ <70 %, SvO_2 <65 % [583, 599]) and too high levels ($ScvO_2$ >85 %/90 %, SvO_2>80 %/85 %) have again recently even been reported to be associated with increased mortality [606, 612]. Thus, as one of our more or less empirically established *macro*circulatory therapeutic targets [563] is to augment in critical illness a diminished central venous oxygen saturation [260, 556], we should carefully watch our measures (i.e. run up CO or increase the oxygen carriers (erythrocytes) if the haematocrit is less than 30 % [260]) aiming for a "normal" $ScvO_2$/resp. SvO_2 level, but are strongly advised to avoid "supra-normal" values which have been demonstrated already years ago to be harmful [613–616]. However, for pathologically high venous oxygen saturations – aetiologically largely disease immanent [193, 606, 609, 611], if not precipitated by "overtreatment" – no actually entitled and validated therapeutic tool is currently available [478, 617].

Finally, it has to be noted that SvO_2 and $ScvO_2$ are in *daily clinical practice* delivering the same adequacy and content of information [603, 618] as confirmed in several experimental and clinical studies under various conditions [597, 598, 603, 619–621]. $ScvO_2$ is usually less than SvO_2 by 2–3 % [619, 622]; in septic shock a difference of about 8 % is described [597]. However, $ScvO_2$ is by far easier to achieve as the critically ill patient will anyway need central venous access, while SvO_2 will require pulmonary artery catheter assessment [623, 624]. Furthermore, it needs to be emphasized that $ScvO_2/SvO_2$ indeed is an indicator of oxygenation, but represents merely *whole-body* tissue oxygenation rather than regional or local relations [7, 625, 626] and as such is a parameter of *global, systemic* tissue perfusion [400, 606, 627, 628].

Altered microcirculation plays a crucial role in the development of *tissue hypoxia/dysoxia* and thus in the formation and the course of shock states and multiorgan failure [27, 90, 91, 148, 555, 629]. Sufficient resuscitation of macrocirculation does not guarantee improvement of microcirculation as persistently high mortality rates after optimalization of macrocirculatory haemodynamics in patients with severe sepsis and septic shock suggest [7, 627, 630], and assessments of microcirculatory haemodynamics confirm [91, 196, 631]. Unfortunately, clinical assessment and ascertainment of global clinical and haemodynamic/oxygenation parameters fail to reflect the microcirculatory condition near enough [91, 178, 566, 632]. In recent years a couple of optical assessment and monitoring methods and devices as well as gastrointestinal and transcutaneous tissue pCO_2 measurement systems have been displayed to examine the microcirculation: The optical systems are able to visualize microcirculation, e.g. sublingually [91, 212, 478, 555]. Furthermore, the systems are feasible to semi-quantify microcirculatory perfusion using a special scoring approach (MFI, microcirculatory flow index) [196]. By the help of these tools, it was possible to detect and establish the typical features of microcirculatory alterations found in the setting of shock: reduced vascular density, particularly of the small vessels (a); large number of non-perfused and intermittently perfused small vessels (b); and marked heterogeneity of blood flow within an area of an organ, an organ and between organs (c) [90, 91]. The *heterogeneity of blood flows within the microcirculation of an area/section of an organ*, within an organ and among organs promoting dysoxia is specific and may be that characteristic hallmark of severe and substantial haemodynamic alteration typically displayed in sepsis, but verified in other shock conditions as well [91, 185, 202, 203, 631, 633].

Optical devices such as OPS (orthogonal polarization spectral) imaging [634] and its further development SDF (sidestream dark-field) imaging [635] as well as NIRS (near infrared spectroscopy) [636] detect the absorption spectrum of oxygenated haemoglobin, deoxygenated haemoglobin and cytochrome aa_3 and as such assess tissue oxygenation [637, 638]. Unfortunately, none of these very useful techniques are ready for routine clinical use due to several technical and applicable problems [639]; also, profound clinical study results including bigger patient populations, using these devices in daily practice are lacking [478, 640].

A different approach to assess and monitor microcirculation are the pCO_2 measurements (tonometry or capnometry) detecting and measuring local/regional tissue concentrations/content of carbon dioxide [641–643]. Initially used by Boda and Muranyi to control mechanical ventilation [644], gastrointestinal tonometry has been considered as being a reliable method to evaluate local tissue perfusion (gastric mucosa) and as such, tissue pCO_2 characteristics are an index of regional tissue metabolism indicative for hypoxia/dysoxia and ensuing or even manifest anaerobic energy production reflecting an imbalance between oxygen supply and oxygen demand [645–651]. Sublingual pCO_2 capnometry, a very attractive approach and further development of gastrointestinal tonometry [652, 653], has given proof, particularly in research issues, as being a very valuable tool to assess microcirculation, to monitor and to adopt treatment, and has confirmed its prognostic quality [626, 654–656] – unfortunately, that technique is not yet validated sufficiently in daily clinical practice [657].

Moreover, the gastro- intestinal tract is understood as being particularly susceptible to systemic hypoperfusion [650, 658, 659] and often is the organ system affected first [553, 660–662]. This is related to its specific anatomy (formation of the vasculature facilitates countercurrent flow and exchange of solutes but unfortunately thereby allows AV shunting and recirculation of carbon dioxide) [553] and the blood flow redistribution [68, 109, 110] causing markedly diminished blood flow to the intestine in the setting of disturbed circulation eliciting low-flow states such as circulatory shock [648, 658, 663–665]. Meanwhile, there is convincing evidence that the circulatory alterations displayed by the gastrointestinal tract may reflect conditions in other organs including the kidneys, brain, liver myocardium, and lingua [666–669].

The physiological basis for this assessment method is that the *local tissue carbon dioxide tension* of the stomach mucosa or other tissues *is largely determined either by the regional, microcirculatory blood flow conditions or actual local energy formation* [670]: In low-flow states, e.g. in case of hypoperfusion or shock situations, the clearance and evacuation of carbon dioxide and thus the CO_2 washout will be diminished causing an increase in local CO_2 content and hence gastric mucosal pCO_2 in the presence of normal, aerobic metabolism [657, 659]. On the other hand, arising or manifest anaerobic energy production as in case of any imbalance between oxygen delivery/local utilization and actual metabolic demand will result in an increase in local CO_2 tissue tension as well [659, 671]. When anaerobic metabolism occurs, hydrogen ions will be incurred, immediately buffered by tissue bicarbonate to maintain a balanced base–acid proportion, but entailing in carbon dioxide accumulation, H^+ + HCO_3^- ↔ $H_2O + CO_2$ [650, 652, 659]. Theoretically, two sources of hydrogen ion formation have to be considered, *glycogenolysis and lactate generation* (glucose + pyruvate → lactic acid + energy + 2 ATP molecules, with lactate acid may dissociate if tissue pH falls below 6.0 [672]; lactate acid ↔ lactate$^-$ + H^+, simplified equations) and/or *excess hydrolysis of energy-rich phosphates*, e.g. ATP (ATP → H^+ + ADP + energy) [659, 671, 673, 674]. As basic research revealed, net hydrogen ion production is due to excess hydrolysis in the setting of disturbed and altered aerobic energy allocation impairing and diluting oxidative phosphorylation and ATP regeneration [673–675].

The above described controversy (↑ pCO_2 either as a result of altered microcirculation or local energy formation) has been recently solved as the studies by Dubin et al. [676, 677] and Vallet et al. [678], further supported by a mathematical model by Gutierrez [679]. The findings strongly suggest that our capnometrically achieved tissue pCO_2 measurements reflect "only" tissue perfusion, and hence, elevated values indicate merely microcirculatory hypoperfusion rather than tissue dysoxia. Nevertheless, Dubin et al. [657] allocates capnometry strong qualities for assessment and evaluation of microcirculatory haemodynamics in critical illness and insistently advises to use this method to address the conditions present by suitable treatment measures in shock state diseases. Moreover, assessing microcirculation is absolutely desirable and needs to become an integrative component in evaluation and management of shock maladies [400, 478, 617, 680] as soon as the current technology is ready for routine use and suitable clinical study results are available [188, 214, 400, 617].

In conclusion, we should still follow the recommendation by Kellum and Pinsky [313] as "the integration of physical examination findings (e.g. delayed capillary refill, oliguria, confusion), hemodynamic variables (e.g. MAP, cardiac index, mixed venous oxygen saturation), and metabolic parameters (e.g. arterial base excess, lactate, glucose) reflects the present-day approach to diagnose shock and monitor response to therapy. Unfortunately, physical examination findings alone are insensitive in the diagnosis of shock and *poorly predict circulating blood volume*".

3.7 Management

3.7.1 General Approach

Since altered perfusion causing tissue hypoxia or even ischaemia/dysoxia is the fundamental and crucial disorder in shock state diseases [1, 3, 4, 29, 90, 91, 188, 203], our basic therapeutic efforts are directed at a rapid, broad-based and sustained restoration and recovery of physiological circulatory conditions to sufficiently provide tissues and even remote cells with oxygen and nutrient amounts essentially and vitally required [91, 681–684]. As such, resuscitative measures primarily aiming for an enhanced oxygenation particularly include the (re-) establishment of normovolaemia and the promotion of a sufficient blood pressure in order to augment tissue perfusion and to prevent organ failure [684–689]. Furthermore, resolving microcirculatory pathophysiology with redirection of blood flow within an organ, readdressing and ensuring that the supply is meeting the actual metabolic requirements of each organ section is obviously essentially necessary to overcome the critical condition of ensuing or manifest shock [1, 193, 204, 212, 213]. While our knowledge of microcirculatory abnormalities occurring is still being developed [29, 88, 90, 91, 555], our approach to manage the macrocirculatory dysfunction is well established and widely adopted in daily practice [563, 681, 682, 690], although recently challenged and consequently modified [563, 687, 691–693].

Although most of the investigations discussed and presented in this paragraph to base treatment options have been undertaken in septic patients as septic shock is the most frequent shock type, the principle knowledge achieved may, generally accepted, be adopted to and transferred to other shock forms as despite natural differences the principal features have been found and confirmed in all entities [478].

Since CO (SV × HR) is a major determinant of oxygen supply [694–696] and the primary parameter by which oxygen delivery to the tissues is varied to match metabolic demands [697], the basic and prevailing intention and purpose to *administer fluids* is to augment cardiac preload and consecutively CO to a significant extent [563, 689, 698], further to replenish absolute and relative hypovolaemia, to replace fluids lost by capillary leakage and to compensate for changes in vascular capacities [243, 689]. Consecutively, this measure may translate into a rise in arterial BP and an increase in urine production, the lactate level is expected to decrease and metabolic acidaemia to be ameliorated [682].

Accordingly, fluid administration is the commonly accepted first-line approach to circulatory disorders, especially in ensuing and manifest shock maladies [12, 684, 690, 699–701]. Meanwhile, volume repletion alone has provided evidence not only to improve macrocirculatory haemodynamics including cardiac output (CO) and hence raised systemic oxygen delivery, but indeed thereby significantly ameliorates tissue perfusion [702–707]; moreover, in some studies even a diminished morbidity and mortality is reported [556, 686, 706]. Even in cardiogenic shock with its potential risk for a fluid-unresponsive ventricle (due to a markedly impaired cardiac function) and hence fluid overload, a fluid challenge should not be withheld [31, 334, 708] unless frank pulmonary oedema is obviously existent [42]. Apparently, initial vigorous and thoroughly volume application is applicable independent of which shock type is present and which pathophysiology predominates [688].

The research work within the last decade after River's [556] landmark study on septic shock, the first systematic approach and tailored treatment strategy to the macrohaemodynamic alterations occurring in severe sepsis and septic shock, generated plenty of new insights and was able to specify our practice much more: we have come to the conclusion that (a) shock maladies are attributed to a failure of macro- *and microcirculation* [90, 91, 212, 213], both may, beyond that, be dissociated [7, 212] – and perhaps a failure of the mitochondria as well [709]; consecutively we have to pay attention to both in our therapeutic efforts and need *to resuscitate* both, *macro- and microhaemodynamics* [194, 400, 478, 617, 627, 680]. (b) The timing of fluid application is of particular impact [217, 260, 710], and (c) the amount of fluids infused assumes a major role in patients' outcome as fluid overload is associated with adverse results [12, 711–714]. Finally (d), we have learned that it is of essential importance to assess, guide and determine more precisely intravascular volume status [563, 698].

(a) Since microcirculatory alterations are identified being an essential factor in the pathophysiology of severe circulatory failure and shock state diseases [29, 90, 91, 478], playing a crucial role in the development of organ dysfunction and in the course of critical illness [91, 212, 632], it is natural and fundamental at the same time that the microhaemodynamic malfunctions need to be resolved in order to re-establish or maintain adequate tissue perfusion and thus oxygenation; hence, the microcirculation has to be addressed by our therapeutic efforts [400, 478, 617, 627, 680]. Unfortunately, as we are just attempting to understand more about the underlying pathology, no trials and no data – apart from some theoretical considerations and a few experiments – exist how to do so [478].

From animal studies [715, 716], and some clinical trials we have fair evidence that fluid loading alone improves microcirculatory haemodynamics [216, 217, 493, 555]. The effect of improved microcirculatory blood flow (increase in total and small vessel density, higher proportion of small vessels perfused compared to baseline [217]) seems to be more pronounced in cases where the initial conditions are markedly altered [717].

However, the changes observed by macrocirculatory and microcirculatory indicators were disproportionate in some studies, suggesting different mechanisms implicated in the regulation of micro- and macrocirculation [216, 217, 617]. Indeed, tissue perfusion is beyond macrocirculatory features directly influenced by specific *intrinsic* microcirculatory issues (local factors with vasoactive effects, e.g. NO, prostacyclins) [199, 400], triggered by septic injury and/or ischaemia [212, 632]. De Backer suggests that issues like microthrombi, adhered erythrocytes and leukocytes and especially a substantially altered glycocalyx (see below) anatomy and function may barely affect microhaemodynamics rather than macrocirculation [194]. Consistently, microcirculation literally needs to be addressed specifically, e.g. peripheral vasoconstriction by vasopressors in case of peripheral vasodilation [687, 718].

The dissociation between macro- and microcirculatory response to volume loading may be the reason why in some investigations although macrohaemodynamics clearly improved and achieved physiological goals microcirculatory alterations persisted and the mortality rate did not change [91, 195, 212, 213, 630, 719]. The lack of response to fluid administration is dependent on the point of time applied as reported by Ospina-Tascon [217]. They demonstrated that fluid application induced an improvement in microcirculatory surrogates in the early (first 24 h after diagnosis) course of septic shock, but was totally blunted if fluids were given in later stages (more than 48 h after diagnosis) (see as well section b, below). Furthermore, it became evident that macro- and microcirculation have different dose responses [216, 217].

Moreover, if fluid therapy is only guided by macrocirculatory surrogates such as MAP, CO and $ScvO_2$, therapeutic measures indeed may result in improved macrohaemodynamics, but unfortunately, macrocirculatory variables do not display the effects on microhaemodynamics; the latter may even be more altered by our therapeutic measures with subsequent further impairment of regional tissue perfusion [493, 720]. On the other hand, in non-septic shock types suitable tissue oxygenation is frequently restored or maintained by addressing and titrating global haemodynamics to normal ranges [627].

Though to reiterate, microcirculation requires to be specifically assessed, and evaluation of microcirculation and close guidance of the measures clasped are absolutely essential [493, 617, 720]. Although in recent years, tracking sublingual flow by OPS/SDF has emerged being the preferred – at least for research issues – method [400], with impaired sublingual flow heralding the *onset of systemic* hypoperfusion in circulatory shock [721, 722] and of independent prognostic power in clinical trials [91, 212, 632]. Unfortunately, the tool is not yet ready for clinical practice as no clinical end points for microcirculatory resuscitation and therapeutic interventions have been defined [194].

As we have to integrate and incorporate requirements and targets of both macro- and microhaemodynamics [478, 617, 627, 680], and due to the substantial variation of interindividual responses found in studies, *individualized solutions need to be sought after* [617, 723].

(b) Several trials convincingly revealed that fluids given at the right time is an issue special attention has to be paid to [724, 725]. Fluid administration during the early disease phase (first 24 h after diagnosis) of sepsis has been reported to significantly improve microcirculatory haemodynamics [86, 726, 727], whereas in timely advanced (48 h or later after diagnosis) septic stages, microcirculatory response to fluids was found to be completely attenuated [217, 727, 728].

Other studies indicate different and dissociated responses of macro- and microcirculation to therapeutic measures (fluid loading as well as vasopressor application) and their time dependency [199, 213, 216]. Pottcher verified a nonlinear relationship between changes in CO and microcirculation following fluid administration; moreover, the microcirculatory changes even impressed disproportionately compared to the macrohaemodynamic modification [216]. While in advanced shock stages a dissociation between macro- and microhaemodynamics in the presence of normalized macrocirculatory variables has been demonstrated independent of the shock type [632], early on, microcirculation is not *generally* dissociated from macrocirculation, even not in sepsis, despite earlier controversial findings [199, 213, 216, 217].

Furthermore, organ failure early in the disease course is obviously mediated by the abnormal perfusion, while in the later phases the mitochondrial disruption seems to be responsible for the organ malfunction [2, 729, 730].

With *ongoing treatment days, a more restrictive fluid management* is vital as fluid overload and especially a positive cumulative fluid balance have been demonstrated to be harmful with serious complications and worsened outcome

[700, 731–733]. As such, Boyd showed that a more positive fluid balance at 12 h and over 4 days was associated with increased mortality [699], Rosenberg reported in his large review of cumulative fluid balance on outcome in acute lung injury based on clinical series (nearly 100 patients) of a clearly favourable result with independently lower mortality if a negative cumulative fluid balance could be achieved on day 4 onwards [734]. Slightly different time frames are reported in other studies (2 negative balances within 7 treatment days [732], negative fluid balance by 500 ml or more on day 2 or 3 [735]), but they do not change the essential aim *to achieve a cumulative negative balance within roughly 4 days after admission.*

Furthermore, it is beneficial to start therapy as early as possible [555, 556, 692]!

(c) The intention to fluid administration and the amount of volume infused is primarily based on physiological considerations rather than on available evidence [736]: Traditionally, more or less arbitrary treatment goals have been chosen within the physiological ranges or somewhere above taking the pathobiology into account and guided by numeric parameters representing macrocirculation such as MAP, CVP/PCWP (preload determinant), CO/CI and ScvO2 [243, 685, 688, 691, 737]. However, an improper amount of fluids applied (either too much or too little) is obviously a relevant and major factor of morbidity and mortality [260, 736, 738]; particularly "volume overload" has turned out as being a major contributor to poor outcome and needs to be avoided at all [699, 700, 714, 739–742].

It is not only the acute formation of fluid overload, one of the two (aside left heart failure) most frequent aetiologies of pulmonary oedema development, particularly in the setting of capillary leakage [743], but the cumulative fluid balance during the treatment course that needs to be paid attention to: A cumulative positive fluid balance in critical illness is generally associated with worse outcome [733]! In acute lung injury, a clearly positive association between mean cumulative fluid balance and mortality has been described [732, 744, 745]: Wiedemann et al. [700] reported of significant less days of required mechanical ventilation if patients with ARDS were guided by a more restrictive fluid management strategy with a cumulative balance of 136 ml ± 491 ml versus 6,992 ml ± 502 ml during the first 7 days of treatment. A small pilot study in septic shock patients found the survival rate to be superior in patients with a negative fluid balance by 500 ml or more within day 2 or 3 of treatment [735]; this was in principle confirmed in a bigger study by Murphy including 212 patients with septic shock and associated ALI, where patients with negative fluid balances on at least 2 consecutive days during the first 7 days of treatment showed a reduced mortality [732]. The Vasopressin and Septic Shock Trial (VASST) delivered the result that the quartile of patients who hold the highest positive fluid balance at 12 h and at 4 days showed the highest adjusted mortality rate [699]. The big observational SOAP study confirmed a generally increased mortality rate in patients with a positive cumulative fluid balance [12].

Unfortunately, assessing circulating blood volume and predicting volume requirements and response are very difficult and absolutely challenging [313, 552].

Further, we should differentiate between fluid management (therapy of fluid imbalances and losses) and volume therapy, the replacement of intravascular volume deficit to restore or to prevent impaired circulatory function following absolute or relative hypovolaemia – ineffective circulating (blood) volume [736, 746]. Fluid administration in case of acute circulatory failure primarily aims to ensure and to improve insufficient tissue perfusion, hence to compensate for relative or absolute hypovolaemia (volume therapy) [687, 746].

In order to increase CO and thus tissue blood flow, a fluid challenge may be applied (fluid management) [563, 697, 698, 701]. As hypovolaemia is a frequent cause of compromised circulation and hypoperfusion a hallmark of shock maladies, volume replenishment and, by increased preload, an augmented CO are desired [563, 698, 701]. However, only roughly 50 % of patients in shock will exhibit a significant increase in SV and hence CO in response to fluid administration [747, 748]. Furthermore, the pre-challenge vascular and cardiac filling status is not the only determinate of fluid responsiveness [749] since an increase in preload can only translate into a rise in SV via Frank–Starling relationship, if the heart is intact and not already working on the flattened part of the ventricular function curve [563, 701, 750] – physiologically, the normal heart (both ventricles) acts on the steep part of the P–V relationship keeping some myocardial reserve capacity [751], and an increase in preload will result in a significant rise in SV [752]. In a compromised heart, especially if the contractility is impaired, the P–V relationship is considerably flatter and the steep part may be virtually missed resulting in a clearly weaker or even undetectable SV effect: An increase in LV-SV decisively depends – aside from pre-infusion preload – on the cardiac contractility [337, 753]. Fluids given in such a situation may lead to deterioration of cardiopulmonary function [25, 747]. Non-responsiveness still assessed by commonly used parameters such as CVP or MAP bears the risk to overload the patient, particularly in case fluids are given repetitively because a missing positive response is explained by a too short volume administration or by a CVP measurement suggesting to give more fluids [563, 754, 755].

The increase in SV may not translate into a rise in MAP and thus in perfusion pressure as the pressure responses to increases in SV substantially depend on arterial tone [314], defined by both vascular resistance and compliance [315]. Attributed to sympathetic modulation, the arterial tone tends to maintain MAP constant while CO varies; thus, changes in MAP may be dissociated from changes in CO [314, 756, 757].

Hence, it is no doubt that the assessment and the prediction of the patient's response to fluid administration are of crucial importance and that we have to use reliable parameters which can predict what will happen if we infuse bigger amounts (i.e. 500 ml crystalloids) of volume [563].

Traditionally, central venous pressure (CVP) and pulmonary occlusion pressure (PCWP) have been used to guide fluid management [563, 701, 754, 758]. Over 100 publications gave substantial evidence that no relationship at all exists between CVP/change in CVP and fluid responsiveness [754]. CVP reflects RAP [759], a major determinant of RV filling [760], and as such CVP was assumed to be a good indicator of RV preload [563]. Further, as RV-SV determines LV preload, it was believed that CVP indicates LV preload [563]. However, changes in venous tone, intrathoracic pressures, RV/LV geometry, and particularly changes in RV and LV compliance – both ventricles tend to stiffen during shock [9] – no feasible correlation between CVP and ventricular filling volume/intravascular filling could ever be demonstrated [754]. Thus, measuring CVP (and PCWP) and guiding fluid application on the pressure measured is absolutely useless in this context [563, 701].

A CVP >9–10 mmHg is always pathological [25, 761], and CVP ≥10–12 mmHg has to be considered high [761], indicates an elevated RV afterload [762], and most patients will not respond to fluids [762]. Even attempts to install special regimen on how the CVP may change in response to fluids, e.g. modification (see chapter about AKI) of the 5–2 rule [763] by Vincent and Weil, are not really helpful [764], mainly because a change in the magnitude of the CVP of at least 2 mmHg is the minimum necessary for detection with confidence on currently available monitors [762] – apart from relevant imprecisions due to height settings of level zero. Insofar, even the Surviving Sepsis Campaign guidelines [260] need to be challenged [691, 765–767].

Currently, the most validated, sensitive and reproducible predictors of fluid responsiveness are the dynamic parameters of intravascular volume which have been applied in a wide range of clinical settings [747, 748, 768–772]: Unfortunately, they are in generally *applicable only in mechanically ventilated and anaesthetized patients* [559, 698, 701, 773, 774]. Physiologically, they are based on the heart–lung interactions during mechanical ventilation (see chapter on mechanical ventilation) and a concept of Perel et al. [775] who investigated the effects of interruptions of mechanical ventilation on systolic blood pressure characteristics and variation, variations in SV (SVV) and its surrogates, pulse pressure variation (PPV), systolic pressure variation (SPV) and even aortic blood flow, all reflect changes in LV-SV associated with positive pressure ventilation and indicate preload responsiveness of both ventricles [748, 768, 776]. A *variation of SVV >12–13 %* is highly predictive of volume responsiveness and that has been demonstrated with a remarkable consistency [748]. Briefly, positive pressure ventilation with its cyclic increases in intrathoracic pressure and lung volume induces inspiratory reductions in venous return, thus impairs RV preload during inspiration, and increases in RV afterload attributed to the inspiratory increase in transpulmonary pressure [777]. Both dilute RV-SV resulting in a decreased LV preload, affecting the LV after a delay of a few cardiac cycles [698]. As a result, LV-SV is diminished, and these cyclic changes in stroke volume are even more prominent if the ventricles are working on the steep part of the volume–pressure relationship – Frank–Starling dependency [563]. The dynamic swings in LV-SV are the basically underlying physiological mechanism and considered as being the gold standard [778, 779]. However, the PPV turned out to be even more robust and sensitive [747, 768], as influenced by the vascular properties inclusive of filling volume as well – pulse pressure is inversely related to arterial compliance [701, 780]. It is created by the SV ejected into a central arterial compartment as quantified by a transfer function [781]. PVV shows a positive predictive value of 94 %, a negative of 96 % [747, 782].

Limitations of the method [698] (aside from that it is only applicable to mechanically ventilated, paralysed patients) lie in

- cardiac arrhythmias, particularly atrial fibrillation with irregular cardiac cycles and inherent SV variations,
- high-frequency ventilation
- *raised intra-abdominal* pressure [783] – as SVV decreases with increasing intra-abdominal pressure and hypovolaemia [784], higher cut-off level needs to be considered – unfortunately currently ranging from 20 % [785] SVV to 40 % [786] indicating positive fluid responsiveness,
- open chest surgery and
- *applying too low tidal volume* with consecutively lower and obviously insufficient increases in intrathoracic pressure during inspiration not decisively reducing venous return and lung volume [787, 788], at least 6–8 ml/kg tidal volume seems to be necessary [788].

In daily practice either rapid application of a fluid bolus or passive leg raising (PLR) are established procedures to test fluid responsiveness [747, 778, 789–792]. When lifting the legs, both preload of RV [778, 793] and LV [559, 778] will increase [559]. The biggest advantage to use PLR for testing fluid responsiveness may be that PLR is a transient autotransfusion which fully reverses once the legs are tilted down with no need to infuse fluids; thus, there is no risk to overload the patient [559, 778]. Furthermore, PLR is a bedside test and, exerting its effects over several cardiac and respiratory cycles, may be used in spontaneously breathing patients and in case of cardiac arrhythmias [559, 794]. On the other hand, the effect is transient, lasting only for about 30–90 s, and hence needs a real-time measurement [698, 795] such as pulse contour analysis-derived cardiac output [796, 797], aortic flow measured by oesophageal Doppler [559, 790] or cardiac output assessed by bioreactance [798], ascending aortic bloodstream velocity measured by suprasternal Doppler [799] or end-tidal carbon dioxide [800]. The effect has shown to be more intense if PLR is started from a 45° semirecumbent position as the increase in central venous pressure is more pronounced [793, 795]. The effect cannot be assessed properly by monitoring simply arterial BP [698].

Another valuable tool to guide volume and fluid therapy is measuring and evaluating the extravascular lung water index (EVLWI): The lungs are particularly exposed to proinflammatory processes and thus to microcirculatory alterations [741, 801]. *EVLWI*, determined by transpulmonary thermodilution [802, 803], provides accurate measurements of the real amount of the fluid within the interstitial lung tissue [803–805], thus *reflects and quantifies* the *capillary leakiness* and *fluid overload* [806–809] – excess fluids will be lost into the tissues and this even more in the setting of capillary leakage [801]. EVLWI correlates quite well with the mortality in critically ill patients in various clinical settings [739, 804, 810–812] and is undoubtedly a useful tool to guide fluid management [563, 748].

At least if the index exceeds *10 ml/kg*, the mortality rate disproportionally increases; thus, fluid restriction is indicated [804, 812–816]. The recently introduced permeability index (PVPI) even further specifies cardiogenic and/or fluid overload-associated lung water from permeability-induced ones [805, 808, 814]. Indexing EVLW to predicted body weight may be a more accurate assessment of the amount of interstitial lung water and indicator for survival (compared as to indexed to current, admission weight) and thus helpful to guide fluid and volume treatment – restrictive fluid strategy, if EVLW indexed to predicted body weight *>10 ml/kg* as described in an review by Brown et al. [816].

Both hypo- and hypervolaemia decrease tissue perfusion and may lead to organ failure [817–819]. *Ascitic or pleural effusions, in turn, should be recognized as a physiological overflow* provoked by intravascular overhydration [682, 820] since volume "over"-replenishment/"over"-resuscitation in general may promote interstitial fluid accumulation [821, 822]. An increased intra-abdominal pressure following ascites development or even an abdominal compartment syndrome may result – as compartment syndromes in general – in reduced perfusion pressure and thus in jeopardized tissue perfusion [823, 824]. For example, a moderately elevated intra-abdominal pressure (12–20 cmH_2O) affects glomerular filtration pressure and may cause AKI in the presence of altered systemic haemodynamics such as shock maladies [825].

The *type of fluid* is still a matter of debate [382, 826]; however, in recent years, at least in sepsis, crystalloids seem to be the preferred fluids [827] although clinical trials and meta-analysis never showed a significant benefit of one fluid type compared to another – crystalloids, colloids and albumin [89, 828–831]. It is that most colloids may show some undesirable effects; most important, they may compromise the kidney function, often described when applying hydroxyethyl starches or dextrans [832, 833], and individual studies even report a higher mortality when using 6 % HES 130/0.4 solutions [834]. As the hyperchloraemic acidosis associated with the application of N. saline resulting in reduced splanchnic mucosal perfusion, glomerular filtration and coagulopathy [835], Ringer's solution or balanced crystalloids should be the first choice [680, 836].

Further, we have to notify that only roughly 20–30 % of the infused amount of crystalloid fluid will stay intravascularly; the majority is lost immediately [837, 838]. Colloids in healthy persons remain in the vessel system for a couple of hours, but in sick patients, particularly in those who are critically ill, only marginal more fluids (in comparison to crystalloids) will be maintained in the vascular system after an equivalent infusion [829]: this is due to the capillary leakage, but largely to the disrupted glycocalyx, precipitating the loss of intravascular fluids [697, 829, 839]. The glycocalyx, a thin endothelial layer of transmembrane syndecans and membrane-bound glypicans coating the vascular endothelium [840, 841], serves as a kind of vascular barrier retaining proteins and as such enhances the oncotic pressure limiting fluid loss across the vascular barrier [841, 842]. Furthermore, the glycocalyx is binding about 800 ml of blood plasma, hence, plasma volume comprises of a circulating part and a kind of reserve part [841]. Accordingly, the glycocalyx takes part in the fluid interchange between vessel system and tissues [697]. In critical illness filter and "storage" function are substantially disturbed leading to a more or less unrestricted fluid shift towards the tissues provoking tissue oedema [843, 844]. Of decisive importance, hypervolaemia in otherwise healthy persons may likewise promote an impairment of the glycocalyx structure/function, and inadequately high fluid administration will provoke increased transendothelial permeability and thus tissue oedema generation [845, 846].

3.7.2 *Vasopressor Application/Use*

"The treatment of acute haemodynamic impairment should be tailored based on the aetiological mechanisms of the cardiovascular dysfunction" [718]. Acute haemodynamic disorders and shock can be referred as to both cardiac and/or arterial–vascular dysfunction since LV-SV is a function of myocardial contractility and loading conditions, pre- and afterload, the latter decisively affected by the circulatory (vessel) system [718, 847–849]. Therefore, treatment and resuscitation of life-threatening shock states include *promotion and augmentation of cardiac function* and hence CO by inotropic agents [5, 30, 42, 850] *and BP* in case of persistent hypotension and haemodynamic instability by vasopressors [5, 681, 851–853]. Vasopressors are indicated in continued hypoperfusion accompanied with hypotension despite and after sufficient and adequate intravascular volume replenishment/intravascular filling [243, 719].

In distributive shock types volume replenishment in general provokes an increase in CO, but this is accompanied by a decrease in vasomotor tone (which can be addressed by vasopressors) [313, 854].

Fluid application still remains the first step, even in cardiogenic shock with often rather poor response [217], as long as no frank pulmonary oedema is present [42, 855], since the use of vasopressors, particularly in uncorrected hypovolaemia, may further promote ischaemia and organ hypoperfusion following excessive peripheral vasoconstriction [856, 857]. Nevertheless, relative hypovolaemia should be corrected related to the underlying pathomechanism(s) early on by vasopressors [687]; notably, prophylactic application of crystalloids in order to avoid vasodilation-related hypotension does not exhibit the desired effects and is as such senseless, just jeopardizing the patient for fluid overload [858]. Furthermore, early use of vasopressors reduces the incidence of organ failure and may be started overlapping with the initial administration of fluids – with increasing vascular filling the vasopressors may be titrated down [313, 692, 859]. Thooft states that fluid resuscitation alone will not re-establish haemodynamic stability; thus, vasopressors are required early in septic shock [199, 478]. Marik [692] recommends in sepsis syndrome norepinephrine (NA) to be started immediately after the patient has received 2 l crystalloids if a MAP of at least 65 mmHg cannot be achieved as vasoconstrictors early on reduce the incidence of organ failure [859].

Acute life-threatening settings with inacceptable low BP and hence the risk of considerable myocardial ischaemia [280, 296, 860] require vasopressor therapy immediately; concurrently fluids should be commenced [199, 687, 859].

Arterial blood pressure is the driving force to blood flow through the tissues [737], and perfusion pressure (also called driving pressure) is a function of total perfusion and vasomotor tone (often termed as arterial tone; see above) [313] – the latter is one of the major determinants of tissue perfusion [3, 861]. Vasomotor tone is altered or even lost in shock states [199–201, 862], particularly in septic shock, but may also apply in cardiogenic shock situations [90]. A reduced and maldistributed tissue perfusion results from a perfusion pressure beyond the autregulatory threshold [199–201, 863]. Moreover, hypotension alone is found to blunt autoregulation of blood flow since regional blood flow distribution is issued by local dilation of the arterioles triggered by metabolic requirements [199–201, 864].

As such, a reasonable BP level, literally MAP which represents fairly enough the organ perfusion pressure for daily practice [761], is absolute necessary and required: "To guarantee perfusion pressure to all vascular beds, adequate CO *and adequate MAP* are necessary" [596, 865]. Furthermore, as described above, the administration of fluids does not necessarily and consistently translate into an increase in MAP, *as arterial pressure response* to an increase in SV depends on arterial tone [314, 315], and is for a given increase in SV, the larger the arterial tone is [866]. Both, a fluid-responsive ventricle *increasing SV by volume loading and enough arterial tone* are required to boost MAP and thus perfusion pressure [867]. Gracia reports that roughly 60 % of the fluid responders increase MAP [314]. If we further take into consideration that "only" about 50 % of patients generally will respond to fluids with an increase in SV [747], vasopressors may be needed even early in the treatment course to reinstall autoregulation and to modify vasomotor tone [868, 869] – sympathomimetic agents such as NA generally increase systemic vasomotor tone [870], hence to ensure by a sufficient MAP an improved and appreciated tissue perfusion [871]. Furthermore, autoregulatory threshold as a result of intravascular thrombosis, vasoconstrictive mediators and local NO deficits may be shifted to higher values, just as these abnormalities may contribute to the altered vascular reactivity reported [862, 872]. Due to a loss of responsiveness to sympathetic stimuli of the vessels, a complete restoration of arterial vasomotor tone is usually missing [392].

It is well established that by administering vasopressors (and fluids), we are able to restore macrocirculation indicated by CO/CI, MAP, oxygen variables, etc. [243, 313, 478, 682]. Important in this context, Thooft [199] gleaned in his study on "The Effects of Changes in Arterial Pressure on Organ Perfusion During Septic Shock" evidence that the *recovery of vasomotor tone and autoregulation rather than the rise in CO were causative accountable for the improvement of microhaemodynamics* (verified by NIRS and in some patients by SDF) with *increasing vasopressor* (NA) *doses*. Further, noticeable interindividual variations suggest the necessity for a personal titration of dose and MAP, whereupon the therapeutic actions may probably be best guided by direct visualization of microcirculation, i.e. by sublingual SDF technique [199]. Thooft's study results are furthermore remarkable:

(a) Because mechanisms (vasopressors such as NA are supporting vasomotor tone/may re-establish suitable vasomotor tone) independent of an enhanced CO obviously promoting microcirculation could be identified for the first time,

indirectly confirming the relevance of microcirculatory features in the pathobiology of shock states, particularly in sepsis.
(b) The necessity to independently guide and to therapeutically address microcirculatory dysfunction is once more confirmed.
(c) This study challenges previous trials, not finding beneficial effects of tissue perfusion with increasing vasopressor doses [723, 873–876].

These different results may rely upon differences in timing and incomplete fluid repletion at the time vasopressors were commenced [876], patient selection [723], variables used in the study to assess microcirculation (uo, renal perfusion) [874, 875] and the vasopressor drug applied (dopamine and NA, norepinephrine) [873].

Dubin as well demonstrated an improvement of microcirculation (verified by SDF) with increasing amounts of norepinephrine in patients still hypotensive after volume loading, whereupon those with normal blood flow or marginal microcirculatory disorders displayed circulatory abnormalities [723]. However, the study confirmed that vasopressors potentially improve microhaemodynamics, but the circumstances when to apply and how much warrant further precision; again, *one size does not fit all* [478].

Noteworthy, both Vlahakes et al. [876] and Di Giantomasso et al. [877] gave proof of a significant increase in myocardial tissue perfusion while administering NA to treat hypotension in patients with cardiogenic shock.

The ideal target MAP is controversial, however, the level needs to be individualized, but *in all cases*, it should *be well above the autoregulatory threshold*! [199] For reinstallation in case of autoregulatory breakdown, higher levels may be required [874], and, for example, persons with known arteriosclerosis will need a higher level than a previously (cardiocirculatory) healthy person [313]. Furthermore, different thresholds in different organs are reported: For the kidneys, an autoregulatory range between 80 and 180 mmHg MAP is reported [484, 878, 879], for the brain 60–160 mmHg [880, 881], and for the coronary circulation the space is roughly 60–140 mmHg [882, 883]. However, in case of ischaemic heart disease, higher pressures, ≥75 mmHg, are recommended [873, 882, 883]. As the kidney function is always crucial and the presence of even mild renal dysfunction already worsens outcome and increases mortality [884, 885], maintenance or re-establishment of renal function (mainly estimated by uo being easily and instantaneously observable) may give the determining hints [886]: Obviously, a MAP between 70 and 80 mmHg seems to be the most suitable to avoid the development of acute kidney injury (AKI) [493, 737, 887, 888]. If we base our decision upon the results by Lehman, who assessed retrospectively in a large number of ICU patients (nearly 17,000) the main risk factor(s) promoting AKI development, we would tailor the MAP around 80 mmHg [889]: Both, the severity and the duration of hypotension significantly increase the risk of AKI. For each 10 mmHg below MAP, each additional hour of continuous hypotension elevates the risk by more than twofold, and the odds to evolve into AKI swells by 3 % for each mmHg below a MAP of 80 mmHg [889]. However, marked interindividual variations are reported in various trials [723, 875, 876]; thus, the MAP level has to be determined and titrated individually [199, 723].

Vasopressor application with subsequent increase in organ perfusion pressure (indicative by an increase in MAP) has provided proof as being able to meliorate microhaemodynamics (assessed by tissue perfusion analysis, e.g. capnometry, NIRS) [723, 876, 890]. Unfortunately, the target perfusion pressure is not known at all, and an individual approach, ideally monitored by microcirculatory assessment, e.g. sublingual capnometry or SDF analysis, is essential [493, 723, 876]. The application of vasopressor agents may not affect microcirculation [723, 876], but be aware of that they may well compromise microhaemodynamics [891, 892], particularly if macrocirculatory variables (MAP) are normalized and NA is given in order to ameliorate a still disrupted microcirculation [723, 876].

If microcirculation is normal, no interventions should be conducted: A couple of studies found that at least in septic patients with normalized microcirculation after initial resuscitation, no further improvement from any additional increase in MAP (or infusion of fluids) can be expected; in contrary, microcirculation may deteriorate [723], while those with still abnormal microhaemodynamics may very well meliorate from increasing MAP (higher than 65 mmHg) [199, 617, 723, 876].

Noradrenaline (NA) is now considered the vasopressor of choice in shock state diseases: Several trials demonstrated a significant better outcome in septic but other shock states as well when NA was applied compared to dopamine, further with less side effects and complications and was superior in achieving the pressure goals desired [31, 199, 692, 851, 876, 893, 894]. Notably, noradrenaline has the ability to increase peripheral vascular tone [313], a crucial issue in organ perfusion and tissue blood flow distribution (autoregulation) [199]. As such, vasopressors basically enhance perfusion pressure and by increasing MAP above autoregulatory threshold re-establish or preserve regional distribution of blood flow [45]. NA promotes an increase in BP (MAP), CO/CI, renal, splanchnic and cerebral blood flow, and microcirculatory perfusion [876, 895].

Of course, catecholamines in general, particularly at higher dose, may exhibit adverse effects such as derangements in regional blood flow (due to vasoconstriction), and an enhanced likelihood of death is described [11, 852, 877, 894]. Notably, physicians have been concerned about renal and splanchnic tissue perfusion following admittedly inconclusive study results [896–898]. However, supported by several clinical studies, NA is shown to improve both renal function and splanchnic microvascular perfusion in the vast majority of circumstances [852, 899–901]. Bourgoin demonstrated in their publication on "*Increasing Mean Arterial Pressure in Patients with Septic Shock; Effects on Oxygen Variables and Renal Function*" that with *increasing doses of NA* and thus stronger vasoconstriction boosting the MAP from 65 to 85 mmHg; *the increase in vasoconstriction did not adversely affect systemic tissue perfusion* and oxygen utilization [874]. As the SVRI remained well within the normal range, they suggest that the results are allocated to *an increase in vascular tone*, whereupon once autoregulatory threshold is achieved and autoregulation effective, further enhancements in perfusion pressure provide no additional benefit, in this study shown by no further improvement in renal function [874].

Furthermore, a combination of NA and dobutamine has been shown to secure proper tissue perfusion even in the splanchnic vessel bed [259, 902].

Dobutamine (Dob) providing inotropic support is broadly indicated in (a) shock states unresponsive to fluids, (b) CS and (c) postcardiac arrest [4] and (d) may be considered in septic shock in the presence of severe myocardial depression causative for a failing normalization (of CO) after fluid resuscitation [243, 719]. Rabuel and Mebazaa estimate that roughly 10–15 % of patients with septic shock may require inotropes due to inappropriately low CO and $ScvO_2$ [860].

Dobutamine improves oxygen delivery [313] and potentially corrects tissue hypoxia [903, 904]. Again the combination of both NA and Dob may be even more effective and appropriate compensating undesirable effects, i.e. vasodilation induced by Dob [903].

Please find the haemodynamic profile and clinical effects of both drugs in the following box, adopted from Ellender and Skinner [5] and Van Thielen and Price [905]:

Drug	Main receptor activity				Clinical/haemodynamic effects						
	α1	α2	β1	β2	CO	dp/dt	HR	SVR	PVR	PCWP	MVO_2
NA	4+	3+	3+	0(+)	↑	↑	±	↑↑	±	±	↑
Dob	0(+)	0(+)	4+	3+	↑↑↑	↑	↑↑	↓	↓	↓/±	↑

Dose, NA 0.02–1.0 (open end) μg/kg/min
Dob 2–20 μg/kg/min

3.7.3 Cardiogenic Shock

The very complex pathophysiology of CS is macrocirculatory largely characterized by an abrupt, marked impairment of cardiac performance with acute loss of pump force [52, 276, 278], frequently attributed to substantially diminished myocardial contractility [58, 906] in the presence of (in the vast majority) mild to moderately increased afterload attributed to compensatory sympathetic nervous [38, 52] and neuroendocrine activation [52, 907]: The resultant fall in CO and thus actually decreased systemic blood flow is immediately counteracted by a (compensatory) redistribution of blood volume, partly by active vasoconstriction, issuing a decline in peripheral vascular capacitance, thereby amplifying cardiac filling pressures and ventricular filling (risk for pulmonary congestion/oedema) with consecutive increase in CO via Frank–Starling mechanism, provided a cardiac reserve is available [214, 352, 908]. Vasoconstriction may, to some extent, preserve a reasonable coronary perfusion pressure [909]; nevertheless, the compensatory response may, in turn, be maladaptive as vasoconstriction potentially boosts afterload which will be detrimental in the presence of a markedly compromised cardiac pump force [42, 855, 910]. Moreover, as recognized just a decade ago, in most cases the compensatory efforts, particularly the intensity of systemic vasoconstriction achieved, are inappropriate related to the severity of circulatory compromise [40, 911]. This is caused by a systemic inflammatory response following the substantial circulatory injury [41], playing a crucial part in the pathobiology of CS [41, 42, 214]. The mediators released (e.g. proinflammatory cytokines) and the subsequent induction of nitric oxide synthase leading to elevated NO levels may prompt disproportionate vasodilation and attenuation of vasomotor tone and regulation [89, 214, 352], thus lowering systemic and coronary perfusion pressures [912]. This potentiates the compromise of coronary perfusion and subsequently promotes even more myocardial ischaemia in the setting of acute myocardial infarction, the main reason for CS [38, 296]. Furthermore, De Backer provided considerable evidence that in CS *microhaemodynamics* are altered as well [90] and the microcirculatory vessel malfunction further compromises myocardial and other organ tissue perfusion [90, 209, 310, 913] which may result in multiorgan dysfunction and failure [42, 914, 915]. The considerable impact of these systemic response mechanisms and features affecting macrocirculation but *microcirculation* as well in the acute disease course are not at least substantiated by the fact that patients with CS show on average an LV-EF of about 30 % [274] – patients with CS may even demonstrate preserved EF [286] – which often does not change in the further course [287] and which is reported in many *stable* chronic heart failure patients [916–918], but obviously the functional status differs fundamentally [42, 908]. Comparably to septic shock, resuscitation and optimalization of macrocirculation may not translate into clinical improvement and survival of the patient suggestive for ongoing microcirculatory failure and the impact of microhaemodynamics on the disease course [20, 310, 919]. However, in spite of these undoubtful microhaemodynamic disorders in CS, a complete dissociation from macrohaemodynamics as found in septic shock has not been reported; in contrast, the correlation between macro- and microcirculations seems to be quite close – an important issue for the management [920–922].

Accordingly, the management of these patients includes specific measures, *predominantly percutaneous coronary intervention/angioplasty* in case of AMI as reperfusion is still the *keystone of a successful treatment of CS* [274, 923]. Furthermore, the *maintenance of a sufficient BP* is crucial to break the vicious cycle of perpetuating ischaemia (ischaemia begets ischaemia [52]) as progressive hypotension causes further myocardial ischaemia and thus aggravates diminished cardiac performance [11, 42, 319]. In addition, myocardial contractility may need support, loading conditions should be optimized, clinically relevant rhythm disturbances must be addressed and we have to keep in mind that microcirculation is a noteworthy element in the disease course that needs discrete monitoring and resuscitation [19, 42, 855, 908, 924].

3.7.3.1 Specific Measures

Early invasive coronary intervention in case CS is associated with AMI

Early invasive revascularization to restore myocardial perfusion is the only approach that has consistently shown to reduce mortality in patients suffering from CS complicating AMI [42, 274, 855, 924, 925]. In the SHOCK trial, an absolute increase in 1-year survival of 13 %, accordingly <8 patients have to be treated to save 1 life (NNT <8), confirmed by the Swiss Multicenter Study of Angioplasty for Shock [926], could be achieved [274, 925].

Evidence level I B [925], ACC/AHA I B [927]

Furthermore, the evidence-based guidelines of AHA/ACC and ESC treatment recommendations for acute coronary syndromes apply [42, 928].

Surgical revascularizations probably have a survival benefit, but this intervention is not common [928, 929].

Emergency operation (timely repair) for special mechanical complications is critical for survival, e.g. acute rupture of the ventricular septum following AMI [42, 928, 930].

Emergency valve replacement/repair in case of acute/acutely decompensated mitral or aortic regurgitation attributed to, e.g. acute endocarditis [931, 932].

Evidence level I B [932]

Thrombolysis/thrombus fragmentation or even operation for acute severe and fulminant pulmonary embolism with CS due to combined acute right and consecutive left heart failure [923]

3.7.3.2 Adjunctive Treatment

(i) Re-establishing and/or maintaining appropriate coronary and systemic perfusion

Critical hypoperfusion reduces not only myocardial oxygen supply but aggravates any myocardial perfusion deficit [860], exacerbating myocardial ischaemia and implementing a vicious cycle leading to a more and more severe myocardial ischaemia [280, 296]: “Ischemia begets ischemia” [52].

As outlined above, a sufficient perfusion pressure indicative by the MAP is a crucial prerequisite for a working autoregulation and an appropriate vasomotor tone to ensure or improve/restore proper microcirculation and thus tissue perfusion [199, 201, 863]. Hence, in order to provide an appropriate coronary perfusion, particularly in patients with coronary artery disease, a MAP of 70 (75) –80 mmHg is usually advised [883, 933].

In case of other aetiologies than ischaemic heart disease, lower values may be sufficient, e.g. 65 mmHg in case CS is related to acute myocarditis [873, 874].

On the other hand, impaired tissue perfusion associated with CS is always a sign of inadequate CO and flow – a determinant of perfusion [30].

(ii) Fluid application as described above is basically recommended although its effect is questionable, just 20 % are fluid responsive, and with the risk of fluid overload, particularly pulmonary oedema [199, 478]. However, a monitored bolus of 250–500 ml crystalloid seems to be reasonable [334, 855, 934] although a relevant and sustained effect on BP increase cannot be expected [718].

(iii) In the event of ongoing or critical hypoperfusion, NA is now the vasopressor of choice [11, 31]. The study results by De Backer clearly challenge the guidance of the AHA/ACC [272] still preferring dopamine as first-line vasopressor; the ESC guidelines advocate to prefer NA over dopamine in case of persistent hypotension [928].

Start up sufficiently early while addressing the underlying reason [42]!

In cases of CS with low BP (sBP <90 mmHg, MAP <65–70 mmHg), the application of NA is essentially indicated to increase BP, particularly MAP, to a level where autoregulation of the coronary system should work and myocardial perfusion – *a microcirculatory property* – can be assured: As Vlahakes et al. [839] and Di Giantomasso et al. ([877]) demonstrated, a significant increase in myocardial perfusion could be shown with administration of NA to treat CS-associated hypotension, aiming for a MAP well above 70 mmHg.

Evidence level for vasopressor use, NA: II b B/C [928, 935]
The graduation needs to be challenged!

(iv) If a MAP >70 mmHg under NA has been restored, Dob may be added [272]. In cases of cardiogenic pre-shocks (ensuing or manifest hypoperfusion in the presence of adequate or ongoing optimalization of volume loading and a reasonable BP [308], sBP >90 mmHg, MAP 70 mmHg (65)) and in low output settings with still reasonable BP, early Dob alone may be indicated and is the agent of choice [4, 30, 907, 922, 928]. Dob will increase cardiac contractility (and as such augment CO, which means flow) and BP, and by trend reduce PCWP [907, 922]. Its beneficial *effects* on *microhemodynamics* (displayed independently of its macrocirculatory changes) and hence *tissue perfusion* [627, 936] improve the situation and may even prevent transition to manifest shock [316, 907].

However, the effects of vasoactive drugs on the microcirculation are more or less unpredictable and thus application need to be monitored and individualized [907].

As a possible alternative, especially in case of a clinically relevant diastolic disorder [937], a marked right heart problem [938], in patients who are on chronic β-blockers [928] or acute renal impairment related to cardiac dysfunction [939], levosimendan (*Lev*), a calcium sensitizer with positive inotropic and vasodilative properties [940–942], may be applied.

Further, due to current practice, Lev may be given on top of Dob [943].

Although initial studies were very promising, suggesting that Lev had a lower mortality rate as compared to Dob [852, 902], but unfortunately could not be confirmed in other trials such as SURVIVE [944].

Like Dob [945, 946], Lev [942, 947] combined with NA is a suitable and preferable drug combination in the setting of CS and other severe circulatory disruptions exploiting the beneficial effects of both drugs thereby trying to use the lowest dosages possible diminishing the potentially harmful side effects of catecholamines [5, 30, 42, 924] and to withdraw as soon as possible [42, 946, 948].

Evidence level for inotropic support (in general) in CS/pre-CS/low output due to cardiac dysfunction: II a C [928]

Evidence level for Dob IIa C, for Lev IIb C [928, 935]

(v) Intra-aortic balloon counterpulsation (IABP)

The appliance of the IAPB has been a standard component in the therapy of CS for a long time [38, 271, 272]. Intra-aortic balloon pumping has been used as an adjuvant measure in the context of AMI complicated by CS due to its associated favourable haemodynamic effects such as the increase in diastolic coronary perfusion pressure and coronary blood flow, the rise CO and the diminution of LV afterload and, not at least, because of the attenuation of the systemic inflammatory response [910, 949–951].

Indications included medically refractory CS complicating AMI, refractory angina, postcardiac surgery, refractory malignant arrhythmias, severe acute mitral regurgitation or ventricular septal rupture [272, 952].

Unfortunately, a just recently published high-quality study revealed that the application of IABP in patients with CS complicating AMI *who have undergone interventional revascularization* did not (further) improve outcome – the 30-day mortality remained about 40 % [928, 953]. Thus, the ECS already downgraded their recommendation from IC to II B; hence, "routine use" of IABP in the discussed setting is no longer an issue [935].

However, the use of the pump as an amplifier of the coronary perfusion, supporter of CO and reducer of afterload is still a subject of considerable debate and its use may well be indicated in individual circumstances [954].

Evidence level II b B [928], ACC/AHA still IIa B [927]

(vi) There are inconsistent results using LV assist devices in CS lacking clear evidence of a better outcome [42]. Nevertheless, the ESC guidelines rate this tool with an evidence of II b C [928]

(vii) CPAP in case of serious pulmonary congestion/pulmonary oedema reduces the need for intubation [955], augments oxygenation and dilutes afterload [956] and should be considered in quite dyspnoeic and/or moderate hypoxic patients, patients with relevant pulmonary congestion or oedema, patients who are exhausted due to respiratory distress or patients who do not improve under pharmacological treatment [935, 957].

Indications for endotracheal intubation include therapy-resistant hypoxaemia with contraindications for non-invasive ventilatory support, hypercapnia and acidosis, further physical exhaustion and compromised consciousness [928, 958].

Evidence level for non-invasive ventilator support I a C [928]

(viii) All patients need thromboembolism prophylaxis, preferably with low molecular weight heparins (LMWHs) [959, 960].

Evidence level Ia A [935]

(ix) Buffering of metabolic acidosis, often found in shock state diseases, is highly controversial [961, 962] as there is little evidence of any beneficial effect in case of application [963]; some authors even refuse to buffer at all [964].

However, application of buffering agents, preferably tromethamine, should only be considered if the pH is <7.1 or if there is evidence that vasopressor or inotropic medication is not effective associated with the low pH [965, 966]; target is a pH of 7.2–7.25.

(x) In special circumstances and desperate situations, individual solutions should be sought after; accordingly, the application of vasodilators, namely, nitroglycerin (GTN), may and should be considered: GTN is effective in attenuating symptoms (dyspnoea) in case of elevated filling pressures and improving pulmonary haemodynamics by diluting filling pressures and congestion, especially indicated in patients with low CO [967, 968], and has well demonstrated salutary effects on altered microcirculation [190, 196, 969]. Low-dose GTN (33–50 μg/min) on top of basic, adjuvant therapy illustrated improvement of microhaemodynamics and ameliorated myocardial and other tissue perfusion; unfortunately, a statistically beneficial effect on mortality could not be demonstrated [196, 968, 969].

3.7.4 Haemorrhagic Shock

Haemorrhage is one of the main aetiologies of hypovolaemic shock [33] and a major cause of death in trauma patients: Around 40 % of trauma patients die from haemorrhage and its complications [970–972]; roughly 45 % of prehospital and about 55 % of in-hospital deaths in the trauma shock group are provoked by exsanguination [973]. Being a failure of the peripheral macrocirculation [6, 32, 35] with reduced venous return and impaired CO [6, 35], the MAP will, due to an intense sympathetic compensatory response [35, 67], not fall until 20–25 % of total blood volume is lost [63, 365, 974]. The microhaemodynamics are only somewhat affected [198, 245], not substantially dissociated from macrohaemodynamics as in sepsis [975], and will only be considerably compromised in advanced stages [240, 241, 379, 380]. Hence, the treatment may be suitably guided by global haemodynamic parameters, at least in early stages. However, an increased capillary permeability leading to fluid losses into the interstitial space is found in the initial stages as well [367]. Decompensation of this “blood tank disorder” occurs following marked hypoxia/ischaemia of the central nervous system causing failure of sympathetic activity (sympathetic escape), often induced by exsanguination [32, 224, 976]. As such, macrocirculatory targets, in particular an appropriate MAP, are suitable key elements of the therapeutic approach with *timely control of the bleeding source* and maintenance of a reasonable BP/MAP for sufficient organ and tissue perfusion [367]. However, haemorrhage secondary to *penetrating torso injury* needs a special management route, the so-called permissive hypotension or hypotensive resuscitation [977].

The favourable effects of hypotensive resuscitation in trauma-related haemorrhage have been demonstrated in multitude animal and clinical studies [367, 978–983]. Disruption of soft clots and dilutional coagulopathy and thrombocytopaenia are obviously avoided [367, 984–986]; a decreased loss of blood, a better splanchnic perfusion and tissue oxygenation, less acidaemia and haemodilution and finally significant less apoptotic cell deaths and tissue/organ injuries as well as improved survival rates are seen [987–990]. Major concerns, hypotensive resuscitation may provoke organ hypoperfusion (particularly of the splanchnic vessel network), are meanwhile completely destroyed [976, 991]; moreover, overzealous fluid administration in this setting is definitely harmful [823, 992] and is in case of trauma-associated haemorrhage considered being a major cause of coagulopathy [993].

Permissive resuscitation in the setting of penetrating torso trauma is a well-established acute treatment approach [976, 994] and can be achieved by two different strategies, goal-directed resuscitation or by predetermined fixed rates of infusion, the latter also called controlled resuscitation [367, 987]. A recent Cochrane review [995] found no evidence for or against early volume resuscitation in those trauma patients; in contrary, withholding or only marginal fluids early on until the bleeding is definitely controlled – known as delayed resuscitation – showed significant organ hypoperfusion and poorer outcomes [996, 997]. Accordingly to overall data, hypotensive resuscitation at a *fixed rate of 60–80 ml/kg/h aiming for a MAP between 40 and 60* [990, 998, 999] and *sBP of 80–90 mmHg* seems to be an appropriate way to maintain controlled hypotension and empiric control of infusion [983, 987, 1000–1002]. The US military guidelines recommend to monitor fluid resuscitation by the patient's mentation and peripheral pulses; in turn, fluid administration should aim for a proper mental status and palpable peripheral pulses [1003]. Crystalloids, particularly Ringer's lactate, are the preferred fluid [987, 994].

However, intense search of the bleeding source remains the first priority and all efforts should be undertaken on that issue [987].

Later in the course, after control of the bleeding, preload-directed resuscitation with repetitive fluid boluses is the recommended standard therapy [994].

Notably, in patients with head injury, hypotensive resuscitation is strongly forbidden (contraindicated) – just a single episode of hypotension doubles the mortality rate [1004]. In patients suffering from shock in the setting of *blunt trauma*, hypotensive resuscitation seems not to achieve better results compared to other strategies, e.g. preload-controlled repetitive boluses [1005].

Additional treatment issues:

(a) Blood products constitute a vital element in haemorrhage; unfortunately, no substantial evidence exists regarding the appliance of RBCs (when and how much) and other blood compounds, e.g. FFP [367, 987, 1006]. In general, a restrictive regimen is advisable – less complication such as transfusion-associated lung injury [1007, 1008], infections and renal failure [1009–1011] are seen; some studies even reported a lower mortality [1007, 1012–1015]. The ATLS protocol recommends administration of RBCs if fluid administration fails to reverse signs of shock [1016]; the most recent European guideline goes for an Hb target between 70 and 90 g/l [1006]; Klein et al. recommend to administer RBCs in case of acute blood loss >1,500 ml or >30 % of total blood volume [1017]. However, contemporarily RBC transfusions should be based on the patient's clinical conditions and the avoidance of organ ischaemia and not on a laboratory value (Hb), e.g. signs of organ dysfunction due to insufficient oxygen delivery such as ischaemic ECG marks or angina symptoms, and mental disorders [1017–1019]. A very well evidence-based review by Shander et al. published in 2011 discussing the hitherto existing data concludes that in general a restrictive regimen to be clearly superior over a liberal one and one can reach a crucial Hb level of around 70–80 (90) g/l [1020].

If several RBC packages are transfused, FFP infusion becomes necessary: Most studies on this issue have shown that in massive RBC transfusions a ratio of RBC–FFP by 3:2 (1.5:1.0) significantly reduces mortality rate compared to patients having received less FFPs [1021, 1022].

Platelets are recommended if the platelet count falls below 50×10^9/l [1006].

(b) The results from the CRASH-2 trial demonstrate an attenuation of the bleeding activity and a significant better outcome if tranexamic acid (TXA), an antifibrinolytic agent inhibiting plasminogen activation and plasminogen activity, was applied to patients with major bleedings requiring RBC transfusions undergoing surgery; an increase in vascular clotting could not be detected [1023].

(c) Impaired ionized Ca^{2+} should be replaced during massive transfusion or if typical ECG changes are present [1006].

3.7.5 Septic Shock

The cornerstones in the treatment of severe sepsis and septic shock are early antibiotics (less than an hour after diagnosis [827, 1024]) and early haemodynamic resuscitation [260, 692, 693]: "Besides removal of the septic focus, effective resuscitation combined with specific vasoconstrictor therapy is considered the standard adjunctive therapy, aiming at optimized microvascular perfusion" [1, 193].

The Surviving Sepsis Campaign advocates resuscitation measures in severe sepsis and septic shock tending to normalize haemodynamic and metabolic disorders indicated by *global parameters* such as central venous pressure, MAP, CO, uo, $ScvO_2$ and lactate [260]. However, severe sepsis and septic shock are maladies of disturbed *micro*haemodynamics [44, 189] with heterogeneous tissue perfusion as their major characteristic feature [214] resulting in tissue dysoxia and hypoxia [91, 203, 204, 632]. The microhaemodynamic alterations may occur independently of macrocirculatory derangements [91, 913, 936] and thus cannot be assessed, monitored and guided by global variables [91, 178, 632]. Nevertheless, fluid administration tailored to global haemodynamic and metabolic parameters accompanied by the immediate supply with antibiotics [1024–1028] are the paramount first steps in the treatment of severe sepsis and septic shock [691, 1029–1031]. Furthermore, recent publications encourage the clinicians in case of substantial hypotension to commence with vasopressors, noradrenaline (NA), very soon [859], if not concomitant to fluids [313, 552, 563]. This is a pathophysiologically related approach [687, 858] (attenuated vasomotor tone) and emphasized by clinical experience that often fluids alone will not restore a sufficient BP and thus autoregulation and normalization of a metabolic adjusted tissue blood distribution [199, 314, 718, 858, 1025].

As to a lack of profound evidence applying to many elements of the current Surviving Sepsis Campaign guidelines [691, 692, 765, 767], we suppose to follow in general a modified resuscitation regimen by Marik [692]:

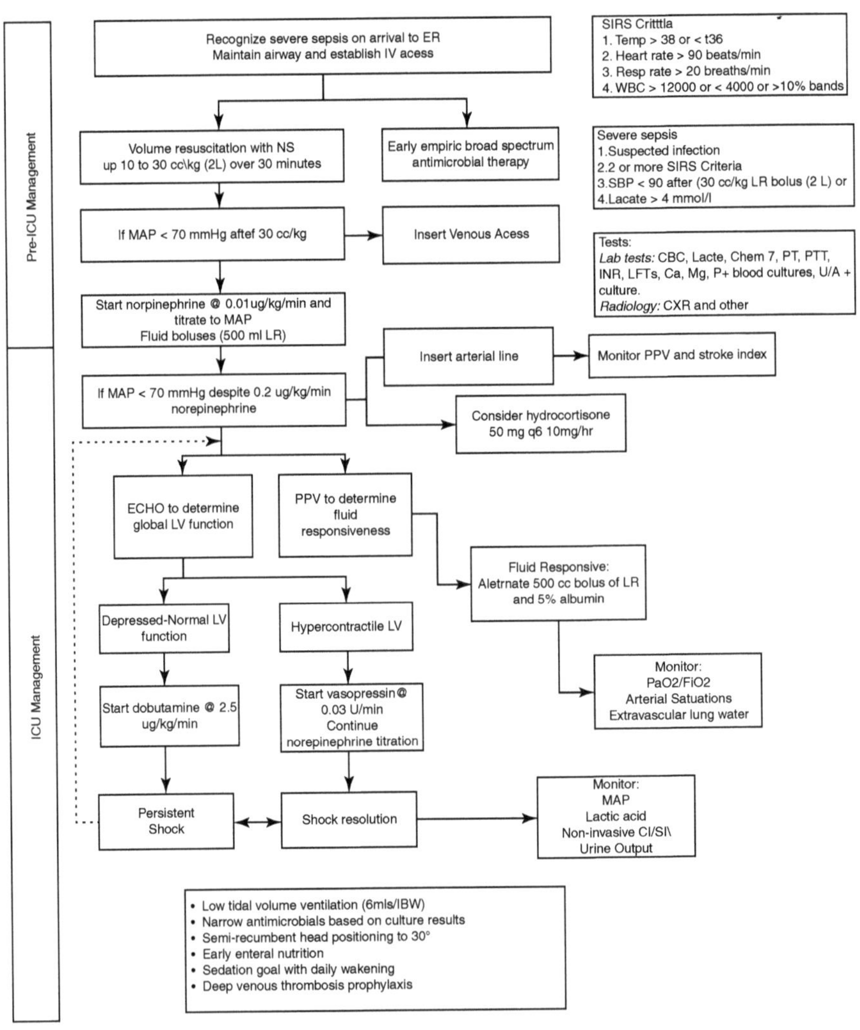

Modified from Marik [662], with permission

(i) Antibiotics

The choice of antimicrobial agent(s) depends on several factors including the source or focus of infection, the immunological condition of the patient, whether the infection is community or hospital (nosocomial) acquired, local microbial flora, susceptibilities and likelihoods as well as local antimicrobial resistance patterns, indwelling catheters and the possible presence of other foreign material [388, 693, 827, 1032]. There is plenty of strong evidence that inappropriate initial antimicrobial treatment yields severe consequences on outcome as it will substantially augment mortality rate [1033–1036].

As such, broad-spectrum antibiotics most likely effectively hitting the underlying microbe(s) considering the above-mentioned factors are commonly advised as initial empirical approach [388, 693, 827, 1024, 1032, 1036]. Aside from an immediate commencement of substance [1024, 1037], it recently seems to turn out that a proper dosage may be another important component: An agent concentration at the site of infection well above (>100 %) the minimum inhibitory concentration (MIC) obviously has mortality-reducing potential – perhaps an initial combination of drugs is able to increase microbial clearance [1038–1041].

Hit as soon and as hard as possible!

Evidence level I B (initial empiric agent(s) likely to hit the pathogen(s) with sufficient concentration at the site of source) and I B (commenced within an hour of recognition of septic shock) [827]

(ii) Fluid therapy and vasopressors (NA)

Essentially, however, for early resuscitation is to "aggressively" infuse fluids within the first 24 h, particularly within the first 6 h – "the early golden hours" [30, 693, 827, 1042]:

Start up with a *very rapid infusion of 500–1,000 ml N. saline 0.9%* (or Hartmann's), *2 l* within the *first hour*, titrated to end points such as MAP >65–70 mmHg, HR <60/min, uo >0.5 ml/kg/h [259, 556, 563, 693, 1043].

The Surviving Sepsis Campaign from 2008 asks for 1,000 ml crystalloids within the first 30 min after the diagnosis is made [260]; the most recent modification encourages for *a more rapid infusion of 30 ml/kg crystalloids* [827].

Similarly Marik proposes *up to 30 ml/kg (roughly max 2 l) N. saline within the first 30 min* [692].

Some recommend to infuse 10–20 ml/kg bolus or 1 l–2 l crystalloids as fast as possible [994, 1044].

At least if there is no satisfactorily clinical stabilization with clear signs of improved tissue perfusion and/or not achieved targets after 2 l of fluids, but may be a great deal earlier (in case of ongoing hypotension consecutively not achieving (renal) autoregulatory threshold and a reasonable vascular tone), *start up with vasopressors early on* [692, 859], overlapping the time necessary to infuse fluids, and reduce NA while fluids are restoring vascular filling and improving SV [313, 563]. Increasingly a target MAP >70 mmHg (70–80 mmHg) to ensure renal autoregulation is reported [693, 737]. May be around 80 mmHg is most suitable to provide proper macrocirculatory renal perfusion as a study by Lehman suggests [889]. However, a wider range exists and the target BP and other *resuscitation end points, particularly microcirculatory ones when known in the future, need to be individualized* [563, 617, 692, 723].

Then, after the initial larger bolus, e.g. 2 l, continue with 150–200 ml/h accompanied by boluses of up to 500 ml if the patient is still and further fluid-responsive and clinical, haemodynamic or metabolic variables indicate ongoing global hypoperfusion [1043, 1045].

As already mentioned, the amount of fluids required may be large, up to 5 l in the first 6 h [259, 556, 1043], and up to 10 l crystalloids in the first 24 h are described [702, 703, 829]. But, *avoid fluid overload during the course – day 3* seems to mark in many aspects *a crucial turning point*, as in patients who overcome shock show within 3 days balanced inflammatory and anti-inflammatory mediators, a restoration of plasma oncotic pressure by closure of capillary leak and healing of microcirculation is seen [631, 724, 1046]. Notably, a *negative cumulative fluid balance* with (consecutively) lower extravascular lung water on *day 3* [809] *or at least on day 4* is obviously a wise target as associated with a clearly better survival [699, 734] – oedema formation as cumulative fluid overload seems to represent a mediator of adverse outcome [700, 714, 731, 1047].

Provided that microcirculatory monitoring is available (e.g. sublingual capnometry), adjust fluids and vasopressors to actual microhaemodynamic targets – improvement/normalizing of perfusion disorders [400, 617, 627]. If *microhaemodynamics have normalized*, be very cautious with any further change in your management as if microcirculatory perfusion is suitable and matches requirements, it has nothing to gain anymore [199, 617, 723].

In case you are not equipped to monitor microhaemodynamics, most of us are not (yet) equipped with SDF or sublingual capnometry, after initial resuscitation "the prevention of fluid shifting", causing tissue oedema impairing even more cell and tissue oxygenation, provoking fluid collection in the pleural and abdominal space, precipitating respiratory problems and abdominal hypertension, thus contributing to organ dysfunction [1048–1050], e.g. renal failure [1051], "should probably be prioritized and fluid accumulation should be prevented by restricting fluids based on fluid responsiveness" [1048].

Evidence level I C (initially at least 30 cm/kg fluids), I B (for crystalloids), I C (to achieve a MAP of 65 mmHg), and I B (NA preferred)

(iii) Inotropic support

Dob is the first choice inotropic agent and indicated in patients with sepsis-associated shortened myocardial function as indicated by persistently (unresponsive to volume loading) low CO accompanied by elevated filling pressures in the presence of suitable vascular filling [4, 260, 693, 719, 1052]. Furthermore, with ongoing hypoperfusion despite proper intravascular filling and reasonable BP, Dob may be considered [827] as it has demonstrated to enhance tissue perfusion [936]. Rabuel and Mebazaa [850] recommend inotropic agents *only* in patients with "decompensated" septic cardiomyopathy reflected by both, a low CO (inappropriate low CO in relation to SVR) accompanied by a low $ScvO_2$ (<70 % [63]), respectively, a hypokinetic profile as determined by Vieillard-Baron [495].

Since catecholamines in general and thus Dob are potentially harmful, they should not be given unnecessarily [945, 948, 1053]!

Furthermore, the administration of Dob or other inotropes "just" to achieve the haemodynamic goals (increase MAP by a rise in CO and thus enhance tissue oxygen supply) coined by Rivers [556] is neither based on any evidence [691, 692, 765, 767] nor does it meet the Joint Commission and Institute for Healthcare Improvement criteria [1054, 1055].

In case of hypotension requiring vasopressor treatment and ongoing low output, a combination of NA and Dob is advisable [719, 903].

Levosimendan may be a valuable alternative to Dob as it may benefit both, macrocirculation as well microcirculation, the latter described as being the fundamental disorder in sepsis [1056–1058].

Evidence level I C for Dob [827]

(iv) Low/moderate doses of hydrocortisone

The role of low/moderate doses (200–300 μg/24 h) of hydrocortisone as part of the management of severe sepsis and septic shock still lies in its ability to improve vasomotor tone [1059]: Initially based on the results of clinical trials demonstrating a more rapid resolution of septic shock when hydrocortisone was given [1060–1062], and the pathophysiological verification that septic patients frequently display a relative adrenal insufficiency [1061, 1063, 1064] – one study even showed a significant reduction of mortality in the hydrocortisone group [1060] – hydrocortisone was affiliated to the sepsis care bundle [259, 260]. The CORTICUS study, indeed, confirmed the favourable effects of cortison application on shock reversal, but, unfortunately, could not demonstrate a clear survival benefit in this cohort, within which the patients received 50 μg hydrocortisone every 6 hours regardless of their blood pressure response to vasopressors [1065]. Thus, given potentially serious side effects like the promotion of infections or myopathy, the gerenral use of hydrocortisone in sepsis and septic shock cannot be advised [260, 1066]. However, Buechele and co-workers revealed that hydrocortisone (50 μg×4/24 h) administered within the first 12 h of diagnosis to patients with septic shock requiring vasopressors to maintain/achieve a reasonable BP significantly improved microvascular perfusion [1066]. These results destroyed concerns that hydrocortisone may blunt and further compromise microhaemodynamics and as such tissue perfusion [1066], but may, in fact, be beneficial to the crucial underlying pathobiology and may open new therapeutic indications and options for steroids, e.g. to address microcirculatory disorders. Furthermore, this result supports the use of hydrocortisone in the most severe septic patients, not responding appropriately to vasopressors in order to attempt to re-establish sufficient BP, as recommended in current guidelines [827].

Evidence II C [827]

(v) RBC transfusions

Since the landmark study by Rivers [556] the standard guideline advocates RBC transfusions to increase oxygen carriage (Hct ≥30 %) and consecutively to support the oxygen delivery to the tissues indicated by $ScvO_2$ ≥70 %. Accumulating study results highly challenge that approach which is not evidence-based at all [691, 692, 765]. Moreover, beneficial effects seem to be more than doubtful as we have quite substantial and resilient data that more restrictive transfusion regimen should be preferred to liberal ones and that clinical aspects determined by organ dysfunction attributed to hypoxia/ischaemia rather than more or less arbitrarily defined haemodynamic and oxygen content/saturation markers should be used to decide whether to transfuse or not [1020]. RBC transfusions in critical illness bear a marked risk for infections, and to trigger ALI/ARDS, multiorgan failure and even death [1010, 1067]. Accordingly, the Joint Commission and Institute for Healthcare Improvement criteria do not include RBC transfusions for septic patients as applied to the early goal-directed advices coined by Rivers [1054, 1055]. Even the latest recommendations of the international guidelines for the management of severe sepsis and septic shock advocate RBC packs if Hb is <70 g/l independent of signs of organ dysfunction – graded I B [827] and thus deviate somehow from current transfusion practice as the clinical condition should be treated, not a laboratory value [1018, 1020].

(vi) rhAPC

Following the results of the large (nearly 1,700 patients were included) PROWESS SHOCK trial [1068] and a Cochrane review [1069], demonstrating no survival benefit in the cohorts receiving rhAPC, but a higher risk of serious bleeding, the drug cannot any longer be recommended as an element in the management of severe sepsis and septic shock and meanwhile has been withdrawn from the market [827].

(vii) Early goal-directed therapy (EGDT)

The integral merit of Rivers' landmark study on the management of severe sepsis and septic shock [556] probably lies in the structured, designated and systemized approach. Most of the original haemodynamic advices and measures aim to achieve arbitrarily determined targets such as *CVP*, *Hb*, *BP and* $ScvO_2$*, target, Dob application to increase BP and CO* (the latter to augment oxygen delivery to peripheral tissues), and *RBC transfusions* to increase the oxygen carriers thereby promoting oxygen supply to the tissues. These recommendations are largely unchanged since then and still advocated [827], however are in fact are in fact debatable and lacking clear scientific evidence [691, 692, 755, 765, 767, 1070]. The combination of measures, the bundled care, may have had a relevant beneficial impact on survival in Rivers' relatively small study population comprising just 263 patients together for both groups [691, 1055, 1071, 1072]. Anyway, there is no convincing data and evidence of positive influence on outcome for the care bundles [691, 1073,

1074]. Furthermore, as alluded to above, most of the single components of the bundle and particularly their interdependent and independent efficacy has not been proven [1075] – may be, another combination and/or other elements would be more successful?

There is strong and clear evidence that *early treatment onset*, exploiting the first 6, "golden" hours after diagnosis, comprising an appropriate *antibiotic regimen accompanied by haemodynamic support including fluids and vasopressors* and, where applicable, surgical infection source control, e.g. surgical treatment of gut perforation, drainage of abscesses, etc., is paramount for beneficial outcome and makes the cornerstones in the treatment [691–693, 827, 1024, 1076].

Addressing the patients protocol guided, with tight monitoring of all therapeutic measures is common sense [557, 692, 827]. Taking into consideration the current evidence and data body, we would advise to follow in general the protocol suggested by Marik [692], but do not forget to individualize [767] and, if possible, address the microcirculation as the fundamental pathobiology [29, 63].

(viii) Intubation and mechanical ventilation

In patients with severe septic shock being in respiratory distress, profound hypoxaemia or acidosis, and those who are comatose, early intubation should be envisaged [4, 1077]. The respirator settings and the proceedings of mechanical ventilation should strictly comply with the lung protective ventilation strategy [1078–1082]; for detailed information see chapter on Mechanical Ventilation.

Evidence level I A for low tidal volumes (6 ml/kg), I B for plateau pressure limitation ($\leq$30 cmH_2O), and I B for adequate PEEP (avoidance of alveolar collapse) [827].

(ix) Thromboembolism prophylaxis

ICU patients in general bear a marked risk for deep vein thrombosis (DVT) and pulmonary embolism (PE) [1083, 1084]. A substantial body of studies including meta-analyses demonstrated a significantly reduced incidence of DVT and PE when septic patients received thromboembolism prophylaxis [1085–1087]. Since the occurrence of PE in a person already vitally jeopardized may be fatal [827] and as the risk of complications, e.g. particularly bleedings, associated with the prophylaxis is very small [260], the evidence to apply antithrombotic medication is very strong [14].

Low molecular weight heparins should be preferred, in case of markedly reduced renal function (creatinine clearance <30 ml/min), dalteparin or unfractionated heparins are advocated [827].

Level of evidence I B [827]

(x) Blood glucose control

Hyperglycaemia displays procoagulant activity [1088], impairs neutrophil function and wound healing [1032], increases the risk of infection [1032], induces apoptosis [1089] and is associated with worse outcome [1032].

A large study (roughly 1,550 patients were included) by van der Berghe and co-workers on surgical critically ill patients revealed a significant 3.4 % absolute mortality reduction among patients with tight glucose control (keeping the blood glucose range during their ICU stay between 80 and 110 mg/dl, compared to 200–280 mg/dl in the other group) [1090]. Unfortunately, this could not be confirmed when applying this concept to nonsurgical patients [1091]. Furthermore, a systematic review including nearly 30 randomized controlled studies with a total of nearly 8,500 patients found no benefit in the tight controlled cohort; all the more, significantly more hypoglycaemic episodes were recorded [1092]. Moreover, the large (more than 6,000 medical and surgical patients included) NICE-SUGAR study (Normoglycemia in Intensive Care Evaluation-Survival Using Glucose Algorithm Regulation) revealed a significantly higher risk of mortality in the tight glucose control group (80–110 mg/dl) attributed to hypoglycaemic events (blood glucose level <70 mg/dl, moderate hypoglycaemia, or <40 mg/dl, severe hypoglycaemia) occurring significantly more frequently in that study pool compared to the conventionally treated cohort (glucose blood level kept 180 mg/dl or less) [1093, 1094]. The logical summary of those comprehensive data sets has been to advocate keeping the blood glucose level <180 mg/dl but ≥145 mg/dl and (thus) avoiding numerous hypoglycaemic episodes in all critically ill patients of whatever underlying disease [827, 1094, 1095].

Moreover, since studies show that moderate hyperglycaemia, defined as blood glucose levels between 180 and 220 mg/dl, is, under stress conditions, a valuable source of energy extraction for the immune system and other organs/cells utilizing glucose independent of insulin like the central nervous system, myocardial cells, brain and red blood cells [1096–1099], even higher levels than the currently used ones, may be more beneficial improving the outcome [1095]. More studies examining this issue are warranted [1095].

The Surviving Sepsis Campaign graded this issue by:

Evidence level I A [827]

(xi) Treatment of acidosis, buffers

The use of buffering agents is in general not advocated and should only considered in selected patients having exhibited severe acidosis (pH <7.1/7.0) as most but not all experts believe [1100–1102]. This is based on the opinion that acidaemia may cause haemodynamic instability, and thus, buffers can stabilize the condition. However, studies on that issue have been unable to demonstrate significantly beneficial effects [961, 963, 1103, 1104]. If buffering agents are given, hydroxymethyl aminomethane (THAM) may be preferred as in animal studies where improvement of acidosis and a reduction in mortality without provoking an intracellular acidosis are reported [1105] – unfortunately, older clinical studies in humans had not shown any difference [1104, 1106].

Level of evidence II B [827]

(xii) Special treatment issues: Agents with vasodilative properties

(a) Dobutamine: Already in the 1980s, dobutamine was used to improve microhaemodynamics in severe sepsis and septic shock. As systemic oxygen consumption (VO_2) increase under the administration of Dob, it was suggested that microvessels which were disease associated poorly or non-perfused could be recruited by Dob effects [1107]. De Backer recently substantiated that suggestion as he could confirm by OPS that Dob indeed facilitates capillary perfusion in septic patients independently of its effects on macrocirculation [936]. As underlying mechanisms, vasodilatory and rheological effects are supposed [936].

(b) GTN (dosages between 0.5 mg/h and up to 4.0 mg/h [196] or 2.0 mg/h [1108]) was shown in several studies as being able to improve microcirculation [196, 968, 969, 1102] and thus may be a treatment option at least in individual cases of septic shock [399, 617]. Noteworthy, the effect was not associated with changes of macrohaemodynamics such as CO or MAP [196, 907, 968, 969, 1102], is dose dependent [968, 969] and suggests that the endothelium is still NO responsive [91]. Thus, GTN/NO may have a role in the treatment of endothelial dysfunction, a crucial component in the pathobiology of sepsis and septic shock [399]. However, as intravenous GTN may potentially diminish BP in sepsis [196], topical appliance which does not significantly affect systemic arterial BP may be preferable, such as inhaled nitric oxide [1109]. Certainly, potential beneficial effects on microcirculation and particularly on patient outcome need to be evaluated and assessed in bigger clinical trials [399, 617].

(xiii) Vasopressin

The place of vasopressin in the treatment of hypotension is not determined: Although a relative vasopressin deficiency is commonly found in sepsis and septic shock [1110], the routine use of vasopressin (or its analogue terlipressin) compared to NA could not deliver any benefit as several study results revealed [463, 1111, 1112]. Moreover, higher doses (>0.04 U/min) are associated with intestinal, cardiac and digital ischaemia and a trend of a higher risk of death [827, 1113, 1114]. The international Surviving Sepsis Campaign merely recommends that vasopressin (max 0.03 U/min) may be added to NA in order to further increase MAP or to decrease NA doses [827] – conspicuously this recommendation is not graded! Furthermore, low-dose vasopressin (0.03–0.04 U/min) is not recommended as single vasopressor and should be reserved for salvage therapy in case BP target cannot be achieved by other vasopressors [827] – again not graded!

References

1. Wheeler AP, Bernard GR. Treating patients with severe sepsis. N Engl J Med. 1999;340:207–14.
2. Fink MP. Bench-to-bedside review: cytopathic hypoxia. Crit Care. 2002;6:491–9.
3. Kumar A, Parrillo JE. Shock: classification, pathophysiology, and approach to management. In: Parrillo JE, Dellinger RP, editors. Critical care medicine. Principles of diagnosis and management in the adult. London: Mosby; 2001. p. 291–339.
4. Graham CA, Parke TRJ. Critical care in the emergency department: shock and circulatory support. Emerg Med J. 2005;22:17–21.
5. Ellender TJ, Skinner JC. The use of vasopressors and inotropes in the emergency medical treatment of shock. Emerg Med Clin North Am. 2008;26:759–86.
6. Sethi AK, Sharma P, Motha M. Shock – a short review. Indian J Anaesth. 2003;47:345–59.
7. Oroto RM, Nguygen HB, Huang DT. Early-goal directed therapy in severe sepsis and septic shock revisited: concepts, controversies, and contemporary findings. Chest. 2006;130:1579.
8. Parrillo JE. Approach to the patient with shock. In: Goldman L, Ausiello DA, editors. Cecil textbook of medicine. 22nd ed. Philadelphia: WB Saunders; 2004. p. p 609.
9. Jones AE, Kline JA. Shock. In: Marx J, editor. Rosen`s emergency medicine: concepts and clinical practice, vol. 1. 6th ed. Philadelphia: Mosby; 2006. p. 230–42.
10. Barber AE. Cell damage after shock. New Horiz. 1996;4:161.
11. DeBacker D, Biston P, Devriendt J. Comparison of dopamine and norepinephrine in the treatment of shock. N Engl J Med. 2010;362:779–89.
12. Vincent JL, Sakr Y, Sprung CL. Sepsis in European intensive care units: results of the SOAP study. Crit Care Med. 2006;34:344–53.
13. Silva E, de Almeida Pedro M, Sogayar ACB. Brazilian sepsis epidemiological study (BASES Study). Crit Care. 2004;8:R251–60.
14. Angus DC, Linde- Zwirble WT, Lidicker J. Epidemiology of severe sepsis in the United States: analysis of incidence, outcome, and associated costs of care. Crit Care Med. 2001;29:1303–10.
15. McCaig, LF, Burt, CW. National Hospital Ambulatory Medical Care Survey: 2002 emergency department summary. Adv Data. 2004;18(340):1–34.
16. Hochman JS, Boland J, Sleeper LA. Current spectrum of cardiogenic shock and effect of early revascularization on mortality. Results of n International Registry. SHOCK Registry Investigators. Circulation. 1995;91:873–81.
17. Osborn TM, Tracy JK, Dunne JR. Epidemiology of sepsis in patients with traumatic injury. Crit Care Med. 2004;32:2234.
18. Prasad A, Lennon RJ, Rihal CS. Outcomes of elderly patients with cardiogenic shock treated with early percutaneous revascularization. Am Heart J. 2004;147:1066.
19. Hochman JS, Buller CE, Sleeper LA. Cardiogenic shock complicating acute myocardial infarction – etiologies, management, and outcome: a report from the SHOCK Trail Registry. J Am Coll Cardiol. 2000;36(3 Suppl A):1063–70.
20. Babev A, Frederick PD, Pasta DJ. Trends in management and outcome of patients with acute myocardial infarction complicated by cardiogenic shock. JAMA. 2005;294:448–54.
21. Molyneux EM, Maitland K. Intravenous fluids – getting the balance right. N Engl J Med. 2005;353:941–4.
22. Weil MH, Shubin H. Shock following acute myocardium infraction: current understanding of hemodynamic mechanisms. Prog Cardiovasc Dis. 1968;11:1–17.
23. Weil M, Shubin H. Proposed reclassification of shock states with special reference to distributive defects. Adv Exp Med Biol. 1971;23:13–23.
24. Tintinalli JE. Emergency medicine: a comprehensive study guide (emergency medicine (tintinalli)). New York: McGraw-Hill Companies; 2010. pp 168. ISBN 0-07-148480-9.
25. Pinsky MR. Hemodynamic evaluation and monitoring in ICU. Chest. 2007;132:2020–9.
26. Hadian M, Punsky MR. Functional hemodynamic monitoring. Curr Opin Crit Care. 2007;13:318–23.

27. Walley KR, Wood LDH. Shock. In: Hall JB, Schmidt GA, Wood LDH, editors. Principles of critical care. 2nd ed. New York: McGraw-Hill; 1998. p. p 286.
28. Shoemaker WC, Beez M. Pathophysiology, monitoring and therapy fo shock with organ failure. Appl Cardiopulm Pathophysiol. 2010;14:5–15.
29. Bonanno FG. Shock – a reappraisal: the holistic approach. J Emerg Trauma Shock. 2012;5:167–77.
30. Hollenberg SM. Vasoactive drugs in circulatory shock. Am J Respir Crit Care Med. 2011;183:847–55.
31. Herget-Rosenthal S, Saner F, Chawla LS. Approach to hemodynamic shock and vasopressors. Clin J Am Soc Nephrol. 2008;3:546–53.
32. Worthley LIG. Shock: a review of pathophysiology and management. Part I. Crit Care Resusc. 2000;2:55–65.
33. Heckbert SR. Outcome after hemorrhagic shock in trauma patients. J Trauma. 1998; 45:545.
34. Holmes CL, Walley KR. The evaluation and management of shock. Clin Chest Med. 2003;24:775–89.
35. Groeneveld ABJ. Hypovolemic shock: chapter 27. In: Parriullo JE, Dellinger RP, editors. Critical care medicine. 3rd ed. Philadelphia: Mosby Elsevier; 2008. p. pp 485.
36. Bressack MA, Raffin TA. Importance of venous return, venous resistance and mean circulatory pressure in the physiology and management of shock. Chest. 1987;92:906.
37. Elbers PWG, Ince C. Bench-to-bedside review: mechanisms of critical illness – classifying microcirculatory abnormalities in distributive shock. Crit Care. 2006;10:221–8.
38. Califf RM, Bentgson JR. Cardiogenic shock. N Engl J Med. 1994;330:1724–30.
39. Cotter G, Moshkovitz Y, Kaluski E. The role of cardiac power and systemic vascular resistance in the pathophysiology and diagnosis of patients with acute congestive heart failure. Eur J Heart Fail. 2003;5:443–51.
40. Kohsaka S, Menon V, Lowe AM. Systemic inflammatory response syndrome after acute myocardial infarction complicated by cardiogenic shock. Arch Intern Med. 2005;165:1643–50.
41. Hochman JS. Cardiogenic shock complicating acute myocardial infarction: expanding the paradigm. Circulation. 2003;107:2998–3002.
42. Reynolds HR, Hochman JS. Cardiogenic shock: current concepts and improving outcomes. Circulation. 2008;117:686–97.
43. Landry DW, Oliver JA. The pathophysiology of vasodilatory shock. N Engl J Med. 2001;345:588–95.
44. Ince C. The microcirculation is the motor of sepsis. Crit Care. 2005;9 Suppl 4:S13–9.
45. Holmes CL. Vasoactive drugs in the intensive care unit. Curr Opin Crit Care. 2005;11:413–7.
46. Beale RJ, Hollenberg SM, Vincent JL. Vasopressor and inotropic support in septic shock: an evidence-based review. Crit Care Med. 2004;32:S455–65.
47. Gkisioti S, Mentzelopoupus SD. Vasogenic shock physiology. Open Access Emerg Med. 2011;3:1–6.
48. Rackow ME, Astiz EC. Septic shock. Lancet. 1998;351:1501–5.
49. Ellis CG, Jagger J, Sharpe M. The microcirculation as a functional system. Crit Care. 2005;9 Suppl 4:S3–8.
50. Ince C, Sinaappel M. Microcirculatory oxygenation and shunting in sepsis and shock. Crit Care Med. 1999;27:1369–77.
51. Steingrub JS. Shock in the intensive care unit. In: Higgins TL, Steingrub JS, Kacmarek RM, Stoller JK, editors. Cardiopulmonary critical care. Oxford: Bios Scientific Publishers; 2002. p. 81.
52. Hollenberg SM, Kavinsky CJ, Parrillo JE. Cardiogenic shock. Ann Intern Med. 1999;131:47–59.
53. Gaieski D, Parson PE, Wilson KC. Shock in adults: types, presentation, and diagnostic approach. Up-to-Date. 2012. http://www.uptodate.com/contents/atropine-drug-information?source=search_result&search=atropine&selectedTitle=1~150.
54. Lappin E, Ferguson AJ. Gram-positive toxic shock syndromes. Lancet Infect Dis. 2009;9:281–90.

55. Bonanno F. Extending damage control philosophy to non-hemorrhagic situations: implications for a re-classification of shock states. ANZ J Surg. 2008;78:634–7.
56. Peng J, Du B. Sepsis-related stress response: known knowns, known unknowns, and unknown unknowns. Crit Care. 2010;14:179.
57. Dewachter P, Mouton-Faivre C, Nace L. Treatment of anaphylactic reaction in pre-hospital and in the emergency room: literature review. Ann Fr Anesth Reanim. 2007;26:218–28.
58. Wan S, LeClerc JL, Vincent JL. Inflammatory response to cardiopulmonary bypass: mechanisms involved and possible therapeutic strategies. Chest. 1997;112:676–92.
59. Jung C, Figulla HR, Ferrari M. High frequency of organ failures during extracorporeal membrane oxygenation: is microcirculation the answer? Ann Thorac Surg. 2010;89:345–6.
60. Dondorp AM, Angus BJ, Chotivanich K. Red cell deformability as a predictor of anemia in severe falciparum malaria. Am J Trop Med Hyg. 1999;60:733–7.
61. Tsai AG, Cabrales P, Winslow RM. Microvascular oxygen distribution in awake hamster window chamber model during hyperoxia. Am J Physiol Heart Circ Physiol. 2003;285:H 1537–45.
62. Schwarte LA, Fournell A, van Bommel J. Redistribution of intestinal microcirculatory oxygenation during acute hemodilution in pigs. J Appl Physiol. 2005;98:1070–5.
63. Bonanno FG. Clinical pathology of shock syndromes. J Emerg Trauma Shock. 2011;4:233–43.
64. Hinshaw LB. Sepsis/septic shock: participitation of microcirculation: an abbreviated review. Crit Care Med. 1996;24:1072–8.
65. Singer M. System interactions: chapter 17. In: Chadwick DJ, Goode J, editors. Sepsis: new insights, new therapies: Novartis Foundation symposium 280. Chichester: Wiley; 2008. p. 252–65.
66. Kristensen SR. Mechanisms of cell damage and enzyme release. Dan Med Bull. 1994;41:423.
67. Abboud FM. Pathophysiology of hypotension and shock. In: Hurst JW, editor. The heart. New York: McGraw-Hill; 1982. p. p 452.
68. Coimbra R, Doucet JJ, Bansal V. Resuscitation strategies in hemorrhagic shock – hypertonic saline revisited. Emerg Med Crit Care Rev. 2007, vols. Reference section, Touch briefings 2007, p. 1–9.
69. Mela L, Bacalaz OL, Miller L. Defective oxidative metabolism of rat mitochondria in hemorrhage and endotoxin shock. Am J Physiol. 1971;220:571.
70. Chaudry IH. Cellular mechanisms in shock and ischemia and their repair. Am J Physiol. 1983;245:R117.
71. Shoemaker WC. Relations of oxygen transport patterns to the pathophysiology and therapy of shock states. Intensive Care Med. 1987;13:230.
72. Mizock BA, Falk JL. Lactic acidosis in critical illness. Crit Care Med. 1992;20:80.
73. Gutierrez G, Reines HD, Wulf-Gutierrez ME. Clinical review: hemorrhagic shock. Crit Care. 2004;8:373–81.
74. Hobler KE, Carey LC. Effect of acute progressive hypoxemia on cardiac output and plasma excess lactate. Ann Surg. 1973;177:199–202.
75. Groeneveld ABJ. Relation of arterial blood lactate to oxygen delivery and hemodynamic variables in human shock states. Circ Shock. 1987;22:35.
76. Rixen D, Siegel JH. Bench-to- bedside review: oxygen debt and its metabolic correlates as quantifiers of the severity of hemorrhagic and posttraumatic shock. Crit Care. 2005;9:441.
77. Brodsky SV, Yamamoto T, Tada T. Endothelial dysfunction in ischemic acute renal failure: rescue by transplanted endothelial cells. Am J Physiol Renal Physiol. 2002;282:F1140–9.
78. Vetterlein F, Petho A, Schmidt G. Distribution of capillary blood flow in rat kidney during postischemic renal failure. Am J Physiol. 1986;251:H510–9.
79. Vogt MT, Fraber E. The effects of ethionine treatment on the metabolism of mitochondria. Arch Biochem Biophys. 1970;141:162.
80. Stacpoole PW. Lactate acidosis and other mitochondrial disorders. Metabolism. 1997;46:306–21.
81. Oakes SA, Opferman JT, Pozzan T. Regulation of endoplasmic reticulum Ca by proapoptotic BCL-2 family members. Biochem Pharmacol. 2003;66:1335–40.

82. Boutilier RG. Mechanisms of cell survival in hypoxia and hypothermia. J Exp Biol. 2001;204:3171–81.
83. Gutierraz G. Cellular effects. In: Crystal RG, West JB, Weibel ER, Barnes PJ, editors. The lung: scientific foundations. 2nd ed. New York: Raven; 1996. p. 1969–79.
84. Al Aseri Z. Vital indices to be used in resuscitation of patients with shock in the emergency department setting. Emerg Med. 2012;2:108. doi:10.4172/2165-7548.1000108.
85. Rogers KG. Cardiovascular shock. Emerg Med Clin North Am. 1995;13:793–810.
86. Shoemaker WC, Appel PL, Kram HR. Tissue oxygen debt as a determinant of lethal and non-lethal postoperative organ failure. Crit Care Med. 1988;16:1117–20.
87. Cheatham ML, Block EFJ, Smith HG, Promes JT. Shock: an overview. Orlando: s.n. www.surgicalcriticalcare.net/Lectures/shock_overview.pdf.
88. Guyton A, Hall J. Circulatory shock and physiology of its treatment: chapter 24. In: Gruliow R, editor. Textbook of medical physiology. 11th ed. Philadelphia: Elsevier; 2006. p. 278–88.
89. Tintinalli JE. Emergency medicine: a comprehensive study guide. New York: McGraw-Hill Comp; 2010. p. 165–72. ISBN 0-07-148480-9.
90. De Backer D, Creteur J, Dubois MJ. Microvascular alterations in patients with acute severe heart failure and cardiogenic shock. Am Heart J. 2004;147:91–9.
91. De Backer D, Creteur J, Preiser JC. Microvascular blood flow is altered in patients with sepsis. Am J Respir Crit Care Med. 2002;166:98–104.
92. Kovach ABG, Lefer M. Endothelial dysfunction in shock states. Physiology. 1993;8:145–8.
93. Saotome T, Ishikawa K, May CN. The impact of experimental hypoperfusion on subsequent kidney function. Intensive Care Med. 2010;36:533–40.
94. Johannes T, Mik EG, Ince C. Nonresuscitated endotoxemia induces microcirculatory hypoxic areas in the renal cortex in the rat. Shock. 2009;31:97–103.
95. Yassin MM, Harkin DW, Barros D´Sa AA. Lower limb ischemia-reperfusion injury triggers a systemic inflammatory response and multiple organ dysfunction. World J Surg. 2002;26:115–21.
96. Yassin MM, Barros D`Sa AA, Parks G. Mortality following lower limb ischemia-reperfusion: a systemic inflammatory response ? World J Surg. 1996;20:961–7.
97. Tsukamoto T, Chanthaphavong S, Pape HC. Current theories on pathophysiology of multiple organ failure after trauma. Injury. 2010;41:21–6.
98. Malinowski DJ, Slater MS, Mullins RJ. Crush syndrome: a review. Crit Care Clin. 2004;20:171–92.
99. Siegmund M, Stegenga ME, Mathura K. Changes in the porcine microcirculation of the ileum after supra-mesenteric aortic cross-clamping. Intensive Care Med. 1999;25 Suppl 1:S10.
100. Kologu M, Yorganci K, Renda N. Effect of local and remote ischemia-reperfusion injury on healing of colonic anastomoses. Surgery. 2000;128:99–104.
101. Koike K, Moore EE, Moore EA. Gut ischemia/reperfusion produces lung injury independent of endotoxin. Crit Care Med. 1994;9:1438–44.
102. Reilly PM, Wilkins KB, Fuh KC. The mesenteric hemodynamic response to circulatory shock: an overview. Shock. 2001;15:329–43.
103. Saini MS, Liberati DM, Diebel LN. Sequential changes in mucosal immunity after hemorrhagic shock. Ann Surg. 2001;67:797–801.
104. Baylor AE, Diebel LN, Liberati DM. The synergistic effects of hypoxia/reoxygenation or tissue acidosis and bacteria on intestinal apoptosis. J Trauma. 2003;55:241–8.
105. Niu CY, Hou YL, Zhao ZG. Role of intestinal lymphatic pathway in pathogenesis of intestine derived bacteria/endotoxin translocation in rats in shock. Zhongguo Wei Zhong Bing Ji Jiu Yi Xue. 2007;19:266–9.
106. Feinman R, Deitch EA, Watkins AC. Hypoxia inducible factor (HIF-1) mediates pathogenic inflammatory responses to intestinal ischemia reperfusion injury. Am J Physiol Gastrointest Liver Physiol. 2010;299:G833–43.
107. Garrison RN, Spain DA, Wilson MA. Microvascular changes explain the “two-hit” theory of multiple organ failure. Ann Surg. 1998;227:851–60.
108. Chien S. Role of the sympathetic nervous system in hemorrhage. Physiol Rev. 1967;47:214.

109. Nakai K, Ikatura T, Naka Y. The distribution of adrenergic receptors in cerebral vessels: an autoradiographic study. Brain Res. 1986;381:148–52.
110. Bond RF, Green HD. Cardiac output redistribution during bilateral common carotid artery occlusion. Am J Physiol. 1969;216:393.
111. Haljamäe H. Microcirculation and hemorrhagic shock. Am J Emerg Med. 1984;2:100–7.
112. Clark JA, Coppersmith CM. Intestinal crosstalk: a new paradigm for understanding the gut as the "motor" of critical illness. Shock. 2007;28:384–93.
113. Hassoun HT, Kone BC, Mercer DW. Post-injury multiple organ failure: the role of the gut. Shock. 2001;15:1–10.
114. Alverdy JC, Laughlin RS, Wu L. Influence of the critically ill state on host – pathogen interactions within the intestine: gut-derived sepsis redefined. Crit Care Med. 2003;31:598–607.
115. Rutherford RB, Arora S, Fleming PW. Delayed onset pulmonary insufficiency in primates resuscitated from hemorrhagic shock. J Trauma. 1979;19:422–31.
116. Regel G, Grotz M, Weltner T. Pattern of organ failure following severe trauma. World J Surg. 1996;20:422–9.
117. Fink MP, Delude RL. Epithelial barrier dysfunction: a unifying theme to explain the pathogenesis of multiple organ dysfunction at the cellular level. Crit Care Clin. 2005;21:177–96.
118. Souza DG, Vieira AT, Soares AC. The essential role of the intestinal microbiota in facilitating acute inflammatory responses. J Immunol. 2004;173:4137–46.
119. Moore FA, Moore EE, Poggetti R. Gut bacterial translocation via the portal vein: a clinical perspective with major torso trauma. J Trauma. 1991;31:629–36.
120. Tamion F, Richard V, Lyoumi S. Gut ischemia and mesenteric synthesis of inflammatory cytokines after hemorrhagic or endotoxic shock. Am J Physiol. 1997;273:314–21.
121. Perl M, Chung CS, Perl U. Fas induced pulmonary apoptosis and inflammation during indirect lung injury. Am J Respir Crit Care Med. 2007;176:591–601.
122. Moore FA, Moore EE. Evolving concepts in the pathogenesis of postinjury multiple organ failure. Surg Clin North Am. 1995;75:257–77.
123. Chow CC, Clermont G, Kumar R. The acute inflammatory response in diverse shock states. Shock. 2005;24:74–84.
124. Singer M. System interactions. Novartis Found Symp. 2007; p. 252–62.
125. Ferrero-Miliani L, Nielsen OH, Andersen PS. Chronic inflammation: importance of NOD2 and NALP3 in interleukin – 1 beta generation. Clin Exp Immunol. 2007;147:227–35. 0611270 1532 7006.
126. Abbas AK, Lichtman AH. Innate immunity: chapter 2. In: Abbas AK, Lichtman AH, editors. Functions and disorders of the immune system. Philadelphia: Saunders; 2009.
127. Brame AL, Singer M. Stressing the obvious? An allostatic look at critical illness. Crit Care Med. 2010;38(Suppl):S600–7.
128. Broechner A, Toft P. The pathophysiology of the systemic inflammatory response after major accidental trauma. Scand J Trauma Resus Emerg Med. 2009;17:43.
129. Astiz AE, Rachow EC, Weil MH. Circulatory Shock. In: Carlson RW, Geheb MA, editors. Principles and practice of medical intensive care. Philadelphia: Saunders; 1993. p. 978–88.
130. Cordemans C, De Laet I, Van Regenmortel N. Fluid management in critically ill patients: the role of extravascular lung water, abdominal hypertension, capillary leak, and fluid balance. Ann Intensive Care. 2012;2 Suppl 1:S1.
131. Fleck A, Hawker F, Wallace PI. Increased vascular permeability: a major cause of hypoalbuminaemia in disease and injury. Lancet. 1985;325:781–4.
132. Lefer AM, Lefer DJ. Pharmacology of the endothelium in ischemia – reperfusion and circulatory shock. Annu Rev Pharmacol Toxicol. 1993;33:71–90.
133. Nathan C. Oxygen and the inflammatory cell. Nature. 2003;17:675–6.
134. Cryer HG, Leong K, McArthur DL. Multiple organ failure: by the time you predict it, it is already there. J Trauma. 1999;46:596–604.
135. Douzinas EE, Andrianakis I, Livaditi O. The level of hypotension during hemorrhagic shock is a major determinant of the post-resuscitation systemic inflammatory response: an experimental study. BMC Physiol. 2008;18:8–15.

136. Fink MP, et al. Shock: An overview. In: Rippe JM, editor. Intensive care medicine. Boston: Little and Brown; 1996. p. 1857–77.
137. Barbee RW, Reynolds PS, Ward KR. Assessing shock resuscitation strategies by oxygen debt repayment. Shock. 2010;33:113–22.
138. Nathan C. Points of control in inflammation. Nature. 2002;420:846–52.
139. Jarrar D, Chaudry IH, Wang P. Organ dysfunction following hemorrhage and sepsis: mechanisms and therapeutic approaches. Int J Mol Med. 1999;4:575–83.
140. Marshall JC. Inflammation, coagulopathy, and the pathogenesis of multiple organ dysfunction syndrome. Crit Care Med. 2001;29:S99–106.
141. Quilez ME, Lopez-Aguilar J, Blanch L. Organ crosstalk during acute lung injury, acute respiratory distress syndrome, and mechanical ventilation. Curr Opin Crit Care. 2012; 18:23–8.
142. Pannu N, Mehta RL. Effect of mechanical ventilation on the kidney. Best Pract Res Clin Anaesthesiol. 2004;18:189–203.
143. Ranieri VM, Giunta F, Suter PM. Mechanical ventilation as a mediator of multisystem organ failure in acute respiratory distress syndrome. JAMA. 2000;284:43–4.
144. Aird WC. The role of the endothelium in severe sepsis and multiple organ dysfunction syndrome. Blood. 2003;101:3765–77.
145. Bonanno FG. Physiopathology of shock. J Emerg Trauma Shock. 2004;4:222–32.
146. Sutton TA, Fisher CJ, Molitoris BA. Microvascular injury and dysfunction during ischemic acute renal failure. Kidney Int. 2002;62:1539–49.
147. Flores J, DiBona DR, Beck CH. The role of cell swelling in ischemic renal damage and the protective effect of hypertonic solute. J Clin Invest. 1972;51:118–26.
148. Vallet B. Endothelial cell dysfunction and abnormal tissue perfusion. Crit Care Med. 2002;30:S229–34.
149. Shapiro N, Schuetz P, Yano K. The association of endothelial cell signaling , severity of illness, and organ dysfunction in sepsis. Crit Care. 2010;15:R182.
150. Devarjan P. Update on mechanisms of ischemic acute kidney injury. J Am Soc Nephrol. 2006;17:1503–20.
151. Lameire N, Vanholder R. Pathophysiology of ischaemic acute renal failure. Best Pract Res Clin Anaesth. 2004;18:21–36.
152. Ten VS, Pinsky DJ. Endothelial response to hypoxia: physiologic adaption or pathologic dysfunction. Curr Opin Crit Care. 2002;8:242–50.
153. Fisk JE, Yan MS, Matouk CC. Hypoxic repression of endothelial nitric-oxide synthase transcription is coupled with eviction of promotor histones. J Biol Chem. 2010;285:810–27.
154. Chi J-T, Wang Z, Nuyten DSA. Gene expression programs in response to hypoxia: cell type specificity and prognostic significance in human cancers. PLoS Med. 2006;3(3):e47.
155. Galley HF, Webster NR. Physiology of the endothelium. Br J Anaesth. 2004;93:105–13.
156. Wolinsky H. A proposal linking clearance of circulating lipoproteins to tissue metabolic activity as a basis for understanding atherogenesis. Circ Res. 1980;47:301–11.
157. Savoia C, Sada L, Zezza L. Vascular inflammation and endothelial dysfunction in experimental hypertension. Intern J Hypertens. 2011; 2011:281240, 8p, doi: 10.4061/2011/281240.
158. Cines DB, Pollak ES, Buck CA. Endothelial cells in physiology and in the pathophysiology of vascular disorders. Blood. 1998;91:3527–61.
159. Gross PL, Aird WC. The endothelium and thrombosis. Semin Thromb Hemost. 2000;26:463–78.
160. Bombeli T, Mueller M, Haeberli A. Anticoagulant properties of the vascular endothelium. Thromb Haemost. 1997;77:408–23.
161. Janeway Jr CA, Medzhitov R. Innate immune recognition. Annu Rev Immunol. 2002;20:197–216.
162. Danese S, Dejana E, Fiocchi C. Immune regulation by microvascular endothelial cells: directing innate and adaptive immunity, coagulation, and inflammation. J Immunol. 2007;178:6017–22.
163. Sandoo A, van Zanten JJCS, Metsios GS. The endothelium and its role in regulating vascular tone. Open Cardiovasc Med J. 2010;4:302–12.

164. Bateman RM, Sharpe MD, Ellis CG. Bench-to-bedside review: microvascular dysfunction in sepsis-hemodynamics, oxygen transport, and nitric oxide. Crit Care. 2003;7:359–73.
165. Hauser B, Matejovic M, Radermacher P. Nitric oxide, leukocytes and microvascular permeability: causality or bystanders? Crit Care. 2008;12:104.
166. Lidington D, Tyml K, Ouellette Y. Lipopolysaccharide-induced reductions in cellular coupling correlate with tyrosine phosphorylation of connexin 43. J Cell Physiol. 2002;193:373–9.
167. Munford RS, Pugin J. Normal responses to injury prevent systemic inflammation and can be immuno-suppressive. Am J Respir Crit Care Med. 2001;163:316–21.
168. Pinsky MR. Immune balance in critically ill patients. Arch Immunol Ther Exp (Warsz). 2000;48:439–42.
169. van der Poll T, van Deventer SJ. Cytokines and anticytokines in the pathogenesis of sepsis. Infect Dis Clin North Am. 1999;13:413–26.
170. Pottmeyer E, Vassar MJ, Holcroft JW. Coagulation, inflammation and response to injury. Crit Care Clin. 1986;2:683–703.
171. Vanhoutte PM, Shimokawa H, Tang EHC. Endothelial dysfunction and vascular disease. Acta Physiol (Oxf). 2009;196:193–222.
172. Durier S, Fassot C, Laurent S. Physiological genomics of human arteries: quantitative relationship between gene expression and arterial stiffness. Circulation. 2000;108:1845–50.
173. Dvorak HF, Nagy JA, Feng D. Vascular permeability factor/vascular endothelial growth factor and the significance of microvascular hyperpermeability in angiogenesis. Curr Top Microbiol Immunol. 1999;237:97–132.
174. De Backer D, Ortiz JA, Salgado D. Coupling microcirculation to systemic hemodynamics. Curr Opin Crit Care. 2010;16:250–4.
175. Tyml K, Wang K, Lidington D. Lipopolysaccharide reduces intracellular coupling in vitro and arteriolar conducted response in vivo. Am J Physiol Heart Circ Physiol. 2001;287:1397–H1406.
176. Morin MJ, Unno N, Hodin RA. Differential expression of inducible nitric oxide synthase messenger RNA along the longitudinal and crypt-villus axes of the intestine in endotoxemic rats. Crit Care Med. 1998;26:1258–64.
177. Revelly JP, Ayuse T, Brienza N. Endotoxic shock alters distribution of blood flow within the intestinal wall. Crit Care Med. 1996;24:1345–51.
178. Trzeciak S, Clinel I, Dellinger PR, Microcirculatory Alterations in Resuscitation and Shock (MARS) Investigators. Resuscitation the microcirculation in sepsis: the central role of nitric oxide, emerging concepts for novel therapies, and challenges for clinical trails. Acad Emerg Med. 2008;15:399–413.
179. De Backer D, Donadello K, Taccone FS. Microcirculatory alterations: potential mechanisms and implications for therapy. Ann Intensive Care. 2011;1:27.
180. Price SA, Spain DA, Wilson MA, Harris PD, Garrison RN. Subacute sepsis impairs vascular smooth muscle contractile machinery and alters vasoconstrictor and dilator mechanisms. J Surg Res. 1999;83:75–80.
181. Chen SJ, Li SY, Shih CC. NO contributes to abnormal vascular calcium regulation and reactivity induced by peritonitis-associated septic shock in rats. Shock. 2010;33:473–8.
182. Baker CH, Willmoth FR. Microvascular responses to E. coli endotoxin with altered adrenergic activity. Circ Shock. 1984;12:165–76.
183. Tooke JE, Hannemann MM. Adverse endothelial function and the insulin resistance syndrome. J Intern Med. 2000;247:425–31.
184. Manjo G. Chronic inflammation: links with angiogenesis and wound healing. Am J Pathol. 1998;153:1035–9.
185. McCuskey RS, Urbaschek R, Urbaschek B. The microcirculation during endotoxemia. Cardiovasc Res. 1996;32:752–63.
186. Jhanji S, Lee C, Watson D. Microvascular flow and tissue oxygenation after major abdominal surgery: association with postoperative complications. Intensive Care Med. 2009; 35:671–7.

187. Bauer A, Kofler S, Thiel M. Monitoring of sublingual microcirculation in cardiac surgery using orthogonal polarization spectral imaging: preliminary results. Anesthesiology. 2007;197:939–45.
188. Vincent J-L, Ince C, Bakker J. Clinical review: circulatory shock – an update: a tribute to Professor Max Henry Weil. Crit Care. 2012;16:239.
189. Spronk PE, Zandstra DF, Ince C. Bench-to-bedside review: sepsis is a disease of microcirculation. Crit Care. 2004;8:462–8.
190. den Uil CA, Caliskan K, Lagrand WK. Dose-dependent benefit of nitroglycerin on microcirculation of patients with severe heart failure. Intensive Care Med. 2009;35:1893–9.
191. Uil D, Lagrand WK, Spronk PE. The effects of intra-aortic balloon pump support on macrocirculation and tissue microcirculation in patients with cardiogenic shock. Cardiology. 2009;114:42–6.
192. Kerger H, Waschke KF, Ackern KV. Systemic and microcirculatory effects of autologous whole blood resuscitation in severe hemorrhagic shock. Am J Physiol. 1999;276:H2035–43.
193. Lam C, Tmyl K, Martin C. Microvascular perfusion is impaired in a rat model of normotensive sepsis. J Clin Invest. 1994;94:2077–83.
194. De Backer D. Monitoring the microcirculation. J Clin Monit Comput. 2012;26:361–6.
195. Edul CS, Enrico C, Laviolle B. Quantitative assessment of the microcirculation in healthy volunteers and in patients with septic shock. Crit Care Med. 2012;40:1443–8.
196. Spronk PE, Ince C, Gardien MJ. Nitoglycerin in septic shock after intravacsular volume resuscitation. Lancet. 2002;360:1395–6.
197. Vallet B. Vascular reactivity and tissue oxygenation. Intensive Care Med. 1998;24:3–11.
198. Boczkowski J, Vicaut E, Aubier M. In vivo effects of Escherichia coli endotoxemia on diaphragmatic microcirculation in rats. J Appl Physiol. 1992;72:2219–24.
199. Thooft A, Favory R, Salgado-Ribeiro D. Effects of changes in arterial pressure on organ perfusion during septic shock. Crit Care. 2011;15:R222.
200. Ruokonen E, Takala J, Uusaro A. Effect of vasoactive treatment on the relationship between mixed venous and regional oxygen saturation. Crit Care Med. 1991;19:1365–9.
201. Bellomo R, Kellum JA, Wisnieski SR. Effects of norepinephrine on the renal vasculature in normal and endotoxemic dogs. Am J Respir Crit Care Med. 1999;159:1186–92.
202. Menger MD, Vollmer B. Role of microcirculation in transplantation. Microcirculation. 2000;7:291–306.
203. Tsai AG, Kerger H, Intaglietta M. Oxygen distribution and consumption by the microcirculation and the determinants of tissue survival. In: Sibbald WJ, Messmer K, Fink MP, editors. Tissue oxygenation in acute medicine. Berlin/Heidelberg/New York: Springer; 1998. p. 55–62.
204. Ellis CG, Bateman RM, Sharpe MD. Effect of maldistribution of microvascular blood flow on capillary O_2 extraction in sepsis. Am J Physiol Heart Circ Physiol. 2002;282:156–64.
205. Walley KR. Heterogeneity of oxygen delivery impairs oxygen extraction by peripheral tissues: theory. J Appl Physiol. 1996;81:885–94.
206. Farquhar I, Martin CM, Lam C. Decreased capillary density in vivo in bowel mucosa of rats with normotensive sepsis. J Surg Res. 1996;61:190–6.
207. Goldman D, Bateman RM, Ellis CG. Effect of decreased O_2 supply on skeletal muscle oxygenation and O_2 consumption during sepsis: role of heterogenous capillary spacing and blood flow. Am J Physiol Heart Circ Physiol. 2006;290:H2277–85.
208. Eipel C, Bordel R, Nickels RM. Impact of leukocytes and platelets in mediating hepatocyte apoptosis in a rat model of systemic endotoxemia. Am J Physiol Gastrointest Liver Physiol. 2004;286:G769–76.
209. den Uil CA, Lagrand WK. Impaired microcirculation predicts poor outcome of patients with acute myocardial infarction complicated by cardiogenic shock. Eur Heart J. 2010;31:3032–9.
210. Shapiro NI, Arnold R, Sherwin R. The association of near-infrared spectroscopy- derived tissue oxygenation measurements with sepsis syndromes, organ dysfunction and mortality in emergency department patients with sepsis. Crit Care. 2011;15:R223.

211. Top AP, Ince C, de Meij N. Persistent low microcirculatory vessel density in nonsurvivors of sepsis in pediatric intensive care. Crit Care Med. 2011;39:8–13.
212. Sakr Y, Dubois MJ, De Backer D. Persistent microcirculatory alterations are associated with organ failure and death in patients with septic shock. Crit Care Med. 2004;32:1825–31.
213. Treciak S, Dellinger RP, Parrillo JE, Microcirculatory Alterations in Resuscitation and Shock (MARS) Investigators. Early microcirculatory perfusion derangements in patients with severe sepsis and septic shock: relationship to hemodynamics, oxygen transport, and survival. Ann Emerg Med. 2007;49:88–98.
214. den Uil CA, Klijn E, Lagrand WK. The microcirculation in health and critical disease. Prog Cardiovasc Dis. 2008;51:161–70.
215. Munsterman LDH, Elbers PW, Ozdemir A. Withdrawing intra-aortic balloon pump support paradoxically improves microvascular flow. Crit Care. 2010;14:R161.
216. Pottecher J, Deruddre S, Teboul JL. Both passive leg raising and intravascular volume expansion improve sublingual microcirculatory perfusion in severe sepsis and septic shock patients. Intensive Care Med. 2010;36:1867–74.
217. Ospina-Tascon G, Neves AP, Occhipinti G. Effects of fluids on microvascular perfusion in patients with severe sepsis. Intensive Care Med. 2010;36:949–55.
218. McKenna TM, Reusch DW, Simpkins CO. Macrophage conditioned medium and interleukin-1 suppress vascular contractility. Circ Shock. 1988;25:187.
219. Hollenberg SM, Cunnion SE, Parrillo JE. The effect of tumor necrosis factor on vascular smooth muscle. In vitro studies using rat aortic rings. Chest. 1991;100:1133.
220. Koyama S, Aibiki M, Kanai K. Role of the central nervous system in renal nerve activity during prolonged hemorrhagic shock in dogs. Am J Physiol. 1988;254:R761–9.
221. Thiemermann C, Szab C, Mitchell JA. Vascular hyporeactivity to vasoconstrictor agents and hemodynamic decompensation in hemorrhagic shock is mediated by nitric oxide. Proc Natl Acad Sci. 1993;30:267.
222. Zingarelli B, Squadrito F, Altavilla D. Evidence for a role of nitric oxide in hypovolemic hemorrhagic shock. J Cardiovasc Pharmacol. 1992;19:982.
223. Guyton AC, Hall JE. Textbook of medical physiology. 11th ed. Philadelphia: Saunders Elsevier; 2006. p. 278.
224. Klabunde RE. Chapter 9. In: Cardiovascular Integration, Adaptation, and Pathophysiology. 2nd ed. Philadelphia: Lippincott Williams and & Wilkens; 2012. p. 198.
225. Haupt MT. The use of crystalloidal and colloidal solution for volume replacement in hypovolemic shock. Crit Rev Clin Lab Sci. 1989;27:1.
226. Davis JO, Freman RH. Mechanisms regulating rennin release. Physiol Rev. 1976;56:1.
227. Gettings LG, Reynolds HN, Scalea T. Outcome in post-traumatic acute renal failure when continuous renal replacement therapy is applied early vs late. Intensive Care Med. 1999;25:805–13.
228. Bagshaw SM, George C, Bellomo R. Early acute kidney injury and sepsis: a multicentre evaluation. Crit Care. 2008;12:R47.
229. Sander-Jensen K, Secher NH, Bie P. Vagal slowing of the heart during haemorrhage: observations from 20 consecutive hypotensive patients. Br Med J. 1986;292:364–6.
230. Barriot P, Riou B. Hemorrhagic shock with paradoxial bradycardia. Intensive Care Med. 1987;13:203–7.
231. Schadt JC, Ludbrook J. Hemodynamic and neurohormonal responses to acute hypovolemia in conscious mammals. Am J Physiol. 1991;260:H305.
232. Bone RC. Gram – negative sepsis. Background, clinical features, and intervention. Chest. 1991;100:802–8.
233. Fuster V, Alexander RW, O´Rourke RA. Hurst`s the heart book. 11th ed. New York: McGraw–Hill; 2004. p. 513.
234. Fung YC. Biomechanics: circulation. 2nd ed. New York: Springer; 1997. p. 10–1.
235. Pries AR, Secomb TW. Blood flow in the microcirculatory networks. In: Tuma RF, Duran, WN, LeyK, editors. Comprehensive physiology- handbook of physiology, supplement 9:

handbook of physiology, the cardiovascular system, microcirculation. s.l.: Wiley Online Library for the American Physiological Society, 2008, online published 1 Jan 2011, p. 3–36.
236. Popel A, Johnson P. Microcirculation and hemorheology. Annu Rev Fluid Mech. 2005;37:43.
237. Cholley B, Payen P. Left ventricular afterload and ventriculoarterial coupling. In: Pinsky M, editor. Applied cardiovascular physiology. Berlin: Springer; 1997. p. 14–27.
238. Nichols WW, O` Rourke M. McDonald's blood flow in arteries. Theoretical, experimental, and clinical principles. London: Oxford University Press; 2005.
239. Zweifach BW, Fronek A. The interplay of central and peripheral factors in irreversible hemorrhagic shock. Prog Cardiovasc Dis. 1975;18:147.
240. Baker CH. Microvascular responses of intact and adrenal medullectomized rats in hemorrhagic shock. Circ Shock. 1988;26:203.
241. Sakai H. Changes in resistance vessels during hemorrhagic shock and resuscitation in conscious hamster model. Am J Physiol. 1999;276:H563.
242. Taylor AE, Moore TM. Capillary fluid exchange. Am J Physiol. 1999;277:S203–10.
243. Dellinger RP. Cardiovascular management of septic shock. Crit Care Med. 2003;31:946–55.
244. Ellis CG, Wrigley SM, Groom AC. Heterogeneity of red blood cell perfusion in capillary networks supplied by a single arteriole in resting skeletal muscle. Cric Res. 1994;75:357–68.
245. Nakajima Y, Baudry N, Duranteau J. Microcirculation in intestinal villi: a comparison between hemorrhagic and endotoxin shock. Am J Respir Crit Care Med. 2001;164:1526–30.
246. Sylvester JT, Scharf SM, Gilbert RD. Hypoxic and CO hypoxia in dogs: hemodynamics, carotid reflexes, and catecholamines. Am J Physiol. 1979;236:H22–8.
247. Benedict CR, Rose JA. Arterial norepinephrine changes in patients with septic shock. Circ Shock. 1992;38:165–72.
248. Cumming AD, Driedger A, McDonald JWD. Vasoactive hormones in the renal response to systemic sepsis. Am J Kidney Dis. 1988, Vol. 11, 23–32, Erratum. Am J Kidney Dis. 1988;11:363.
249. Young JD. The heart and circulation in severe sepsis. Br J Anaesth. 2004;93:114–20.
250. Young JD, Cameron EM. Dynamics of skin blood flow in human sepsis. Intensive Care Med. 1995;21:669–74.
251. Stephan F, Novara A, Tournier B. Determination of total effective vascular compliance in patients with sepsis syndrome. Am J Respir Crit Care Med. 1998;157:50–6.
252. Astiz M, Rackow EC, Weil MH. Early impairment of oxidative metabolism and energy production in severe sepsis. Circ Shock. 1988;26:516–23.
253. Anning PB, Sair M, Winlove CP. Abnormal tissue oxygenation and cardiovascular changes in endotoxemia. Am J Respir Crit Care Med. 1999;159:1710–5.
254. Rachow EC, Kaufmann BS, Falk JL. Hemodynamic response to fluid repletion in patients with septic shock: evidence for early depression od cardiac performance. Circ Shock. 1987;22:11–22.
255. Yang S, Cioffi WG, Bland KI. Differential alterations in systemic and regional oxygen delivery and consumption during the early and late stages of sepsis. J Trauma. 1999;47:706–12.
256. Cuthbertson DP. Post shock metabolic response. Lancet. 1942;1:433–42.
257. Partrick DA, Bensard DD, Janik JS. Is hypotension an reliable indicator of blood loss from traumatic injury in children? Am J Surg. 2002;184:555–9.
258. Worthley LIG. Shock: a review of pathophysiology and management. Part II. Crit Care Resusc. 2000;2:66–84.
259. Dellinger RP, Carlet JM, Masur H. Surviving Sepsis Campaign guidelines for the management of severe sepsis and septic shock. Intensive Care Med. 2004;30:536–55.
260. Dellinger RP, Levy MM, Carlet JMS. Sepsis campaign: international guidelines for management of severe sepsis and septic shock. Intensive Care Med. 2008;34:17–60.
261. Myers JM, Neighbors M, Tannehille-Jones R. Pathophysiology of shock: chapter 6. In: Principles of pathophysiology and emergency medical care. New York: Delmar; 2002. p. 95.
262. Moore FA, Moore EE, Read RA. Postinjury multiple organ failure: role of extrathoracic injury and sepsis in adult respiratory distress syndrome. New Horiz. 1993;1:538–49.

263. Collard CD, Gelman S. Pathophysiology, clinical manifestations, and prevention of ischemia-reperfusion injury. Anesthesiology. 2001;94:1133–8.
264. Landry DW, Oliver JA. The ATP-sensitive K+ channel mediates hypotension in endotoxemia and hypoxic lactic acidosis in dogs. J Clin Invest. 1992;89:2071.
265. Barroso-Aranda J, Schmid-Schönbein GW, Zweifach BW. Granulocytes and no- reflow phenomenon in irreversible hemorrhagic shock. Cric Res. 1988;63:437.
266. Thiemermann C, Szabo` C, Mitchell JA. Vascular hyporeactivity to vasoconstrictor agents and hemodynamic decompensation in hemorrhagic shock is mediated by nitric oxide. Proc Natl Acad Sci U S A. 1993;90:267.
267. Hierholzer C, Harbrecht B, Menezes JM. Essential role of induced nitric oxide in the initiation of the inflammatory response after hemorrhagic shock. J Exp Med. 1998;187:917.
268. Bauer M, Feucht K, Ziegenfuss T. Attenuation of shock-induced hepatic microcirculatory disturbances by the use of a starch-deferoxamine conjugate for resuscitation. Crit Care Med. 1995;23:316.
269. Robin JK, Oliver JA, Landry DW. Vasopressin deficiency in the syndrome of irreversible shock. J Trauma. 2003;54:S149.
270. Ullian ME. The role of corticosteroids in the regulation of vascular tone. Cardiovasc Res. 1999;41:55–64.
271. Van de Werf F, Bax J, Betriu A. Management of acute myocardial infarction in patients presenting with persistent ST-segment elevation: the Task Force on the Management of ST-Segment Elevation Acute Myocardial Infarction of the European Society of Cardiology. Eur Heart J. 2008;29:2909–45.
272. Antman EM, Anbe DT, Armstrong PW. ACC/AHA guidelines for the management of patients with ST-elevation myocardial infarction: executive summary: a report of the American College of Cardiology/American Heart Association Task Force on Practice Guidelines. Circulation. 2004;110:588–636.
273. Menon V, Hochman JS. Management of cardiogenic shock complicating acute myocardial infarction. Heart. 2002;88:531–7.
274. Hochman JS, Sleeper LA, Webb JG. Early revascularisation in acute myocardial infraction complicated by cardiogenic shock. SCHOCK Investigators. Should we emergently revascularize occluded coronaries for cardiogenic shock. N Engl J Med. 1999;341:625–34.
275. Nieminen MS, Brutsaert D, Dickstein K. EuroHeart Failure Survey II (EHFS II): a survey on hospitalized acute heart failure patients: description of population. Eur Heart J. 2006;27:2725–36.
276. Topalian S, Ginsberg F, Parrillo JE. Cardiogenic shock. Crit Care Med. 2008;36(1 Suppl):S66–74.
277. Alpert S. Pathophysiology, diagnosis and management of cardiogenic shock. In: Blunt RC, Alexander RC, editors. Hurst`s the heart: arteries and veins. New York: McGraw-Hill; 1994. p. 907.
278. McGhie AI, Goldstein RA. Pathogenesis and management of acute heart failure and cardiogenic shock: role of inotropic therapy. Chest. 1992;102(5 Suppl 2):S626–32.
279. Steingrub S. Cardiac physiology. In: Steingrubb H, Stoller K, editors. Cardiopulmonary critical care. Milton Park/Oxfordshire: BIOS Scientific Pub Ltd; 2002. p. 17.
280. Califf RM, Bengtson RF. Current concepts: cardiogenic shock. N Engl J Med. 1994;330:1724–30.
281. Webb JG. Interventional management of cardiogenic shock. Can J Cardiol. 1998;14:233–44.
282. Greenberg MA, Menegus MA. Ischemia-induced diastolic dysfunction: new observations, new questions. J Am Coll Cardiol. 1989;13:1071–2.
283. Gillebert TC, Leite-Moreira AF, De Hert SG. Load dependent diastolic dysfunction in heart failure. Heart Fail Rev. 2000;5:345–65.
284. Braunwald E. Mechanisms of cardiac contraction and relaxation. In: Braunwald E, editor. Braunwald heart disease. 5th ed. Philadelphia: Elsevier Saunders; 1997. p. 434.
285. Ramanathan K, Harkness SM, Nayar AC. Cardiogenic shock in patients with preserved left ventricular systolic function: characteristics and insight into mechanisms. J Am Coll Cardiol. 2004;43:241A.

286. Reynolds HR, Anand SK, Fox JM. Restrictive physiology in cardiogenic shock: observations from echocardiography. Am Heart J. 2006;151:890.e9–15.
287. Yehudai L, Reynolds HR, Schwarz SA. Serial echocardiograms in patients with cardiogenic shock: analysis of the SHOCK trial. Poster session 12.03.2006. J Am Coll Cardiol. 2006;47(Suppl A):111A.
288. Picard MH, Davidoff R, Sleeper LA. Echocardiographic predictors of survival and response to early revascularization in cardiogenic shock. Circulation. 2003;107:279–84.
289. Ellis SG, Hentschke CI, Sandor T. Time course of functional and biochemical recovery of myocardium salvaged by reperfusion. J Am Coll Cardiol. 1983;1:1047–55.
290. Bolli R, Zhu W-X, Thornby JI. Time course and determinants of recovery of function after reversible ischemia in conscious dogs. Am J Physiol. 1988;254:H102–14.
291. Bolli R. Myocardial 'stunning' in man. Circulation. 1992;86:1671–91.
292. Bolli R. Basic and clinical aspects of myocardial stunning. Prog Cardiovasc Dis. 1998;40:477–516.
293. Kloner RA, Przyklenk K. Stunned and hibernating myocardium. Annu Rev Med. 1991; 42:1–8.
294. Wijns W, Vatner SF, Camici PG. Hibernating myocardium. N Engl J Med. 1998;339:173–81.
295. Boffard KD. Resuscitative physiology. In: Boffard KD, editor. Manual of definitive trauma care. London: Arnold Publishers; 2003. p. 7–38.
296. Antman EM, et al. Acute myocardial infarction. In: Braunwald E, Fauci E, Kasper D, editors. Harrison`s principles of internal medicine. New York: McGraw-Hill; 2001. p. p1395.
297. Cotter G, Moshkovitz Y, Milovanov O. Acute heart failure: a novel approach to its pathogenesis and treatment. Eur J Heart Fail. 2002;4:227–34.
298. Geppert A, Steiner A, Zorn G. Multiple organ failure in patients with cardiogenic shock is associated with high plasma levels of interleukin-6. Crit Care Med. 2002;30: 1987–94.
299. Kohsaka S, Menon V, Lange M. High incidence of sepsis complicating cardiogenic shock following acute myocardial infarction. Circulation. 2001;104(Suppl II):II–483.
300. Li H, Förstermann VH. Nitric oxide in the pathogenesis of vascular disease. J Pathol. 2000;190:344–254.
301. Frangogigannis NG, Smith CW, Entman ML. The inflammatory response in myocardial infarction. Cardiovasc Res. 2002;53:31.
302. Finkel T, Holbrook NJ. Oxidants, oxidative stress and the biology of ageing. Nature. 2000;408:239–47.
303. de Werra I, Jaccard C, Corradin SB. Cytokines, nitrite/nitrate, soluble tumor necrosis factor receptors, and procalcitonin concentrations: comparisons in patients with septic shock, cardiogenic shock, and bacterial pneumonia. Crit Care Med. 1997;25:607–13.
304. Theroux P, Armstrong PW, Mahaffey KW. Prognostic significance of blood markers of inflammation in patients with ST-segment elevation myocardial infarction undergoing primary angioplasty and effects of pexelizumab, a C5 inhibitor: a substudy of the COMMA trial. Eur Heart J. 2005;26:1964–70.
305. Zhang C, Xu X, Potter BJ. TNF-α Contributes to Endothelial Dysfunction in Ischemia/ Reperfusion Injury. Arterioscler Thromb Vasc Biol. 2006;26:475–80.
306. Feng O, Lu X, Jones DL. Increased inducible nitric oxide synthase expression contributes to myocardial dysfunction and higher mortality after myocardial infraction in mice. Circulation. 2001;104:700–4.
307. Investigators TRIUMPH. Effect of tilarginine acetate in patients with acute myocardial infarction and cardiogenic shock. JAMA. 2007;297:1657–66.
308. Dzavik V, Cotter G, Reynolds HR. Effect of nitric oxide synthase inhibition on haemodynamics and outcome of patients with persistent cardiogenic shock complicating acute myocardial infarction: a phase II dose-ranging study. Eur Heart J. 2007;28:1109–16.
309. Smith HJ, Oriol A, Morch J. Hemodynamic studies in cardiogenic shock: treatment with isoproterenol and metaraminol. Circulation. 1967;35:1084–91.
310. Lim N, Dubois MJ, De Backer D. Do all nonsurvivors of cardiogenic shock die with a low cardiac index? Chest. 2003;124:1885–91.

311. Vlachopoulos C, Dima I, Aznaouridis K. Acute Systemic Inflammation Increases Arterial Stiffness and Decreases Wave Reflections in Healthy Individuals. Circulation. 2005;112:2193–200.
312. Reichek N, Wilson J, Sutton MSJ. Noninvasive determination of left ventricular end-systolic stress: validation of the method and initial application. Circulation. 1982;65:99–108.
313. Kellum JA, Pinsky MR. Use of vasopressor agents in critically ill patients. Curr Opin Crit Care. 2002;8:236–41.
314. Garcia MI, Cano AG, Romero MG. Dynamic arterial elastance to predict arterial pressure response to volume loading in preload-dependent patients. Crit Care. 2011;15:R15.
315. Pinsky MR. Defining the boundaries of bedside pulse contour analysis: dynamic arterial elastance. Crit Care. 2011;15:120.
316. Neumann FJ, Ott I, Gawaz M. Cardiac release of cytokines and inflammatory responses in acute myocardial infarction. Circulation. 1995;92:748–55.
317. Menon V, Slater JN, White HD. Acute myocardial infarction complicated by systemic hypoperfusion without hypotension: report of the SHOCK trail registry. Am J Med. 2000;108:374–80.
318. Liuzzo G, Buffon A, Biasucci LM. Enhanced inflammatory response to coronary angioplasty in patients with severe unstable angina. Circulation. 1998;98:2370–6.
319. Cotter G, Kaluski E, Milo O. LINCS: L-NAME (a NO synthase inhibitor) in the treatment of refractory cardiogenic shock: a prospective randomized study. Eur Heart J. 2003;24: 1287–95.
320. Mueller-Werdan U, Schumann H, Fuchs R. Tumor necrosis factor(TNFalpha) is cardiodepressant in pathophysiologically relevant concentrations without inducing inducible nitric oxide (NO-) synthase (iNOS) or triggering serious cytotoxicity. J Mol Cell Cardiol. 1997;29:2915–33.
321. Parrillo JE, Burch C, Shelhamer JH. A circulating myocardial depressant substance in humans with septic shock. Septic shock patients with a reduced ejection fraction have a circulating factor that depresses in vitro myocardial cell performance. J Clin Invest. 1985;76:1539–53.
322. Li H, Förstermann U. Nitric oxide in the pathogenesis of vascular disease. J Pathol. 2000;190:244–54.
323. Valencia R, Theroux P, Granger CB. Congestive heart failure and cardiogenic shock complicating acute myocardial infarction have a high mortality and are associated with intense inflammatory response: results from the CARDINAL trials (abstract). J Am Coll Cardiol. 2004;43(5, Suppl A):291A.
324. Hochman JS. Cardiogenic shock. Annual Scientific Session of the AHA, Dallas; 1998.
325. Harizi RC, Bianco JA, Alpert JS. Diastolic Function of the Heart in Clinical Cardiology. Arch Intern Med. 1988;148:99–109.
326. Banka VS, Helfant RH. Temporal sequence of dynamic contractile characteristics in ischemic and nonischemic myocardium after acute coronary ligation. Am J Cardiol. 1974; 34:158–63.
327. Figueras J, Singh BN, Ganz W. Mechanism of rest and nocturnal angina: observations during continuous hemodynamic and electrocardiographic monitoring. Circulation. 1979;59:955–68.
328. Holubarsch C. Existence of the Frank-Starling mechanism in the failing human heart: investigations on the organ, tissue, and sarcomere levels. Circulation. 1996;94:683–9.
329. Ware LB, Matthay MA. Acute pulmonary edema. N Engl J Med. 2005;353:2788–96.
330. Diamond G, Forrester JS. Effect of coronary artery disease and acute myocardial infarction on left ventricular compliance in man. Circulation. 1972;45:11–9.
331. Calvin JE, Driedger AA, Sibbald WJ. Does the pulmonary capillary wedge pressure predict left ventricular preload in critically ill patients? Crit Care Med. 1981;9:437–43.
332. Tousignant CP, Walsh F, Mazer CD. The use of transesophageal echocardiography for preload assessment in critically ill patients. Anesth Analg. 2000;90:351–5.
333. Kumar A, Anel R, Bunnell E. Pulmonary artery occlusion pressure and central venous pressure fail to predict ventricular filling volume, cardiac performance, or the response to volume infusion in normal subjects. Crit Care Med. 2004;32:691–9.

334. Menon V, White H, LeJemtel T. The clinical profile of patients with suspected cardiogenic shock due to predominant left ventricular failure: a report from the SHOCK Trail Registry. J Am Coll Cardiol. 2000;36:1071–6.
335. Rothbaum DA, Linnemeier TJ, Landin RJ. Emergency percutaneous transluminal coronary angioplasty in acute myocardial infarction: a 3 year experience. J Am Coll Cardiol. 1987;10:264–72.
336. Bindels J, van der Hoeven G, MeindersBindels AE. Pulmonary artery wedge pressure and extravascular lung water in patients with acute cardiogenic pulmonary edema requiring mechanical ventilation. Am J Cardiol. 1999;84:1158–63.
337. Michard F, Alaya S, Zarka V. Global end-diastolic volume as an indicator of cardiac preload in patients with septic shock. Chest. 2003;124:1900–8.
338. Atherton JJ, Moore TD, Lele SS. Diastolic ventricular interaction in chronic heart failure. Lancet. 1997;349:1720–4.
339. Santamore WP, Lynch PR, Meier G. Myocardial interaction between the ventricles. J Appl Physiol. 1976;41:362–8.
340. Janicki JS, Weber KT. The pericardium and ventricular interaction, distensibility, and function. Am J Physiol. 1980;238:H494–503.
341. Jacobs AK, Leopold JA, Bates E. Cardiogenic shock caused by right ventricular infarction: a report from the SHOCK registry. J Am Coll Cardiol. 2003;41:1273–9.
342. Kinch JW, Ryan TJ. Current concepts: right ventricular infarction. N Engl J Med. 1994;330:1211–7.
343. Stolzfus DP. Right ventricular function and failure in the perioperative period. Anesthesiol Clin North America. 1997;15:797–822.
344. Bleasdale RA, Frenneraux M. Prognostic importance of right ventricular dysfunction. Heart. 2002;88:323–4.
345. Jardin F. Ventricular interdependence: how does it impact on hemodynamic evaluation in clinical practice ? Intensive Care Med. 2003;29:361–3.
346. Dell'Italia LJ, Pearce DJ, Blackwell GG. Right and left ventricular volumes and function after acute pulmonary hypertension in intact dogs. J Appl Physiol. 1995;78:2320–7.
347. Sibbald WJ, Driedger AA. Right ventricular function in acute state disease: pathophysiologic considerations. Crit Care Med. 1983;11:339–45.
348. Feneley MP, Olsen CO, Glower DD. Effect of acutely increased right ventricular afterload on work output from the left ventricle in conscious dogs. Systolic ventricular interaction. Circ Res. 1989;65:135–45.
349. Dell`Italia LJ. Comparative effects of volume loading, dobutamine, and nitroprusside in patients with predominant right ventricular infarction. Circulation. 1985;72: 1327–35.
350. Brookes C, Ravn H, White P. Acute right ventricular dilatation in response to ischemia significantly impairs left ventricular systolic performance. Circulation. 1999;100:761–7.
351. Jung C, Ferrari M, Roediger C. Evaluation of sublingual microcirculation in cardiogenic shock. Clin Hemorheol Microcirc. 2009;42:141–8.
352. Kirschenbaum LA, Astiz ME, Rackow EC. Microvascular response in patients with cardiogenic shock. Crit Care Med. 2000;28:1290–194.
353. Freelander SO, Lenhart CH. Clinical observation on the capillary circulation. Arch Intern Med. 1992;29:12–32.
354. Wroblewski H, Kastrup J, Norgaard T. Evidence of increased microvascular resistance and arteriolar hyalinosis in skin in congestive heart failure secondary to idiopathic dilated cardiomyopathy. Am J Cardiol. 1992;69:769–74.
355. Humer MF, Phang PT, Friesen BP. Heterogeneity of gut capillary transit times and impaired gut oxygen extraction in endotoxemic pigs. J Appl Physiol. 1996;81:895–904.
356. Drost EM, Kassabian G, Meiselman HJ. Increased rigidity and priming of polymorphonuclear leukocytes in sepsis. Am J Respir Crit Care Med. 1999;159:1696–702.
357. Hebert MT, Marshall JM. Direct observation of responses of mesenteric microcirculation of the rat to circulating noradrenaline. J Physiol. 1985;368:393–407.

358. Vicant E, Hou X, Payen D. Acute effects of tumor necrosis factor on the microcirculation in rat cremaster muscle. J Clin Invest. 1991;87:1537–40.
359. Levine B, Kalman J, Mayer L. Elevated circulating levels of tumor necrosis factor in severe chronic heart failure. N Engl J Med. 1990;323:236–41.
360. Farquharson CA, Buttler R, Hill A. Allopurinol improves endothelial dysfunction in chronic heart failure. Circulation. 2002;106:221–6.
361. Chen ZM, Pan HC, Chen YP. Early intravenous then oral metoprolol in 45,852 patients with acute myocardial infarction: randomised placebo-controlled trail. Lancet. 2005;366:1622–32.
362. Group, ISIS-4 Collaborative. ISIS-4: a randomised factorial trial assessing early captopril, oral mononitrate, and intravenous magnesium sulphate in 58,050 patients with suspected acute myocardial infarction. Lancet. 1995;345:669–85.
363. Meine TJ, Roe MT, Chen AY. Association of intravenous morphine use and outcomes in acute coronary syndromes: results from the CRUSADE Quality Improvement Initiative. Am Heart J. 2005;149:1043–9.
364. Cosgriff N, Moore EE, Sauaia A. Predicting life-threatening coagulopathy in the massively transfused trauma patient: hypothermia and acidosis revisited. J Trauma. 1997;42:857–61.
365. Hinshaw LB, Peterson M, Huse WM. Regional blood flow in hemorrhagic shock. Am J Surg. 1961;102:224–34.
366. Trauma, Committee on, American College of Surgeons. Advanced trauma life support manual. Chicago: American College of Surgeons; 1997. p. 103–12.
367. Cherkas D. Traumatic hemorrhagic shock: advances in fluid management. Emerg Med Pract. 2011;13:1–19.
368. Bonanno G. Hemorrhagic shock: the "physiology approach". J Emerg Trauma Shock. 2012;5:285–95.
369. Shoemaker WC, Appel PL, Kram HB. Measurement of tissue perfusion by oxygen transport pattern in experimental shock and high -risk surgical patients. Intensive Care Med. 1990;16:S135.
370. Wong DH. Changes in cardiac output after acute blood loss and position changes in man. Crit Care Med. 1989;17:979.
371. McGee S, Abernethy WB, Simel DL. Is this patient hypovolemic? JAMA. 1999;281:1022.
372. Birkhahn RA. Shock index in diagnosing early acute hypovolemia. Am J Emerg Med. 2003;23:323.
373. Bond RF, Johnson III G. Vascular adrenergic interactions during hemorrhagic shock. Fed Proc. 1985;44:281.
374. Victorino GP, Battistella FD, Wisner DH. Does tachycardia correlate with hypotension after trauma. J Am Coll Surg. 2003;196:697.
375. Schlichtig R, Kramer DJ, Pinsky MR. Flow redistribution during progressive hemorrhage is a determinant of critical O-2 delivery. J Appl Physiol. 1991;70:169.
376. Cooke WH, Ryan KL, Convertino VA. Lower body negative pressure as a model to study progression to acute hemorrhagic shock in humans. J Appl Physiol. 2004;96:1249.
377. Lieberthal W. Nitric oxide inhibition in rats improves blood pressure and renal function during hypovolemic shock. Am J Physiol. 1991;261:F868.
378. Chang H. Plasma endothelin level changes during hemorrhagic shock. J Trauma. 1993;35:825.
379. Connolly HV, Maginniss LA, Schumacker PT. Transit time heterogeneity in canine small intestine: significance of oxygen transport. J Clin Invest. 1997;99:228.
380. Corso CO. Hypertonic saline dextran attenuates leukocyte accumulation in the liver after hemorrhagic shock and resuscitation. J Trauma. 1999;46:417.
381. Bone RC, Sibbald WJ, Sprung CL. The ACCP-SCCM consensus conference on sepsis and organ failure. Chest. 1992;101:1481–38.
382. Levy MM, Fink MP, Marshall JC. SCCM/ESICM/ACCP/ATS/SIS international sepsis definitions conference. Crit Care Med. 2003;31:1250–6.
383. Hurtado FJ, Buroni M, Tenzi J. Sepsis: clinical approach, evidenced-based at the bedside. In: Gullo A, Besso J, Lumb PD, Williams GF, editors. Intensive and critical care medicine. Milan: Springer; 2009.

384. Kumar A. Sepsis and septic shock: old concepts, new precepts. http://www.slideshare.net/fergua/sepsis-and-septic-shock?utm_source=slideshow03&utm_medium=ssemail&utm_campaign=share_slideshow_loggedout.
385. Conference, American College of Chest Physicians/Society of Critical Care Medicine. American College of Chest Physicians/Society of Critical Care Medicine Consensus. Definitions for sepsis and organ failure and guidelines for the use of innovative therapies in sepsis. Crit Care Med. 1992;20:864–74.
386. Annane D, Bellissant E, Cavaillon J-M. Septic shock. Lancet. 2005;365:63–78.
387. Marshall JC, Cook DJ, Christou NV. Multiple organ dysfunction score: a reliable descriptor of a complex clinical outcome. Crit Care Med. 1995;23:1638–52.
388. Nguyen HB, Rivers EP, Abrahamian FM. Severe sepsis and septic shock: review of the literature and emergency department management guidelines. Ann Emerg Med. 2006;48:28–54.
389. Nduka OO, Parrillo JE. The pathophysiology of septic shock. Crit Care Clin. 2009;25:677–702.
390. Han Y, Carcillo J, Dragotta M. Early reversal of pediatric- neonatal septic shock by community physicians is associated with improved outcome. Pediatrics. 2003;112:793–9.
391. Zanotti- Cavazzoni SL, Dellinger RP. Hemodynamic optimization of sepsis-induced tissue hypoperfusion. Crit Care. 2006;10 Suppl 3:S2.
392. Parrillo JE. Pathogenetic mechanisms of septic shock. N Engl J Med. 1993;328:1471–7.
393. Rangel-Frausto MS, Pittet D, Costigan M. The natural history of the systemic inflammatory response syndrome (SIRS): a prospective study. JAMA. 1995;273:117–23.
394. Brun-Buisson C, Doyon F, Carlet J. Incidence, risk factors, and outcome of severe sepsis and septic shock in adults: a multicenter prospective study in intensive care units: French ICU Group for Severe Sepsis. JAMA. 1995;274:968–74.
395. Estenssoro E, Gonzalez F, Laffaire E. Shock on admission day is the best predictor of prolonged mechanical ventilation in the ICU. Chest. 2005;127:598–603.
396. Volk T, Kox WJ. Endothelium function in sepsis. Inflamm Res. 2000;49:185–98.
397. Parrillo JE, Parker MM, Natanson C. Septic shock in humans. Advances in the understanding of pathogenesis, cardiovascular dysfunction, and therapy. Ann Intern Med. 1990;113: 227–42.
398. Sherwin RL, Garcia AJ, Bilkovski R. Do low dose corticosteroids improve mortality or shock reversal in patients with septic shock? A systematic review and position statement prepared for the American Academy of Emergency Medicine. J Emerg Med. 2012;43:7–12.
399. Parrillo JE. Cardiovascular dysfunction in septic shock: new insights into a deadly disease. Int J Cardiol. 1985;7:314–21.
400. Trzeciak S, Cinel I, Dellinger RP. Resuscitating the microcirculation in sepsis: the central role of nitric oxide, emerging concepts for novel therapies, and challenges for clinical trails. Acad Emerg Med. 2008;15:399–413.
401. Vincent J-L, De Backer D. Microvascular dysfunction as cause of organ dysfunction in severe sepsis. Crit Care. 2005;9 Suppl 4:S9–12.
402. Schumacker PT, Wood LDH. Limitations of aerobic metabolism in critically illness. Chest. 1984;85:453–4.
403. Cumming AD, Driedger A, McDonald JWD. Vasoactive hormones in the renal response to systemic sepsis. Am J Kidney Dis. 1988;11:23–32. Erratum 1988;11: 363.
404. Carleton SC. The cardiovascular effects of sepsis. Cardiol Clin. 1995;13:249–56.
405. Thijs LG, Schneider AJ, Groeneveld AB. The hemodynamics of septic shock. Intensive Care Med. 1990;16 Suppl 3:S182–6.
406. Grocott-Mason RM, Shak AM. Cardiac dysfunction in sepsis: new theories and clinical implications. Intensive Care Med. 1998;24:286–95.
407. Levy B, Mansart A, Montemont C. Myocardial lactate deprivation is associated with decreased cardiovascular performance, decreased myocardial energetics, and early death in endotoxic shock. Intensive Care Med. 2007;33:495–502.
408. Bersten AD, Hersch M, Cheung H, Rutledge FS, Sibbald WJ. The effect of various sympathomimetics on the regional circulations in hyperdynamic sepsis. Surgery. 1992;112:549–61.

409. Bellomo R, Wan L, May C. Vasoactive drugs and acute kidney injury. Crit Care Med. 2008;36(Suppl):S179–86.
410. Dennen P, Douglas IS, Anderson R. Acute kidney injury in the intensive care unit: an update and primer for the intensivist. Crit Care Med. 2010;38:261–75.
411. Farand P, Hamel M, Lauzier F. Review article: organ perfusion/permeability-related effects of norepinephrine and vasopressin in sepsis. Can J Anaesth. 2006;53:934–46.
412. Astiz ME, DeGent GE, Lin RY. Microvascular function and rheologic changes in hyperdynamic sepsis. Crit Care Med. 1995;23:265–71.
413. Price LC, Wort SJ, Finney SJ. Pulmonary vascular and right ventricular dysfunction in adult critical care: current and emerging options for management: a systematic literature review. Crit Care. 2010;14:R169.
414. Zerbe RL, Henry DP, Robertson GL. Vasopressin response to orthostatic hypotension: etiologic and clinical implications. Am J Med. 1983;74:265–71.
415. Kaufmann H, Oribe E, Oliver JA. Plasma endothelin during upright tilt: relevance for orthostatic hypotension? Lancet. 1991;338:1542–5.
416. Arnauld E, Czernichow P, Fumoux F. The effects of hypotension and hypovolaemia on the liberation of vasopressin during haemorrhage in unanaesthetized monkey (Macaca mulatta). Pflugers Arch. 1977;371:193–200.
417. Mohrman DE, Heller LJ. The peripheral vascular system: chapter 7. In: Cardiovascular physiology. 4th ed. New York: McGraw-Hill; 1997. p. 91–108.
418. Feihl F, Liaudet L, Levy BI. Hypertension and microvascular remodeling. Cardiovasc Res. 2008;78:274–85.
419. Sparks H, Lorenz J, Rooke TW. Essentials of cardiovascular physiology. Minneapolis: University Press; 1987. p. 95–8.
420. Christensen KL, Mulvany MJ. Location of resistance arteries. J Vasc Res. 2001;38:1–12.
421. Levy BI, Ambrosio G, Pries AR. Microcirculation in hypertension – a new target for treatment. Circulation. 2001;104:735–40.
422. Mulvancy MJ. Structural changes in the resistance vessels in human hypertension. In: Laragh JH, Brenner BRG, editors. Hypertension: pathophysiology, diagnosis and management. 2nd ed. New York: Raven; 1995. p. 508–15.
423. Feihl F, Liaudet L, Waeber B. Hypertension: a disease of the microcirculation? Hypertension. 2006;48:1012–7.
424. Polanco PM, Pinsky MR. Practical issues of hemodynamic monitoring at the bedside. Surg Clin North Am. 2006;86:1431–56.
425. Berster, AD, Holt, AW. Vasoactive drugs and the importance of renal perfusion pressure. Vol 3, New Horizons; 1995. pp 650–661
426. Lang RM, Borow KM, Neumann A. Systemic vascular resistance: an unrealiable index of left ventricular afterload. Circulation. 1986;74:1114–23.
427. Weber KT, Janicki JS, Hunter WC. The contractile behaviour of the heart and its functional coupling to the circulation. Prog Cardiovasc Dis. 1982;24:375.
428. O`Rourke MF. Vascular impedance in studies of arterial and cardiac function. Physiol Rev. 1982;62:570.
429. Nichols WW, O`Rourke M. McDonald's blood flow in arteries. In: Theoretical, experimental and clinical principles. London: Oxford University Press; 2005.
430. Kelly RP, Ting CT, Yang TM. Effective arterial elastance as index of arterial vascular load in humans. Circulation. 1992;86:513–21.
431. Westerhof N, Stergropulos N, Noble MIM. Snapshots of hemodynamics: an aid for clinical research and graduate education. New York: Springer; 2005.
432. Sunagawa K, Maughan WL, Burkhoff D. Left ventricular interaction with arterial load studied in isolated canine ventricle. Am J Physiol. 1983;245:H773–80.
433. Kass DA. Age-related changes in ventriculo-arterial coupling: pathophysiologic implications. Heart Fail Rev. 2002;7:51–62.
434. Ignarro LJ, Buga GM, Wood KS. Endothelium-derived relaxing factor produced and released from artery and vein is nitric oxide. Proc Natl Acad Sci U S A. 1987;84:9265–9.

435. Kotsovolis G, Kallaras K. The role of endothelium and endogenous vasoactive substances in sepsis. Hippokratia. 2010;14:88–93.
436. Hutcheson IR, Whittle BJ, Boughton-Smith NK. Role of nitric oxide in maintaining vascular integrity in endotoxin-induced acute intestinal damage in the rat. Br J Pharmacol. 1990;101:815–20.
437. Ochoa JB, Udekwu AO, Billiar TR. Nitrogen oxide levels in patients after trauma and during sepsis. Ann Surg. 1991;214:621–6.
438. Titheradge MA. Nitric oxide in septic shock. Biochim Biophys Acta. 1999;1411:437–55.
439. Taylor BS, Geller DA. Molecular regulation of the human inducible nitric oxide synthase (iNOS) gene. Shock. 2000;13:413–24.
440. Ikeda U, Kurosaki K, Ohya K. Adenosine stimulates nitric oxide synthase in vascular smooth muscle cells. Cardiovasc Res. 1997;35:168–74.
441. Hollenberg SM, Cinel I. Bench-to-bedside review: nitric oxide in critical illness-update 2008. Crit Care. 2009;13:218.
442. Mosi R, Seguin B, Cameron B. Mechanistic studies on AMD6221: a ruthenium- based nitric oxide scavenger. Biochem Biophys Res Commun. 2002;29:519–29.
443. Chandra A, Enkhbaatar P, Nakano Y. Sepsis: emerging role of nitric oxide and selections. Clinics. 2006;61:71–6.
444. Matsuda N, Hattori Y. Vascular biology in sepsis: pathophysiological and therapeutic significance of vascular dysfunction. J Smooth Muscle Res. 2007;43:117–37.
445. Ignarro LJ, Harbison RG, Wood KS. Activation of purified soluble guanylate cyclase by endothelium-derived relaxing factor from intrapulmonary artery and vein: stimulation by acetylcholine, bradykinin, and arachidonic acid. J Pharmacol Exp Ther. 1986;237:893–900.
446. Jones KA, Wong GY, Jankowski CJ. cGMP modulation of Ca sensitivity in airway smooth muscle. Am J Physiol Lung Cell Mol Physiol. 1999;276:L35–40.
447. Collins P, Griffith TM, Henderson AH. Endothelium-derived relaxing factor alters calcium fluxes in the rabbit aorta: a cyclic guanosine monophosphate-mediated effect. J Physiol. 1986;381:427–37.
448. Cornwell TL, Przywansky KB, Wyatt TA. Regulation of sarcoplasmatic reticulum protein phosphorylation by localized cyclic GMP-dependent protein kinase in vascular smooth muscle cells. Mol Pharmacol. 1991;40:923–31.
449. Jackson WE. Ion channels and vascular tone. Hypertension. 2000;35:173–8.
450. Davies NW. Modulation of ATP-sensitive K-channels in skeletal muscle by intracellular protons. Nature. 1990;343:375–7.
451. Quayle JM, Nelson MT, Standen NB. ATP-sensitive and inwardly rectifying potassium channels in smooth muscle. Physiol Rev. 1997;77:1165–232.
452. Buckley J, Singer M, Clapp L. Role of KATP channels in sepsis. Cardiovasc Res. 2006;72:220–30.
453. Keung EC, Li Q. Lactate activates ATP-sensitive potassium channels in guinea pig ventricular myocytes. J Clin Invest. 1991;88:1772–7.
454. Nelson MT, Patlak JB, Worley JF. Calcium channels, potassium channels, and voltage dependence of arterial smooth muscle tone. Am J Physiol. 1990;259:C3–18.
455. Nelson MT, Quayle JM. Physiological roles and properties of potassium channels in arterial smooth muscle. Am J Physiol. 1995;268:C799–822.
456. Geisen K, Vegh A, Krause E. Cardiovascular effects of conventional sulfonylureas and glimepiride. Horm Metab Res. 1996;28:496–507.
457. Gardiner SM, Kemp PA, March JE. Regional hemodynamic responses to infusion of lipopolysaccharide in conscious rats: effects of pre- or post-treatment with glibenclamide. Br J Pharmacol. 1999;128:1772–8.
458. Murphy ME, Brayden JE. Nitric oxide hyperpolarizes rabbit mesenteric arteries via ATP-sensitive potassium channels. J Physiol. 1995;486:47–58.
459. Archer SL, Huang JMC, Hampl V. Nitric oxide and cGMP cause vasorelaxation by activation of a charybdotoxin-sensitive K channel by cGMP-dependent protein kinase. Proc Natl Acad Sci U S A. 1994;91:7583–7.

460. Warrillow S, Egi M, Bellomo R. Randomized, double-blind, placebo-controlled crossover pilot study of a potassium channel blocker in patients with septic shock. Crit Care Med. 2006;34:980–5.
461. D`Emmanuele di Villa d Bianca R, Lippolis R, Autore G. Dexamethasone improves vascular hyporeactivity induced by LPS in vivo by modulation ATP-sensitive potassium channel activity. Br J Pharmacol. 2003;140:91–6.
462. Japiassu AM, Salluh JI, Bozza PT. Revisiting steroid treatment for septic shock: molecular actions and clinical effects- a review. Mem Inst Oswaldo Cruz. 2009;104:531–48.
463. Russell JA, Walley KR, Singer J. Vasopressin versus norepinephrine infusion in patients with septic shock. N Engl J Med. 2008;358:877–87.
464. Baker CH, Sutton ET, Zhou Z. Microvascular vasopressin effects during endotoxin shock in the rat. Circ Shock. 1990;30:81–95.
465. Mentzelopoulos SD, Zakynthinos SG, Tzoufi M. Vasopressin, epinephrine, and corticosteroids or in-hospital cardiac arrest. Arch Intern Med. 2009;169:15–24.
466. Holmes CL, Patel BM, Russell JA. Physiology of vasopressin relevant to the management of septic shock. Chest. 2001;120:989–1002.
467. Holmes CL, Landry DW, Grandon J. Science review: vasopressin and the cardiovascular system part 1 – receptor physiology. Crit Care. 2003;7:427–34.
468. Wilson MF, Brackett DJ, Tompkins P. Elevated plasma vasopressin concentrations during endotoxin and E. coli shock. Adv Shock Res. 1981;6:15–26.
469. Morales D, Madigan J, Cullinane S. Reversal by vasopressin of intractable hypotension in the late phase of hemorrhagic shock. Circulation. 1999;100:226–9.
470. Errington ML, Rocha e Silva Jr M. Vasopressin clearance and secretion during hemorrhage in normal dogs and in dogs with experimental diabetes insipidus. J Physiol. 1972;227:395–418.
471. Landry DW, Levin HR, Gallant EM. Vasopressin deficiency contributes to vasodilation of septic shock. Circulation. 1997;95:1122–5.
472. Holmes CL, Landry DW, Grandon J. Science review: vasopressin and the cardiovascular system part 2 – clinical physiology. Crit Care. 2004;8:15–23.
473. Reid IA. Role of nitric oxide in the regulation of renin and vasopressin secretion. Neuroendocrinology. 1994;15:351–83.
474. Day TA, Randle JC, Renaud LP. Opposing alpha- and beta- adrenergic mechanisms mediate dose-dependent actions of noradrenaline on supraoptic vasopressin neurones in vivo. Brain Res. 1985;358:171–9.
475. Sharshar T, Carlier R, Blanchard A. Depletion of neurophyseal content of vasopressin in septic shock. Crit Care Med. 2002;30:497–500.
476. Trzeciak S, Rivers EP. Clinical manifestations of disordered microcirculatory perfusion in severe sepsis. Crit Care. 2005;9 Suppl 4:S20–6.
477. Hack CE, Zeerleder S. The endothelium in sepsis: source of and a target for inflammation. Crit Care Med. 2001;29(7 Suppl):S21–7.
478. Bezemer R, Bartels SA, Bakker J, Ince C. Clinical review: clinical imaging of the sublingual microcirculation in the critically ill – where do we stand? Crit Care. 2012;16:224.
479. Ferrer M, Encabo A, Conde MV. Heterogeneity of endothelium-dependent mechanisms in different rabbit arteries. J Vasc Res. 1995;32:339–246.
480. Thorin E, Shatos MA, Shreeve SM. Human vascular endothelium heterogeneity. A comprehensive study of cerebral and peripheral cultured vascular endothelial cells. Stroke. 1997;28:375–81.
481. Hill CE, Phillips JK, Sandow SL. Heterogeneous control of blood flow amongst different vascular beds. Med Res Rev. 2001;21:1–60.
482. Ttero RM, Nguyen HB, Huang DT. Early goal-directed therapy in severe sepsis and septic shock revisited: concepts, controversies, and contemporary findings. Chest. 2006;130:1579–95.
483. Legrand M, Payen D. Understanding urine output in critically ill patients. Ann Intensive Care. 2011;1:13–20.
484. Bellomo R, Di Giantomasso D. Noradrenaline and the kidney: friends or foes? Crit Care. 2001;5:294–8.

485. Fink MP, Fiallo V, Stein KL. Systemic and regional hemodynamic changes after intraperitoneal endotoxin in rabbits: development of a new model of the clinical syndrome of hyperdynamic sepsis. Circ Shock. 1987;22:73–81.
486. Siegemund M, van Bommel J, Schwarte LA. Inducible nitric oxide synthase inhibition improves intestinal microcirculatory oxygenation and CO2 balance during endotoxemia in pigs. Intensive Care Med. 2005;31:985–92.
487. Nelson DP, Samsel RW, Wood LD. Pathological supply dependence of systemic and intestinal O_2 uptake during endotoxemia. J Appl Physiol. 1988;64:2410–9.
488. Brealey D, Brand M, Hargreaves I. Association between mitochondrial dysfunction and severity and outcome of septic shock. Lancet. 2002;360:219–23.
489. McCoy J, Trzeciak S, Parrillo J. Improved organ function at 24 hours is associated with increased microcirculatory flow during the early resuscitation of patients with sepsis. Acad Emerg Med. 2007;14(5 Suppl 1):S10–1.
490. Doerschug KC, Delsing AS, Schmidt GA. Impairments in microvascular reactivity are related to organ failure in human sepsis. Am J Physiol Heart Circ Physiol. 2007;293:H1065–71.
491. Hunter JD, Doddi M. Sepsis and the heart. Br J Anaesth. 2010;104:3–11.
492. Abraham E, Shoemaker WC, Bland RD. Sequential cardiorespiratory patterns in septic shock. Crit Care Med. 1983;11:799–803.
493. Jhanji S, Vivan-Smith A, Lucena-Amaro S. Hemodynamic optimisation improves tissue microvascular flow and oxygenation after major surgery: a randomised controlled trial. Crit Care. 2010;14:R151.
494. Wilson RF, Sarver EJ, LeBlanc PL. Factors affecting hemodynamics in clinical shock with sepsis. Ann Surg. 1971;174:939–43.
495. Vieillard-Baron A. Septic cardiomyopathy. Ann Intensive Care. 2011;1:6.
496. Landesberg G, Gilon D, Meroz Y. Diastolic dysfunction and mortality in severe sepsis and septic shock. Eur Heart J. 2012;33:895–903.
497. Merx MW, Weber C. Sepsis and the heart. Circulation. 2007;116:793–802.
498. Vieillard-Baron A, Schmitt JM, Beauchet A. Early preload adaptation in septic shock? A transesophageal echocardiographic study. Anesthesiology. 2001;94:400–6.
499. Ellrodt AG, Riedinger MS, Kimchi A. Left ventricular performance in septic shock: reversible segmental and global abnormalities. Am Heart J. 1985;110:402–9.
500. Vieillard-Baron A, Caille V, Charron C. Actual incidence of global left ventricular hypokinesia in adult septic shock. Crit Care Med. 2008;36:1701–6.
501. Etchecopar-Chevreuil C, Francois B, Clavel M. Cardiac morphological and functional changes during early septic shock: a transesophageal echocardiographic study. Intensive Care Med. 2008;34:250–6.
502. Parker M, Shelamer J, Bacharach S. Profound but reversible myocardial depression in patients with septic shock. Ann Intern Med. 1984;100:483–90.
503. Bouhemad B, Nicolas-Robin A, Arbelot C. Acute left ventricular dilatation and shock-induced myocardial dysfunction. Crit Care Med. 2009;37:441–7.
504. Jardin F, Brun-Ney D, Auvert B. Sepsis-related cardiogenic shock. Crit Care Med. 1990;18:1055–60.
505. Monsalve F, Rucabado L, Salvador A. Myocardial depression in septic shock caused by meningococcal infection. Crit Care Med. 1984;12:1021–3.
506. Price S, Anning PB, Mitchell JA. Myocardial dysfunction in sepsis: mechanisms and therapeutic implications. Eur Heart J. 1999;20:715–24.
507. Poelaert J, Declerck C, Vogelaers D. Left ventricular systolic and diastolic function in septic shock. Intensive Care Med. 1997;23:553–60.
508. Munt B, Gin K, et al. Diastolic filling in human severe sepsis: an echocardiographic study. Crit Care Med. 1998;26:1829–33.
509. Pirracchio R, Cholley B, De HS. Diastolic heart failure in anaesthesia and critical care. Br J Anaesth. 2007;98:701–21.
510. Opie LH. Mechanisms of cardiac contraction and relaxation. In: Braunwald J, editor. Heart disease. 7th ed. Philadelphia: WB Saunders Company; 2005. p. 457.

511. Zile MR, Baicu CF, Gaasch WH. Diastolic heart failure- abnormalities in active relaxation and passive stiffness of the left ventricle. N Engl J Med. 2004;350:1953–9.
512. Jafri SM, Lavine S, Field BE. Left ventricular diastolic function in sepsis. Crit Care Med. 1990;18:709–14.
513. Barraud D, Faivre V, Damy T. Levosimendan restores both systolic and diastolic cardiac performance in lipopolysaccharide-treated rabbits: comparison with dobutamine and milrinone. Crit Care Med. 2007;35:1376–82.
514. Pilz G, McGinn P, Boekstegers P. Pseudomonas sepsis does not cause more severe cardiovascular dysfunction in patients than non-Pseudomonas sepsis. Circ Shock. 1994;42:174–82.
515. Reilly JM, Cunnion RE, Burch-Whitman C. A circulating myocardial depressant substance is associated with cardiac dysfunction and peripheral hypoperfusion (lactic acidemia) in patients with septic shock. Chest. 1989;95:1072–80.
516. Tavener SA, Kubes P. Is there a role for cardiomyocyte toll-like receptor 4 in endotoxemia? Trends Cardiovasc Med. 2005;15:153–7.
517. Natanson C, Eichenholz PW, Danner RL. Endotoxin and tumor necrosis factor challenges in dogs stimulate the cardiovascular profile of human septic shock. J Exp Med. 1989;169:823–32.
518. Hollenberg SM, Cunnion RE, Lawerence M. Tumor necrosis factor depress myocardial cell function: results using an in vitro assay of myocyte performance. Clin Res. 1989;37:528–34.
519. Kumar A, Thota V, Dee L. Tumor necrosis factor – alpha and interleukin – 1-beta are responsible for depression of in-vitro myocardial cell contractility induced by serum from human septic shock. J Exp Med. 1996;183:949–58.
520. Brady AJ, Poole-Wilson PA. Circulatory failure in septic shock Nitric oxide: too much of a good thing? Nitric oxide: too much of a good thing? Br Heart J. 1993;70:103–5.
521. Brady AJ, Warren JB, Poole-Wilson PA. Nitric oxide attenuates cardiac myocyte contraction. Am J Physiol. 1993;265:H176–82.
522. Rassaf T, Poll LW, Brouzos P. Positive effects of nitric oxide on left ventricular function in humans. Eur Heart J. 2006;27:1699–705.
523. Kawaguchi H, Shin WS, Wang Y. In vivo gene transfection of human endothelial cell nitric oxide synthase in cardiomyocytes causes apoptosis-like cell death. Identification using Sendai virus-coated liposomes. Circulation. 1997;95:2441–7.
524. Pacher P, Beckman JS, Liaudet L. Nitric oxide and peroxynitrite in health and disease. Physiol Rev. 2007;87:315–424.
525. Ferdinandy P, Danial H, Ambrus I. Peroxynitrite is a major contributor to cytokine-induced myocardial contractile failure. Circ Res. 2000;87:241–7.
526. Shepherd RE, Lang CH, McDonough KH. Myocardial adrenergic responsiveness after lethal and nonlethal doses of endotoxin. Am J Physiol. 1987;252:H410–6.
527. Flierl MA, Rittirsch D, Huber-Lang MS. Molecular events in the cardiomyopathy of sepsis. Mol Med. 2008;14:327–36.
528. Rudiger A, Singer M. Mechanisms of sepsis-induced cardiac dysfunction. Crit Care Med. 2007;35:1599–608.
529. Levy RJ, Piel DA, Acton PD. Evidence of myocardial hibernation in the septic heart. Crit Care Med. 2005;33:2752–6.
530. Cariou A, Pinsky MR, Monchi M. Is myocardial adrenergic responsiveness depressed in human septic shock ? Intensive Care Med. 2008;34:917–22.
531. Silverman HJ, Penaranda R, Orens JB. Impaired beta-adrenergic receptor stimulation of cyclic adenosine monophosphate in human septic shock: association with myocardial hyporesponsiveness to catecholamines. Crit Care Med. 1993;21:31–9.
532. Cunnion RE, Schaer GL, Parker MM. The coronary circulation in human septic shock. Circulation. 1986;76:637–44.
533. Dhainaut JF, Huyghebaert MF, Monsallier JF. Coronary hemodynamics and myocardial metabolism of lactate, free fatty acids, glucose, and ketones in patients with septic shock. Circulation. 1987;75:533–41.
534. Solomon MA, Correa R, Alexander HR. Myocardial energy metabolism and morphology in a canine model of sepsis. Am J Physiol. 1994;266:H757–68.

535. Hoffmann R. Tissue Doppler echocardiography: already of clinical significance. Z Kardiol. 2002;91:677–84.
536. Parker MM, McCarthy KE, Ognibene FP. Right ventricular dysfunction and dilatation, similar to left ventricular changes, characterize the cardiac depression of septic shock in humans. Chest. 1990;97:126–31.
537. Vincent J-L, Reuse C, Frank N. Right ventricular dysfunction in septic shock: assessment by measurements of right ventricular ejection fraction using thermodilution technique. Acta Anaesthesiol Scand. 1989;33:34–8.
538. Kimichi A, Ellrodt AG, Berman DS. Right ventricular performance in septic shock: a combined radionuclide and hemodynamic study. J Am Coll Cardiol. 1984;4:945–51.
539. Schneider AJ. Right ventricular performance in sepsis and septic shock. Neth J Med. 1988;33:187–204.
540. Jardin F, Fourme T, Page B. Persistent preload defect in severe sepsis despite fluid loading: a longitudinal echocardiographic study in patients with septic shock. Chest. 1999; 116:1354–9.
541. Ognibene FP, Parker MM, Natanson C. Depressed left ventricular performance. Response to volume infusion in patients with sepsis and septic shock. Chest. 1988;93:903–10.
542. Westermann D, Kasner M, Steendijk P. Role of left ventricular stiffness in heart failure with normal ejection fraction. Circulation. 2008;117:2051–60.
543. Schneider AJ, Teule GJ, Groeneveld AB. Biventricular performance during volume loading in patients with early septic shock, with emphasis on the right ventricle: a combined hemodynamic and radionuclide study. Am Heart J. 1988;116:103–12.
544. Marshal JC. Complexity, chaos, and incomprehensibility: parsing the biology of critical illness. Crit Care Med. 2000;28:2646.
545. Stein PK, Kleiger RE. Insights from the study of heart rate variability. Annu Rev Med. 1999;50:249–61.
546. Vieillard-Baron A, Prin S, Chergui K. Hemodynamic instability in sepsis. Bedside assessment by Doppler echocardiography. Am J Respir Crit Care Med. 2003;168: 1270–6.
547. Glower DD, Spratt JA, Snow ND. Linearity of the Frank-Starling relationship in the intact heart: the concept of preload recruitable stroke work. Circulation. 1985;71:994–1009.
548. Robotham JL, Takata M, Berman M. Ejection fraction revisited. Anesthesiology. 1991;74:172–83.
549. Fincke R, Hochman JS, Lowe AM. Cardiac power is the strongest hemodynamic correlate of mortality in cardiogenic shock: a report from the SHOCK Trail Registry. J Am Coll Cardiol. 2004;44:340–8.
550. Parker MM, Shelhamer JH, Natanson C. Serial cardiovascular variables in survivors and nonsurvivors of human septic shock: heart rate as an early predictor of prognosis. Crit Care Med. 1987;15:923–9.
551. Groeneveld AB, Nauta JJ, Thijs LG. Peripheral vascular resistance in septic shock: its relation to outcome. Intensive Care Med. 1988;14:141–7.
552. Mello PMV, Sharma VK, Dellinger RP. Shock overview. Semin Respir Crit Care Med. 2004;25:619–28.
553. Bajwa EK, Malhotra A, Thompson BT. Methods of monitoring shock. Semin Respir Crit Care Med. 2004;25:629–44.
554. Cowell C, Marx JA, Grayzel J. Initial evaluation and management of shock in adult trauma. UpToDate 2012. 2012 vols. Last topic updated: 18 May 2012, www.uptodate.com.
555. Trzeciak S, McCoy JV, Dellinger PR. Microcirculatory alterations in resuscitation and shock (MARS): early increases in microcirculatory perfusion during protocol – directed resuscitation are associated with reduced multi-organ failure at 24 h in patients with sepsis. Intensive Care Med. 2008;34:2210–7.
556. Rivers E, Nguyen B, Havstad S. Early-goal – directed therapy collaborative group: early goal-directed therapy in the treatment of severe sepsis and septic shock. N Engl J Med. 2001;345:1368–77.

557. Lin SM, Huang CD, Lin HC. A modified goal-directed protocol improves clinical outcomes in intensive care unit patients with septic shock: a randomised control trail. Shock. 2006;26:551–7.
558. Nohria A, Lewis E, Stevenson LW. Medical management of advanced heart failure. J Am Med Assoc. 2002;287:628–40.
559. Monnet X, Rienzo M, Osman D. Passive leg raising predicts fluid responsiveness in the critically ill. Crit Care Med. 2006;34:1402–7.
560. Joly HR, Weil MH. Temperature of the great toe as an indication of the severity of shock. Circulation. 1969;39:131–8.
561. Schriger DL, Baraff L. Defining normal capillary refill: variation with age, sex, and temperature. Ann Emerg Med. 1988;17:932–5.
562. Champion HR, Sacco WJ, Carnazzo AJ. Trauma score. Crit Care Med. 1981;9:672–6.
563. Marik PE, Monnet X, Teboul J-L. Hemodynamic parameters to guide fluid therapy. Ann Intensive Care. 2011;1:1. doi:10.1186/2110-5820-1-1.
564. Jimenez E. Shock. In: Civetta JM, Taylor RW, Kirby RR, editors. Critical care. 3rd ed. Philadelphia: Lippincott, Williams & Wilkins; 1997. p. 359–87.
565. Colwell, C, Marx, JA, Grayzel, J. Initial evaluation and management of shock in adult trauma. UpToDate. 18th May 2012; 2012.
566. Rady MY, Rivers EP, Nowak RM. Resuscitation of the critically ill in the ED: responses of blood pressure, heart rate, shock index, central venous oxygen saturation, and lactate. Am J Emerg Med. 1996;14:218–25.
567. Vincent JL, De Backer D. Oxygen uptake/ oxygen supply dependency: fact or fiction. Acta Anaesthesiol Scand Suppl. 1995;107:229–974.
568. Cohen ES, Hollenberg SM. Tissue oxygenation and sepsis. Crit Care Med. 2001;29:1479–80.
569. Donnino MW, Nguyen B, Jacobson G. Cryptic septic shock: a sub-analysis of early, goal-directed therapy. Chest. 2003;124:90S.
570. Puskarich MA, Trzeciak S, Shapiro NI. Outcome of patients undergoing early sepsis resuscitation for cryptic shock compared with overt shock. Resuscitation. 2011;82:1289–93.
571. Cho CS, Rothrock SG. Circulatory emergencies: shock. In: Baren JM, editor. Pediatric emergency medicine. Philadelphia: Saunders Elsevier; 2008. p. 78–94.
572. Trauma, Am College of Surgeons Committee of. Advanced trauma life support of doctors: student course manual. 7th ed. Chicago: Am College of Surgeons; 2004.
573. Parks JK, Elliott AC, Gentiletto LM. Systemic hypotension is a late marker of shock after trauma: a validation study of Advanced Trauma Life Support principles in a large national sample. Am J Surg. 2006;192:727.
574. Mizushina Y, Ueno M, Watanabe H. Discrepancy between heart rate and markers of hypoperfusion is a predictor of mortality in trauma patients. J Trauma. 2011;71:789.
575. Vayer JS, Henderson JV, Bellamy RF. Absence of tachycardiac response to shock in penetrating intraperitoneal injury. Am Emerg Med. 1988;17:227.
576. Hasdai D, Holmes DR, Califf RM. Cardiogenic shock complicating acute myocardial infarction: predictors of death. Am Heart J. 1999;138:21–31.
577. Rice IR, Kirkeby OJ. Effect of cerebral ischemia on the cerebrovascular and cardiovascular response to hemorrhage. Acta Neurochir (Wien). 1998;140:699–705.
578. Girard TD, Opal SM, Ely EW. Insights into severe sepsis in older patients: from epidemiology to evidence-based management. Clin Infect Dis. 2005;40:719–27.
579. Chauhan K, Schocken DD. Shock with jugular venous distension. Hosp Physician. 2005;41:17–24.
580. Castle SC, Norman DC, Yeh M. Fever response in elderly nursing home residents: are the older truly colder? J Am Geriatr Soc. 1991;39:853–7.
581. Lind K, Bucht E, Ljunghall S. Pronounced elevations in circulating calcitonin in critical care patients is related to the severity of illness and survival. Intensive Care Med. 1995;21:63–6.
582. Brunckhorst FM, Wegscheider K, Forycki ZF. Procalcitonin for early diagnosis and differentiation of SIRS, sepsis, severe sepsis, and septic shock. Intensive Care Med. 2000;26: S148–52.

583. Rady MY, Rivers EP, Martin GB. Continuous central venous oximetry and shock index in the emergency department. Am J Emerg Med. 1992;10:538–41.
584. Rady MY, Nightingale P, Little RA. Shock-index: a re-evaluation in acute circulatory failure. Resuscitation. 1992;23:227–34.
585. Vincent J-L. Shock. In: Kuhlen R, Moreno R, Ranieri M, Rhodes A, editors. 25 years of progress and innovation in intensive care medicine. Berlin: Medizinisch Wissenschaftliche Verlagsgesellschaft; 2007. p. 261.
586. De Backer D. Lactic acidosis. Intensive Care Med. 2003;29:699–702.
587. Matejovic M, Radermacher R, Fontaine E. Lactate in shock: a high-octane fuel for the heart? Intensive Care Med. 2007;33:406–8.
588. Connett RJ, Gayesky TE, Honig CR. Lactate efflux is unrelated to intracellular PO-2 in working red muscle in situ. J Appl Physiol. 1986;61:402–8.
589. Levy B, Gibot S, Frank P. Relation between muscle Na+K+ATPase activity and raised lactate concentrations in septic shock: a prospective study. Lancet. 2005;365:871–5.
590. Bundgaard H, Kjeldsen H, Krabbe KS. Endotoxemia stimulates skeletal muscle Na+-K+-ATPase and raises blood lactate under aerobic conditions in humans. Am J Physiol Heart Circ Physiol. 2003;284:H1028–34.
591. Bakker J, Coffernils M, Leon M. Blood lactate levels are superior to oxygen-derived variables in predicting outcome in human septic shock. Chest. 1991;99:956–62.
592. Vincent J-L, Dufaye P, Berre J. Serial lactate determinations during circulatory shock. Crit Care Med. 1983;11:449–51.
593. Vincent J-L, Roman A, Kahn RJ. Dobutamine administration in septic shock: addition to a standard protocol. Crit Care Med. 1990;18:689–93.
594. Waxman K, Nolan LS, Shoemaker WC. Sequential perioperative lactate determination: physiological and clinical implications. Crit Care Med. 1982;10:96–9.
595. Bakker J, Gris P, Coffernils M. Serial blood lactate levels can predict the development of multiple organ failure following septic shock. Am J Surg. 1996;171:221–6.
596. Antonelli M, Levy M, Andrews PJ. Hemodynamic monitoring in shock and implications for management: international consensus conference, Paris, France, 27–28 April 2006. Intensive Care Med. 2007;33:575–90.
597. Reinhart K, Kuhn HJ, Hertog C. Continuous central venous and pulmonary artery oxygen saturation monitoring in the critically ill. Intensive Care Med. 2004;30:1572–8.
598. Ladakis C, Myrianthefs P, Karabinis A. Central venous and mixed venous oxygen saturation in critically ill patients. Respiration. 2001;68:279–85.
599. Bloos F, Reinhart K. Venous oximetry. In: Pinsky MR, Brochard L, Mancebo J, editors. Applied physiology in intensive care medicine. Berlin/Heidelberg: Springer; 2006. p. 37.
600. Baele PL, McMichan JC, Marsh HM. Continuous monitoring of mixed venous oxygen saturation in critically ill patients. Anesth Analg. 1982;61:513–7.
601. Kasnitz P, Druger GL, Yorra F. Mixed venous oxygen tension and hyperlactatemia severe cardiopulmonary disease. JAMA. 1976;236:570.
602. Richard C, Thuilliez C, Pezzano M. Relationship between mixed venous saturation and cardiac index in patients with chronic congestive heart failure. Chest. 1989;95:1289.
603. Reinhardt K, Rudolph T, Bredle DL. Comparison of central-venous to mixed – venous oxygen saturation during changes in oxygen supply/demand. Chest. 1989;95:1216.
604. Reinhart K. Monitoring O_2 transport and tissue oxygenation in critically ill patients. In: Reinhart K, Eyrich K, editors. Clinical aspects of O_2 transport and tissue oxygenation. Berlin/Heidelberg/New York: Springer; 1989. p. 195–211.
605. Reinhart K, Bloos F. The value of venous oximetry. Curr Opin Crit Care. 2005;11:259–63.
606. Pope JV, Jones AE, Gaieski DF. Multicenter study of central venous oxygen saturation (ScvO-2) as a predictor of mortality in patients with sepsis. Ann Emerg Med. 2010;55:40–6.
607. Marx G, Reinhart K. Venous oximetry. Curr Opin Crit Care. 2006;12:63–8.
608. Textoris J, Fouché L, Wiramus S. High central venous oxygen saturation in the latter stages of septic shock is associated with increased mortality. Crit Care. 2011;15:R176.

609. Fink M. Cytopathic hypoxia in sepsis. Acta Anaesthesiol Scand Suppl. 1997;110:87–95.
610. Singer M. Mitochondrial function in sepsis: acute phase versus multiple organ failure. Crit Care Med. 2007;35(9 Suppl):S441–8.
611. Levy RJ, Deutschmann CS. Cytochrome c oxidase dysfunction in sepsis. Crit Care Med. 2007;35(9 Suppl):S468–75.
612. Varpula M, Tallgren M, Saukkonen K. Hemodynamic variables related to outcome in septic shock. Intensive Care Med. 2005;31:1066–71.
613. Shoemaker WC, Appel PL, Kram HB. Prospective trail of supranormal values of survivors as therapeutic goals in high-risk surgical patients. Chest. 1988;94:1176–86.
614. Hayes MA, Timmins AC, Yau EH. Elevation of systemic oxygen delivery in the treatment of critically ill patients. N Engl J Med. 1994;330:1717–22.
615. Yu M, Levy MM, Smith P. Effect of maximizing oxygen delivery on morbidity and mortality rates in critically ill patients: a prospective, randomized, controlled study. Crit Care Med. 1993;21:830–8.
616. Krafft P, Steltzer H, Hiesmayr M. Mixed venous oxygen saturation in critically ill septic shock patients. The role of defined events. Chest. 1993;103:900–6.
617. Harrois A, Duranteau J. Targeting the microcirculation in resuscitation of acutely unwell patients. Curr Opin Crit Care. 2011;17:303–7.
618. Scalea TM, Holman M, Fuortes M. Central venous blood oxygen saturation: an early, accurate measurement of volume during hemorrhage. J Trauma. 1988;88:725–32.
619. Scheinman MM, Brown MA, Rapaport E. Critical assessment of use of central venous oxygen saturation as a mirror of mixed venous oxygen in severely ill cardiac patients. Circulation. 1969;40:165–72.
620. Turnaoglu S, Tugrul M, Camci E. Clinical applicability of the substitution of mixed venous oxygen saturation with central venous oxygen saturation. J Cardiothorac Vasc Anesth. 2001;15:574–679.
621. Rivers E. Mixed vs central venous oxygen saturation may be not numerically equal, but both are still clinically useful. Chest. 2006;129:507–8.
622. Baratt-Boyes BG, Wood EH. The oxygen saturation of blood in the vena cavae, right heart chambers and pulmonary vessels of healthy subjects. J Lab Clin Med. 1957;50:93–106.
623. Walley KR. Use of central venous oxygen saturation to guide therapy. Am J Respir Crit Care Med. 2011;184:514–20.
624. Reinhart K, Bloss F. Central venous oxygen saturation ($ScvO_2$). In: Pinsky MR, Payen D, editors. Functional hemodynamic monitoring in: update in intensive care and emergency medicine, vol. 42. Heidelberg: Springer; 2005. p. 241–50.
625. Lobo SM, de Oliveira NE. Clinical review: what are the best hemodynamic targets for noncardiac surgical patients. Crit Care. 2013;17:210.
626. Marik PE, Bankov A. Sublingual capnometry versus traditional markers of tissue oxygenation in critically ill patients. Crit Care Med. 2003;31:818–22.
627. Nencioni A, Trzeciak S, Shapiro NI. The microcirculation as a diagnostic and therapeutic target. Intern Emerg Med. 2009;4:413–8.
628. Russell JA, Phang PT. The oxygen delivery/consumption controversy: approaches to management of critically ill. Am J Respir Crit Care Med. 1994;149:533–7.
629. Deitch EA. Multiple organ failure: pathophysiology and potential future therapy. Ann Surg. 1992;216:117–34.
630. Otero RM, Nguyen HB, Huang DT. Early goal- directed therapy in sever sepsis and septic shock revisited: concepts, controversies, and contemporary findings. Chest. 2006;130:1579–95.
631. Boerma EC, van den Spronk PE, Ince C. Relationship between sublingual and intestinal microcirculatory perfusion in patients with abdominal sepsis. Crit Care Med. 2007;35:1055–60.
632. Trzeciak S, Dellinger RP, Parrillo JE. Early microcirculatory perfusion derangements in patients with severe sepsis and septic shock. Ann Emerg Med. 2007;49:88–98.e1–2.
633. Drazenovic R, Samsel RW, Wylam ME. Regulation of perfused capillary density in canine intestinal mucosa during endotoxemia. J Appl Physiol. 1992;72:259–65.

634. Groner W, Winkelmann JW, Harris AG. Orthogonal polarization spectral imaging: a new method for study of microcirculation. Nat Med. 1995;5:1209–12.
635. Goedharr PT, Khalilzada M, Bezemer R. Sidestream dark field (SDF) imaging: a novel stroboscopic LED ring-based imaging modality for clinical assessment of the microcirculation. Opt Express. 2007;15:15101–14.
636. Liu H, Boas DA, Zhang Y. Determination of optical properties and blood oxygenation in tissues using continuous NIRS light. Phys Med Biol. 1995;40:1983–93.
637. Lima A, Bakker J. Noninvasive monitoring of peripheral perfusion. In: Pinsky MR, Brochard L, Mancebo J, editors. Applied physiology in intensive care medicine. Berlin/Heidelberg: Springer; 2006. p. 131–41.
638. Flewelling R. Noninvasive optical monitoring. In: Bronzino JD, editor. The biomedical engineering handbook. Berlin/Heidelberg/New York: Springer; 2000. p. 1–10.
639. Lindert J, Werner J, Redlin M. OPS imaging in human microcirculation: a short technical report. J Vasc Res. 2002;39:368–72.
640. De Backer D. Monitoring microcirculation: the next frontier. In: Kuhlen R, Moreno R, Ranieri M, Rhodes A, editors. 25 years of progress and innovation in intensive care medicine. Berlin: Medizinisch Wissenschaftliche Verlagsgesellschaft; 2007. p. 301–8.
641. Haslibeder W, Haisjackl M, Sparr H. Factors influencing transcutaneous oxygen and carbon dioxide measurements in adult intensive patients. Intensive Care Med. 1991;17:272–5.
642. Carter B, Hochmann M, Osborne A. A comparison of two transcutaneous monitors for the measurement of arterial PO_2 and PCO_2 in neonates. Anaesth Intensive Care. 1995;23:708–14.
643. Phan CQ, Tremper KK, Lee SE. Noninvasive monitoring of carbon dioxide: a comparison of the partial pressure of transcutaneous and end-tidal carbon dioxide with the partial pressure of arterial carbon dioxide. J Clin Monit. 1987;3:149–54.
644. Boda D, Muranyi L. Gastrotonometry: an aid to the control of ventilation during artificial respiration. Lancet. 1959;73:181–2.
645. Grum CM, Fiddian-Green RG, Pittenger GL. Adequacy of tissue oxygenation in intact dog intestine. J Appl Physiol. 1984;56:1065–9.
646. Marik PE. Gastric intramucosal pH. A better predictor of multiorgan dysfunction and death than oxygen-derived variables in patients with sepsis. Chest. 1993;104:225–9.
647. Doglio GR, Pusajo JF, Egurrola MA. Gastric mucosal pH as a prognostic index of mortality in critically ill patients. Crit Care Med. 1991;19:1037–40.
648. Maynard N, Bihari D, Beale R. Assessment of splanchnic oxygenation by gastric tonometry in patients with acute circulatory failure. JAMA. 1993;270:1203–10.
649. Mythen MG, Woolf R, Noone RB. Gastric mucosal tonometry: towards new methods and applications. Anasthesiol Intensivmed Notfallmed Schmerzther. 1998;33 Suppl 2:S85–90.
650. Heard SO. Gastric tonometry: the hemodynamic monitor of choice. Chest. 2003;123:469S–74474.
651. Walley KR, Friesen BP, Humer MF. Small bowel tonometry is more accurate than gastric tonometry in detecting gut ischemia. J Appl Physiol. 1998;85:1770–7.
652. Weil MH, Nakagawa Y, Tang W. Sublingual capnometry: a new noninvasive measurement for diagnosis and quantification of severity of circulatory shock. Crit Care Med. 1999;27: 1225–9.
653. Toledo MA, Createur J, Vincent JL. Tissue capnometry: does the answer lie under the tongue? Intensive Care Med. 2004;30:2157–65.
654. Povoas HP, Weil MH, Tang W. Comparisons between sublingual and gastric tonometry during hemorrhagic shock. Chest. 2000;118:1127–32.
655. Rackow EC, O`Neil P, Astiz ME. Sublingual capnometry and indexes of tissue perfusion in patients with circulatory failure. Chest. 2001;120:1633–8.
656. Marik PE. Sublingual capnometry: a clinical validation study. Chest. 2001;120:923–7.
657. Dubin A, Edul VSK, Ince C. Determinants of tissue pCO-2 in shock and sepsis: relationship to the microcirculation. In: Vincent JL, editor. Yearbook of intensive care and emergency medicine. Berlin/Heidelberg: Springer; 2009. p. 195–204.

658. Kivilaakso E, Ahonen J, Aronsen KF. Gastric blood flow, tissue gas tension and microvascular changes during hemorrhage-induced stress ulceration in the pig. Am J Surg. 1982;143:322–30.
659. Cerny V, Cvachovec K. Gastric tonometry and intramucosal pH – theoretical principles and clinical application. Physiol Res. 2000;49:289–97.
660. Chiara O, Pelosi P, Segala M. Mesenteric and renal oxygen transport during hemorrhage and reperfusion: evaluation of optimal goal for resuscitation. J Trauma. 2001;51:356–62.
661. Fink MP. Gastrointestinal mucosal injury in experimental models of shock, trauma, and sepsis. Crit Care Med. 1991;19:627–41.
662. Nelson DO, King CE, Dodd SL. Systemic and intestinal limits of O_2 extraction in the dog. J Appl Physiol. 1987;63:387–94.
663. Gutierrez G, Bismar H, Dantzker DR. Comparison of gastric intramucosal pH with measures of oxygen transport and consumption in critically ill patients. Crit Care Med. 1992;20:451–7.
664. Fiddian-Green RG. Studies in splanchnic ischemia and multiple organ failure. In: Marston A, Bulkley GB, Fiddian-Green RG, editors. Splanchnic ischemia and multi-organ failure. St. Louis: Mosby; 1989. p. 349–63.
665. Björck M, Hedberg B. Early detection of major complications after abdominal aortic surgery: predictive value of sigmoid colon and gastric intramucosal pH monitoring. Br J Surg. 1994;81:25–30.
666. Desai VS, Weil MH, Tang W. Hepatic, renal, and cerebral tissue hypercarbia during sepsis and shock in rats. J Lab Clin Med. 1995;125:456–61.
667. Kette F, Weil MH, Gazmuri RJ. Intramyocardial hypercarbic acidosis during cardiac arrest and resuscitation. Crit Care Med. 1993;21:901–6.
668. Xie J, Weil MH, Jin X. Sublingual capnometry for quantification of the severity of the septic shock. Shock. 1997;7:13–4.
669. Desai VS, Weil MH, Tang W. Gastric intramural pCO_2 during peritonitis and shock. Chest. 1993;104:1254–8.
670. Schlichtig R, Bowles S. Distinguishing between aerobic and anaerobic appearances of dissolved CO_2 in intestine during low flow. J Appl Physiol. 1994;76:2443–51.
671. Sato Y, Weil MH, Tang W. Tissue hypercarbic acidosis as a marker of acute circulatory failure (shock). Chest. 1998;114:263–74.
672. Fiddian-Green RG. Gastric intramucosal pH, tissue oxygenation and acid base balance. Br J Anaesth. 1995;74:591–606.
673. Johnson BA, Weil MH, Tang W. Mechanisms of myocardial hypercarbic acidosis during cardiac arrest. J Appl Physiol. 1995;78:1579–84.
674. Turner T, Weil MH, Yang L. Source of myocardial hypercarbia during cardiac arrest (abstract). Crit Care Med. 1994;22:A223.
675. Guttierez G, Brown SD. Gastrointestinal tonometry: a monitor of regional dysoxia. New Horiz. 1996;4:413–9.
676. Dubin A, Murias G, Estenssoro E. Intramucosal-arterial PCO-2 gap fails to reflect intestinal dysoxia in hypoxic hypoxia. Crit Care. 2002;6:514–20.
677. Dubin A, Estenssoro E, Murias G. Intramucosal-arterial PCO-2 gradient does not reflect intestinal dysoxia in anemic hypoxia. J Trauma. 2004;57:1211–7.
678. Vallet B, Teboul JL, Cain S. Venoarterial CO-2 difference during regional ischemic or hypoxic hypoxia. J Appl Physiol. 2000;89:1317–21.
679. Gutierrez G. A mathematical model of tissue -blood carbon dioxide exchange during hypoxia. Am J Respir Crit Care Med. 2004;169:525–33.
680. Bateman RM, Walley KR. Microvascular resuscitation as a therapeutic goal in severe sepsis. Crit Care. 2005;9 Suppl 4:S27–32.
681. Myburgh JA. Catecholamines for shock: the quest for high-quality evidence. Crit Care Resus. 2007;9:352–6.
682. Groeneveld ABJ, Polderman KH. Fluid resuscitation: the good, the bad, and the ugly. Crit Care Shock. 2005;8:52–4.
683. Moore FA, McKinley BA, Moore EE. The next generation in shock resuscitation. Lancet. 2004;363:1988–96.

684. Girbes ARJ, Groeneveld ABJ. Circulatory optimalization of the patient with or at risk for shock. Clin Intensive Care. 2000;11:77–88.
685. Vincent J-L, Gerlach H. Fluid resuscitation in severe sepsis and septic shock: an evidence – based review. Crit Care Med. 2004;32:S451–4.
686. McKendry M, McGloin H, Saberi D. Randomised controlled trail assessing the impact of a nurse delivered, flow monitored protocol for optimalization of circulatory status after cardiac surgery. BMJ. 2004;329:258–61.
687. Zarbock A, Ertmer C, Van Aken H. Current controversies in volume therapy in critically ill patients. Adv Anaesthesiol Crit Care. 2009;1:3–10.
688. Martin G. Fluid management in shock. Semin Respir Care Med. 2004;25:683–93.
689. Al-Khafaji AH, Webb AR. Fluid resuscitation. Con Educ Anaesth Crit Care Pain. 2004;4:127–31.
690. Finfer S, Liu B, Taylor C. Resuscitation fluid use in critically ill adults: an international cross-sectional study in 391 intensive care units. Crit Care. 2010;14:R185.
691. Barochia A, Cui X, Vitberg D. Bundled care for septic shock: an analysis of clinical trails. Crit Care Med. 2010;38:668–78.
692. Marik PE. Surviving sepsis: going beyond the guidelines. Ann Intensive Care. 2011;1:17.
693. Raghavan M, Marik PE. Management of sepsis during the early “golden hours”. J Emerg Med. 2006;31:185–99.
694. Leach RM, Treacher DF. The pulmonary physician in critical care, 2. oxygen delivery and consumption in the critically ill. Thorax. 2002;57:170–7.
695. Vincent J-L, DeBacker D. Oxygen transport – the oxygen delivery controversy. In: Pinsky MR, editor. Applied physiology in intensive care medicine. Heidelberg: Springer; 2009. p. 409.
696. Schumacker PT, Samsel RW. Analysis of oxygen delivery and uptake relationships in the Krogh tissue model. J Appl Physiol. 1989;67:1234–44.
697. Strunden MS, Heckel K, Goetz AE. Perioperative fluid and volume management: physiological basis, tools, and strategies. Annals Intensive Care. 2011;1:2.
698. Monnet X, Teboul J-L. Assessment of volume responsiveness during mechanical ventilation: recent years. Crit Care. 2013;17:217.
699. Boyd JH, Forbes J, Nakada TA. Fluid resuscitation in septic shock: a positive fluid balance and elevated central venous pressure are associated with increased mortality. Crit Care Med. 2011;39:259–65.
700. Wiedemann HP, Wheeler AP, Bernard GR. Comparison of two fluid-management strategies in acute lung injury. N Engl J Med. 2006;354:2564–75.
701. Zochios V, Wilkinson J. Assessment of intravascular fluid status and fluid responsiveness during mechanical ventilation in surgical and intensive care patients. J Intensive Care Soc. 2011;12:295–300.
702. Ognibene FP. Hemodynamic support during sepsis. Clin Chest Med. 1996;17:279–87.
703. Rackow EC, Falk JL, Fein IA. Fluid resuscitation in circulatory shock: a comparison of the cardiorespiratory effects of albumin, hetastarch, and saline solutions in patients with hypovolemic and septic shock. Crit Care Med. 1983;11:839–50.
704. Flancbaum L, Dick M, Dasta J. A dose-response study of phenylephrine in critically ill, septic surgical patients. Eur J Clin Pharmacol. 1997;51:461–5.
705. Haupt MT, Gilbert EM, Carlson RW. Fluid loading increases oxygen consumption in septic patients with lactic acidosis. Am Rev Respir Dis. 1985;131:912–6.
706. Levy MM, Macias WL, Russell JA. Failure to improve during the first day of therapy is predictive of 28-day mortality in severe sepsis. Chest. 2004;124(Suppl):120S.
707. Scharper J, Ahmed R, Schafer T. Volume therapy with colloid solutions preserves intestinal microvascular perfusion in endotoxaemia. Resuscitation. 2008;76:120–8.
708. Hunt SA, Baker DW, Marshall H. ACC/AHA guidelines for the evaluation and management of chronic heart failure in the adult: executive summary. J Am Coll Cardiol. 2001;38:2101–13.
709. Watts JA, Kline JA. Bench to bedside: the role of mitochondrial medicine in the pathogenesis and treatment of cellular injury. Acad Emerg Med. 2003;10:985–97.

710. Axler O, Tousignant C, Thompson CR. Small hemodynamic effect of typical rapid volume infusions in critically ill patients. Crit Care Med. 1997;25:965–70.
711. Schuller D, Mitchell JP, Calandrino FS. Fluid balance during pulmonary edema. Is fluid gain a marker or a cause of poor outcome? Chest. 1991;100:1068–75.
712. Goldstein SL, Somers MJ, Baum MA. Pediatric patients with multi-organ dysfunction syndrome receiving continuous renal replacement therapy. Kidney Int. 2005;67:653–8.
713. Brandstrup B, Tonnesen H, Beier-Holgersen R. Effects of intravenous fluid restriction on postoperative complications: comparison of two perioperative fluid regimen: a randomized assessor-blinded multicenter trail. Ann Surg. 2003;238:641–8.
714. Bouchard J, Soroko SB, Chertow GM. Fluid accumulation, survival and recovery of kidney function in critically ill patients with acute kidney injury. Kidney Int. 2009;76:422–7.
715. Anning PB, Finney SJ, Singh S. Fluids reverse the early lipopolysaccharide – induced albumin leakage in rodent mesenteric venules. Intensive Care Med. 2004;30:1944–9.
716. Walley KR, McDonald TE, Wang Y. Albumin resuscitation increases cardiomyocyte contractility and decreases nitric oxide synthase II expression in rat endotoxemia. Crit Care Med. 2003;31:187–94.
717. Silva E, De Backer D, Creteur J. Effects of fluid challenge on gastric mucosal PCO-2 in septic patients. Intensive Care Med. 2004;30:423–9.
718. Guarracino F, Baldassari R, Pinskc MR. Ventriculo-arterial decoupling in acutely altered hemodynamic states. Crit Care. 2013;17:213.
719. Cavazzoni SLZ, Dellinger RP. Hemodynamic optimalization of sepsis-induced tissue hypoperfusion. Crit Care. 2006;10 Suppl 3:S2.
720. Dubin A, Pozo MO, Casabella CA. Comparison of 6 % hydroxyethyl starch 130/0.4 and saline solution for resuscitation of the microcirculation during early goal-directed therapy of septic patients. J Crit Care. 2009;25:659. e1–8.
721. Guzman JA, Lacoma FJ, Kruse JA. Relationship between systemic oxygen supply dependency and gastric intramucosal PCO-2 during progressive hemorrhage. J Trauma. 1998;44:696–700.
722. DeBacker D, Creteur J, Noordally O. Does hepato-splanchnic VO_2/DO_2 dependency exist in critically ill septic patients? Am J Respir Crit Care Med. 1998;157(4 Pt 1):1219–25.
723. Dubin A, Pozo MO, Casabella CA. Increasing arterial blood pressure with norepinephrine does not improve microcirculatory blood flow: a prospective study. Crit Care. 2009;13:R92.
724. Rivers EP. Fluid management strategies in acute lung injury: liberal, conservative, or both? N Engl J Med. 2006;355:2564–75.
725. Levy MM, Macias WL, Vincent J-L. Early changes in organ function predict eventual survival in severe sepsis. Crit Care Med. 2005;33:2194–201.
726. Boyd O, Grounds RM, Bennett ED. A randomized clinical trail of the effect of deliberate perioperative increase of oxygen delivery on mortality in high-risk surgical patients. JAMA. 1993;270:2699–707.
727. Kern JW, Shoemaker WC. Meta-analysis of hemodynamic optimization in high-risk patients. Crit Care Med. 2002;30:1176–86.
728. Gattinoni L, Brazzi L, Pelosi P. A trail of goal-orientated hemodynamic therapy in critically ill patients. SvO_2 Collaborative Group. N Eng. J Med. 1995;333:1025–32.
729. Hotchkiss RS, Karl IE. Reevaluation of the role of cellular hypoxia and bioenergetic failure in sepsis. JAMA. 1992;267:1503–10.
730. Singer M, De Santis V, Vitale D. Multiorgan failure is an adaptive, endocrine-mediated, metabolic response to overwhelming systemic inflammation. Lancet. 2004;364:545–8.
731. Payen D, de Pont AC, Sakr Y. A positive fluid balance is associated with a worse outcome in patients with acute renal failure. Crit Care. 2008;12:R74.
732. Murphy CV, Schramm GE, Doherty JA. The importance of fluid management in acute lung injury secondary to septic shock. Chest. 2009;136:102–9.
733. Uchino S, Bellomo R, Morimatsu H. Pulmonary artery catheter versus pulse contour analysis: a prospective epidemiological study. Crit Care. 2006;10:R174.
734. Rosenberg AL, Dechert RE, Park PK. Review of a large clinical series: association of cumulative fluid balance on outcome in acute lung injury: a retrospective review of the ARDSnet tidal volume study cohort. J Intensive Care Med. 2009;24:35–46.

735. Alsous F, Khamiees M, DeGirolamo A. Negative fluid balance predicts survival in patients with septic shock: a retrospective pilot study. Chest. 2000;117:1749–54.
736. Hilton AK, Pellegrino VA, Scheinkestel CD. Avoiding common problems associated with intravenous fluid therapy. Med J Aust. 2008;189:509–13.
737. Correa TD, Vuda M, Takala J. Increasing mean arterial pressure in sepsis: effects on fluid balance, vasopressor load and renal function. Crit Care. 2013;17:R21.
738. Perel P, Roberts I, Pearson M. Colloids versus crystalloids for fluid resuscitation in critically ill patients. Cochrane Database Syst Rev. 2007;4:CD000567.
739. Sakka SG, Klein M, Reinhart K. Prognostic value of extravascular lung water in critically ill patients. Chest. 2002;122:2080–6.
740. Jozwiak M, Silva S, Persichini R. Extra-vascular lung water is an independent prognostic factor in patients with acute respiratory distress syndrome. Crit Care Med. 2013;41: 472–80.
741. Groeneveld AB. Vascular pharmacology of acute lung injury and acute respiratory distress syndrome. Vascul Pharmacol. 2002;39:247–56.
742. Groeneveld AB, Polderman KH. Acute lung injury, overhydration or both. Crit Care. 2005;9:136–7.
743. Gluecker T, Capasso P, Schnyder P. Clinical and radiologic features of pulmonary edema. Radiographics. 1999;19:1507–31.
744. de Almeida JP, Palomba H, Galas FR. Positive fluid balance is associated with reduced survival in critically ill patients with cancer. Acta Anaesthesiol Scand. 2012;56:712–7.
745. Bellomo R, Cass A, Cole L. An observational study fluid balance and patient outcomes in the Randomized Evaluation of Normal vs Augmented Level of Replacement Therapy trail. Crit Care Med. 2012;40:1753–60.
746. Ertmer C, Rehberg S, Van Aken H. Physiologische Grundlagen der perioperativen Flüssigkeitstherapie. Intensivmed up2date. 2009;4:9–20.
747. Michard F, Teboul JL. Predicting fluid responsiveness in ICU patients: a critical analysis of the evidence. Chest. 2002;121:2000–8.
748. Marik PE, Cavallazzi R, Vasu T. Dynamic changes in arterial waveform derived variables and fluid responsiveness in mechanically ventilated patients. A systematic review of literature. Crit Care Med. 2009;37:2642–7.
749. Michard F, Reuter DA. Assessing cardiac preload or fluid responsiveness? It depends on the question we want to answer. Intensive Care Med. 2003;29:1396.
750. Michard F, Lopes MR, Auler Jr JOC. Pulse pressure variation: beyond the fluid management of patients with shock. Crit Care. 2007;11:131.
751. Braunwald E, Sonnenblick EH, Ross J. Mechanisms of cardiac contraction and relaxation. In: Braunwald E, editor. Heart disease. Philadelphia: WB Saunders; 1988. p. 383–425.
752. Nixon JV, Murray RG, Leonhard PD. Effect of large variations in preload on left ventricular characteristics in normal subjects. Circulation. 1982;65:698–703.
753. Cohn JN, Franciosa JA. Vasodilator therapy of cardiac failure. N Engl J Med. 1977;297:27–31.
754. Marik PE, Baram M, Vahid B. Does the central venous pressure predict fluid responsiveness? A systematic review of the literature and the tale of seven mares. Chest. 2008;134:172–8.
755. McIntyre LA, Hebert PC, Fergusson D. A survey of Canadian intensivists' resuscitation practices in early septic shock. Crit Care. 2007;11:R74.
756. Monnet X, Letierce A, Hamzaoui O. Arterial pressure allows monitoring the changes in cardiac output induced by volume expansion but not by norepinehrine. Crit Care Med. 2011;39:1394–9.
757. Pierrakos C, Velissaris D, Scoletta S. Can changes in arterial pressure be used to detect changes incardiac index during fluid challenge in patients with septic shock? Intensive Care Med. 2012;38:422–8.
758. Kastrup M, Markewitz A, Spies C. Current practice of hemodynamic monitoring and vasopressor and inotropic therapy in post-operative cardiac surgery patients in Germany: results from a postal survey. Acta Anaesthesiol Scand. 2007;51:347–58.
759. Alzeer A, Arora S, Ansari A. Central venous pressure from common iliac vein reflects right atrial pressure. Can J Anaesth. 1998;45:798–801.

760. Hamilton DR, Dani RS, Semlacher RA. Right atrial and right ventricular transmural pressures in dogs and humans. Effects of the pericardium. Circulation. 1994;90:2492–500.
761. Mirsky MR, Payen D. Functional hemodynamic monitoring. Crit Care. 2005;9:566.
762. Madger S. Central venous pressure monitoring. Curr Opin Crit Care. 2006;12:219–27.
763. Weil MH, Henning RJ. New concepts in the diagnosis and fluid treatment of circulatory shock. Thirteenth annual Becton, Dickinson and Company Oscar Schwidetsky memorial Lecture. Anesth Analg. 1979;58:124–32.
764. Vincent JL, Weil MH. Fluid challenge revisited. Crit Care Med. 2006;34:1333–7.
765. Hicks P, Cooper S, Webb J. The surviving sepsis campaign: international guidelines for management of severe sepsis and septic shock: 2008. An assessment by the Australian and New Zealand Intensive Care Society. Anaesth Intensive Care. 2008;36:149–51.
766. Marik PE, Varon J. Early goal-directed therapy (EGDT): on terminal life support? Am J Emerg Med. 2010;28:243–5.
767. Perel A. Bench-to-bedside review: the initial hemodynamic resuscitation of the septic patient according to Surviving Sepsis Campaign guidelines – does one size fit all? Crit Care. 2008;12:223.
768. Michard F, Boussat S, Chemla D. Relation between respiratory changes in arterial pulse pressure and fluid responsiveness in septic patients with acute circulatory failure. Am J Respir Crit Care Med. 2000;162:134–8.
769. Reuter DA, Kirchner A, Felbinger TW. Usefulness of left ventricular stroke volume variation to assess fluid responsiveness in patients with reduced cardiac function. Crit Care Med. 2003;31:1399–404.
770. Berkenstadt H, Margalit N, Hadani M. Strike volume variation as a predictor of fluid responsiveness in patients undergoing brain surgery. Anesth Analg. 2001;92:984–9.
771. Feissel M, Michard F, Mangin I. Respiratory changes in aortic blood velocity as an indicator of fluid responsiveness in ventilated patients with septic shock. Chest. 2001;119:867–73.
772. Reuter DA, Felbinger TW, Schmidt C. Stroke volume variations for assessment of cardiac responsiveness to volume loading in mechanically ventilated patients after cardiac surgery. Intensive Care Med. 2002;28:392–8.
773. Heenen S, DeBacker D, Vincent JL. How can the response to volume expansion in patients with spontaneous respiratory movements be predicted? Crit Care. 2006;10:R102.
774. Soubrier S, Saulnier F, Hubert H. Can dynamic indicators help the prediction of fluid responsiveness in spontaneously breathing critically ill patients? Intensive Care Med. 2007;33:1117–24.
775. Perel A, Pizov R, Cotev S. Systolic Blood Pressure Variation is a Sensitive Indicator of Hypovolemia in Ventilated Dogs Subjected to Graded Hemorrhage. Anesthesiology. 1987;67:498–502.
776. Michard F, Teboul JL. Using heart lung interactions to assess fluid responsiveness during mechanical ventilation. Crit Care. 2000;2:282–9.
777. Theres H, Binkau J, Laule M. Phase-related changes in right ventricular cardiac output under volume-controlled mechanical ventilation with positive end-expiratory pressure. Crit Care. 1999;27:953–8.
778. Boulain T, Achard JM, Teboul JL. Changes in BP induced by passive leg raising predict response to fluid loading in critically ill patients. Chest. 2002;121:1245–52.
779. Preisman S, Kogan S, Berkenstadt H. Predicting fluid responsiveness in patients undergoing cardiac surgery: functional haemodynamic parameters including the Respiratory Systolic Variation Test and static preload indicators. Br J Anaesth. 2005;95:746–55.
780. Chemla D, Hebert JL, Coirault C. Total arterial compliance estimated by stroke volume-to-aortic pulse pressure ratio in humans. Am J Physiol. 1998;274:H500–5.
781. Hamilton WF, Remington W. The measurement of stroke volume from the pressure pulse. Am J Physiol. 1947;148:14–24.
782. Marx G, Cope T, McCrossan L. Assessing fluid responsiveness by stroke volume variation in mechanically ventilated patients with severe sepsis. Eur J Anaesthesiol. 2004;21:132–8.
783. Malbrain ML, de Laet I. Functional hemodynamics and increased intraabdominal pressure: same thresholds for different conditions …? Crit Care. 2009;37:781–3.

784. Vivier E, Metton O, Piriou V. Effects of increased intra-abdominal pressure on central circulation. Br J Anaesth. 2006;96:701–7.
785. Renner J, Gruenewald M, Quaden R. Influence of increased intra-abdominal pressure on fluid responsiveness predicted by pulse pressure variation and stroke volume variation in a porcine mode. Crit Care Med. 2009;37:650–8.
786. Jacques D, Bendjelid K, Duperret S. Pulse pressure variation and stroke volume variation during increased intra-abdominal pressure: an experimental study. Crit Care. 2011; 15:R33.
787. Candela D, Zoric L, deLa Coussaye JE. The influence of the airway driving pressure on pulsed pressure variation as a predictor of fluid responsiveness. Intensive Care Med. 2010;36:496–503.
788. DeBacker D, Heenen S, Piagnerelli M. Pulse pressure variation to predict fluid responsiveness: influence of tidal volume. Intensive Care Med. 2005;31:517–23.
789. Michard F. Changes in arterial pressure during mechanical ventilation. Anesthesiology. 2005;103:419–28.
790. Lafanechere A, Pene F, Goulenok C. Changes in aortic blood flow induced by passive leg raising predict fluid responsiveness in critically ill patients. Crit Care. 2006;10:R132.
791. Maizel J, Airapetian N, Lorne E. Diagnosis of central hypovolemia by using passive leg raising. Intensive Care Med. 2007;33:1133–8.
792. Cavallaro F, Sandroni C, Marano C. Diagnostic accuracy of passive leg raising for prediction of fluid responsiveness in adults: systematic review and meta-analysis of clinical studies. Intensive Care Med. 2010;36:1475–83.
793. Jabot J, Teboul JL, Richard C. Passive leg raising for predicting fluid responsiveness: importance of postural change. Intensive Care Med. 2009;35:85–90.
794. Monnet X, Osman D, Ridel C. Predicting volume responsiveness by using the end-expiratory occlusion in mechanically ventilated intensive care unit patients. Crit Care Med. 2009;37:951–6.
795. Monnet X, Teboul JL. Passive leg raising. Intensive Care Med. 2008;34:659–63.
796. Monnet X, Bleibtreu A, Ferre A. Passive leg raising and end-expiratory occlusion tests perform better than pulse pressure variation in patients with low respiratory system compliance. Crit Care Med. 2012;40:152–7.
797. Biais M, Vidil L, Sarrabay P. Changes in stroke volume induced by passive leg raising in spontaneously breathing patients: comparison between echocardiography and Vigileo/FloTrac device. Crit Care. 2009;13:R195.
798. Benomar B, Ouattara A, Estagnasie P. Fluid responsiveness predicted by noninvasive bioreactance-based passive raise test. Intensive Care Med. 2010;36:1875–81.
799. Thiel SW, Kollef MH, Isakow W. Non-invasive stroke volume measurement and passive leg raising predict volume responsiveness in medical ICU patients: an observational cohort study. Crit Care. 2009;13:R111.
800. Monge Gracia MI, Cano GA, Gracia Romero M. Non-invasive assessment of fluid responsiveness by changes in partial end-tidal CO-2 pressure during a passive leg – raising maneuver. Ann Intensive Care. 2012;2:9.
801. Schrier RW, Wang W. Acute renal failure and sepsis. N Engl J Med. 2004;351:159–69.
802. Tagami T, Kushimoto S, Yamamoto Y. Validation of extravascular lung water measurement by single transpulmonary thermodilution: human autopsy study. Crit Care. 2010;14:R162.
803. Fernandez-Mondejar E, Rivera-Fernandez R, Gracia-Delgado M. Small increases in extravascular lung water are accurately detected by transpulmonary thermodilution. J Trauma. 2005;59:1420–3.
804. Sturm JA. Development and significance of lung water measurement in clinical and experimental practice. In: Lewis FR, Pfeiffer UJ, editors. Practical applications of fibreoptics in critical care monitoring. Berlin/Heidelberg/New York: Springer; 1990. p. 129–39.
805. Katzenelson R, Perel A, Berkenstadt H. Accuracy of tranpulmonary thermodilution versus gravimetric measurement of extravascular lung water. Crit Care Med. 2004;32:1550–4.
806. Fernandez-Mondejar E, Gurrero-Lopez F, Colmenero M. How important is the measurement of extravascular lung water? Curr Opin Crit Care. 2007;13:79–83.

807. Berkowitz DM, Danai PA, Eaton S. Accurate characterization of extravascular lung water in acute respiratory distress syndrome. Crit Care Med. 2008;35:1803–9.
808. Monnet X, Anguel N, Osman D. Assessing pulmonary permeability by transpulmonary thermodilution allows differentiation of hydrostatic pulmonary edema from ALI/ARDS. Intensive Care Med. 2007;33:448–53.
809. Kuzkov W, Kirov MY, Sovershaev MA. Extravascular lung water determined with single transpulmonary thermodilution correlates with the severity of sepsis-induced acute lung injury. Crit Care Med. 2006;34:1647–53.
810. Eisenberg P, Hansbrough JR, Anderson D. A prospective study of lung water measurements during patient management in an intensive care unit. Am Rev Res Dis. 1987;136:662–8.
811. Chung FT, Lin SM, Lin SY, Lin HC. Impact of extravascular lung water index on outcomes of severe sepsis patients in a medical intensive care unit. Respir Med. 2008;102:956–61.
812. Phillips CR, Chesnutt MS, Smith SM. Extravascular lung water in sepsis-associated acute respiratory distress syndrome: indexing with predicted body weight improves correlation with severity of illness and survival. Crit Care Med. 2008;36:69–73.
813. Boussat S, Jacques T, Levy B. Intravascular volume monitoring and extravascular lung water in septic patients with pulmonary edema. Intensive Care Med. 2002;28:712–8.
814. Isakow W, Schuster DP. Extravascular lung water measurements and hemodynamic monitoring in the critically ill: bedside alternatives to the pulmonary artery catheter. Am J Physiol Lung Cell Mol Physiol. 2006;291:L1118–31.
815. Maharaj R. Extravascular lung water and acute lung injury. Cardiol Res Pract. 2012. doi:10.1155/2012/407035.
816. Brown LM, Liu KD, Matthay MA. Measurement of extravascular lung water using the single indicator method in patients: research and potential clinical value. Am J Physiol Lung Cell Mol Physiol. 2009;297:L547–58.
817. Grocott MP, Mythen MG, Gan TJ. Perioperative fluid management and clinical outcomes in adult. Anesth Analg. 2005;100:1093–106.
818. Hachenberg T, Grundling M. Acute failure of the intestinal barrier – pathophysiology, diagnosis, prophylaxis and therapy. Anaesthesiol Reanim. 1999;24:4–12.
819. Holte K, Sharrock NE, Kehlet H. Pathophysiology and clinical implications of perioperative fluid excess. Br J Anaesth. 2002;89:622–32.
820. Mayberry JC, Welker KJ, Goldman RK. Mechanisms of acute ascites formation after trauma resuscitation. Arch Surg. 2003;138:773–6.
821. Guyton AC. Interstitial fluid pressure. II. Pressure-volume curves of interstitial space. Circ Res. 1965;16:452–60.
822. Staub NC. The pathogenesis of pulmonary edema. Prog Cardiovasc Dis. 1980;23:53–80.
823. Balogh Z, McKinley BA, Cocanour CS. Supranormal trauma resuscitation causes more cases of abdominal compartment syndrome. Arch Surg. 2003;138:637–43.
824. Balogh Z, McKinley BA, Cocanour CS. Secondary abdominal compartment syndrome is an elusive early complication of traumatic shock resuscitation. Am J Surg. 2002;184:538–44.
825. Jacobs FM, Pinot G, Pilorge C. Intra abdominal pressure elevation and its link with renal failure in medical intensive care patients. Intensive Care Med. 2009;223(Suppl):35 S (Poster).
826. Astiz ME, Rackow EC. Crystalloid-colloid controversy revisited. Crit Care Med. 1999;27:34–5.
827. Dellinger RP, Levy MM, Rhodes A. Surviving sepsis campaign: international guidelines for the management of severe sepsis and septic shock: 2012. Crit Care Med. 2013;41:580–637.
828. Schierhout G, Roberts I. Fluid resuscitation with colloid or crystalloid solutions in critically ill patients: a systematic review of randomised trails. BMJ. 1998;316:1080–7.
829. Finfer S, Bellomo R, Boyce N. A comparison of albumin and saline for fluid resuscitation in the intensive care unit. N Engl J Med. 2004;350:2247–56.
830. Choi PT, Yip G, Quinonez LG. Crystalloids vs colloids in fluid resuscitation: a systemic review. Crit Care Med. 1999;27:200–10.
831. Perel P, Roberts I. Colloids versus crystalloids for fluid resuscitation in critically ill patients. Cochrane Database Syst Rev. 2011;(3):CD000567.

832. Jungheinrich C, Scharpf R, Wargenau M. The pharmacokinetics and tolerability on an intravenous infusion of the new hydroxyethyl starch 130/0.4 (6 %, 500 ml) in mild-to severe renal impairment. Anesth Analg. 2002;95:544–51.
833. Kato A, Yonemura K, Matsushima H. Complication of oliguric acute renal failure in patients treated with low -molecular weight dextran. Ren Fail. 2001;23:679–84.
834. Perner A, Haase N, Guttormsen AB, 6S Trail Group; Scandinavian Critical Care Trails Group. Hydroxyethyl starch 130/0.4 versus Ringer`s acetate in severe sepsis. N Engl J Med. 2012;367:124–34.
835. Kellum JA, Bellomo R, Kramer DJ. Etiology of metabolic acidosis during saline resuscitation in endotoxemia. Shock. 1998;9:364–8.
836. Wilkes NJ, Woolf R, Mutch M. The effect of balanced versus saline-based hetastarch and crystalloid solutions on acid–base and electrolyte status and gastric mucosal perfusion in elderly surgical patients. Anesth Analg. 2001;93:811–6.
837. Larsen R. Anästhesie. München: Urban & Fischer Elsevier GmbH; 2008.
838. Rossaint R, Werner C, Zwißler B. Die Anästhesiologie. Heidelberg: Springer; 2008.
839. Upadhyay M, Singhi S, Murlidharan J. Randomized evaluation of fluid resuscitation with crystalloid (saline) and colloid (polymer from degraded gelatin in saline) in pediatric septic shock. Indian Pediatr. 2005;42:223–31.
840. Pries AR, Kuebler WM. Normal endothelium. Handb Exp Pharmacol. 2006;1:1–40.
841. Pries AR, Secomb TW, Gaehtgens P. The endothelial surface layer. Pflugers Arch. 2002;440:653–66.
842. Adamson RH, Lenz JF, Zhang X. Oncotic pressures opposing filtration across non- fenestrated rat microvessels. J Physiol. 2004;557:889–907.
843. Rehm M, Orth V, Kreimeier U. Changes in blood volume during acute normovolemic hemodilution with 5% albumin or 6% hydroxethylstarch and intraoperative retransfusion. Anaesthesist. 2001;50:569–79.
844. Rehm M, Orth V, Kreimeier U. Changes in intravascular volume during acute normovolemic hemodilution and intraoperative retransfusion in patients with radical hysterectomy. Anesthesiology. 2000;92:657–64.
845. Chappell D, Jacob M, Hofmann-Kiefer K. Hydrocortisone preserves the vascular barrier by protecting the endothelial gylcocalyx. Anesthesiology. 2007;107:776–884.
846. Bruegger D, Jacob M, Rehm M. Atrial natriuretic peptide induces shedding of endothelial glycocalyx in coronary vascular bed of guinea pig hearts. Am J Physiol Heart Circ Physiol. 2005;289:H1993–9.
847. Cotter G, Felker M, Adams KF. The pathophysiology of acute heart failure: is it all about fluid accumulation? Am Heart J. 2008;155:9–18.
848. Metra M, Teerlink JR, Voors AA. Vasodilators in the treatment of acute heart failure: what we know, what we don't. Heart Fail Rev. 2009;15:299–307.
849. Kawaguchi M, Hay I, Fetics B. Combined ventricular systolic and arterial stiffening in patients with heart failure and preserved ejection fraction: implications for systolic and diastolic reserve limitations. Circulation. 2003;107:714–20.
850. Rabuel C, Mebazaa A. Septic shock: a heart story since the 1960s. Intensive Care Med. 2006;32:799–807.
851. DeBacker D, Biston P, Devriendt J. Comparison of dopamine and norepinephrine in the treatment of shock. N Engl J Med. 2010;362:779–89.
852. Dünser MW, Ruokonen E, Pettilä V. Association of arterial blood pressure and vasopressor load with septic shock mortality: a post hoc analysis of a multicenter trial. Crit Care. 2009;13:R181.
853. Sandroni C, Vacallaro F, Antonelli M. Is there still a place for vasopressors in the treatment of cardiac arrest? Crit Care. 2012;16:213.
854. Weil MH, Shubin H. Shock following acute myocardium infarction: current understanding of hemodynamic mechanisms. Prog Cardiovasc Dis. 1968;11:1–17.
855. Hollenberg SM. Recognition and treatment of cardiogenic shock. Semin Respir Crit Care Med. 2004;25:661–71.

856. Murakawa K, Kobayashi A. Effects of vasopressors on renal tissue gas tensions during hemorrhagic shock in dogs. Crit Care Med. 1988;16:789–92.
857. Silva S, Teboul JL. Defining the adequate arterial pressure target during septic shock: not a "micro" issue but the microcirculation can help. Crit Care. 2011;15:1004.
858. Jackson R, Reid JA, Thorburn J. Volume preloading is not essential to prevent spinal-induced hypotension at cesarean section. Br J Anaesth. 1995;75:262–5.
859. Pottecher S, Calvat H, Dupont J. Haemodynamic management of severe sepsis: recommendations of the French Intensive Care Societies (SFAR/SRLF) Consensus Conference, 13 October 2005, Paris, France. Crit Care. 2006;10:311.
860. Lefer AM. Properties of cardioinhibitory factors produced in shock. Fed Proc. 1978;37:2734–40.
861. Gupta S. Shock and hypotension in the newborn. s.l.: Medscape Sep 2012. http://emedicine.medscape.com/article/979128-overview.
862. Terborg C, Scgummer W, Albrecht M. Dysfunction of vasomotor reactivity in severe sepsis and septic shock. Intensive Care Med. 2001;27:1231–4.
863. Guyton AC, Hall JE. Textbook of medical physiology. Philadelphia: Saunders; 2005.
864. Stainsby WN. Local control of regional blood flow. Annu Rev Physiol. 1973;35:151–68.
865. Pinsky MR. Both perfusion pressure and flow are essential for adequate resuscitation. Sepsis. 2000;4:143–6.
866. Pinsky MR. Protocolized cardiovascular management based on ventricular-arterial coupling. In: Pinsky MR, Payen D, editors. Functional hemodynamic monitoring. Berlin: Springer; 2006. p. 381–95.
867. García MIM, Cano AG, Romero MG. Dynamic arterial elastance to predict arterial pressure response to volume loading in preload-dependent patients. Crit Care. 2011;15:R15.
868. Arnold RC, Shapiro NI, Jones AE. Multicenter Study of Early Lactate Clearance As A Determinant of Survival in Patients With Presumed Sepsis. Shock. 2009;32:35–9.
869. Biais M, Cottenceau V, Petit L. Impact of norepinephrine on the relationship between pleth variability index and pulse pressure variations in ICU adult patients. Crit Care. 2011;15:R168.
870. Landin L, Lorente JA, Renes E. Inhibition of nitric oxide synthesis improves the vasoconstrictive effect of noradrenaline in sepsis. Chest. 1994;106:250–6.
871. Grazia MIN, Cano AG, Romero MG. Dynamic arterial elastance to predict arterial pressure response to volume loading in pre-load – dependent patients. Crit Care. 2011;15:R15.
872. Avontuur JA, Bruining HA, Ince C. Nitric oxide causes dysfunction of coronary autoregulation in endotoxemic rats. Cardiovasc Res. 1997;35:368–76.
873. LeDoux D, Astiz ME, Carpati CM. Effects on perfusion pressure on tissue perfusion in septic shock. Crit Care Med. 2000;28:2729–32.
874. Bourgoin A, Leone M, Delmas A. Increasing mean arterial pressure in patients with septic shock: effects on oxygen variables and renal function. Crit Care Med. 2005;33:780–6.
875. Derudde S, Cheisson G, Mazoit JX. Renal arterial resistance in septic shock: effects of increasing mean arterial pressure with norepinephrine on the renal resistive index assessed with Doppler ultrasonography. Intensive Care Med. 2007;33:1557–62.
876. Vlahakes G, Turley K, Hoffman JI. The pathophysiology of failure in acute right ventricular hypertension: hemodynamic and biochemical correlations. Circulation. 1981;63:87–95.
877. Di Giantomasso D, May CN, Bellomo R. Norepinephrine and vital organ blood flow. Intensive Care Med. 2002;28:1804–9.
878. Perrson PB. Renal blood flow autoregulation in blood pressure control. Curr Opin Nephrol Hypertens. 2002;11:67–72.
879. Navar LG. Integrating multiple paracrine regulators of renal microvascular dynamics. Am J Physiol. 1998;274:F433–44.
880. Chan KH, Miller JC, Dearden MN. The effect of changes in cerebral perfusion pressure upon middle cerebral artery blood flow velocity and jugular bulb venous oxygen saturation after severe brain injury. J Neurosurg. 1992;77:55–61.
881. Juul N, Morris GF, Marshall SB. Intracranial hypertension and cerebral perfusion pressure: influence on neurological deterioration and outcome in severe head injury. J Neurosurg. 2000;92:1–6.

882. Pijls NHJ, Van Gelder B, Van der Voort P. Fractional flow reserve: a useful index to evaluate the influence of an epicardial coronary stenosis on myocardial blood flow. Circulation. 1995;92:3183–93.
883. Bourdarias JP. Coronary reserve: concept and physiological variations. Eur Heart J. 1995;16(Suppl I):2–6.
884. Loef BG, Epema AH, Smilde TD. Immediate postoperative renal function deterioration in cardiac surgical patients predicts in-hospital mortality and long-term survival. J Am Soc Nephrol. 2005;16:195–200.
885. Chertow GM, Burdick M, Honour M. Acute kidney injury, mortality, length of stay, and costs in hospitalized patients. J Am Soc Nephrol. 2005;16:3365–70.
886. Esson ML, Schrier RW. Diagnosis and Treatment of Acute Tubular Necrosis. Ann Intern Med. 2002;137:744–52.
887. Badin J, Boulain T, Ehrmann S. Relation between mean arterial pressure and renal function in the early phase of shock: a prospective, explorative cohort study. Crit Care. 2011;15:R135.
888. Benes J, Chvojka J, Sykora R. Searching for mechanisms that matter early septic acute kidney injury: an experimental study. Crit Care. 2011;15:R256.
889. Lehman L, Saeed M, Moody G. Hypotension as a risk factor for acute kidney injury in ICU patients. Comput Cardiol (2010). 2010;37:1095–8.
890. Georger JF, Hamzaoui O, Chaari A. Restoring arterial pressure with norepinephrine improves muscle tissue oxygenation assessed by near-infrared spectroscopy in severely hypotensive septic patients. Intensive Care Med. 2010;36:1882–9.
891. Boerma EC, van der Voort PH, Ince C. Sublingual microcirculatory flow is impaired by the vasopressin-analogue terlipressin in a patient with catecholamine-resistant septic shock. Acta Anaesthesiol Scand. 2005;49:782–8.
892. Maier S, Hasibeder WR, Hengl C. Effects of phenylephrine on the sublingual microcirculation during cardiopulmonary bypass. Br J Anaesth. 2009;102:485–91.
893. Vasu TS, Cavallazzi A, Hirani A. Norepinephrine or dopamine for septic shock: systematic review of randomized clinical trials. J Intensive Care Med. 2012;27:172–8.
894. De Backer D, Aldecoa C, Njimi H. Dopamine versus norepinephrine in the treatment of septic shock: a meta-analysis. Crit Care Med. 2012;40:725–30.
895. Treggiari MM, Romand D, Burgener PM. Effect of increasing norepinephrine dosage on regional blood flow in a porcine model of endotoxin shock. Crit Care Med. 2002;30:1334–9.
896. Schaer GL, Find MP, Parrillo JE. Norepinephrine alone versus norepinephrine plus low dose dopamine: enhanced renal blood flow with combination pressor therapy. Crit Care Med. 1985;13:492.
897. Redl-Wenzl EM, Armbruster C, Edelmann G. The effects of norepinephrine on hemodynamics and renal function in severe septic shock states. Intensive Care Med. 1993;19:151.
898. Marik PE, Mohedin M. The contrasting effects of dopamine and norepinephrine on systemic and splanchnic oxygen utilization in hyperdynamic sepsis. JAMA. 1994;272:1354.
899. Martin C, Econ B, Saux P. Renal effects of norepinephrine used to treat septic shock patients. Crit Care Med. 1990;18:282–5.
900. Desjars P, Pinaud M, Bugnon D. Norepinephrine therapy has no deleterious renal effects in human septic shock. Crit Care Med. 1990;18:1048–9.
901. Martin C, Papazian L, Perrin G. Norepinephrine or dopamine for the treatment of hyperdynamic septic shock? Chest. 1993;103:1826–31.
902. Duranteau J, Sitbon P, Teboul JL. Effects of epinephrine, norepinephrine, or the combination of norepinephrine and dobutamine on gastric mucosa in septic shock. Crit Care Med. 1999;28:893–900.
903. Levy B, Bollaert PE, Charpentier C. Comparison of norepinephrine and dobutamine to epinephrine for hemodynamics, lactate metabolism, and gastric tonometric variables in septic shock: a prospective, randomized study. Intensive Care Med. 1997;23:282–7.
904. Silverman HJ, Tuma P. Gastric tonometry in patients with sepsis. Effects of dobutamine infusions and packed redblood cell transfusions. Chest. 1992;102:184–8.
905. Van Thielen G, Price S. Ischemic cardiogenic shock. Anaesth Intensive Care Med. 2010;11:519–22.

906. Heyndrickx GR, Wijns W, Vogelaers D. Recovery of regional contractile function and oxidative metabolism in stunned myocardium induced by 1-hour circumflex coronary artery stenosis in chronically instrumented dogs. Circ Res. 1993;72:901–13.
907. Francis GS, Benedict C, Johnstone DE. Comparison of neuroendocrine activation in patients with left ventricular dysfunction with and without congestive heart failure. A substudy of the Studies of Left Ventricular Dysfunction (SOLVD). Circulation. 1990;82:1724–9.
908. den Uil CA, Lagrand WK, Valk SDA. Management of cardiogenic shock: focus on tissue perfusion. Curr Probl Cardiol. 2009;34:330–49.
909. Alpert JS, Becker RC. Cardiogenic shock: elements of etiology, diagnosis, and therapy. Clin Cardiol. 1993;16:182–90.
910. Boehmer JP, Popjes E. Cardiac failure: mechanical support strategies. Crit Care Med. 2006;34(Suppl):S268–77.
911. Cotter G, Moshkovitz Y, Kaluski E. The role of cardiac power and systemic vascular resistance in the pathophysiology and diagnosis of patients with acute congestive heart failure. Eur J Heart Fail. 2003;5:443–51.
912. Wildhirt SM, Dudek RR, Suzuki H. Involvement of inducible nitric oxide synthase in the inflammatory process of myocardial infarction. Int J Cardiol. 1995;50:253–61.
913. den Uil CA, Lagrand WK, Spronk PE. Impaired sublingual microvascular perfusion during surgery with cardiopulmonary bypass: a pilot study. J Thorac Cardiovasc Surg. 2008;136:129–34.
914. Reilly PM, Wilkins KB, Fuh KC, Haglund U, Bulkley GB. The mesenteric hemodynamic response to circulatory shock: an overview. Shock. 2001;5:329–43.
915. Hasper D, Hummel M, Kleber FX. Systemic inflammation in patients with heart failure. Eur Heart J. 1998;19:761–5.
916. Pfeffer MA, McMurray JL, Velazquez EJ. Valsartan, captopril, or both in myocardial infarction complicated by heart failure, left ventricular dysfunction, or both. N Engl J Med. 2003;349:1893–906.
917. Pitt B, Poole-Wilson PA, Segal R. Effect of losartan compared with captopril on mortality in patients with symptomatic heart failure: randomised trail: the Losartan Heart Failure Survival Study ELITE III. Lancet. 2000;355:1582–7.
918. Pitt B, White H, Nicolau J. Eplerone reduces mortality 30 days after randomization following acute myocardial infarction in patients with left ventricular systolic dysfunction and heart failure. J Am Coll Cardiol. 2005;46:425–31.
919. Mann HJ, Nolan PE. Update on the management of cardiogenic shock. Curr Opin Crit Care. 2006;12:431–6.
920. Fries M, Weil MH, Chang YT. Microcirculation during cardiac arrest and resuscitation. Crit Care Med. 2006;34:S454–7.
921. Fries M, Tang W, Chang YT. Microvascular blood flow during cardiopulmonary resuscitation is predictive of outcome. Resuscitation. 2006;71:248–53.
922. Erol-Yilmaz A, Atasever B, Mathura K. Cardiac resynchronization improves microcirculation. J Card Fail. 2007;13:95–9.
923. Kushner FG, Hand M, Smith Jr SC. 2009 focused updates: ACC/AHA guidelines for the management of patients with ST-elevation myocardial infarction (updating the 2004 guideline and 2007 focused update) and ACC/AHA/SCAI guidelines on percutaneous coronary intervention (updating the 2005 Guid.). Circulation. 2009;120:2271–306.
924. Shabana A, Moustafa M, El-Menyar A. Cardiogenic shock complicating myocardial infarction: an updated review. Br J Med Med Res. 2013;3:622–53.
925. Hocjman JS, Sleeper LA, Webb JG. Early revascularization and long-term survival in cardiogenic shock complicating acute myocardial infarction. JAMA. 2006;295:2511–5.
926. Urban P, Stauffer J, Bleed D. A randomized evaluation of early revascularization to treat shock complicating acute myocardial infarction. The Swiss Multicenter Trail of Angioplasty for Shock ((S)MASH). Eur Heart J. 1999;20:1030–8.
927. O'Gara PT, Kushner FG, Ascheim DD. 2013 ACCF/AHA guideline for the management of ST-elevation myocardial infarction: a report of the American College of Cardiology

Foundation/American Heart Association Task Force on Practice Guidelines. Circulation. 2013;2013(127):e362–425.
928. Steg PG, James SK, Atar D. ESC Guidelines for the management of acute myocardial infarction in patients presenting with ST-segment elevation. Eur Heart J. 2012;33:2569–619.
929. Dar MI, Manan AU, Rasheed B. Outcome of patients after coronary bypass grafting in cardiogenic shock. Ann Thorac Cardiovasc Surg. 2007;13:247–50.
930. Lemery R, Smith HC, Giuliani ER. Prognosis in rupture of the ventricular septum after acute myocardial infarction and role of early surgical intervention. Am J Cardiol. 1992;70:147–51.
931. Guidelines, ACC/AHA. Guidelines for the Evaluation and Management of Heart Failure. J Am Coll Cardiol. 1995;26:1376–98.
932. Habib G, Hoen B, Tornos P. Guidelines on the prevention, diagnosis, and treatment of infective endocarditis (new version. Eur Heart J. 2009;30:2369–413.
933. Nico HJ, Pijls B, Van Gelder P. Fractional flow reserve: a useful index to evaluate the influence of an epicardial coronary stenosis on myocardial blood flow. Circulation. 1995;92:3183–93.
934. Anbe DT, Armstrong PW, Bates ER. ACC/AHA guidelines for the management of patients with ST-elevation myocardial infarction . http:// www.cardiosource.com/guidelines/guidelines/stemi/index.pdf. [Online] ACC/AHA 15 Dec 2007.
935. McMurray JJ, Adamopoulos S, Anker SD. ESC guidelines for the diagnosis and treatment of acute and chronic heart failure. Eur Heart J. 2012;33:1787–847.
936. De Backer D, Creteur J, Dubois MJ. The effects of dobutamine on microcirculatory alterations in patients with septic shock are independent of its systemic effects. Crit Care Med. 2006;34:403–8.
937. Sonntag S, Sundberg S, Lehtonen LA. The calcium sentisizer levosimendan improves the function of stunned myocardium after percutaneous transluminal coronary angioplasty in acute myocardial ischemia. J Am Coll Cardiol. 2004;43:2177–82.
938. Morelli A, De Castro S, Teboul JL. Effects of levosimendan on right ventricular afterload in patients with acute respiratory distress syndrome: a pilot study. Crit Care Med. 2006;34:2287–93.
939. Yilmaz MB, Yalta K, Yontar C. Levosimendan improves renal function in patients with acute decompensated heart failure: comparison with dobutamine. Cardiovasc Drugs Ther. 2007;21:431–5.
940. Dominguez-Rodriguez A, Samimi-Fard S, Garcia-Gonzalez MJ. Effects of levosimendan versus dobutamine on left ventricular diastolic function in patients with cardiogenic shock after primary angioplasty. Intern J Cardiol. 2008;128:214–7.
941. Cleland J, McGowan J. Levosimendan: a new era for inodilator therapy for heart failure. Curr Opin Cardiol. 2002;17:257–65.
942. Delle Karth G, Buberl A, Geppert A. Hemodynamic effects of continuous infusion of levosimendan in critically ill patients with cardiogenic shock requiring catecholamines. Acta Anaesthesiol Scand. 2003;47:1251–6.
943. Nanas JN, Papazoglou PP, Terrovitis JV. Hemodynamic effects of levosimendan added to dobutamine in patients with decompensated advanced heart failure refractory to dobutamine alone. Am J Cardiol. 2004;94:1329–32.
944. Mebazaa A, Nieminen MS, Packer M. Levosimendan vs dobutamine for patients with acute decompensated heart failure: the SURVIVE randomized trail. JAMA. 2007;297:1883–91.
945. Silver M, Horton DP, Ghali JK. Effect of nesiritide versus dobutamine on short-term outcomes in the treatment of patients with acutely decompensated heart failure. J Am Coll Cardiol. 2002;39:798–803.
946. Felker GM, O`Connor CM. Inotropic therapy for heart failure: an evidence-based approach. Am Heart J. 2001;142:393–401.
947. Greenberg B, Borghi C, Perrone S. Pharmacotherapeutic approaches for decompensated heart failure: a role for the calcium sensitiser, levosimendan? Eur J Heart Fail. 2003;5:13–21.
948. Abraham WT, Adams KF, Fonarow GC. In-hospital mortality in patients with acute decompensated heart failure requiring intravenous vasoactive medications: an analysis from the

acute decompensated heart failure national register (ADHERE). J Am Coll Cardiol. 2005;46:57–64.
949. Ohman EM, George BS, White CJ. Use of aortic counterpulsation to improve sustained coronary artery patency during acute myocardial infarction: results of a randomized trail. Circulation. 1994;90:792–9.
950. Barron HV, Every NR, Parsons LS. The use of intra-aortic balloon counterpulsation in patients with cardiogenic shock complicating acute myocardial infarction: data from the National Registry of Myocardial Infarction 2. Am Heart J. 2001;141:933–9.
951. Bur A, Bayegan K, Holzer M. Intra-aortic balloon counterpulsation in the emergency department: a 7-year review and analysis of predictors of survival. Resuscitation. 2002;53:259–64.
952. Papaioannou T, Stefanadis C. Basic principles of the intraaortic balloon pump and mechanisms affecting its performance. ASAIO J. 2005;51:296–300.
953. Thiele H, Zeymer U, Neumann F-J. Intraaortic balloon support for myocardial infarction with cardiogenic shock. N Engl J Med. 2012;367:1287–96.
954. Kleber F. Intraaortic balloon support for cardiogenic shock. N Engl J Med. 2013;368:80–1.
955. Pang D, Keenan SP, Cook DJ. The effect of positive pressure airway support on mortality and the need for intubation in cardiogenic pulmonary edema: a systematic review. Chest. 1998;114:1185–92.
956. Sin DD, Logan AG, Fitzgerald FS. Effects of continuous positive airway pressure on cardiovascular outcomes in heart failure patients with and without Cheyne-Stokes respiration. Circulation. 2000;102:61–6.
957. Masip J, Roque M, Sanchez B. Noninvasive ventilation in acute cardiogenic pulmonary edema: systematic review and meta-analysis. JAMA. 2005;294:3124–30.
958. Gray A, Goodacre S, Newby DE. Noninvasive ventilation in acute cardiogenic pulmonary edema. N Engl J Med. 2008;359:142–51.
959. Tebbe U, Schellong SM, Haas S. Certoparin versus unfractionated heparin to prevent venous thromboembolic events in patients hospitalized because of heart failure: a subgroup analysis of the randomized, controlled CERTIFY study. Am Heart J. 2011;161:322–8.
960. Alikhan R, Cohen AT, Combe S. Prevention of venous thromboembolism in medical patients with enoxaparin: a subgroup analysis of the MEDENOX study. Blood Coagul Fibrinolysis. 2003;14:341–6.
961. Cooper DJ, Walley KR, Wiggs BR. Bicarbonate does not improve hemodynamics in critically ill patients who have lactic acidosis: a prospective, controlled clinical trail. Ann Intern Med. 1990;112:492–8.
962. Shapiro JI. Functional and metabolic responses of isolated hearts to acidosis: effects of sodium bicarbonate and Carbicarb. Am J Physiol Heart Circ Physiol. 1990;258:H1835–9.
963. Nahas GG, Sutin KM, Fermon C. Guidelines for the treatment of acidaemia with THAM. Drugs. 1998;55:191–224.
964. Forsythe SM, Schmidt GA. Sodium bicarbonate for the treatment of lactic acidosis. Chest. 2000;117:260–7.
965. Robertson C, Steen P, Adgey J. The European Resuscitation Council guidelines for adult advanced life support: a statement from the Working Group on Advanced Life Support, and approved by the executive committee of the European Resuscitation Council. Resuscitation. 1998;37:81–90.
966. Weil MH, Wanchun T. Management of acidosis: the role of buffer agents. Crit Care. 1997;1:51–2.
967. Dickstein K, Cohen-Solal A, Filippatos GS. ESC guidelines for the diagnosis and treatment of acute and chronic heart failure 2008. Eur Heart J. 2008;29:2388–442.
968. HFSA. Comprehensive heart failure practice guideline. J Card Fail. 2006;12:e 1–2.
969. den Uil CA, Lagrand WK, Spronk PE. Low-dose nitroglycerin improves microcirculation in hospitalized patients with acute heart failure. Eur J Heart Fail. 2009;11:386–90.
970. WHO. World Health Organisation: cause-specific mortality and morbidity (p. 59–71). http://www.who.int/whosis/whostat/EN WHS10 Full.pdf. [Online] 2010. [Cited: 26th Apr 2013.]

971. Cothren CC, Moore EE, Hedegaard HB. Epidemiology of urban trauma deaths: a comprehensive reassessment 10 years later. World J Surg. 2007;31:1507–11.
972. Geeraedts Jr LM, Kaasjager HA, van Vugt AB. Exsanguination in trauma: a review of diagnostics and treatment options. Injury. 2009;40:11–20.
973. Curry N, Hopewell S, Doree C. The acute management of trauma hemorrhage: a systematic review of randomised controlled trails. Crit Care. 2011;15:R92.
974. Hardaway RM. Monitoring of the patient in a state of shock. Surg Gynecol Obstet. 1979;148:339–52.
975. Fang X, Tang W, Sun S. Comparison of buccal microcirculation between septic and hemorrhagic shock. Crit Care Med. 2006;34:S447–53.
976. Schmidt BM, Rezende-Neto JB, Andrade MV. Permissive hypotension does not reduce regional organ perfusion compared to normotensive resuscitation: animal study with fluorescent microspheres. World J Emerg Surg. 2012;7 Suppl 1:S9.
977. Revell M, Greaves I, Porter K. Endpoints for fluid resuscitation in hemorrhagic shock. J Trauma. 2003;54(5 Suppl):S63–7.
978. Cotton BA, Reddy N, Hatch QM. Damage control resuscitation is associated with reduction in resuscitation volumes and improvement survival in 390 damage control laparotomy patients. Ann Surg. 2011;254:598–605.
979. Morrison CA, Carrick MM, Norman MA. Hypotensive resuscitation strategy reduces transfusion requirements and severe postoperative coagulopathy in trauma patients with hemorrhagic shock: preliminary results of a randomized controlled trail. J Trauma. 2011;70:652–63.
980. Burris D, Rhee P, Kaufmann C. Controlled resuscitation for uncontrolled hemorrhagic shock. J Trauma. 1999;46:216–23.
981. Leppaniemi A, Soltero R, Burris D. Fluid resuscitation in a model of uncontrolled hemorrhage: too much too early or too little too late. J Surg Res. 1996;63:413–9.
982. Li T, Zhu Y, Li L. Ideal permissive hypotension to resuscitate uncontrolled hemorrhagic shock and the tolerance time in rats. Anesthesiology. 2011;114:111–9.
983. Mapstone J, Roberts I, Evans P. Fluid resuscitation strategies: a systematic review of animal trails. J Trauma. 2003;55:571–89.
984. Sakles JC, Sena MJ, Knight DA. Effect of immediate fluid resuscitation on the rate, and duration of pulmonary vascular hemorrhage in a sheep model of penetration thoracic trauma. Ann Emerg Med. 1997;29:392–9.
985. Holmes JF, Sakles JC, Lewis G. Effects of delaying fluid resuscitation on an injury to the systemic arterial vasculature. Acad Emerg Med. 2002;9:267–74.
986. Rezende-Neto JB, Rizoli SB, Andrade MV. Permissive hypotension and desmopressin enhance clot formation. J Trauma. 2010;68:42–51.
987. Santry HP, Alam HB. Fluid resuscitation: past, present, and the future. Shock. 2010;33:229–41.
988. Lu YQ, Cai XJ, Gu LH. Experimental study of controlled fluid resuscitation in the treatment of severe and uncontrolled hemorrhagic shock. J Trauma. 2007;63:798–804.
989. Xiao N, Wang XC, Diao YF. Effect of initial fluid resuscitation on subsequent treatment in uncontrolled hemorrhagic shock in rats. Shock. 2004;21:276–80.
990. Shah KJ, Chiu WC, Scalea TM. Detrimental effects of rapid fluid resuscitation on hepatocellular function and survival after hemorrhagic shock. Shock. 2002;18:242–7.
991. Varela E, Cohn SM, Diaz I. Splanchnic perfusion during delayed, hypotensive, or aggressive fluid resuscitation from uncontrolled hemorrhage. Shock. 2003;20:476–80.
992. Cotton BA, Guy JS, Morris JA. The cellular metabolic and systemic consequences of aggressive fluid resuscitation strategy. Shock. 2006;26:115–21.
993. Hoyt DB, Dutton RP, Hauser CJ. Management of coagulopathy in the patients with multiple injuries: results from an international survey of clinical practice. J Trauma. 2008;65:755–65.
994. Surgeons, American College of. Advanced trauma life support (ATLS) for doctors. Chicago: Am College of Surgeons; 2008.
995. Kwan I, Brunn F, Roberts I. Timing and volume of fluid administration for patients with bleeding. Cochrane Database Sys Rev. 2003;(3):CD002245.

996. Meregalli A, Oliveira RP, Friedman G. Occult hypoperfusion is associated with increased mortality in hemodynamic stable high risk surgical patients. Crit Care. 2004;8:R60–5.
997. Nguyen HB, Rivers EP, Knoblich BP. Early lactate clearance is associated with improved outcome in severe sepsis and septic shock. Crit Care Med. 2004;32:1637–42.
998. Knoferl MW, Angele MK, Ayala A. Do different rates of fluid resuscitation adversely or beneficially influence immune responses after trauma-hemorrhage? J Trauma. 1999;46:23–33.
999. Santbanez-Gallerani AS, Barber AE, Williams SJ. Improved survival with early fluid resuscitation following hemorrhagic shock. World J Surg. 2001;25:592–7.
1000. Lier H, Krep H, Schroeder S. Preconditions of hemostasis in trauma: a review. the influence of acidosis, hypocalcemia, anemia, and hypothermia on functional hemostasis in trauma. J Trauma. 2008;65:951–60.
1001. Dutton RP, Mackenzie CF, Scalea TM. Hypotensive resuscitation during active hemorrhage: impact on in-hospital mortality. J Trauma. 2002;52:1141–6.
1002. Sondeen JL, Coppes VG, Holcomb JB. Blood pressure at which rebleeding occurs after resuscitation in swine with aortic injury. J Trauma. 2003;54(5 Suppl):S110–7.
1003. Butler F. Fluid resuscitation in tactical combat casualty care: brief history and current status. J Trauma. 2011;70:s11–2.
1004. Badjatia N, Carney N, Crocco TJ. Guidelines for prehospital management of traumatic brain injury, 2nd edition. Prehosp Emerg Care. 2008;12 Suppl 1:S1–52.
1005. Turner J, Nicholl J, Webber L. A randomised controlled trail of prehospital intravenous fluid replacement therapy in serious trauma. Health Technol Assess. 2000;4:1–57.
1006. Rossaint R, Bouillon B, Cerny V. Management of bleeding following major trauma, an updated European guideline. Crit Care. 2010;14:R52.
1007. Croce MA, Tolley EA, Claridge JA. Transfusions result in pulmonary morbidity and death after a moderate degree of injury. J Trauma. 2005;59:19–23. discussion 23–24.
1008. Chaiwat O, Lang JD, Vavilala MS. Early packed red blood cell transfusion and acute respiratory distress syndrome after trauma. Anesthesiology. 2009;110:351–60.
1009. Claridge JA, Sawyer RG, Schulman AM. Blood transfusions correlate with infections in trauma patients in a dose-dependent manner. Am Surg. 2002;68:566–72.
1010. Marik PE, Corwin HL. Efficacy of red blood cell transfusion in the critically ill: a systematic review of the literature. Crit Care Med. 2008;36:2667–74.
1011. Weinberg JA, McGwin Jr G, Marques MB. Transfusions in the less severely injured: does age of transfused blood affect outcome? J Trauma. 2008;65:794–8.
1012. Malone DL, Dunne J, Tracy JK. Blood transfusion, independent of shock severity, is associated with worse outcome in trauma. J Trauma. 2003;54:898–905.
1013. Charles A, Shaikh AA, Walters M. Blood transfusion is an independent predictor of mortality after blunt trauma. Am Surg. 2007;73:1–5.
1014. Robinson WP, Ahn J, Stiffler A. Blood transfusion is an independent predictor of increased mortality in nonoperatively managed blunt hepatic and splenic injuries. J Trauma. 2005;58:437–44. discussion 444–445.
1015. Hebert PC, Wells G, Blajchmann MA. A multicenter, randomized, controlled clinical trail of transfusion requirements in critical care. Transfusion Requirements in Critical Care Investigators, Canadian Critical Care Trails Group. N Engl J Med. 1999;340:409–17.
1016. Surgeons, Am College of. Advanced trauma life support for doctors. Chicago: American College of Surgeons Committee on Trauma; 1997.
1017. Klein HG, Spahn DR, Carson JL. Red blood cell transfusions in clinical practices. Lancet. 2007;370:415–26.
1018. Napolitano LM, Kurek S, Luchette FA. Clinical practice guideline: red blood cell transfusion in adult trauma and critical care. Crit Care Med. 2009;37:3124–57.
1019. Anesthesiologists, Am Society of. Practice guidelines for perioperative blood transfusion and adjuvant therapies: an updated report by the American Society of Anesthesiologists Task Force on Perioperative Blood Transfusion and Adjuvant Therapies. Anesthesiology. 2006;105:198–208.

1020. Shander A, Javidroozi M, Ozawa S, Hare GMT. What is really dangerous: anemia or transfusion? Br J Anaesth. 2011;107:i41–59.
1021. Rose AH, Kotze A, Doolan D. Massive transfusion – evaluation of current clinical practice and outcome in two large teaching hospital trusts in Northern England. Vox Sang. 2009;97:247–53.
1022. Gonzalez EA, Moore FA, Holcomb JB. Fresh frozen plasma should be given earlier to patients requiring massive transfusions. J Trauma. 2007;62:112–9.
1023. Shakur H, Roberts I, Bautista R. Effects of tranexamic acid on death, vascular occlusive events, and blood transfusion in trauma patients with significant haemorrhage (CRASH-2): a randomised, placebo-controlled trail. Lancet. 2010;376:23–32.
1024. Kumar A, Roberts D, Wood KE. Duration of hypotension before initiation of effective antimicrobial therapy is the critical determinant of survival in human septic shock. Crit Care Med. 2006;34:1589–96.
1025. Kreger BE, Craven DE, McCabe WR. Gram-negative bacteremia. Am J Med. 1980;68:344–55.
1026. Bryan CS, Reynolds KL, Brenner ER. Analysis of 1,186 episodes of gram-negative bacteremia in non-university hospitals: the effects of antimicrobial therapy. Rev Infect Dis. 1983;5:629–38.
1027. Ferrer R, Fraser VJ, Kollef MH. Effectiveness of treatments for severe sepsis: a prospective, multicenter, observational study. Am J Respir Crit Care Med. 2009;180:861–6.
1028. Barie PS, Hydo LJ, Shou J. Influence of antibiotic therapy on mortality of critical surgical illness caused or complicated by infection. Surg Infect (Larchmt). 2005;6:41–54.
1029. Castellanos-Ortega A, Suberviola B, Gracia-Astudillo LA. Impact of Surviving Sepsis Canpaign protocols on hospital length of stay and mortality in septic shock patients: results of a three-year follow-up quasi-experimental study. Crit Care Med. 2010;38:1036–43.
1030. El Solh AA, Akinnusi ME, Alsawalha LN. Outcome of septic shock in older adults after implementation of the sepsis "bundle". J Am Geriatr Soc. 2008;56:272–8.
1031. Gurani PK, Patel GP, Crank CW. Impact of the implementation of a sepsis protocol for the management of fluid-refractory septic shock: a single-center, before-and-after study. Clin Ther. 2010;32:1285–93.
1032. Russell JA. Management of sepsis. N Engl J Med. 2006;355:1699–713.
1033. Ibrahim EH, Sherman G, Ward S. The influence of inadequate antimicrobial treatment of bloodstream infections on patient outcome in the ICU setting. Chest. 2000;118:146–55.
1034. Kollef MH, Sherman G, Ward S. Inadequate antimicrobial treatment of infections: a risk factor for hospital mortality among critically ill patients. Chest. 1999;115:462–74.
1035. Harbarth S, Garbino J, Pugin J. Inappropriate initial antimicrobial therapy and its effect on survival in a clinical trail of immunomodulating therapy for severe sepsis. Am J Med. 2003;115:529–35.
1036. Leone M, Bourgoin A, Cambon S. Empirical antimicrobial therapy of septic shock patients: adequacy and impact on the outcome. Crit Care Med. 2003;31:462–7.
1037. Houk PM, Bratzler DW, Nsa W. Timing of antibiotic administration and outcomes for Medicare patients hospitalized with community- acquired pneumonia. Arch Intern Med. 2004;164:637–44.
1038. Kumar A, Kethireddy S. Emerging concepts in optimizing antimicrobial therapy of septic shock: speed is life but a hammer helps too. Crit Care. 2013;17:104.
1039. Diaz-Martin A, Martinez-Gonzalez ML, Ferrer R. Antibiotic prescription patterns in the empirical therapy of severe sepsis: combination of antimicrobials with different mechanisms of action reduce mortality. Crit Care. 2012;16:R223.
1040. Kumar A, Zarychanski R, Light B. Cooperative Antimicrobial Therapy of Septic Shock (CATSS) Database Research Group: early combination antibiotic therapy yields improved survival compare with monotherapy in septic shock: a propensity – matched analysis. Crit Care Med. 2010;38:1773–85.
1041. Craig WA. Pharmacokinetic/pharmacodynamic parameters: rationale for antibacterial dosing of mice and men. Clin Infect Dis. 1998;16:1–12.

1042. Institute for Wealth Care Improvement. Campaign, sepsis. www.IHI.org. [Online].
1043. Marik PE. Assessment of intravascular volume: a comedy of errors. Crit Care Med. 2001;29:1635–6.
1044. Kontra JM. Evidence-based management of severe sepsis and septic shock. J Lancester Gen Hosp. 2000;1:39–46.
1045. Marik PE, Varon J. Sepsis: state of the art. Dis Mon. 2001;47:465–532.
1046. Bone RC. Immunologic dissonance: a continuing evolution in our understanding of the systemic inflammatory response syndrome (SIRS) and the multiple organ dysfunction syndrome (MODS). Ann Intern Med. 1996;125:680–7.
1047. Foland JA, Fortenberry JD, Warshaw BL. Fluid overload before continuous hemofiltration and survival in critically ill children: a retrospective analysis. Crit Care Med. 2004;32:1771–6.
1048. Bouchard J, Metha RL. Fluid balance issues in the critically ill patient. In: Costanzo MR, Bellomo R, Maisel AS, Ronco C, editors. Fluid overload: diagnosis and management, Contributions to nephrology, vol. 164. Basel: Karger; 2010. p. 69–78.
1049. Maerz L, Kaplan LJ. Abdominal compartment syndrome. Crit Care Med. 2008;36:S212–5.
1050. Nohria A, Hasselbad V, Stebbins A. Cardiorenal interactions: insights from the ESCAPE trail. J Am Coll Cardiol. 2008;51:1268–74.
1051. Andreucci M, Frederico S, Andreucci VE. Edema and acute renal failure. Semin Nephrol. 2001;21:251–6.
1052. Hollenberg SM, Ahrens TS, Annane D. Practice parameters for hemodynamic support of sepsis in adult patients: 2004 update. Crit Care Med. 2004;32:1928–48.
1053. Gattinoni L, Brazzi L, Pelosi P. A trail of goal-oriented hemodynamic therapy in critically ill patients. SvO2 Collaborative Group. N Engl J Med. 1995;333:1025–32.
1054. Haraden C. What is a bundle? [Online] Institute for Healthcare Improvement 27 Feb 2008. http://www.ihi.org/IHI/Topics/CriticalCare/IntensiveCare/ImprovementStories/WhatIsaBundle.htm.
1055. Safety, Joint Commission Perspectives on Patient. Raising the bar with bundles: improving the quality of care by improving the work environment. [Online] April 2006; 6:5–6. [Cited: 2 Sept 2009.] http://www.ingentaconnect.com/content/jcaho/jcpps/2006/0000006/00000004/art0003.
1056. De BD, Taccone FS, Radermacher P. Levosimendan in septic shock: another piece in the puzzle, but many pieces are still lacking. Intensive Care Med. 2007;33:403–5.
1057. Morelli A, De CS, Teboul JL. Effects of levosimendan on systemic and regional hemodynamics in septic myocardial depression. Intensive Care Med. 2005;31:638–44.
1058. Pinto BB, Rehberg S, Ertmer C. Role of levosimendan in sepsis and septic shock. Curr Opin Anaesthesiol. 2008;21:168–77.
1059. Annane D, Bellissant E, Sebille V. Impaired pressor sensitivity to noradrenaline in septic shock patients with and without impaired adrenal function reserve. Br J Clin Pharmacol. 1998;46:589–97.
1060. Ananne D, Sebille V, Charpentier C. Effect of treatment with low doses of hydrocortisone and fludrocortisone on mortality in patients with septic shock. JAMA. 2002;288:862–71.
1061. Schelling G, Kilger E, Kuprat G. Stress doses of hydrocortisone reverse hyperdynamic septic shock: a prospective, randomized, double-blind, single-center study. Crit Care Med. 1999;27:723–32.
1062. Bollaert P-E, Charpentier C, Levy B. Reversal of late septic shock with supraphysiologic doses of hydrocortisone. Crit Care Med. 1998;26:645–50.
1063. Annane D, Sébille V, Troché G. A 3-level prognostic classification in septic shock based on cortisol levels and cortisol response to corticotropin. JAMA. 2000;283:1038–45.
1064. Prigent H, Maxime V, Annane D. Science review: mechanisms of impaired adrenal function in sepsis and molecular actions of glucocorticoids. Crit Care. 2004;8:243–52.
1065. Sprung CL, Annane D, Keh D. Hydrocortisone therapy for patients with septic shock. N Engl J Med. 2008;358:111–24.

1066. Buechele GL, Silva E, Ospina-Tascon GA. Effects of hydrocortisone on microcirculatory alterations in patients with septic shock. Crit Care Med. 2009;37:1341–7.
1067. Marik PE. The hazards of blood transfusions. Br J Hosp Med. 2009;70:12–5.
1068. Ranieri VM, Thompson BT, Barie PS. Drotrecogin alfa (activated) in adults with septic shock. N Engl J Med. 2012;366:2055–64.
1069. Martí-Carvajal AJ, Solà J, Lathyris D. Human recombinant activated protein C for severe sepsis. [Online] Cochrane Database Syst Rev. [Cited: 13 Apr 2011]. doi: 10.1002/14651858. CD004388.pub4.
1070. Peake S, Webb S, Delaney A. Early goal-directed therapy of septic shock: we honestly remain skeptical. Crit Care Med. 2007;35:994.
1071. Fullbrook P, Mooney S. Care bundles in critical care: a practical approach to evidence-based practice. Nurs Crit Care. 2003;8:249–55.
1072. Pronovost PJ, Berenholtz SM, Ngo K. Developing and testing quality indicators in the intensive care unit. J Crit Care. 2003;18:145–55.
1073. DeLeon E. Improving sepsis care. [Online] [Cited: 17 Dec 2009] http://www.socalpatient-safety.org/materials/docs/02-03-2009/05-Improving%20Sepsis%20Care%20-%20 SCPSC%20pdf.pdf.
1074. Aboela SW, Stone PW, Larson EL. Effectiveness of bundled behavioral interventions to control healthcare-associated infections: a systematic review of the literature. J Hosp Infect. 2007;66:101–8.
1075. Eichacker PQ, Natanson C, Danner RL. Surviving sepsis – practice guidelines, marketing campaigns, and Eli Lilly. N Engl J Med. 2006;355:1640–2.
1076. Natanson C, Danner RJ, Reilly JM. Antibiotics versus cardiovascular support in a canine model of human septic shock. Am J Physiol. 1990;259:1440–H1447.
1077. Manthous CA, Hall JB, Kushner R. The effect of mechanical ventilation on oxygen consumption in critically ill patients. Am J Respir Crit Care Med. 1995;151: 210–4.
1078. Network, Acute Respiratory Distress Syndrome. Acute Respiratory Distress Syndrome Network: ventilation with lower tidal volumes as compared with traditional tidal volumes for acute lung injury and the acute respiratory distress syndrome. N Engl J Med. 2000;342:1301–8.
1079. Marini JJ, Gattinoni L. Ventilatory management of acute respiratory distress syndrome: a consensus of two. Crit Care Med. 2004;32:250–5.
1080. Putensen C, Theuerkauf N, Zinserling J. Meta-analysis: ventilation strategies and outcomes of the acute respiratory distress syndrome and acute lung injury. Ann Intern Med. 2009;151:566–76.
1081. Checkley W, Brower R, Korpak A. Acute Respiratory Distress Syndrome Network Investigators: effects of a clinical trail on mechanical ventilation practices in patients with acute lung injury. Am J Respir Crit Care Med. 2008;177:1215–22.
1082. Schultz MJ. Lung-protective mechanical ventilation with lower tidal volumes in patients not suffering from acute lung injury: a review of clinical studies. Med Sci Monit. 2008;14:RA22–6.
1083. Cade JF. High risk of the critically ill for venous thromboembolism. Crit Care Med. 1982;10:448–50.
1084. Hirsch DR, Ingenito EP, Goldhaber SZ. Prevalence of deep venous thrombosis among patients in medical intensive care. JAMA. 1995;274:335–7.
1085. Gardlund B. Randomized, controlled trail of low- dose heparin for prevention of fatal pulmonary embolism in patients with infectious disease: the Heparin Prophylaxis Study Group. Lancet. 1996;347:1357–61.
1086. Samama MM, Cohen AT, Darmon JY. A comparison of enoxaparin with placebo fro the prevention of venous thromboembolism in acutely ill medical patient. N Engl J Med. 1999;341:793–800.
1087. Geerts W, Cook D, Shelby R. Venous thromboembolism and its prevention in critical care. J Crit Care. 2002;17:95–104.

1088. Carr ME. Diabetes mellitus: a hypercoagulable state. J Diabetes Complications. 2001;15:50–6.
1089. Ortiz A, Ziyadeh FN, Neilson EG. Expression of apoptosis-regulatory genes in renal proximal tubular epithelial cells exposed to high ambient glucose and in diabetic kidneys. J Invest Med. 1997;45:50–6.
1090. van den Berghe G, Wouters P, Weekers F. Intensive insulin therapy in the critically ill patients. N Engl J Med. 2001;345:1359–67.
1091. van den Berghe G, Wilmer A, Hermans G. Intensive insulin therapy in the medical ICU. N Engl J Med. 2006;354:449–61.
1092. Wiener RS, Wiener DC, Larson RJ. Benefits and risks of tight glucose control in critically ill adults: a meta-analysis. JAMA. 2008;300:933–44.
1093. Investigators, The NICE-SUGAR Study. Hypoglycemia and risk of death in critically ill patients. N Engl J Med. 2012;367:1108–18.
1094. Finfer S, Chittock DR, Su SY, NICE-SUGAR Study Investigators. Intensive versus conventional glucose control in critically ill patients. N Engl J Med. 2009;360:1283–97.
1095. Marik PE, Bellomo R. Stress hyperglycemia: an essential survival response! Crit Care. 2013;17:305.
1096. Shephard PR, Kahn BB. Glucose transporters and insulin action – implications for insulin resistance and diabetes mellitus. N Engl J Med. 1994;341:248–57.
1097. Losser MR, Darmoisel C, Payen D. Bench-to-bedside review: glucose and stress conditions in the intensive care unit. Crit Care. 2010;14:231.
1098. Malfitano C, Alba Loureio T, Rodrigues B, et al. Hyperglycemia protects the heart after myocardial infarction: aspects of programmed cell survival and cell death. Eur J Heart Fail. 2010;12:659–67.
1099. Ma G, Al-Shabrawey M, Johnson JA. Protection against myocardial ischemia/reperfusion injury by short-term diabetes: enhancement of VEGF-formation, capillary density, and activation of cell survival signaling. Naunyn Schmiedebergs Arch Pharmacol. 2006;373:415–27.
1100. Wiederkehr M, Emmett M. Bicarbonate therapy in lactic acidosis. [Online] Up-to-Date. http://www.uptodate.com/contents/bicarbonate-therapy-in-lactic-acidosis.
1101. Boyd JH, Walley KR. Is there a role for sodium bicarbonate in treating lactic acidosis. Curr Opin Crit Care. 2008;14:379–83.
1102. Kraut JA, Kurtz I. Use of base in the treatment of acute severe organic acidosis by nephrologists and critical care physicians: results of an online survey. Clin Exp Nephrol. 2006;10:111–7.
1103. Mathieu D, Neviere R, Billard V. Effects of bicarbonate therapy on hemodynamics and tissue oxygenation in patients with lactic acidosis: a prospective, controlled clinical study. Crit Care Med. 1991;19:1352–6.
1104. Hoste EA, Colpaert K, Vanholder RC. Sodium bicarbonate versus THAM in ICU patients with mild metabolic acidosis. J Nephrol. 2005;18:303–7.
1105. Kraut JA, Madias NE. Treatment of acute metabolic acidosis: a pathophysiologic approach. Nat Rev Nephrol. 2012;8:589–601.
1106. Rhee KH, Toro LO, McDonald GG. Carbicarb, sodium bicarbonate, and sodium chloride in hypoxic lactic acidosis. Effect on arterial blood gases lactate concentrations, hemodynamic variables, and myocardial intracellular pH. Chest. 1993;104:913–8.
1107. Shoemaker WC, Appel PL, Kram HB. Comparison of hemodynamic and oxygen transport effects of dopamine and dobutamine in critically ill surgical patients. Chest. 1989;96:120–6.
1108. Boerema EC, Koopmans M, Konijn A. Effects of nitroglycerin on sublingual microcirculatory blood flow in patients with severe sepsis/septic shock after strict resuscitation protocol: a double-blind randomized placebo controlled trail. Crit Care Med. 2008; 38:93–100.
1109. Hataishi R, Zapol WM, Bloch KD. Inhaled nitric oxide does not reduce systemic vascular resistance in mice. Am J Physiol Heart Circ Physiol. 2006;290:H1826–9.

1110. Patel BM, Chittock DR, Russell JA. Beneficial effects of short-term vasopressin infusion during severe septic shock. Anesthesiology. 2002;96:576–82.
1111. Dünser MW, Mayr AJ, Ulmer H. Argenine vasopressin in advanced vasodilatory shock: A prospective, randomized, controlled study. Circulation. 2003;107:2313–19.
1112. Lauzier F, Levy B, Lamarre P. Vasopressin or norepinephrine in early hyperdynamic septic shock: A randomized clinical trail. Intensive Care Med. 2006;32:1782–89.
1113. van Haren EM, Rozendaal FW, van der Hoeven JG. The effect of vasopressin on gastric perfusion in catecholamine-dependent patients in septic shock. Chest. 2003;124:2256–60.
1114. Holmes CL, Walley KR, Chittock DR. The effects on vasopressin on hemodynamics and renal function in severe septic shock: a case series. Intensive Care Med. 2001;27:1416–21.

Chapter 4
Sepsis

Sepsis is defined as a "systemic illness caused by microbial invasion of normally sterile parts of the body" [1]. The original (traditional) definition consists of two components:

I. Evidence of a systemic inflammatory response syndrome (SIRS) as characterized by at least two of the following so-called SIRS criteria [2–4]:

   (a) Body temperature of either ≥38 °C or ≤36 °C
   (b) Leukocyte count (WBC) of either ≥12,000 cells/µL or ≤4,000 cells/µL, or >10 % immature cells
   (c) Tachycardia with a heart rate of ≥90 beats per minute
   (d) Tachypnoea with a respiratory rate of ≥20 breaths per minute or proof of hyperventilation by $PaCO_2$ ≤4.3 kPa (<32 mmHg) and

II. Evidence or suspicion of a microbial infection either by microbiological proof or by clinical criteria [4, 5]

Thus, sepsis depicts a clinically relevant systemic response to a presumed or confirmed not necessarily bacterial, but microbial infection [6, 7]. The reference to a microbial agent as the underlying cause differentiates sepsis from a variety of other, non-microbial aetiologies such as trauma, burns, haemorrhage, ischaemia, anaphylaxis and pancreatitis [1, 4, 8–10], which precipitate a similar systemic reaction and clinical syndrome referred to as SIRS [1, 3, 4]. SIRS represents a non-specific, coordinated set of basically physiological systemic measures largely attributed to and launched by the innate immune response [11–13] and the vascular endothelium [14–16] to a wide variety of potentially threatening stimuli causing an allied and comparable clinical picture independent of the deploying insult [1, 17]. Mediated essentially by cytokines, host-derived peptides released in conditions of stress, a widespread bundle of neurohormonal, metabolic, endocrine, and immunological responses with affiliated haemodynamic changes apply [18–20]. Sepsis is generally associated with cardiovascular abnormalities [21–25] potentially displaying tissue malperfusion [26–28].

W. Krüger, A.J. Ludman, *Core Knowledge in Critical Care Medicine*,
DOI 10.1007/978-3-642-54971-7_4, © Springer-Verlag Berlin Heidelberg 2014

In 2003, a revised definition by the North American and European Intensive Care Societies was published [3]. The physicians particularly addressed the diagnosis and the recognition of inflammatory response being a non-specific feature [29] and until then determined by just a few parameters as above [2] and not comprising the broad spectrum of possible signs and symptoms. As such, sepsis should be diagnosed if an infection is documented or suspected and some of the following apply:

General Variables

Fever (>38.3 °C)
Hypothermia (core temperature <36.0 °C)
Heart rate >90/min or >2 SD above the normal value
Tachypnoea (respiratory rate >20/min)
Altered mental status
Significant oedema or positive fluid balance (>20 mL/kg over 24 h)
Hyperglycaemia (plasma glucose >140 mg/dL or 7.7 mmol/L) in the absence of diabetes

Inflammatory Variables

Leukocytosis (WBC count >12,000 μL^{-1})
Leukopaenia (WBC count <4,000 μL^{-1})
Normal WBC with greater than 10 % immature forms
Plasma C-reactive protein >2 standard deviations (SD) above the normal value
Plasma procalcitonin >2 SD above the normal value

Haemodynamic Variables

Arterial hypotension as defined as a

- Systolic blood pressure (SBP) <90 mmHg or
- Mean arterial pressure (MAP) <70 mmHg, or
- SBP decrease >40 mmHg in adults or less than 2 SD below normal SBP for age

Organ Dysfunction Variables

Arterial hypoxaemia (PaO_2/FiO_2 <300)
Acute oliguria (urine output <0.5 mL/kg/h for at least 2 h despite adequate fluid resuscitation)
Creatinine increase >0.5 mg/dL or >44.2 µmol/L
Coagulation abnormalities (INR >1.5 or aPTT >60 s)
Ileus (absent bowl sounds)
Thrombocytopaenia (platelet count <100,000 μL^{-1})
Hyperbilirubinaemia (plasma total bilirubin >4 mg/dL or >70 µmol/L)

Tissue Perfusion Variables

Hyperlactataemia of >1 mmol/L
Decreased capillary refill or mottling

With this revision, although enlisting more possible clinical signs and symptoms to identify an inflammatory response/process, the definition has become more vague, as these features are quite unspecific for diagnosing sepsis. However, the diagnosis of sepsis has become more indicative [6]. Accordingly, **sepsis** is defined as a systemic inflammatory response syndrome in the setting of confirmed or suspected microbial infection [1, 3, 30]. Furthermore, **severe sepsis** is defined as sepsis plus sepsis-induced organ dysfunction or sepsis-induced tissue hypoperfusion. Sepsis-induced tissue hypoperfusion or organ dysfunction is determined by any of the following [1, 3, 30]:

Sepsis-induced hypotension:

- SBP <90 mmHg and/or
- MAP <70 mmHg and/or
- SBP drop of >40 mmHg or <2 SD below age-adopted normal BP in the absence of other causes

Lactate above upper limits of normal laboratory range
Urine output <0.5 mL/kg/h for more than 2 h despite adequate fluid resuscitation
Acute lung injury with PaO_2/FiO_2 <250 in the absence of pneumonia as the source of infection
Acute lung injury with PaO_2/FiO_2 <200 in the presence of pneumonia as the infection source
Creatinine >2.0 mg/dL (176.8 μmol/L)
Bilirubin >2 mg/dL (34.2 μmol/L)
Platelet count <100,000 μL
Coagulopathy (international normalized ratio >1.5)

Finally, **septic shock** applies in the setting of severe sepsis where the sepsis-related hypotension persists despite adequate volume resuscitation [1, 3, 30].

Thus, septic shock comprises sepsis-related, *fluid-unresponsive hypotension* (SBP ≤90 mmHg, MAP ≤65 mmHg) or the requirement of catecholamines to ensure a reasonable perfusion pressure (SBP >90 mmHg, MAP >65 mmHg) [26] *in the presence of malperfusion*, e.g. signs of hypoperfusion or organ dysfunction [6, 31–33]. To note, unfortunately, the definition of septic shock differs to a considerable degree in the literature [1, 3, 30, 31], with Annane [32] specifying septic shock as severe sepsis and one of the following:

- MAP <60 mmHg (<80 mmHg if hypertension is known) after 20–30 mL/kg starch or 40–60/kg mL normal saline or with a pulmonary capillary wedge pressure (PCWP) between 12 and 20 mmHg
- Need for dopamine >5 μg/kg/min or norepinephrine or epinephrine <0.25 μg/kg/min to maintain MAP >60 mmHg (>80 mmHg in hypertensive patients)

4.1 Epidemiology

The incidence of sepsis is approximately 750,000 patients per year in the United States or 3 cases per 1,000 persons per year [34]. In the elderly population with an average age of 63 years [35] to 68 years [34], a much higher incidence is noted at 26 cases per 1,000 persons per year [34, 35] with both sepsis incidence and prevalence increasing with age [34].

Sepsis accounts for 2 % of all hospital admissions [32], with the yearly incidence of sepsis on the rise by up to 9 % [36]. Nine percent of the septic patients progress to severe sepsis, the incidence of this has also increased by 1.7 % [37] in recent years (1993–2003) [37, 38] and a further 3 % of those will develop septic shock [39].

The overall mortality has not substantially changed over the past 20–30 years with reported rates varying from 18 to 50 % [36, 38, 40, 41]. A large observational US study found an overall hospital mortality of 28 % [34], and a more recent European study recognized 36 % [42]. A recent review of an international registry confirmed the 50 % upper limit of mortality in severe septic cases [43].

As mortality analysis from trials conducted between 1990 and 2000 show the overall outcome appears significantly improved since the implementation of the surviving sepsis campaign (SSC) to standardize care, with mortality rates declining from 44 to 35 % [44, 45]. Results published by Levy et al. in December 2012 evaluating the mortality of patients admitted to intensive care units in the Unites States and Europe during a period between 2005 and 2010 suffering from severe sepsis and septic shock are very encouraging. While the raw hospital mortality was 41 and 28 % in Europe and the United States, respectively, the adjusted mortality did not differ at 31 and 32 % in Europe and the United States, respectively [46].

Increasing severity correlates with increasing mortality; indeed, older data described a mortality of about 25–30 % in severe sepsis that may rise to 40–70 % in septic shock [1, 47, 48]. In the long run, survivors of sepsis and septic shock are shown to have a higher 6- and 12-month mortality rate and a significant lower health-related quality of life [49, 50]. Annane suggests up to 20 % of hospital survivors might die within their first year after discharge [51].

Martin et al. acknowledged in their large international register, including more than 11,000 patients from 37 countries, that the lung was the primary source of infection in 47 %, the abdomen in 23 % and the urinary tract in 8 % of cases [43]. Gram-positive infections were found in 57 % of all cases and Gram-negative infections occurred in 44 %, while fungal infections were confirmed in 11 % with some patients obviously suffering from mixed infections [43]. Relevant comorbidities included diabetes (23 %), chronic lung disease (16 %) and heart failure (14 %); impaired renal function was detected in 11 % [43]. Chest, abdominal, genitourinary tract and primary bloodstream infections are accountable for more than 80 % of all sepsis cases [36, 52, 53]. Martin earlier described that in the elderly, the respiratory tract is still the most common site of infection followed by the genitourinary system, while abdominal infections are less frequent [35]. The proportion of viruses and parasites is estimated at 2–4 %, but this may be underrated [52].

4.2 Pathophysiology and Pathogenesis of Sepsis

If microbial organisms like bacteria or fungi invade the human body or tissue, the competent residential immune cells initiate a local inflammatory response [12, 19]: The so-called innate immune system is employed. Key elements are comprised of tissue macrophages and their circulating dependent monocytes [18], other myeloid-derived cells, polymorphonuclear leukocytes and natural killer cells [18] and, in part, the endothelial cells [11, 54, 55]. The residential macrophages phagocytose the bacteria and activate a number of proinflammatory cytokines which provoke a response from the local innate immune system and the endothelium [13, 56, 57]. These cytokines influence the cells of their local microenvironment [56, 57] via cell signalling [58] and as such affect the endothelial cells too [56, 57].

The endothelial cells cover an area totalling between 4,000 and 7,000 m^2; their main functions include controlling vascular tone and blood flow by adjusting microperfusion according to the local metabolic requirements, balancing local proinflammatory and anti-inflammatory activity, allowing appropriate exchange of macromolecules and fluids between blood and tissues and maintaining blood rheology thus preventing thrombosis [59].

The endothelium plays a predominant role, as it appears to be a target and concomitantly a source of the inflammatory process [60]. The endothelial cells may be causally activated directly by the pathogens themselves [61]; commonly stimulation emerges from lipopolysaccharides (LPS) or other components of the bacterial wall via pattern recognition receptors localized on the endothelial cells' surface [62, 63]. Further to this they can be activated by a broad range of host-derived factors and substances such as cytokines (e.g. released by the phagocyting macrophages), other proinflammatory agents, complement, serine proteases, activated platelet factor and alerted leukocytes, even hyperglycemia or altered blood flow and oxygenation may trigger the endothelial cells and insofar are markedly involved and jointly responsible for the encouragement of the endothelium [11, 61–63]). Pathogen-associated molecular patterns allow to specifically recognize invading microbiological pathogens or components of microbiological material (e.g. LPS, fungal molecules, and viral material), called pathogen-associated molecular patterns, via pattern recognition receptors (e.g. toll-like receptors) that are expressed on the surface of innate immune competent cells (e.g. macrophages, dendritic cells, endothelial cells), a special property of those cells [54, 64–66]. In contrast but in addition as well, in the adaptive immune response pathway, antigen receptors may bind to all kind of antigens: bacterial, environmental or even of self-antigenic origin [64]. However, the cells of the innate immune system (including the endothelial cells with their distinctive properties) will immediately display their functions, e.g. prompt release of a variety of mediators, especially cytokines rather than after they have proliferated and thus offer a rapid response [64]. The interaction between cells with innate immune expertise and bacteria, bacterial and microbial components and the crosstalk between the elements of innate immunity result in local release of cytokines and a myriad of other mediators including tissue factor, platelet-activating factor, chemokines, arachidonic acid metabolites and nitric oxide (NO) [12, 18, 19, 67, 68] and will trigger specific inflammation and coagulation cascades [11]. The mediators released and engaged act in an autocrine and paracrine manner, further impelling the areal cells like tissue macrophages/monocytes and/or endothelial cells [11, 58]. In addition to the local process, circulating white blood cells are recruited by binding to the endothelial cells as cytokines stimulate synthesis of adhesion molecules [69] and then transferred through the vascular wall to the site of injury [12, 70].

Inflammation as a local and/or systemic process is basically a vascular response to any insult [14–16] with the endothelial cells playing a major role integrating the multifaceted signals arriving [71] and triggering, perpetuating and modulating the inflammatory process [11, 72, 73]. Inflammation is largely characterized by increased capillary leakiness (disturbed permeability resulting in tissue oedema), enhanced leukocyte adhesion, activated procoagulability and altered microhaemodynamics [74–77]. This activation of the endothelial cells with the aim of re-establishing homeostasis [59] has to be distinguished from endothelial cell dysfunction which describes an impaired response or even a failure to achieve this goal [59]. Increasing evidence indicates cellular injury rather than implications of cellular involvement in the process being the basic feature provoking dysfunction during sepsis [78, 79]. As such, endothelial cell activation and dysfunction are substantially involved in the pathology of sepsis [80, 81], and dysfunction engaged in microcirculatory impairment and organ failure [59]. Interestingly, the response of the endothelial cells (structural and/or functional alterations including nuclear vacuolization, cytoplasm swelling, changes in vascular tone or loss of barrier function) differs somewhat depending on the nature of the pathogen, age, gender, host genetics, underlying comorbidities and the location within the vascular bed [11, 82–86]. Furthermore, pathogens and the interactions of these pathogens may set in motion coagulation [87], contact and complement cascades [11]. Moreover, pathogens and the interactions between inflammatory and coagulation pathways may amplify the host's response and thus aggravate the clearance and the response processes [11, 88–90]. Accordingly, crosstalk between the pathways is imperative to coordinate the response and thus facilitate survival [12, 20].

As a matter of fact, even local invasion of bacteria or fungi results in a "stress response" [19, 91]; thus, the sympathetic nervous system, namely, the brain stem and hypothalamus which regulate the response to stress [92], will be aroused and, depending on the nature and severity of the harassment, may get markedly involved [19, 91]. Activation of the sympathetic nervous system is an important feature that allows the body to face and to cope with stressful insults like bacterial infection [93]; noradrenaline is discharged and the adrenal medulla increases the secretion of adrenaline and noradrenaline [19]. As blood vessels and most endocrine cells are richly innervated by adrenergic nerve fibres [19], the brain receives sensory input from the local immune reaction and inflammation via sensory fibres ascending to the nucleus tractus solitarius and may directly affect the microcirculation and the endocrine system (particularly the adrenal and pituitary glands) via the neural route [19]. Concurrently due to our sophisticated mechanisms, the parasympathetic activity will be enhanced and anti-inflammatory hormones and mediators such as cortisol and prostacyclin are secreted in order to balance the pathways and processes activated [19, 91].

A suitable inflammatory response indicated by the release of mediators, particularly cytokines, adequate in amount and duration and with limited effect characterizes the physiological reply [94]. Further, although the defence mechanisms may inflict injury in normal tissue in the near vicinity and cause some collateral damage, the pathology is virtually focal and limited rather than diffuse [11, 95]. Local and systemic negative feedback mechanisms make sure that the response is contained at the remote site [91, 96], and the compartmentalization of the innate immune system helps to limit the collateral damage to the host and to maintain the integrity of the noninvolved endothelium

[11]. Accordingly, the local immunological and inflammatory response is a physiological host-derived, phylogenetically developed, meaningful, necessary and very effective approach to any injury in order to eradicate pathogens, repair and remove damaged tissue, heal wounds, and to limit tissue damage [18, 58, 91, 97].

However, if the host response exceeds the boundaries of the local area [68, 98], the process may become generalized and a clinically relevant, poorly compensated, unbalanced and potentially unregulated inflammation may occur [11, 12, 91, 99, 100]. This is characterized by marked pathophysiological features [12] and the widespread involvement of the endothelium [68, 101, 102] causing a significant and serious illness representing the systemic inflammatory response syndrome (SIRS) and potentially leading to multiorgan dysfunction (MODS) [11, 12, 17, 103–105]. The pathophysiological properties of the cytokines released define the clinical syndrome [1, 91, 106, 107], since if they turn up in the bloodstream, where they are usually not detectable [107], in a significant amount, they will exhibit a similar clinical picture independent of the type of primary insult [1, 17, 91]. Occasions where detectable amounts of proinflammatory mediators are found in the circulation may include the following: primary bloodstream infection [1, 32], if the local site is overrun by cytokines following a critical volume of predominantly cell-associated mediators reaching the limit of the area's capacity [67, 68, 99, 108], failure of local control mechanisms [32, 100], literally production of pro-inflammatory substances in excess although reasonable in regard to the severity of infection [12, 17, 100], if the infection is sustained and/or prolonged [19, 20, 100], due to genetic disposition [109], and in case of disturbed regulation/crosstalk [104, 109–111].

Sepsis applies if the initial threat is of a microbial nature, and sepsis develops when the initial, appropriate response to infection becomes amplified, and then unregulated [12, 112, 113], or in conditions of direct penetration of the pathogen(s) into the bloodstream creating a septic state [1]. Consequently, the basically physiological and normally localized host reaction to an insult [12, 68, 95] will, in the case of overspill into the systemic circulation [114], result in a widespread activation of our extremely well-developed defence [105] and response to affliction pathways [19], provoking a generalized inflammation [19, 103, 104, 115].

In order to achieve a balanced response, concurrent with the inflammatory initiation, anti-inflammatory pathways termed compensatory anti-inflammatory response syndrome (CARS) [116] are elicited [99, 116–119]. Their task is to dampen the inflammatory reaction and to equilibrate the processes triggered [11, 18, 99, 117–119]. Anti-inflammatory cytokines (e.g. IL-6, IL-10) and soluble receptors (IL-1 receptor antagonist, TNF-α) counteract proinflammatory mediators protecting the organism from endotoxin shock [32, 99, 107]. Lymphocytes exert a major input by modulating SIRS, they are qualified to interact with the innate and adoptive immune response and may coordinate, mitigate or boost the inflammatory reaction [119]. Monocyte functions will be down-regulated [120], and the action of neutrophils (polymorphonuclear cells) will be attuned [99, 121]. CARS is a physiological feature [91, 99] and obviously does not result in a generalized reduction of all immune functions. Rather, it results in an adaptive and compartmentalized (regional tissue, blood, organ) reply that is adjusted to the type of insult, the cytokines released, the cells involved and the function affected, mainly reprogramming the leukocytes [99, 122, 123]. Overall, CARS is considered being an adopted reply to SIRS intending to well balance the inflammatory response to injury by silencing some active proinflammatory genes and as such limiting tissue damage while not interfering with the elimination of pathogen/destructed cells and tissues by keeping other genes involved in the inflammatory and finally repair processes running [99, 119]. As this dampening may be poorly timed or even inappropriate, this compensatory setup has the risk of jeopardizing the desired and necessary immunological effects and defence [99, 119], thus allowing for a secondary hit as demonstrated in several conditions [124–127] and sepsis development in critically ill patients (the concept of second hits is expanded below) [121, 128, 129]. Furthermore, deficient activation and release of anti-inflammatory substances may promote systemic effects (preferentially lungs and kidneys are affected) and systemic complexity, as the necessary diluting effect of CARS mediators onto the pro-inflammatory responses of circulating leukocytes and the hindering effect of remote endothelial cell involvement by counteracting cytokines breaching in the circulation will be insufficient [91, 100].

Moreover, neuroendocrine input (hypothalamic–pituitary–adrenal axis [92]) is of major impact on the control of the inflammation [32]: Firstly, catecholamines released interfere with cytokine production [32]. Noradrenaline promotes TNF-α production [130], while epinephrine suppresses both TNF-α [130, 131] and NO generation [132]. Secondly and of crucial importance is a well operating crosstalk between cytokines and neurohormones in order to re-establish homeostasis during stress [133]. As such,

- Circulating TNF-α and other cytokines, locally expressed IL-1 beta and NO, as well as afferent vagal fibres all will augment the production and release of vasopressin and corticotropin-releasing hormone, and
- Cortisol synthesis is modulated by locally released IL-6 and TNF-α [32] exerting an anti-inflammatory effect [1, 91].

Thus, the alerted neuroendocrine system contributes to maintain cardiovascular homeostasis and cellular metabolism and to clear the initial microbial focus [32]. A blunted or an inappropriate reply related to neuronal apoptosis, metabolic or ischaemic disorder of the hypothalamic, pituitary or adrenal glands or interferences from drugs administered and the cytokines acting [134] may deploy a relative or absolute shortage of adrenal hormones (50 % of septic shock patients) [135] and vasopressin [136] (occurring in 33 % of septic shock cases). This results in alteration of the required homeostatic balance contributing to hypotension and multiorgan dysfunction syndrome (MODS) [135, 137, 138].

As the host's reaction has a relevant impact on the pathways set up and is not necessarily a direct effect of the pathogen(s) [11, 12, 19], SIRS is traditionally considered a disease largely incited and exacerbated by uncontrolled inflammatory response emerging in and provoked by the affected patient [104, 139]. The theory postulated by Thomas [140] signified sepsis representing an uncontrolled, exaggerated inflammatory response associated with an over-stimulated immune system [141–143]. Pinsky and Matuschak described the uncontrolled hyper-inflammatory response as a generalized malignant intravascular inflammation [101], stressing the special and decisive involvement of the endothelium in the process and the potentially serious consequences of being responsible for the host's critical clinical condition, associated

MODS and the substantial mortality found [12, 144]. However, the theory was based on animal studies neither reflecting the clinical picture in humans [145, 146] nor was the expected frequency of an exaggerated systemic inflammation consistent with the study results-a clinically relevant SIRS was demonstrated in a markedly lower proportion in septic humans [147–149]. Moreover, histopathological examinations show normal cell and tissue structures even in the presence of markedly disturbed function in the vast majority of patients with sepsis syndrome and septic shock [109, 150, 151]. Further, the high potential of complete organ recovery, if the critical phase is survived, [152, 153] strengthens even more the challenge of the theory [20, 91, 154, 155]. However, the pathophysiology will help to understand.

The pathophysiological characteristics of sepsis include altered microhaemodynamics and mitochondrial dysfunction, with maldistributed and heterogeneous microcirculatory blood flows being a hallmark of sepsis and septic shock promoting cellular and tissue dysoxia [27, 156–159]. Disproportionate micro-perfusion is associated with disturbed oxygen extraction capabilities causing more severely perturbed tissue oxygenation than homogeneously diminished perfusion [160, 161]. This may lead to cellular injury [68, 162], albeit the oxygen supply, reflected by a rise in tissue oxygen tension, appears to be sufficient [155, 163, 164]. However, although cardiovascular changes and abnormalities seen in sepsis and septic shock may definitely result in net global tissue hypoxia [165, 166], the crucial issue is a dysfunctional cellular oxygen uptake and energy metabolism referred to as mitochondrial dysfunction [144, 164, 167–170]. Mitochondrial dysfunction may shape up to be the central process in the development of MODS [168, 171, 172] rather than the circulatory changes observed [155]. Nevertheless, altered microhaemodynamics are a typical feature of sepsis and septic shock pathobiology [165, 173] and may facilitate SIRS, thus having a relevant impact on the disease process. "Both regional flow alterations and mitochondrial dysfunction undoubtedly coexist in the septic state" [109] impairing tissue oxygen use [166, 168, 174]. This may potentially result in multiorgan dysfunction with the cells merging into a dormant, hibernating-like state precipitating occasionally severe organ malfunction but no structural organ damage of concern [151, 152, 155]. This provides a survival option, if the acute phase is outlived, as the histopathological examinations show and the high recovery rates support [20, 104, 175, 176]. Hence, SIRS is definitely essential for survival although sometimes it may be somewhat dysregulated, uncoordinated and not always well interlinked [11, 12, 68, 91, 105, 133]. The subsequent MODS is arguably quite adaptive rather than maladaptive, and the body's sophisticated, absolutely necessary well-structured response needs to be seen and interpreted in a completely different way [20, 155].

Conditions discussed and associated with the incidence of SIRS are a severe injury/infection or a direct bloodstream affliction by infectious material/organisms [100, 104]. Furthermore, initially local reactions may exceed the regional site, signalling that systemic help/response is needed to face the incurred threat properly [12, 100], or the inflammatory response is very strong, thus crossing the local site, although actually considered adequate in relation to the severity of the threat [11, 12]. The collateral damage caused in such a situation might not be compensated, facilitating the systemic inflammatory reply [12]. Often the generalized response precipitated is multifactorial [12, 17, 68], and features and circumstances may be related to:

I. Microbial-related issues.
   - An infection with a high microbial load, with highly virulent pathogens [12, 98].

II. Host-related properties.
   - Prolonged and sustained inflammation [12, 19, 67, 95, 105] with systemic overspill occurs [12, 68, 97].
   - Genetic factors may be of substantial significance [109, 112, 177] as genetic polymorphism causes a considerable variability in the individual response and hence may be associated with raised susceptibility and a strong reaction to infection including a genetically up-regulated expression of inflammatory mediators [109, 144, 178, 179].
   - Genetically determined abnormalities and deficits [1, 180, 181].
   - Imbalanced release of pro- and anti-inflammatory measures and mediators (poorly timed CARS, very strong CARS, etc.) [12, 91, 98, 99].
   - Especially noteworthy are the so-called second hits

     (a) Second hits related to circulatory issues:
     The translocation of gut bacteria/bacterial material into the bloodstream attributed to compromised gut barrier function. In the setting of SIRS, the gut as a remote organ may get pre-primed [182, 183]. Further due to redistribution and/or hypoperfusion of the intestinal vascular bed ascribed to the circulatory changes arising in systemic inflammatory states, especially septic conditions [68, 109], the barrier function weakens. This allows for translocation of bacteria/bacterial material into the bloodstream, boosting the inflammatory pathways and potentially promoting the development of MODS [183–186], "the gastrointestinal tract is being/is becoming the motor of critical illness" [187].

     (b) Second hits may be related to immunosuppressive states:
     Intense, prolonged and persistent inflammation (frequently termed "excess inflammation") may over time create an anti-inflammatory immunosuppressive state [17, 91, 99, 104, 188, 189] – CARS may be one of the mechanisms and/or contributors. This will impede the clearance of infection and may predispose to a new nosocomial infection (II. hit) while concurrently feeding the inflammatory process [12, 91, 104, 188–190].

     Patients with pre-existing serious illnesses such as diabetes, liver cirrhosis, chronic kidney disease, and neoplastic or specific immunodeficiency disorders and the elderly are at an increased risk of developing an unbalanced response and thus SIRS and consecutively MODS [67, 191–193]. These patients are considered as being pre-primed, as most of these conditions are associated with abnormal levels of cytokines [100, 115].

III. Consequence of therapeutic interventions, for example, antibiotics and other drugs affecting cytokine release [194]

Furthermore, the inflammatory process tends to be self-sustaining [68], since at a critical level, proinflammatory mediators locally overspill into the systemic circulation [99, 114] and will potentially affect healthy tissues as demonstrated by Nathan [195] and Jarrar [196]. This process can aggravate the inflammation, probably ending in organ failure and even death [197].

Mechanisms specifically and particularly provoking and leading to multiorgan dysfunction (MODS) during the septic malady may include the up-regulation of inflammatory response and/or of the neuroendocrine system [155, 198, 199]. Inflammatory mediators are reported to disrupt the crosstalk between organs (altered modulation of biological homeostatic functions of the autonomic nervous system) [110]. TNF-α, IL-1 α and NO may precipitate a reduced energy production by disturbing the mitochondrial respiratory chain, inducing cellular dysoxia [155]. The formation of a procoagulant state may interact with the inflammatory mediators, creating a self-perpetuating vicious cycle [11, 200]. Initially triggered by a microbial infection or other initiators of the SIRS, MODS can develop independently of the infective process [109] with widespread endothelial affliction and injury [68, 102]. The lungs followed by the kidneys are the most common remote organs developing ALI/ARDS or AKI, respectively [102, 109, 152, 199, 201].

"Today it is increasingly recognized that diseases stem from genetic interactions among genetics, environment, threat and response" [19] and a broad variety of factors, conditions and circumstances may overcharge, interact and interfere with these very intricate, complex, cross-linked and sophisticated, crosstalking and regulated pathways [12, 91, 99, 109, 110, 112, 133].

Cardiovascular abnormalities associated with sepsis include a broad spectrum of derangements [21, 26]. Key characteristic features are altered microcirculation, hypovolaemia, compromised myocardial performance, hypo-responsiveness of the cardio-vascular system to catecholamines, and peripheral vasodilation [22, 24, 27, 28, 157, 158, 202–204]. As they often earmark the clinical picture, they need to be addressed separately:

1. Altered microhaemodynamics are the hallmark of (severe) sepsis and septic shock [27, 28, 157, 158, 205], typified by mal-perfusion and particularly maldistribution as a result of diminished vascular tone, shunting and blunted microvascular (metabolic) autoregulation. Furthermore, the coexistence of a limited number of well-perfused and well-functioning capillaries, although often of limited quantity, in close vicinity with non-perfused or even over-perfused microvessels leads to heterogenic regional blood flows provoking hypoxia/dysoxia aside normoxic areas. This results in altered mitochondrial oxygen extraction/utilization and thus cytopathic hypoxia [6, 113, 157, 159, 167, 173, 206–211].
2. Hypovolaemia may be caused by true fluid losses related to increased permeability found in SIRS but may be related to the reduced peripheral resistance and resultant venous pooling [21, 56, 203].
3. Myocardial depression is a common finding [22, 24] and may be present at least to a mild degree in all septic patients [212]. Causative are cardiodepressant factors diluting myocardial contractility such as cytokines with cardiodepressive properties e.g. TNF alpha [213–215], nitric oxide which is overproduced in sepsis [216–218] and its metabolite peroxynitrate [219, 220]. Both systolic and diastolic properties are compromised with impaired intrinsic contractility (systolic function) and diminished compliance (diastolic property) [25, 221–223].

4. Hypo-responsiveness of the vasculature and myocardium to catecholamines provokes an inappropriate regional circulation and subsequently inadequate distribution of blood flow among the organs [68, 224].
5. Peripheral vasodilation, often with consecutive hypotension, is the key feature of the macrohaemodynamics in sepsis [204, 225, 226]: Proinflammatory substances released both locally and systemically trigger the expression of the inducible nitric oxide synthase resulting in an excessive production of NO by the endothelial cells [226–228]. NO has been identified as probably the most important mediator of organ blood flow regulation, thus decisively affects vascular tone and blood pressure [158, 229]. NO interferes with vasoconstriction of and impairs the ability of the vascular smooth muscle cells to constrict and as such induces vasodilation in cases of an activated systemic inflammatory response where the NO concentrations are excessively and inappropriately high [224, 226, 230–232]. This condition in its extreme is called vasoplegia by some authors [224, 230, 232].

Well treated and "compensated" in the early disease course, sepsis is after suitable fluid replenishment macro-circulatory characterized by an elevated cardiac output/index (CO/CI) in the presence of profoundly reduced systemic vascular resistance obscuring any myocardial dysfunction, termed the hyperdynamic phase [24, 109, 233]. Ongoing profound vasoplegia allowing for a good/excellent LV systolic function, traditionally indicated by an LVEF well above normal range, is associated with a poor prognosis [234]. This hyperkinetic cardiac performance is driven by a persistently low peripheral resistance due to ongoing sepsis and as such is suggestive that the body is not capable to get control over the disease process [235]. Septic patients with systolic dysfunction and normal or reduced LVEF indicating normalization of SVR, are demonstrated having a significant better prognosis than those with a good LVEF [233, 236], as an increase in afterload determines a corresponding reduction in EF and vice versa [237].

Of further note and for understanding, acute systemic inflammation exhibits different impacts on different vascular beds: Large, elastic vessels like the aorta are getting stiffened as a result of blunted wave reflections following peripheral vasodilation [238], while medium and small peripheral muscular arteries and arterioles are obviously hampered and compromised in their ability to constrict and thus vasodilate. This is attributed to pro-inflammatory mediators provoking excessive NO synthesis [238] and release of vasodilative prostanoids [239] both facilitating vasodilation, further vascular adrenorceptor hypo-reactivity to catecholamines (partly triggered by functional modification of potassium channels) is reported [240]. The effect of this aortic and big arteries' stiffening is a rise in LV-load (via ventriculo-arterial coupling) and LV oxygen demand, although the peripheral resistance is low [238] since the heart is in general primarily affected by the directly adjacent vessels [249]. As a stiff aorta will stretch the arteries, this will result in mechanical fatigue risking vessel dissection [241]. Moreover, these results by Vlachopoulos explain why in sepsis, despite profound peripheral vasodilation, normotension in the central vessel system (aorta and artery tree) may be present and as such may misguide the clinician to assume suitable peripheral and tissue perfusion [238].

Acute kidney injury (AKI) is the most common complication in sepsis as 40–50 % of all septic patients develop renal dysfunction [242–244]. Sepsis is the predominant cause of all AKI cases in critically ill patients [242, 245, 246], and the severity and duration of hypotension are both significant risk factors for AKI development [247]. Interestingly, the cut-off mean arterial pressure (MAP) was 80 mmHg, and for every mmHg drop below 80 mmHg, the odds of developing AKI increased by 3 %. Further, every hour of continuous hypotension more than doubled the risk (to develop AKI) for each 10 mmHg drop in MAP below 80 [247]. Should we aim for a MAP of 80 mmHg in septic patients?

Haemodynamic and inflammatory pathways have been identified as contributing to the formation of septic AKI (S-AKI) [248, 249] with renal vascular dysfunction as the major issue [250–252]. Firstly, there is an uncoupling between systemic macrocirculation (peripheral vasodilation indicated by ↓ SVR) and regional organ perfusion (elevated renal vascular resistance) resulting in reduced renal blood flow despite maintained or even enhanced CO [253, 254]. Secondly, altered microhaemodynamics with heterogenic intra-renal blood flow distribution [27, 250] and disturbed intraglomerular haemodynamics [255, 256] generate functional AKI that may precede to manifest (intrinsic) AKI [257] – for further information, read Chap. 5.

In summary, the human body during the evolutionary process over the millennia has evolved an intricate, integrative, multi-faceted, sophisticated, and vital response system to any injury, called the systemic inflammatory response syndrome [17, 20, 97, 105, 155, 232]. To cope with such sometimes challenging threats, a broad spectrum of pathways including neural, innate immune, neuroendocrine and metabolic processes are available and may be deployed [18, 20, 27, 155]. Proinflammatory cytokines released by innate immunocompetent cells like macrophages and endothelial cells are central mediators in this reply system [32, 39, 71, 100] and jointly responsible for the typical clinical picture characterizing SIRS [1, 17, 94, 107]. The endothelium is playing a fundamental role in the inflammatory reaction integrating the signals receiving and orchestrating the defence mechanisms [11, 71–73]. SIRS is called sepsis, if the triggering threat is of microbial nature [2] and occurs when microbes or microbial material primarily affect the bloodstream [1, 100] or, most commonly, if the inflammatory response to an initially localized infection exceeds the local boundaries and pro-inflammatory cytokines submerge into the systemic circulation [12, 68, 107] provoking a generalized response within the host [12, 17]. With overspill into the systemic circulation, a widespread activation of the cross-linked pathways [19, 20, 32, 92, 103, 133] and particularly of the endothelium [11, 71, 73, 232] ensues. This reaction is basically not harmful to the host, as even a strong systemic reaction might be quite appropriate in response to the threat [12] and/or the local site is just signalling that the regional microenvironment is not able to get control over the insult and is calling for systemic help [17, 100] (e.g. a very virulent microbe or a very high bacterial load requires a powerful response) [12, 98]. Thus, in the vast majority, SIRS is beneficial for the host [19, 32, 105], and a response is appropriate and successful if the reactions launched and the mediators released are balanced, of appropriate magnitude, of limited duration and effect and lead to control of infection and the restoration of homeostasis resulting in tissue repair and healing [94, 100]. In order to suitably balance the response and to avoid a clinically apparent SIRS being the ongoing disease course, a complex network of dampening and modifying tools and cascades are concomittantly employed, called, at large, CARS [19, 32, 99, 112, 117–119]. However, sepsis is a critical illness and bears a high risk of mortality [1, 32, 144]. Accordingly, if sepsis applies, besides the desired control of the infectious site, collateral damage may occur [11, 12]: Due to their large intricacy the cross-linked pathways and their regulation, particularly the cross-talk between several elements (e.g. neuroendocrine hormones and cytokines [32], between endothelium and the organs and between and within organs [20, 110]), may get disturbed, uncoordinated and unregulated [11, 95, 99, 104, 105]. The underlying features generating a clinically relevant systemic reply are diverse, yet not completely known and understood [112], however include genetic polymorphism creating a broad variety of responses within SIRS [18, 144, 178]. Pathogen-associated causes (virulence) [12, 98] as well as improper neuroendocrine input [32] are of note, while pre-existing illnesses such as diabetes or chronic kidney failure will pre-prime the patient [67, 192]. Further, second hits [95, 185, 186] may amplify and perpetuate the inflammation [183, 185, 187]. As such, a prolonged and sustained inflammation may merge into immunosuppression [99, 189], jeopardizing and exposing the patient to new/further microbial threats [105, 258], promoting critical illness and MODS [91]. Moreover, as prolonged immune depressed states are associated with an increased risk of death, this may be the main reason for a poor outcome [104, 128, 259, 260], while SIRS per se is beneficial and a physiological answer [19, 20, 155, 261]. **Impaired cellular and tissue oxygen utilization attributed to coexisting microhemodynamic** (characterized by heterogenous and maldistributed microcirculatory blood flows [27, 157, 159] associated with disturbed oxygen extraction capabilities [160, 161]) **and mitochondrial dysfunction is a hallmark of septic pathology** [109, 166, 168, 169], and blunted mitochondrial energy metabolism appears to be basic in the process resulting in MODS/MOF [170, 172]. This setting obviously generates a milieu which facilitates the cells moving into a dormant, hibernating-like state [20, 151, 152, 155, 175]. Consecutively, severe multiorgan dysfunction may occur particularly affecting the lung and kidneys. However, this disruption turned out to be of functional rather than of structural nature and can be adaptive, apparently allowing the body the possibility to fully recover without reduction in organ function if the patient survives the critical illness phase [20, 150, 152, 155].

4.3 Clinical and Diagnostic Issues

Clinical signs and symptoms are greatly variable, non-specific, subtle but should be suggestive of a systemic illness [1, 144, 262]. Typical features such as fever, tachycardia (>90/min) and tachypnoea (>20/min) representing systemic inflammation [6, 11] may not be present [6], particularly in the elderly [263, 264]. Fever, although an attribute of infection and common in sepsis [6, 265], is somewhat difficult since body temperature varies from human to human and depends on time of day. Accordingly, an oral temperature of >37.2 °C, or rectally >37.5 °C, means fever in the elderly, and in general, temperatures <36.0 °C may indicate severe sepsis

[192, 266]. Altered consciousness indicating involvement of the central nervous system is frequently seen although the brain is rarely the original site of infective focus but may be the only clinical conspicuity [1, 6, 68].

The straightforward definition of the 1992 consensus document only allowed for the core features and did unfortunately not include a number of subtle clinical signs, thus did not comprehensively represent the clinical picture in daily practice. A follow-up conference developed a long list of symptoms and signs, as depicted above, indicative for and consistent with sepsis [3]. From this, evidence of organ dysfunction that has no other explanation may point towards sepsis as the diagnosis [265]. However, the clinical features listed are highly unspecific and thus may dilute the diagnosis of sepsis [267]. Consequently, the concept has been challenged, whether sepsis is a specific diagnosis and a discrete entity or not? Carlet suggests to focus on the infection as the underlying process, which may in some cases be complicated by organ failures [268]. Accordingly, we have to address and align to the specific infection when considering the cause of malady and when applying the clinical picture to a specific diagnosis rather than call it by the somewhat imprecise term sepsis, e.g. peritonitis or pneumonia complicated by kidney dysfunction [268].

Features indicating organ hypoperfusion, even in the presence of normal or near normal blood pressure, should be imperatively looked for [1] as haemodynamic alterations associated with sepsis are common, often earmarking the clinical picture [21, 24, 27, 204].

Signs of hypoperfusion include [1, 32, 262, 269, 270]:

- Mottled skin
- Cold extremities
- Altered consciousness
- Thready pulse
- Concentrated urine or oliguria and elevated creatinine levels
- Thirst
- Narrowed pulse pressure
- Delayed capillary refill

The aging immune system of the elderly more so than other adults predispose them to infections [271, 272], and the diagnosis can be difficult as symptoms potentially suggestive for sepsis, although non-specific, comprise altered mental status, weakness, falls, delirium, malaise and urinary incontinence [264].

The physical examination should pay particular attention to identifying the infected organ/site. Premorbid conditions predisposing to infection such as immunocompromise (e.g. HIV), prosthetic devices, intravenous and urinary lines and artificial heart valves need to be ascertained [48]. Repeat examinations will ensure that new signs emerging are recognized [265].

Laboratory results may give some clues; unfortunately, they are in general non-specific and poorly sensitive [6]. Complete blood cell counts with differentiation although demonstrating leukocytosis, neutrophilia and several young and immature cells, all suggestive for bacterial infection, are of poor sensitivity, highly unspecific and are not able to confirm bacterial infection [273–275]. Moreover, as stated in the 1992 consensus guideline [141], leukopaenia and/or neutropaenia may be the predominant finding [6].

CRP is still the classic inflammatory marker, but too slow and non-specific [276]. Meanwhile, more than 80 (bio)markers have been investigated, assessing their diagnostic and prognostic capabilities in septic conditions: Regrettably, they are imprecise, not distinct, and lack sensitivity [277, 278]. Procalcitonin (PCT) concentrations are usually increased in septic patients but will not discriminate between infection and inflammation [279]. Anyhow, PCT may be quite helpful to identify sepsis [278], and Werra estimated a diagnostic sensitivity of 100 % and specificity of 72 % for sepsis if the PCT concentration is ≥1.5 μgL [280]. Notably, PCT shows a high negative predictive value and may exclude a serious septic process by a serum level of <0.25 μg/L [6]. Furthermore, PCT has demonstrated its potential to guide antibiotic prescribing and a level below 0.25 μg/L indicates that antibiotics are not needed or maybe discontinued [281].

Elevated concentrations of lactic acid may represent abnormal tissue and organ perfusion as a result of deficient tissue and organ oxygen supply and support a suggested diagnosis of sepsis [6, 32]. Unfortunately, hyperlactataemia is highly unspecific and may be elevated due to comorbidities, e.g. liver dysfunction [282]. Where the diagnosis of sepsis is confirmed, lactate levels and further increases in lactate may be used to guide therapeutic measures [283, 284] and are prognosticative as elevated, and increasing concentrations indicate a poor prognosis [285, 286].

Blood cultures are important for the diagnosis of sepsis and have a major impact on its treatment. As infection is "a conditio sine qua non of sepsis" [265] and bacteraemia "the essential and fundamental pathophysiological determinate of sepsis" [287], blood cultures are a standard and still a crucial element for its diagnosis [30, 265, 288, 289]. Antimicrobial therapy needs to be effective and must be adjusted based on the results of the cultures [265, 290]. Ideally cultures have to be taken prior to initiation of empiric antibiotic therapy [5, 291, 292]. Two sets of cultures from two different sites are in general recommended [5, 293], although recent research revealed that 2 sets taken within 24 h will detect about 90 % of bacteraemic episodes and that as many as four cultures are necessary to achieve a >99 % detection rate [294, 295]. However, sepsis and particularly septic shock are emergencies, and immediate initiation of empiric antimicrobial treatment will be life-saving [296, 297]. Roughly 30 % (Martin found 50 % [36]) of the blood cultures taken on admission are negative, most often in patients with community-acquired infections already treated with antibiotics [32]. Both, the site of infection, with pneumonia and intra-abdominal infections more often positive than other sources [265, 298], and several other factors may narrow the yield. These include suboptimal timing, e.g. later phase of sepsis [287], insufficient blood volume and prior antibiotic treatment [287]. Furthermore, the bacterial infection might remain quite localized as the clinical picture is considerably more related to the cytokine release of SIRS [1, 94, 107].

A trial by Shapiro [299] is the first systematic investigation to determine the predictable likelihood of bacteraemia in adult patients in the ED setting and as such a recommendation when to take blood cultures and when not: The group could identify criteria suggestive for the appearance of microbial pathogens in the bloodstream. In cases not fulfilling the identified criteria, less than 1 % positive blood cultures can be expected. This rule shows a very high negative predictive value of 99 %.

The criteria derived are:

Major Criteria

- Temperature >39.5 °C
- Indwelling vascular catheter
- Clinical suspicion of endocarditis

Minor Criteria

- Temperature 38.3–39.4 °C
- Chills
- Age >65 years
- Vomiting
- Hypotension (SBP <90 mmHg)
- Relative neutrophil >80 %
- White blood cell count >18,000 cells/mm^3
- Bands (young polymorphs) >5 %
- Platelets <150,000 cells/mm^3
- Creatinine >2.0 mg/dL

Blood cultures are indicated if at least one major and two minor criteria apply. Furthermore, the yield of blood cultures rises with greater blood volumes taken rather than *when* samples were taken from *different sites* [300, 301].

It is imperative that cultures should be obtained from any potential source suspected such as indwelling catheters, tissues and body fluids [302]. While evaluating the results of blood cultures, keep in mind that critically ill patients tend to develop colonizations of bacilli which cover a priori no morbidly potential. Thus, microbes verified in cultures should fit in with the overall picture and be typical for the site of infection, e.g. *Bacteroides fragilis* or aerobic Gram-negative bacilli in case of lower gastrointestinal infection or *Escherichia coli*, *Klebsiella pneumoniae* or *Enterococcus faecalis* if the hepatobiliary system is involved [287, 303].

A number of other diseases may mimic the clinical picture of sepsis and have to be considered as differential diagnosis [304]: Common examples are pulmonary embolism, acute myocardial infarction, acute gastrointestinal haemorrhage, acute pancreatitis, and diuretic-induced hypovolaemia. Less common mimics are diabetic ketoacidosis, relative adrenal insufficiency, rectus and other sheath hematoma, and a flare of systemic lupus. Accordingly, to diagnose sepsis, conditions mimicking it should be ruled out, and the overall findings, including history, physical examination and the laboratory results, should be consistent with and meet a sufficient number of signs and symptoms of the extended SSC criteria in the presence of confirmed or suggested infection [6, 265, 287, 305]. It is important to be aware of that in a substantial number of cases, sepsis could be the reason for the patient's presentation, particularly in the elderly [271], and thus clinicians need a high grade of susceptibility [1]. As such, sepsis is basically a clinical diagnosis, quite often not apparent and not easy to discover and, in a considerable proportion of patients, quite challenging [1, 144, 276].

4.4 Management

4.4.1 Overview

Both, immediate empiric but presuming appropriate antimicrobial treatment and effectual (aggressive) early volume resuscitation are strongly evidence-based approaches and cornerstone in the management in septic conditions [5, 144, 262, 287, 296, 306–310]. Immediate recognition of sepsis is paramount to commence treatment and as such is indispensable in reducing mortality [311, 312]. Antibiotics given as early as possible are fundamental [307, 308, 313]; each hour of delay may reduce survival by about 8 % [296]. Furthermore, the choice of antibiotic drug will markedly influence the outcome as improper coverage may increase mortality by between 1.4 and 8 times [314–317], while an adequate choice of antimicrobial agents demonstrated improved survival [318, 319]. "There is evidence showing that the correct initial choice of antibiotic saves more lives than virtually any other intensive care unit intervention" [5, 314, 316, 320]. Moreover, recent study results seem to indicate that high antimicrobial concentrations well above the minimal inhibitory concentration (MIC) could improve outcomes and are attributed to an increase in bacterial clearance. This may be achieved by either combination therapy using drugs with different mechanisms of action at traditional doses, each reaching 60–70 % of MIC, or high dosages of monotherapy with concentrations >100 % above MIC [321–323].

In sepsis, the relatively compromised vascular tone and absolute hypovolaemia are typical haemodynamic features [27, 144, 158, 203, 204, 324]. Volume repletion improves cardiocirculatory function (macrocirculation) and, in early sepsis, organ and tissue perfusion (microcirculation) as well [325–327, 6, 27, 324, 328, 329]. This results in a reduction in mortality [3, 262, 328, 330], and thus aggressive fluid resuscitation is paramount [3, 5, 262, 282, 331, 332]. Further on in the disease process, too extensive or inadequate amounts of fluids are demonstrated to be harmful [333–336].

Timing of therapy is of major importance: Treatment interventions need to be applied at the right time to be effective and beneficial [104, 312, 332, 337, 338]. Rivers was the first to document in his landmark study that early interventions significantly improve survival. With their early goal-directed approach, the authors achieved a 16 % absolute reduction in mortality [332]. Early sepsis fluid resuscitation has shown significant improvements not only of the macrocirculation but also of microhaemodynamics which are important elements of the septic course [326–329, 339]. Fluids given later in the disease course bear demonstrably harmful effects, and fluids not given immediately after diagnosis and early on (first 24–48 h) to replenish the vascular system will not display the known beneficial effects if administered later [333, 335, 336, 340]. The significance of early administration of antibiotics has already been addressed above [296, 307, 313]. Very specific and least theoretically robust treatments have apparently failed to be successful in the vast majority, if not all, since in septic patients we cannot clinically determine the disease's cycle precisely enough and as such do not know how far advanced the disease already is. Accordingly, anti-inflammatory measures applied during the early hyperimmune phase may be beneficial [154, 341, 342], while if applied during the later hypoimmune phase, they are demonstrated to be deleterious [154, 343].

A well-structured, monitored, systematic and straightforward approach, namely, early goal-directed therapy (EGDT), may be the foundation for a successful management of sepsis syndrome [312, 344] as has been confirmed and validated in several studies [345–348]. Although systematic and having a major contribution to the surviving sepsis guidelines (SSGs), the protocol advocated by Rivers and coworkers [332] has been challenged recently as it is a single-centre study of only 263 patients and with the elements and targets contained within it not evidence-based at all [306, 349–352]. An example is that fluid management cannot be guided by CVP and the target range of 8–12 mmHg that was chosen is arbitrary. More than 100 publications have demonstrated no correlation between CVP/change in CVP and fluid responsiveness [353] and that CVP is a poor parameter of intravascular filling [331]. CVP values above 9–10 mmHg have to be considered pathologically high, and the majority of patients with a CVP >9–10 mmHg will not respond to fluids [354–356]. The results of the "Vasopressin in Septic Shock Trial" (VASST) showed that patients with a CVP <8 mmHg had the lowest mortality rates although the data was gathered 12 h or more after admission and may not reflect the very early phase of sepsis [336]. Anyhow, they again confirm that later in the course, volume administrations need to be considerably reduced and restricted [312].

Another intervention recommended by Rivers and the SSC [30, 332] is to achieve a central venous oxygen saturation ($ScvO_2$) of ≥70 %, if necessary by administration of red blood cells and/or inotropic support if not achieved with volume replacement alone. However, the reason for the altered oxygen metabolism is not a deficient oxygen delivery but rather a failure of the mitochondrial oxygen utilization, a cytopathic hypoxia, yet the administration of red blood cells for oxygen carriage is still recommended in the guidelines [167–169]. The administration of red blood cells is accompanied by several serious side effects, particularly in the setting of sepsis: Firstly, transfusion-related immunomodulation (TRIM) that may compromise the ongoing host defence against the initial infection facilitating blunted immune response [357] and promoting new infections [357, 358] and, secondly, the development of transfusion-related acute lung injury (TRALI), with resultant multiorgan failure and death in critically ill patients [359–362]. The recommendations of the SSC do not comply with current transfusion guidelines [361, 363]. Red blood cell transfusion aiming for a $ScvO_2$ ≥70 % is neither evidence based nor pathophysiologically robust and importantly does not currently meet the Joint Commission and Institute for Healthcare Improvement criteria for inclusion in a treatment bundle [364, 365]. The application of inotropes intending to increase the $ScvO_2$ is highly questionable as well, since this measure is again neither evidence based [306, 351, 352] nor recognized by the Joint Commission and Institute for Healthcare Improvement [364, 365]. Inotropes are only indicated in case of confirmed myocardial failure with low CO despite adequate left ventricular filling [22, 306, 366] and may be applied in case of ongoing hypoperfusion [30] as they might ameliorate microcirculation [367]. However, caution needs to be employed as it might cause a deterioration in the microcirculation [367], even in cases where it improves the macrocirculation, thus monitoring of the microcirculation is essential in

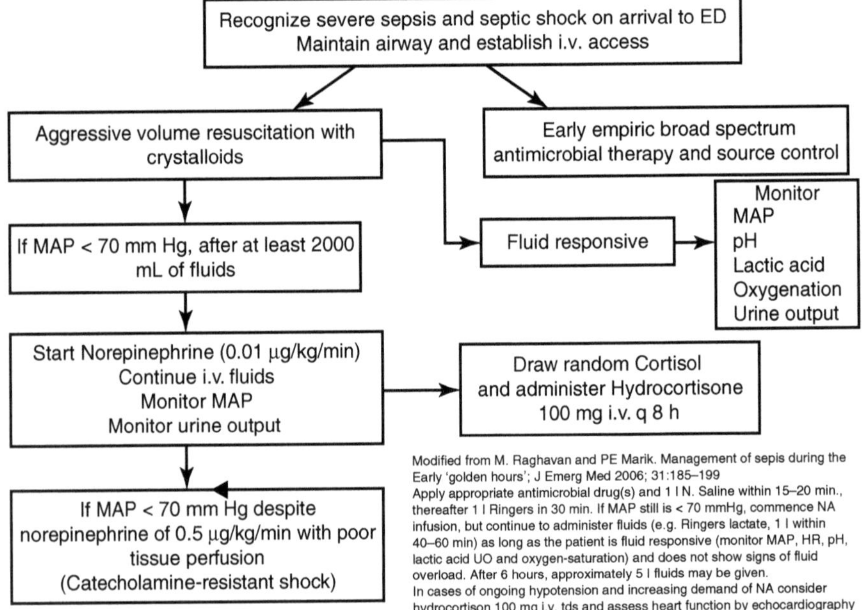

Fig. 4.1 Treatment algorhythm for early sepsis (first 6 hours after admission) modified from Raghavan and Marik [262]

dobutamine use [367, 368] – unfortunately, at present we are not able to systematically and routinely monitor microcirculation. Based on this data and taking all considerations together, we recommend following the EGDT protocol by Raghavan and Marik (Fig. 4.1) exploiting the early golden 6 h [262] in septic patients and, if necessary, the extended schedule by Marik in case the patient is haemodynamically unstable or develops septic shock [306]. However, as expressed by several authors, an individualized approach is necessary to meet the requirements of each single patient; thus, the protocol is only a guide and has to be adapted for each patient [12, 306, 351, 369]. Concerns have always been raised as to the efficacy of EGDT, if it is truly beneficial and which component derives the benefit. Hopefully, the ongoing ARISE, PROCESS and PROMISE multicenter randomized control trials of EGDT across Europe, the United States and Australia can answer this question [370–372].

4.4.2 *Special Issues*

1. In special risk groups, sepsis may be prevented by specific therapeutic measures like vaccination; examples include meningococcal A and C in cases of splenectomy, pneumococcal in case of agammaglobulinaemia and *H. influenza* in children [373–375]. Further, the administration of intravenous gammaglobulin is

advocated if pneumococcal bacteria are the cause of sepsis in patients with deficient gammaglobulin [376].

2. Empiric antimicrobial drug choice has to be targeted to the presumed pathogen(s) which is(are) defined by the organ(s) primarily affected [287]. Current opinion emphasizes and recommends that antimicrobial choice is based on local pathogen surveillance including local patterns of antibiotic susceptibility adopting broad-spectrum agents which have the potential to successfully control the anticipated microbe(s) causing the clinical picture [5, 309]. However, we have to keep in mind that broad-spectrum agents potentially contribute to the development of drug resistance, e.g. MRSA, highly resistant Gram-negative bacilli (GNB), and may promote rise of infections with Clostridium difficile and fungi [377, 378]. De-escalation to narrow-spectrum therapy is strongly advised once the pathogen(s) is(are) identified [379]. Thus, general recommendations should not be given, and institution-specific antibiotic nomograms should strongly be followed [307, 380]. Nevertheless, Table 4.1 below is an overview of commonly used antibiotics. The most common conditions are respiratory [43], hepatobiliary [287], colonic and pelvic infections [43, 287] and urosepsis [43, 287]. Community-acquired pneumonia (CAP) is a very common disease, estimated to cause 2–12 cases/1,000 population per year whereof the majority (80 %) of these will be managed in an outpatient setting [265]. The remaining 20 % will require hospital admission of which 10 % will need ICU care [381]. Etiologically, up to 20 % of CAPs are caused by viruses; the most common bacteria are *S. pneumoniae*, *H. influenzae* and *M. pneumoniae*. The most common nosocomial sepsis source is central venous catheters (CVCs) followed by nosocomial pneumonia (NP), which can be further differentiated into hospital-acquired pneumonia (HAP) and ventilator-associated pneumonia (VAP) [287].

   As community-acquired pneumonia is usually not associated with sepsis syndrome, only severe cases of CAP and particularly nosocomial conditions are indexed. Ventilator-associated pneumonia is a special subgroup of NP with a different formation mechanism, but is caused by the same bacterial species [287]. Concurrent surgical intervention in the case of intra-abdominal perforation, obstruction or abscesses is crucial, if infected devices are present their removal is indicated [287].

   Early and appropriate antibiotics are rated a IC or in the case of septic shock a IB level of evidence by the SSC [30].

3. Several proposals can be found in the literature regarding the amount of fluids:
   (a) Very rapid administration of 500–1000 ml crystalloid bolus (N.saline or Hartman's) within 20 to 30 minutes, may be repeatedly, 2 l within the first hour titrated to endpoints such as MAP >65–70 mmHg, HR <60/min, or uo >0.5 ml/kg/h is recommended by several authors [5, 262, 331, 332].
   (b) Alternatively, 20–30 ml/kg crystalloids within the first 30 min may be required [30, 306].

   Further on, 150–200 ml/h maintenance fluids with boluses of 500 mls as required are advocated [391, 392].

   However, before more fluids are given, the responsiveness to fluids should be evaluated by passive leg raise using the pulse pressure variation method

Table 4.1 Examples of organisms, site and appropriate antibiotics

Pathogen(s) commonly found	Organ/region of infection	Antibiotics commonly effective
	Hepatobiliary	
E. coli *K. pneumoniae* *E. faecalis* (VSE)		Meropenem, piperazillin-tazobactam, levofloxacin, or tigecyclin
	Lower GI-tract: colonic and pelvic	
Bacteroides fragilis (75 %) Aerobe GNBs		Meropenem, tigecyclin, or ertapenem
Accompanied surgical approach in case of intra-abdominal rupture, abscess, or leak essential[a]		
	Severe community-acquired pneumonia	
S. pneumoniae *H. influenza* *M. pneumonia*[b] Syncytial viruses (no specific antibiotic therapy) Influenza (no specific antibiotic therapy)		Ampicillin-sulbactam or ceftriaxone plus azithromycin
If suspicion for special risk factor e.g., P. aeroginosa, MRSA		
P. aeroginosa[c] *S. aureus*, MRSA[c, d] Gram-negative bacilli[c]		Piperacillin-tazobactam, meropenem/imipenem, or cefepime plus one of[d] if MRSA is likely
	Nosocomial pneumonia HAP, VAP	
P. aeroginosa Aerobic GNBs		Meropenem, cefipime, or levofloxacin
	Genitourinary	
Aerobic GNBs *E. faecalis*		Piperacillin-tazobactam, or meropenem
	CVC	
Aerobic GNBs *E. faecalis*		Meropenem, tigecyclin, or piperacillin-tazobactam

All data, if not especially marked are from Cunha, "Antibiotic Essentials" [382]

As community-acquired pneumonia is usually not associated with sepsis syndrome, only nosocomial conditions are indexed. Ventilator-associated pneumonia is a special subgroup of NP with a different formation mechanism, but is caused by the same bacterial species [287]. Concurrent surgical intervention in case of intra-abdominal perforation, obstruction or abscesses is crucial, if infected devices are present, their removal is indicated [287]

GNBs gram-negative bacilli, *VSE* vancomycin sensitive enterococci, *MRSA* methicillin-resistent *S. aureus*, *NP* nosocomial pneumonia, *VAP* ventilator-associated pneumonia, *HAP* hospital-acquired pneumonia

[a]Literature from [32, 383]

[b]Literature from [384–389]

[c]If required ICU care, *P. aeroginosa* and MRSA as well influenza, and gram-negative bacilli need to be considered, hence the antibiotics recommended differ; data from [390]

[d]If MRSA is a known local problem, one of the following needs to be added: vancomycin, daptomycin, linezolid, tigecyclin

[393–395] as described in the shock chapter in detail. Rapidly applied boluses may be preferable to maintenance fluid as they usually give some information about the fluid responsiveness of the patient's circulatory system, and thus fluid overloading may be avoided if proper monitoring is followed (such as pulse pressure variation and clearly defined targets, e.g., increase in blood pressure, drop in heart rate, etc.) [328, 396–398]. Up to 5–6 L within the first 6 h [332, 391, 392] and up to 10 L of crystalloids administered within the first 24 h [329, 399] have been described.

As MAP reasonably reflects the organ perfusion pressure [354] and in order to prevent autoregulation being blunted by a too low MAP jeopardizing blood distribution further compromising microhaemodynamics [400–402], early initiation of vasopressor administration overlapping with initial volume replenishment may be advisable [306, 403]. Although interindividual variations are reported [404, 405] and thus the MAP should to be titrated individually [400, 405], a range between 70 and 80 mmHg which corresponds to renal autoregulatory threshold [406], or around 80 mmHg as Lehman's study suggests [247], should be aimed for [407, 408]. Noradrenaline is the preferred agent and doses up to 0.4 μg/kg/min seem to result in no critical side effects [409–411]. Substantial evidence indicates a marked benefit if cumulative fluid balance is negative by day 3–4 [333, 336, 412, 413]. Thus, fluid administration needs to be turned down and restricted markedly in the days after early resuscitation [334, 335].

It has been a long and still ongoing discussion as to which type of fluid is the most beneficial [414]. Until now, no fluid type has been reported to be superior [415–417], and as such, crystalloids are recommended by the SSC [30]. As most colloids may affect the kidneys to some degree, an organ already at risk of dysfunction in sepsis, potentially crystalloids should be preferred [418, 419]. The overall delivery of fluids is not graded by the SSC, the early aggressive administration of up to 30 mL/kg within 30 min considered as evidence level IC [30].

4. Glycaemic control: The approach to blood glucose control has taken a turn, and due to substantial evidence, a range between 180 and 145 mg/dL is strongly advocated, thus avoiding numerous hypoglycaemic episodes in critically ill patients [30, 420, 421]. Moreover, even a higher range (180–220 mg/dL) might be beneficial under stress and the result of the evolutionary process [422].

   Aiming for a moderate blood glucose range between 145 and 180 mg/dL is graded as a IA evidence by the SSC [30].
5. Antithrombotic prophylaxis: Critically ill patients suffer from a substantial risk of developing deep vein thrombosis and pulmonary embolism [423, 424], particularly septic patients where the coagulation cascade is activated by the septic process. Hence, proper antithrombotic prophylaxis is paramount [425–427]. While in the case of normal or moderately diminished renal function any low molecular weight heparin is applicable (I B evidence [30]), in the case of relevant renal impairment (GFR <30 ml/min), dalteparin (I A evidence [30]) or unfractionated heparin (I A evidence [30]) should be given [30]. The overall use of antithrombotic prophylaxis is graded as IB evidence by the SSC [30].

References

1. Lever I, Mackenzie I. Sepsis: definition, epidemiology, and diagnosis. BMJ. 2007;335: 879–83.
2. Bone RC, Sibbald WJ, Sprung CL. The ACCP-SCCM consensus conference on sepsis and organ failure. Chest. 1992;101:1481–3.
3. Levy MM, Fink MP, Marshall JC. SCCM/ESICM/ACCP/ATS/SIS international sepsis definitions conference. Crit Care Med. 2003;31:1250–6.
4. Conference, American College of Chest Physicians/Society of Critical Care Medicine. American College of Chest Physicians/Society of Critical Care Medicine Consensus. Definitions for sepsis and organ failure and guidelines for the use of innovative therapy. Crit Care Med. 1992;20:864–74.
5. Dellinger RP, Levy MM, Carlet JM. Surviving sepsis campaign: international guidelines for management of severe sepsis and septic shock: 2008. Intensive Care Med. 2008;34:17–60.
6. Nguyen HB, Rivers EP, Abrahamian FM. Severe sepsis and septic shock: review of the literature and emergency department management guidelines. Ann Emerg Med. 2006;48:28–54.
7. Han Y, Carcillo J, Dragotta M. Early reversal of pediatric- neonatal septic shock by community physicians is associated with improved outcome. Pediatrics. 2003;112:793–9.
8. Rippe JM, Irwin RS, Cerra FB. Irwin and Rippe's intensive care medicine. Philadelphia: Lippincott-Raven; 1999. ISBN 0-7817-1425-7.
9. Marino PL. The ICU book. Baltimore: Williams and Wilkins; 1998. ISBN 0-683-05565-8.
10. Ten VS, Pinsky DJ. Endothelial response to hypoxia: physiologic adoption or pathologic dysfunction. Curr Opin Crit Care. 2002;8:242–50.
11. Aird WC. The role of the endothelium in severe sepsis and multiple organ dysfunction syndrome. Blood. 2003;101:3765–77.
12. Stearns-Kurosawa DJ, Osuchowski MF, Valentine C. The pathogenesis of sepsis. Annu Rev Pathol. 2011;6:19–48.
13. Benoit M, Desnues B, Mege JL. Macrophage polarization in bacterial infections. J Immunol. 2008;181:3733–9.
14. Majno G. Chronic inflammation: links with angiogenesis and wound healing. Am J Pathol. 1998;153:1035–9.
15. Biedermann BC. Vascular endothelium: checkpoint for inflammation and immunity. News Physiol Sci. 2001;16:84–8.
16. Cook-Mills JM, Deem TL. Active participation of endothelial cells in inflammation. J Leukoc Biol. 2005;77:487–95.
17. Roberstson CM, Coppersmith CM. The systemic inflammatory response syndrome. Microbes Infect. 2006;8:1382–9.
18. Broecher AC, Toft P. Pathophysiology of the systemic inflammatory response after major accidental trauma. Scan J Trauma Resusc Emerg Med. 2009;17:43–52.
19. Brame AL, Singer M. Stressing the obvious? An allostatic look at critical illness. Crit Care Med. 2010;38(Suppl):S 600–7.
20. Singer M. Systemic interactions. [Buchverf] Novartis foundation symposium 280. Sepsis – new insights, new therapies. Chichester: Wiley/Novartis; 2007. p. 252–65.
21. Zanotti- Cavazzoni SL, Dellinger RP. Hemodynamic optimisation of sepsis-induced tissue hypoperfusion. Crit Care. 2006;10 Suppl 3:S2.
22. Hunter JD, Doddi M. Sepsis and the heart. Br J Anaesth. 2010;104:3–11.
23. Landesberg G, Gilon D, Meroz Y. Diastolic dysfunction and mortality in severe sepsis and septic shock. Eur Heart J. 2012;33:895–903.
24. Merx MW, Weber C. Sepsis and the heart. Circulation. 2007;116:793–802.
25. Poelaert J, Declerck C, Vogelaers D. Left ventricular systolic and diastolic function in septic shock. Intensive Care Med. 1997;23:553–60.
26. Dellinger RP. Cardiovascular management of septic shock. Crit Care Med. 2003;31:946–55.
27. De Backer D, Creteur J, Preiser JC. Microvascular blood flow is altered in patients with sepsis. Am J Respir Crit Care Med. 2002;166:98–104.

28. Spronk PE, Zandstra DF, Ince C. Bench-to-bedside review: sepsis is a disease of microcirculation. Crit Care. 2004;8:462–8.
29. Vincent JL. Dear SIRS, I'm sorry to say that we don't like you. Crit Care Med. 1997;25:372–4.
30. Dellinger RP, Levy MM, Rhodes A. Surviving sepsis campaign: international guidelines for management of severe sepsis and septic shock: 2012. Crit Care Med. 2013;41:580–637.
31. Nduka OO, Parrillo JE. The pathophysiology of septic shock. Crit Care Clin. 2009;25:677–702.
32. Annane D, Bellissant E, Cavaillon J-M. Septic shock. Lancet. 2005;365:63–78.
33. Medicine, American College of Chest Physicians/Society of Critical Care. Definitions for sepsis and organ failure and guidelines for the use of innovative therapies in sepsis. Crit Care Med. 1992;20:864–74.
34. Angus DC, Linde-Zwirble WT, Lidicker J. Epidemiology of severe sepsis in the United States: analysis of incidence, outcome, and associated costs of care. Crit Care Med. 2001;29:1303–10.
35. Martin GS, Mannino DM, Moss M. The effect of age on the development and outcome of adult sepsis. Crit Care Med. 2006;34:15–21.
36. Martin GS, Mannino DM, Eaton S. The epidemiology of sepsis in the United States from 1979 through 2000. N Engl J Med. 2003;348:1546–54.
37. Dombrovsky VY, Martin AA, Sunderram J. Rapid increase in hospitalization and mortality rates for severe sepsis in the United States: a trend analysis from 1993 to 2003. Crit Care Med. 2007;35:1244–50.
38. Padkin A, Godfrad C, Brady AR. Epidemiology of severe sepsis occurring in the first 24 hrs in intensive care units in England, Wales, and Northern Ireland. Crit Care Med. 2003;31:2332–8.
39. Rangel-Fausto MS, Pittet D, Hwang T. The dynamic of disease progression in sepsis: Markov modeling describing the natural history and the likely impact of effective antisepsis agents. Clin Infect Dis. 1998;27:185–90.
40. Linde-Zwirble WT, Angus DC. Severe sepsis epidemiology: sampling, selection, and society. Crit Care. 2004;32:222–6.
41. Friedman G, Silva E, Vincent JL. Has the mortality of septic shock changed with time. Crit Care Med. 1998;26:2078–86.
42. Vincent JL, Sakr Y, Spring CL, Sepsis Occurrence in Acutely Ill Patients Investigators. Sepsis in European intensive care units: results of the SOAP study. Crit Care Med. 2006;34:344–53.
43. Martin GS, Brunckhorst FM, Reinhart K. The international PROGRESS registry of patients with severe sepsis: drotrecogin alfa (activated) use and patient outcomes. Crit Care. 2009;13:R103.
44. Annane D, Bellissant E, Bollaert PE. Corticosteroids in the treatment of severe sepsis and septic shock in adults: a systematic review. JAMA. 2009;301:2362–75.
45. van Ruler O, Schultz MJ, Reitsma JB. Has mortality from sepsis improved and what to expect from new treatment modalities: review of current insights. Surg Infect (Larchmt). 2009;10:339–48.
46. Levy MM, Artigas A, Phillips GS. Outcomes of the surviving sepsis campaign in intensive care units in the USA and Europe: a prospective cohort study. Lancet Infect Dis. 2012;12:919–24.
47. Lundberg JS, Perl TM, Wiblin T. Septic shock: an analysis of outcomes for patients with onset on hospital wards versus intensive care units. Crit Care Med. 1998;26:1020–4.
48. Nguyen HB, Rivers EP, Havstad S. Critical care in the emergency department: a physiologic assessment and outcome evaluation. Acad Emerg Med. 2000;7:1354–61.
49. Teres D, Rapoport J, Lemeshow S. Effects of severity of illness on resource use by survivors and nonsurvivors of severe sepsis at intensive care unit admission. Crit Care Med. 2002;30:2413–9.
50. Heyland DK, Hopman W, Coo H. Long-term health-related quality of live in survivors of sepsis. Short form 36: a valid and reliable measure of health-related quality of life. Crit Care Med. 2000;28:3599–605.

51. Annane D, Sebille C, Carpentier C. Effect of treatment with low doses of hydrocortisone and fludrocortisone on mortality in patients with septic shock. JAMA. 2002;288:862–71.
52. Annane D, Aegerter P, Jars-Guincestre MC. Current epidemiology of septic shock: the CUB-Rea Network. Am J Respir Crit Care Med. 2003;168:165–72.
53. Alberti C, Brun-Buisson C, Burchardi H. Epidemiology of sepsis and infection in ICU patients from an international multicenter cohort study. Intensive Care Med. 2002;28:108–21.
54. Jr Janeway CA, Medzhitov R. Innate immune recognition. Annu Rev Immunol. 2002;20:197–216.
55. Zhao B, Bowden RA, Stavchansky SA. Human endothelial cell response to gram –negative lipopolysaccharide assessed with cDNA microarrays. Am J Physiol Cell Physiol. 2001;281:C1587–95.
56. Parillo JE. Pathogenetic mechanisms of septic shock. N Engl J Med. 1993;328:1471–7.
57. Nathan CF. Secretory macrophages. J Clin Invest. 1987;79:319–26.
58. Licastro F, Candore G, Lio D. Innate immunity and inflammation in ageing: a key for understanding age-related diseases. Immun Ageing. 2005;2:8.
59. Huet O, Harrois A, Duranteau J. Oxidative stress and endothelial dysfunction during sepsis. [Buchverf]. In: Vincent JL, editor. Yearbook of intensive care and emergency medicine. Berlin/Heidelberg/New York: Springer; 2009. p. 59.
60. Hack CE, Zeerleder S. He endothelium in sepsis: source of and a target for inflammation. Crit Care Med. 2001;29(7 Suppl):S21–7.
61. Volk T, Kox WJ. Endothelium function in sepsis. Inflamm Res. 2000;49:185–98.
62. Faure E, Thomas L, Xu H. Bacterial lipopolysaccharide and IFN-gamma induce toll-like receptor 2 and toll-like receptor 4 expression in human endothelial cells: role of NF-kappa B activation. J Immunol. 2001;166:2018–24.
63. Henneke P, Golenbock DT. Innate immune recognition of lipopolysaccharide by endothelial cells. Crit Care Med. 2002;30:S207–13.
64. Medzitov R, Janewy C. Innate immunity. N Engl J Med. 2000;343:338–44.
65. Opal SM, Hiber CE. Bench-to-bedside review: toll-like receptors and their role in septic shock. Crit Care. 2002;6:125–36.
66. Medzhitov R. Toll-like receptors and innate immunity. Nat Rev Immunol. 2001;1:135–45.
67. Bone RC. The pathogenesis of sepsis. Ann Intern Med. 1991;307:1225–30.
68. Neviere R. The pathophysiology of sepsis. [Online] Up to date, 16 Mai 2012. [Zitat vom: 30 June 2013]. http://www.uptodate.com/contents/pathophysiology-of-sepsis.
69. Alcaide P, Auerbach S, Luscinskas FW. Neutrophil recruitment under shear flow: it's all about endothelial cell rings and gaps. Microcirculation. 2009;16:43–57.
70. Movat HZ, Cybulsky MI, Colditz IG. Acute inflammation in gram-negative infection: endotoxin, interleukin 1, tumor necrosis factor, and neutrophils. Fed Proc. 1987;46:97.
71. Aird WC. Endothelium in health and disease. Pharmacol Rep. 2008;60:139–43.
72. Santiago-Delpin EA. The endothelium and early immune activation: new perspectives and interaction. Transplant Proc. 2004;36:1709–13.
73. Shapiro N, Schuetz P, Yano K. The association of endothelial cell signaling, severity of illness, and organ dysfunction in sepsis. Crit Care. 2010;14:R182.
74. Cines DB, Pollak ES, Buck CA. Endothelial cells in physiology and in pathophysiology of vascular disorders. Blood. 1998;91:3527–61.
75. Galley HF, Webster NR. Physiology of the endothelium. Br J Anaesth. 2004;93:105–13.
76. Sandoo A, van Zanten JJCS, Metsios GS. The endothelium and its role in regulating vascular tone. Open Cardiovasc Med J. 2010;4:302–12.
77. Gross PL, Aird WC. The endothelium and thrombosis. Semin Thromb Hemost. 2000;26:463–78.
78. Cerwinka WH, Cooper D, Krieglstein CF. Superoxide mediates endotoxin-induced platelet-endothelial cell adhesion in intestinal venules. Am J Physiol Heart Circ Physiol. 2003;284:H535–41.
79. Chung HY, Yokozawa T, Kim MS. The mechanism of nitric oxide and/or superoxide cytotoxicity in endothelial cells. Exp Toxicol Pathol. 2000;52:227–33.

80. Ince C. Microcirculation in distress: a new resuscitation end point. Crit Care Med. 2004;32:1963–4.
81. Li JM, Shah AM. Endothelial cell superoxide generation: regulation and relevance for cardiovascular pathophysiology. Am J Physiol Regul Integr Comp Physiol. 2004;287: R1014–30.
82. Cariou A, Chiche JD, Charpentier J. The era of genomics: impact on sepsis clinical trial design. Crit Care Med. 2002;30:S341–8.
83. Kumar A, Short J, Parrillon JE. Genetic factors in septic shock. JAMA. 1999;282:579–81.
84. Arbour NC, Lorenz E, Schutte BC. TLR4 mutations are associated with endotoxin hyporesponsiveness in humans. Nat Genet. 2000;25:187–91.
85. Vallet B, Wiel E. Endothelial cell dysfunction and coagulation. Crit Care Med. 2002;29:S36–41.
86. Schuetz P, Jones AE, Aird WC. Endothelial cell activation in emergency department patients with sepsis-related and non-sepsis-related hypotension. Shock. 2011;36:104–8.
87. Aired WC. Vascular bed-specific hemostasis: role of endothelium in sepsis pathogenesis. Arch Immunol Ther Exp. 2000;48:439–42.
88. Pernerstorfer T, Hollenstein U, Hansen JB. Lepirudin blunts endotoxin-induced coagulation activation. Blood. 2000;95:1729–34.
89. Carson SD, Johnson DR. Consecutive enzyme cascades: complement activation at the cell surface triggers increased tissue factor activity. Blood. 1990;76:361–7.
90. Jansen PM, Eisele B, de Jong IW. Effect of C1 inhibitor on inflammatory and physiologic response patterns in primates suffering from lethal septic shock. J Immunol. 1998;160:475c–84.
91. Munford RS, Pugin J. Normal responses to injury prevent systemic inflammation and can be immunosuppressive. Am J Respir Crit Care Med. 2001;163:316–21.
92. Chrousos GP. The hypothalamic-pituitary-adrenal axis and immune- mediated inflammation. N Engl J Med. 1995;332:1351–62.
93. Romanovsky AA, Szekely M. Fever and hypothermia: two adoptive thermoregulatory responses to systemic inflammation. Med Hypotheses. 1998;50:219–26.
94. Tracey KJ. The inflammatory reflex. Nature. 2002;420:853–9.
95. Engelmann L, Petros S, Gundelach K. Die Bedeutung des “second hit” für den Sepsisverlauf. Intensiv Med. 2006;43:189–201.
96. Bochkov VN, Kadl A, Huber J. Protective role of phospholipid oxidation products in endotoxin-induced tissue damage. Nature. 2002;419:77–81.
97. Ayala A, Chung CS, Grutkoski PS. Mechanisms of immune resolution. Crit Care Med. 2003;31(Suppl):558–71.
98. Pugin J. Recognition of bacteria and bacterial products by host immune cells in sepsis. [Buchverf.]. In: Vincent JL, editor. Yearbook of intensive care and emergency medicine. Berlin: Springer; 1996. p. 11.
99. Adib-Conquy M, Cavaillon JM. Compensatory anti-inflammatory response syndrome. Thromb Haemost. 2009;101:36.
100. Bone RC. Immunologic dissonance: a continuing evolution in our understanding of systemic inflammatory response syndrome (SIRS) and the multiple organ dysfunction syndrome (MODS). Ann Intern Med. 1996;125:680–7.
101. Pinsky MR, Matuschak GM. Multiple systems organ failure: failure of host defense homeostasis. Crit Care Clin. 1989;5:199.
102. Ghosh S, Latimer RD, Gray BM. Endotoxin-induced organ injury. Crit Care Med. 1993;21(Suppl):S21.
103. Matsuda N, Hattory Y. Systemic inflammatory response syndrome (SIRS): molecular pathophysiology and gene therapy. J Pharmacol Sci. 2006;101:189–98.
104. Hotchkiss RS, Karl IE. The pathophysiology and treatment of sepsis. N Engl J Med. 2003;348:138–50.
105. Bistrian B. Systemic response to inflammation. Nutr Rev. 2007;65:S170–2.
106. Tracey KJ, Beutler B, Lowry SF. Shock and tissue injury induced by recombinant human cachetin. Science. 1986;234:470.

107. Cavaillon JM, Adib-Conquy M, Fitting C. Cytokine cascade in sepsis. Scand J Infect Dis. 2003;35:535–44.
108. Cavaillon JM, Mu-oz C, Fitting C. Circulating cytokines: the tip of the iceberg? Cric Shock. 1992;38:145–52.
109. Singh S, Evans TW. Organ dysfunction in sepsis. [Buchverf.]. In: Pinsky MR, Brochard L, Mancebo J, editors. Applied physiology in intensive care medicine. Berlin/Heidelberg: Springer; 2006. p. 345–56.
110. Godin PJ, Buchman TG. Uncoupling of biological oscillators: a complementary hypothesis concerning the pathogenesis of multiple organ dysfunction syndrome. Crit Care Med. 1996;24:1107–16.
111. Chrousos GP. The stress response and immune function: clinical implications. The 1999 Novera H Spector Lecture. Ann N Y Acad Sci. 2000;917:38–67.
112. Cohen J. The immunopathogenesis of sepsis. Nature. 2002;420:885–91.
113. Neviere R, Mathieu D, Chagnon JL. Skeletal muscle microvascular blood flow and oxygen transport in patients with severe sepsis. Am J Respir Crit Care Med. 1996;153:191.
114. Fukushima R, Alexander JW, Gianotti L. Isolated pulmonary infection acts as a source of systemic tumor necrosis factor. Crit Care Med. 1994;22:114–20.
115. Bone RC. Toward a theory regarding the pathogenesis of the systemic inflammatory response syndrome: what we do and do not know about cytokine regulation. Crit Care Med. 1996;24:163–72.
116. Bone RC, Grodzin CJ, Balk RA. Sepsis: a new hypothesis for pathogenesis of the disease process. Chest. 1997;112:235–43.
117. van der Poll T, van Deventer SJ. Cytokines and anticytokines in the pathogenesis of sepsis. Infect Dis Clin North Am. 1999;13:413–26.
118. Pinsky MR. Immune balance in critically ill patients. Arch Immunol Ther Exp. 2000;48:438–42.
119. Ward NS, Casserly B, Ayaly A. The compensatory anti-inflammatory response syndrome (CARS) in critilly ill patients. Clin Chest Med. 2008;29:617–viii.
120. Volk HD, Reinke P, Krausch D. Monocyte deactivation-rationale for a new therapeutic strategy in sepsis. Intensive Care Med. 1996;22 Suppl 4:S474–81.
121. Stephan F, Yang K, Tankovic J. Impairment of polymorphonuclear neutrophil functions precedes nosocomial infections in critically ill patients. Crit Care Med. 2002;30:315–22.
122. Cavaillon JM, Adrie C, Fitting C. Reprogramming of circulatory cells in sepsis and SIRS. J Endotoxin Res. 2005;11:311–20.
123. McCall CE, Yoza BK. Gene silencing in severe systemic inflammation. Am J Resipr Crit Care Med. 2007;175:763–7.
124. Prass K, Meisel C, Hoflich C. Stroke-induced immunodeficiency promotes spontaneous bacterial infections and is mediated by sympathetic activation reversal by poststroke T helper cell type 1-like immunostimulation. J Exp Med. 2003;198:725–36.
125. Olszewski MA, Falkowski NR, Surana R. Effect of laparotomy on clearance and cytokine induction in Staphylococcus aureus infected lungs. Am J Respir Crit Care Med. 2007;176:921–9.
126. White JC, Nelson S, Winkelstein JA. Impairment of antibacterial defense mechanisms of the lung by extrapulmonary infection. J Infect Dis. 1986;153:202–8.
127. Lyons A, Kelly JL, Rodrick ML. Major injury induces increased production of interleukin-10 by cells of the immune system with a negative impact on resistance to infection. Ann Surg. 1997;226:450–8.
128. Hensler T, Heidecke CD, Hecker H. Increased susceptibility to postoperative sepsis in patients with impaired monocyte IL-12 production. J Immunol. 1998;161:2655–9.
129. Spolarics Z, Siddiqi M, Siegel JH. Depressed interleukin-12-producing activity by monocytes correlates with adverse clinical course and a shift toward Th2-type lymphocyte pattern in severely injured male trauma patients. Crit Care Med. 2003;31:1722–9.
130. Spengler RN, Allen RM, Remick DG. Stimulation of alpha-adrenergic receptor augments the production of macrophage-derived tumor necrosis factor. J Immunol. 1990;145:1430–4.
131. Severn A, Rapson NT, Hunter CA. Regulation of tumor necrosis factor production by adrenaline and beta-adrenergic agonists. J Immunol. 1992;148:3441–5.

132. Zinyama RB, Bancroft GJ, Sigola LB. Adrenaline suppression of the macrophage nitric oxide response to lipopolysaccharide is associated with differential regulation of tumor necrosis factor-alpha and interleukin-10. Immunology. 2001;104:439–46.
133. Chrousos GP. The stress response and immune function: clinical implications. The 1999 Novera H Spector Lecture. Ann N Y Acad Sci. 2000;917:38–67.
134. Prigent H, Maxime D, Annane D. Mechanisms of impaired adrenal function in sepsis and molecular actions of glucocorticoids. Crit Care. 2004;8:243–52.
135. Annane D, Sebille V, Troche G. A 3-level prognostic classification in septic shock based on cortisol levels and cortisol response to corticotropin. JAMA. 2000;283:1038–45.
136. Sharshar T, Gray F, de la Grandmaison LG. Apoptosis of neurons in cardiovascular autonomic centres triggered by inducible nitric oxide synthase after death from septic shock. Lancet. 2003;362:1799–805.
137. Landry DW, Levin HR, Gallant EM. Vasopressin deficiency contributes to the vasodilation of septic shock. Circulation. 1997;95:1122–5.
138. Sharshare T, Blanchard A, Paillard M. Circulating vasopressin levels in septic shock. Crit Care Med. 2003;31:1752–8.
139. Riedemann NC, Guo R, Ward PA. Novel strategies for the treatment of sepsis. Nat Med. 2003;9:517–24.
140. Thomas L. Germs. N Engl J Med. 1972;287:553–5.
141. Bone RC, Balk RA, Cerra FB. Definitions for sepsis and organ failure and guidelines for the use of innovative therapies in sepsis. Chest. 1992;101:1644–55.
142. Stone R. Search for sepsis drugs goes on despite past failures. Science. 1994;264:365–7.
143. Warren HS. Strategies for the treatment of sepsis. N Engl J Med. 1997;336:952–3.
144. Qureshi K, Rajah A. Septic shock: a review article. Br J Med Pract. 2008;1:7–12.
145. Fink MP, Heard SO. Laboratory models of sepsis and septic shock. J Surg Res. 1990;49:186–96.
146. Deitch EA. Animal models of sepsis and shock: a review and lessons learned. Shock. 1998;9:1–11.
147. Debts JMH, Kampmeijer R, van der Linden MPMH. Plasma tumor necrosis factor and mortality in critically ill septic patients. Crit Care Med. 1989;17:489–94.
148. Oberholzer A, Oberholzer C, Moldawer LL. Cytokine signaling–regulation of the immune response in normal and critically ill states. Crit Care Med. 2000;28(Suppl):N3–12.
149. Pruitt JH, Welborn MB, Edwards PD. Increased soluble interleukin-1 type II receptor concentrations in postoperative patients and in patients with sepsis syndrome. Blood. 1996;87:3282–8.
150. Hotchkiss RS, Swanson PE, Freeman BD. Apoptotic cell death in patients with sepsis, shock, and multiple organ dysfunction. Crit Care Med. 1999;27:1230–51.
151. Weinberg JM, Venkatachalam MA. Guanine nucleotides and acute renal failure. J Clin Invest. 2001;108:1279–81.
152. Noble JS, MacKirdy FN, Donaldson SI. Renal and respiratory failure in Scottish ICUs. Anaesthesia. 2001;56:124–9.
153. Wheeler AP, Bernard GR. Treating patients with severe sepsis. N Engl J Med. 1999;340:207–14.
154. Zeni F, Freeman BF, Natanson C. Antiinflammatory therapies to treat sepsis and septic shock: a reassessment. Crit Care Med. 1997;25:1095–100.
155. Singer M, De Santis V, Vitale D. Multiorgan failure is an adoptive, endocrine-mediated, metabolic response to overwhelming systemic inflammation. Lancet. 2004;364:545–8.
156. Tsai AG, Kerger H, Intaglietta M. Oxygen distribution and consumption by the microcirculation and the determinants of tissue survival. [Buchverf]. In: Messmer WJ, Fink K, Sibbald MP, editors. Tissue oxygenation in acute medicine. Berlin/Heidelberg: Springer; 1998.
157. Ince C. The microcirculation is the motor of sepsis. Crit Care. 2005;9 Suppl 4:S13–29.
158. Trzeciak S, Cinel I, Dellinger RP. Resuscitating the microcirculation in sepsis: the central role of nitric oxide, emerging concepts for novel therapies, and challenges for clinical trials. Acad Emerg Med. 2008;15:399–413.
159. Vincent J-L, De Backer D. Microvascular dysfunction as a cause of organ dysfunction in severe sepsis. Crit Care. 2005;9 Suppl 4:S9–12.

160. Walley KR. Heterogeneity of oxygen delivery impairs oxygen extraction by peripheral tissues: theory. J Appl Physiol. 1996;81:885–94.
161. Goldman D, Bateman RM, Ellis CG. Effect of decreased O2 supply on skeletal muscle oxygenation and O2 consumption during sepsis: role of heterogeneous capillary spacing and blood flow. Am J Physiol Heart Circ Physiol. 2006;290:H2277–85.
162. Eipel C, Bordel R, Nickels RM. Impact of leukocytes and platelets in mediating hepatocyte apoptosis in a rat model of systemic endotoxemia. Am J Physiol Gastrointest Liver Physiol. 2004;286:G769–76.
163. Boeckstegers P, Weidenhofer S, Pilz G. Peripheral oxygen availability within skeletal muscle in sepsis and septic shock: comparison to limited infection and cardiogenic shock. Infection. 1991;19:317–23.
164. Sair M, Etherington PJ, Winlove CP. Tissue oxygenation and perfusion in human skeletal muscle in patients with systemic sepsis. Crit Care Med. 2001;29:1343–9.
165. Nencioni A, Trzeciak S, Shapiro NI. The microcirculation as a diagnostic and therapeutic target in sepsis. Intern Emerg Med. 2009;4:413–8.
166. Karimova A, Pinsky MR. The endothelial response to oxygen deprivation: biology and clinical implications. Intensive Care Med. 2001;27:19–31.
167. Fink MP. Cytopathic hypoxia: Is oxygen use impaired in sepsis as a result of an acquired intrinsic derangement in cellular respiration. Crit Care Clin. 2002;18:165–75.
168. Fink MP. Cytopathic hypoxia: mitochondrial dysfunction as mechanism contributing to organ dysfunction in sepsis. Crit Care Clin. 2001;17:219–37.
169. Crouser ED, Julian MW, Blaho DV. Endotoxin-induced mitochondrial damage correlates with impaired respiratory activity. Crit Care Med. 2002;30:276.
170. Rosengarten B, Hecht M, Auch D. Microcirculatory dysfunction in the brain precedes changes in evoked potentials in endotoxin-induced sepsis syndrome in rats. Cerebrovasc Dis. 2007;23:140.
171. Boulos M, Astiz ME, Barua RS. Impaired mitochondrial function induced by serum from septic shock patients is attenuated by inhibition of nitric oxide synthase and poly (ADP-ribose) synthase. Crit Care Med. 2003;31:353–8.
172. Brealey D, Brand M, Hargreaves I. Association between mitochondrial dysfunction and severity and outcome of septic shock. Lancet. 2002;360:219–23.
173. Ellis CG, Bateman RM, Sharpe MD. Effect of maldistribution of microvascular blood flow on capillary O 2 extraction in sepsis. Am J Physiol Heart Circ Physiol. 2002;282:156–64.
174. Parrillo JE. Cardiovascular dysfunction in septic shock: new insights into a deadly disease. Int J Cardiol. 1985;7:314–21.
175. Sawyer DB, Loscalzo J. Myocardial hibernation: restorative or preterminal sleep? Circulation. 2002;105:1517–9.
176. Khan AU, Delude RL, Han YY. Liposomal NAD(+) prevents diminished O(2) consumption by immunostimulated Caco-2 cells. Am J Physiol Lung Cell Mol Physiol. 2002;282:L1082–91.
177. Sörensen TIA, Nielsen GG, Andersen PK. Genetic and environmental influences on premature death in adult adoptees. N Engl J Med. 1988;318:727–32.
178. Lin MT, Albertson TE. Genomic polymorphism in sepsis. Crit Care Med. 2004;32:569–79.
179. Stuber F. Effects of genomic polymorphisms on the course of sepsis: is there a concept for gene therapy? J Am Soc Nephrol. 2001;12 Suppl 17:S60–4.
180. Read RC, Camp NJ, di Giovine FS. An interleukin-1 genotype is associated with fatal outcome of meningococcal disease. J Infect Dis. 2000;182:1557–60.
181. Eisen DP, Dean MM, Thomas P. Low mannose-binding lectin function is associated with sepsis in adult patients. FEMS Immunol Med Microbiol. 2006;48:274–82.
182. Koike K, Moore FA, Moore EE. Endotoxin after gut ischemia/reperfusion causes irreversible lung injury. J Surg Res. 1992;52:556–62.
183. MacFie J, O'Boyle C, Mitchell CJ. Gut origin of sepsis: a prospective study investigating associations between bacterial translocation, gastric microflora, and septic morbidity. Gut. 1999;45:223–8.
184. Bastian L, Weimann A. Praktische Aspeckte der frühen enteralen Ernährung. Anaesthesiol Reanimat. 1999;24:95–100.

185. Carrico CJ, Meakins JL, Marshall JC. Multiple-organ-failure syndrome: the gastrointestinal tract–the motor of MOF. Arch Surg. 1986;121:197–201.
186. Fine J, Frank ED, Rutenberg SH. The bacterial factor in traumatic shock. N Engl J Med. 1959;260:214–20.
187. Clark J, Coopersmith CM. Intestinal crosstalk – a new paradigm for understanding the gut as the "motor" of critical illness. Shock. 2007;28:384–93.
188. Lederer JA, Rodrick ML, Mannick JA. The effects of injury on the adaptive immune response. Shock. 1999;11:153–9.
189. Oberholzer A, Oberholzer C, Moldawer LL. Sepsis syndromes: understanding the role of innate and acquired immunity. Shock. 2001;16:83–96.
190. Meakins JL, Pietsch JB, Bubenick O. Delayed hypersensitivity: indicator of acquired failure of host defences in sepsis and trauma. Ann Surg. 1977;186:241–50.
191. McCabe WR, Jackson GG. Gram-negative bacteremia. I: etiology and ecology. Ann Intern Med. 1962;110:847–53.
192. Brun-Bulsson C, Doyon F, Carlet J. Incidence, risk factors, and outcome of severe sepsis and septic shock in adults. A multicenter, prospective study in intensive care units. French ICU Group for Severe Sepsis. JAMA. 1995;74:968–74.
193. Bruserud O, Nesthus I, Buhring HJ. Cytokine modulation of interleukin 1 and tumor necrosis factor-alpha secretion from acute myelogenous leukemia blast cells in vitro. Leuk Res. 1995;19:15–22.
194. Singer M, Glynne P. Treating critical illness: the importance of first doing no harm. PLoS Med. 2005;2:e167.
195. Nathan C. Points of control in inflammation. Nature. 2002;420:846–52.
196. Jarrar D, Chaudry IH, Wang P. Organ dysfunction following hemorrhage and sepsis: mechanisms and therapeutic approaches. Int J Mol Med. 1999;4:575–83.
197. Marshall JC. Inflammation, coagulopathy, and the pathogenesis of multiple organ dysfunction syndrome. Crit Care Med. 2001;29:S99–106.
198. Pathan CA, Hemingway CA, Alizadeh AA. Role of interleukin 6 in myocardial dysfunction of meningococcal septic shock. Lancet. 2004;363:203–9.
199. Schrier RW, Wang W. Acute renal failure and sepsis. N Engl J Med. 2004;351:159–69.
200. Levi M, Ten Cate H. Disseminated intravascular coagulation. N Engl J Med. 1999;341:586–92.
201. Bion J. Pathophysiology of acute respiratory distress syndrome and acute lung injury. [Buchverf.]. In: Shapiro MJ, Singer M, Suter PM, Webb AR, editors. Oxford textbook of critical care. Oxford: Oxford University Press; 1999. p. 923–6.
202. Bonanno FG. Shock- a reappraisal: the holistic approach. J Emerg Trauma Shock. 2012;5:167–77.
203. Holmes CL. Vasoactive drugs in the intensive care unit. Curr Opin Crit Care. 2005;11:413–7.
204. Astiz ME, Rackow EC. Septic shock. Lancet. 1998;351:1501–5.
205. Trzeciak S, Rivers EP. Clinical manifestations of disordered microcirculatory perfusion in severe sepsis. Crit Care. 2005;9 Suppl 4:S20–6.
206. Ince C, Sinaasappel M. Microcirculatory oxygenation and shunting in sepsis and shock. Crit Care Med. 1999;27:1369–77.
207. Ruokonen E, Takala J, Kari A. Regional blood flow and oxygen transport in septic shock. Crit Care Med. 1993;21:1296–303.
208. Ellis CG, Jagger J, Sharpe M. The microcirculation as a functional system. Crit Care. 2005;9 Suppl 4:S3–8.
209. Astiz ME, DeGent GE, Lin RY. Microvascular function and rheologic changes in hyperdynamic sepsis. Crit Care Med. 1995;23:265.
210. Schumacker PT, Wood LDH. Limitations of aerobic metabolism in critically illness. Chest. 1984;85:453–4.
211. Bellomo R, Wan L, May C. Vasoactive drugs and acute kidney injury. Crit Care Med. 2008;36(Suppl):S179–86.

212. Vieillard-Baron A. Septic cardiomyopathy. Ann Int Care. 2011;1:6.
213. Reilly JM, Cunnion RE, Burch-Whitman C. A circulating myocardial depressant substance is associated with cardiac dysfunction and peripheral hypoperfusion (lactic acidemia) in patients with septic shock. Chest. 1989;95:1072–80.
214. Tavener SA, Kubes P. Is there a role for cardiomyocyte toll-like receptor 4 in endotoxemia? Trends Cardiovasc Med. 2005;15:153–7.
215. Natanson C, Eichenholz PW, Danner RL. Endotoxin and tumor necrosis factor challenges in dogs stimulate the cardiovascular profile of human septic shock. J Exp Med. 1989;169:823–32.
216. Brady AJ, Poole-Wilson PA. Circulatory failure in septic shock Nitric oxide: too much of a good thing? Nitric oxide: too much of a good thing? Br Heart J. 1993;70:103–5.
217. Rassaf T, Poll LW, Brouzos P. Positive effects of nitric oxide on left ventricular function in humans. Eur Heart J. 2006;27:1699–705.
218. Brady AJ, Warren JB, Poole-Wilson PA. Nitric oxide attenuates cardiac myocyte contraction. Am J Physiol. 1993;265:H176–82.
219. Pacher P, Beckman JS, Liaudet L. Nitric oxide and peroxynitrite in health and disease. Physiol Rev. 2007;87:315–424.
220. Ferdinandy P, Danial H, Ambrus I. Peroxynitrite is a major contributor to cytokine-induced myocardial contractile failure. Circ Res. 2000;87:241–7.
221. Vieillard-Baron A, Caille V, Charron C. Actual incidence of global left ventricular hypokinesia in adult septic shock. Crit Care Med. 2008;36:1701–6.
222. Munt B, Jue J, Gin K, et al. Diastolic filling in human severe sepsis: an echocardiographic study. Crit Care Med. 1998;26:1829–33.
223. Barraud D, Faivre V, Damy T. Levosimendan restores both systolic and diastolic cardiac performance in lipopolysaccharide-treated rabbits: comparison with dobutamine and milrinone. Crit Care Med. 2007;35:1376–82.
224. Gkisioti S, Mentzelopoupus SD. Vasogenic shock physiology. Open Access Emerg Med. 2011;3:1–6.
225. Young JD. The heart and circulation in severe sepsis. Br J Anaesth. 2004;93:114–20.
226. Landry DW, Oliver JA. The pathophysiology of vasodilatory shock. N Engl J Med. 2001;345:588–95.
227. Titheradge MA. Nitric oxide in septic shock. Biochem Biophys Acta. 1999;1411:437–55.
228. Taylor BS, Geller DA. Molecular regulation of the human inducible nitric oxide synthase (iNOS) gene. Shock. 2000;13:413–24.
229. Kotsovolis G, Kallaras K. The role of endothelium and endogenous vasoactive substances in sepsis. Hippokratia. 2010;14:88–93.
230. Ochoa JB, Udekwu AO, Billiar TR. Nitrogen oxide levels in patients after trauma and during sepsis. Ann Surg. 1991;214:612–26.
231. Mosi R, Seguin B, Cameron B. Mechanistic studies on AMD6221: a ruthenium- based nitric oxide scavenger. Biochem Biophys Res Commun. 2002;29:519–29.
232. Matsuda N, Hattori Y. Vascular biology in sepsis: pathophysiological and therapeutic significance of vascular dysfunction. J Smooth Muscle Res. 2007;43:117–37.
233. Parker MM, Shelhamer JH, Bacharach SL. Profound but reversible myocardial depression in patients with septic shock. Ann Intern Med. 1984;100:483–90.
234. Weng L, Liu YT, Du B. The prognostic value of left ventricular systolic function measured by tissue Doppler imaging in septic shock. Crit Care. 2012;16:R71–9.
235. Repesse X, Charron C, Vieillard-Baron A. Evaluation of left ventricular systolic function revisited in septic shock. Crit Care. 2013;17:164.
236. Vieillard-Baron A, Schmidt JM, Beauchet A. Early preload adaption in septic shock: a transesophageal echocardiographic study. Anesthesiology. 2001;94:400–6.
237. Reichek N, Wilson J, Sutton MSJ. Noninvasive determination of left ventricular end-systolic stress: validation of the method and initial application. Circulation. 1982;65:99–108.
238. Vlachopoulos C, Dima I, Aznaouridis K. Acute systemic inflammation increases arterial stiffness and decreases wave reflections in healthy individuals. Circulation. 2005;112:2193–200.
239. Kharbanda RK, Walton B, Allen M. Prevention of inflammation -induced endothelial dysfunction: a novel vasculo-protective action of aspirin. Circulation. 2002;105:2600–4.

240. Pleiner J, Heere-Ress E, Langenberger H. Adrenoreceptor hyporeactivity is responsible for Escherichia coli endotoxin-induced acute vascular dysfunction in humans. Arterioscler Thromb Vasc Biol. 2002;22:95–100.
241. Nichols WW, O'Rourke MF, editors. McDonalds blood flow in arteries. [Buchverf]. 4th ed. London: Arnold E; 1998. p. 170–222, 284–315, 347–95, 450–76.
242. Bagshaw SM, George C, Bellomo R. Early acute kidney injury in sepsis: a multicenter evaluation. Crit Care. 2008;12:R47.
243. Oppert M, Engel C, Brunkhorst FM. Acute renal failure in patients with severe sepsis and septic shock–a significant independent risk factor for mortality: results from the German Prevalence Study. Nephrol Dial Transplant. 2008;23:904–9.
244. Vincent JL, Sakr Y, Sprung CL. Sepsis in European intensive care units: results of the SOAP study. Crit Care Med. 2006;34:344–53.
245. Parmar A, Langenberg C, Wan L. Epidemiology of septic acute kidney injury. Curr Drug Targets. 2009;10:1169–78.
246. Bagshaw SM, Uchino S, Bellomo R. Septic acute kidney injury in critically ill patients: clinical characteristics and outcomes. Clin J Am Soc Nephrol. 2007;2:431–9.
247. Lehman L, Daeed M, Moody G. Hypotension as a risk factor for acute kidney injury in ICU patients. Comput Cardiol. 2010;37:1095–8.
248. Wan L, Bagshaw SM, Langenberg C. Pathophysiology of septic acute kidney injury: what do we really know? Crit Care Med. 2008;36:S198–203.
249. Chvojka J, Sykora R, Karvunidis T. New developments in septic acute kidney injury. Physiol Res. 2010;59:859–69.
250. Langenberg C, Bellomo R, May C. Renal blood flow in sepsis. Crit Care. 2005;9:R363–74.
251. Payen D, Lukaszewicz A-C, Legrand M. A multicentre study os acute kidney injury in severe sepsis and septic shock: association with inflammatory phenotype and HLA genotype. PLoS One. 2012;7:e35838.
252. Sutton TA, Fisher CJ, Molitoris BA. Microvascular endothelial injury and dysfunction during ischemic acute renal failure. J Am Soc Nephrol. 2005;16:117–24.
253. Benes J, Chvojka J, Sykora R. Searching for mechanisms that matter in early septic acute kidney injury: an experimental study. Crit Care. 2011;15:R256.
254. Lerolle N, Guerot E, Faisy C. Renal failure in septic shock: predictive value of Doppler-based renal arterial resistive index. Intensive Care Med. 2006;32:1553–9.
255. Sladen RN, Landry D. Renal blood flow regulation, autoregulation, and vasomotor nephropathy. Anesth Clin North Am. 2000;18:791–807.
256. Yamaguchi N, Jesmin S, Zaedi S. Time-dependent expression of renal vaso-regulatory molecules in LPS-induced endotoxemia in rat. Peptides. 2006;27:2258–70.
257. Wan L, Bellomo R, Giantomaso D. The pathogenesis of septic acute renal failure. Curr Opin Crit Care. 2003;9:496–502.
258. Kox WJ, Volk T, Kox SN. Immunomodulatory therapies in sepsis. Intensive Care Med. 2000;26(Suppl):124–8.
259. Majetschak M, Flach R, Kreuzfelder E. The extent of traumatic damage determines a graded depression of the endotoxin responsiveness of peripheral blood mononuclear cells from patients with blunt injuries. Crit Care Med. 1999;27:313–8.
260. Docke WD, Randow F, Syrbe U. Monocyte deactivation in septic patients: restoration by IFN-gamma treatment. Nat Med. 1997;3:678–81.
261. Bistrian BR. Dietary treatment in secondary wasting and cachexia. J Nutr. 1999;129(Suppl):290–4.
262. Raghavan M, Marik PE. Management of sepsis during the early "golden hours". J Emerg Med. 2006;31:185–99.
263. De Gaudio AR, Rinaldi S, Chelazzi C. Pathophysiology of sepsis in the elderly: clinical impact and therapeutic considerations. Curr Drug Targets. 2009;10:60–70.
264. Girard TD, Opal SM, Ely EW. Insights into severe sepsis in older patients: from epidemiology to evidence-based management. Clin Infect Dis. 2005;40:719–27.
265. Cohen J, Brun-Buisson C, Torres A. Diagnosis of infection in sepsis: an evidence based review. Crit Care Med. 2004;32(Suppl):S466–94.

266. Morris DL, Chambers HF, Morris MG. Hemodynamic characteristics of patients with hypothermia due to occult infection and other causes. Ann Intern Med. 1985;102:153–7.
267. Cohen J. Sepsis and septic shock: inching forward. Clin Med. 2009;9:256–7.
268. Carlet J, Cohen J, Calandra T. Sepsis: time to reconsider the concept. Crit Care Med. 2008;36:964–6.
269. Dellinger RP. Shock overview. Semin Respir Crit Care Med. 2004;25:619–28.
270. Hunter JD, Doddi M. Sepsis and the heart. Br J Anaesth. 2010;104:3–11.
271. Nasa P, Juneja D, Singh O. Severe sepsis and septic shock in elderly: an overview. World J Crit Care. 2012;1:23–30.
272. Opal SM, Girard TD, Ely EW. The immunopathogenesis of sepsis in the elderly patients. Clin Infect Dis. 2005;41 Suppl 7:S504–12.
273. Cornbleet PJ. Clinical utility of the band count. Clin Lab Med. 2002;22:101–36.
274. Wenz B, Gennis P, Canova C. The clinical utility of the leukocyte differential in emergency medicine. Am J Clin Pathol. 1986;86:298–303.
275. Callaham M. Inaccuracy and expense of the leukocyte count in making urgent clinical decisions. Ann Emerg Med. 1986;15:774–81.
276. Gerlach H. Sepsis. [Buchverf]. In: Moreno R, Ranieri R, Rhodes M, Kuhlen A, editors. 25 years of progress and innovation in intensive care medicine. Berlin: Med. Wissenschaftliche Verlagsgesellschaft; 2007. p. 141.
277. Marshall JC, Vincent JL, Fink MP. Measures, markers, and mediators: toward a staging system for clinical sepsis: a report of the fifth Toronto Sepsis Roundtable, Toronto, Ontario, Canada, October 25–26, 2000. Crit Care Med. 2003;31:1560–7.
278. Tang BM, Eslick GD, Craig JC. Accuracy of procalcitonin for sepsis diagnosis in critically ill patients: systematic review and meta-analysis. Lancet Infect Dis. 2007;7:210–7.
279. Gattas DJ, Cook DJ. Procalcitonin as a diagnostic test for sepsis: health technology assessment in the ICU. J Crit Care. 2003;18:52–8.
280. de Werra I, Jaccard C, Corradin SB. Cytokines, nitrite/nitrate, soluble tumor necrosis factor receptors, and procalcitonin concentrations: comparisons in patients with septic shock, cardiogenic shock, and bacterial pneumonia. Crit Care Med. 1997;25:607–13.
281. Christ-Crain M, Jaccard-Stolz D, Bingisser R. Effect of procalcitonin-guided treatment on antibiotic use and outcome in power respiratory tract infections: cluster-randomised, single-blinded intervention trial. Lancet. 2004;363:600–7.
282. Mello PMV, Sharma VK, Dellinger RP. Shock overview. Semin Respir Crit Care Med. 2004;25:619–28.
283. Vincent JL, Roman A, Kahn RJ. Dobutamine administration in septic shock: addition to a standard protocol. Crit Care Med. 1990;18:689–93.
284. Waxman K, Nolan LS, Shoemaker WC. Sequential perioperative lactate determination: physiological and clinical implications. Crit Care Med. 1982;10:96–9.
285. Shapiro NI, Howell MD, Talmor D. Serum lactate as a predictor of mortality in emergency department patients with infection. Ann Emerg Med. 2005;45:524–8.
286. Bakker J, Gris P, Coffernils M. Serial blood lactate levels can predict the development of multiple organ failure following septic shock. Am J Surg. 1996;171:221–6.
287. Cunha BA. Sepsis and septic shock: selection of empiric antimicrobial therapy. Crit Care Clin. 2008;24:313–34.
288. Reimer LG, Wilson ML, Weinstei MP. Update on detection of bacteremia and fungemia. Clin Microbiol Rev. 1997;10:444–65.
289. Munson EL, Diekema DL, Beekmann SE. Detection and treatment of bloodstream infection: laboratory reporting and antimicrobial management. J Clin Microbiol. 2003;41:495–7.
290. Bochud P-Y, Glauser MP, Calandra T. Antibiotics in sepsis. Intensive Care Med. 2001;27 Suppl 1:S33–48.
291. Washington 2nd JA, Ilstrup DM. Blood cultures: issues and controversies. Rev Infect Dis. 1986;8:792–800.
292. Aronson MD, Bor DH. Blood cultures. Ann Intern Med. 1987;106:246–53.
293. Seifert H. The clinical importance of microbiological findings in the diagnosis and management of bloodstream infections. Clin Infect Dis. 2009;48 Suppl 4:S238–45.

294. Lee A, Mirrett S, Reller LB. Detection of bloodstream infections in adults: how many blood cultures are needed? J Clin Microbiol. 2007;45:3546–8.
295. Cockerill III FR, Wilson JW, Vetter EA. Optimal testing parameters for blood cultures. Clin Infect Dis. 2004;38:1724–30.
296. Kumar A, Roberts D, Wood KE. Duration of hypotension before initiation of effective antimicrobial therapy is the critical determinant of survival in human septic shock. Crit Care Med. 2006;34:1589–96.
297. Lodise TP, McKinnon PS, Swiderski L. Outcomes analysis of delayed antibiotic treatment for hospital-acquired Staphylococcus aureus bacteremia. Clin Infect Dis. 2003;36:1418–23.
298. Cruz K, Dellinger RP. Diagnosis and source of sepsis: the utility of clinical findings. [Buchverf.]. In: Carlet J, Opal SM, Vincent JL, editors. The sepsis test. Boston: Kluwer Academic Publishers; 2002. p. 11–28.
299. Shapiro NI. Who needs a blood culture? A prospectively derived and validated prediction rule. J Emerg Med. 2008;35:255.
300. Mermel LA, Maki DG. Detection of bacteremia in adults: consequences of culturing an inadequate volume of blood. Ann Intern Med. 1993;119:270–2.
301. Lamy B, Roy P, Carret G. What is the relevance of obtaining multiple blood samples for culture? A comprehensive model to optimize the strategy for diagnosing bacteremia. Clin Infect Dis. 2002;35:842–50.
302. Levinson AT, Casserly BP, Levy MM. Reducing mortality in severe sepsis and septic shock. Semin Respir Crit Care Med. 2011;32:195–205.
303. Sepsis-Gesellschaft, Deutsche. AWMF-online. Prävention, Diagnose, Therapie und Nachsorge der Sepsis. [Online] Feb 2010. [Zitat vom: 27 Jul 2013].
304. Cunha BA. Sepsis and its mimics in the CCU. [Buchverf.]. In: Cunha BA, editor. Infectious diseases in clinical practice. 2nd ed. New York: Informa Healthcare; 2007.
305. Perman SM, Goyal M, Gaieski DF. Initial emergency department diagnosis and management of adult patients with severe sepsis and septic shock. Scand J Trauma Resusc Emerg Med. 2012;20:41.
306. Marik PE. Surviving sepsis: going beyond the guidelines. Ann Intensive Care. 2011;1:17–29.
307. Gaieski DF, Mikkelson ME, Band RA. Impact of time to antibiotics on survival in patients with severe sepsis and septic shock on whom early goal-directed therapy was initiated in the emergency department. Crit Care Med. 2010;38:1045–53.
308. Bochud P, Bonten Y, Marchetti MO. Antimicrobial therapy for patients with severe sepsis and septic shock: an evidence-based review. Crit Care Med. 2004;32(Suppl):S495–512.
309. Russel F. Management of sepsis. N Engl J Med. 2006;355:1699–713.
310. Natanson C, Danner RJ, Reilly JM. Antibiotics versus cardiovascular support in canine model of human septic shock. Am J Physiol. 1990;259:H1440–7.
311. Tarassenko L, Hann A, Young D. Integrated monitoring and analysis for early warning of patient deterioration. Br J Anaesth. 2006;97:64–8.
312. Perman SM, Goyal M, Gaiseki DF. Initial emergency department diagnosis and management of adult patients with severe sepsis and septic shock. Scan J Trauma Resusc Emerg Med. 2012;20:41.
313. Pushkarich MA, Trzeciak S, Shapiro NI, Emergency Medicine Shock Research Network (EMSHOCKNET). Association between timing of antibiotic administration and mortality from septic shock in patients treated with a quantitative resuscitation protocol. Crit Care Med. 2011;39:2066–71.
314. Garnacho-Montero J, Garcia-Garmendia JL, Barrero-Almodovar A. Impact of inadequate empirical antibiotic therapy on the outcome of patients admitted to the intensive care unit with sepsis. Crit Care Med. 2003;31:2742–51.
315. MacArthur RD, Miller T, Albertson T. Adequacy of early empiric antibiotic treatment and survival in severe sepsis: experience from the MONARCS trial. Clin Infect Dis. 2004;38:285–8.
316. Kollef MH, Sherman G, Ward S. Inadequate antimicrobial treatment of infections: a risk factor for hospital mortality among critically ill patients. Chest. 1999;115:462–74.

317. Ibrahim EH, Sherman G, Ward S. The influence of inadequate antimicrobial treatment of bloodstream infections on patient outcomes in the ICU setting. Chest. 2000;118:146–55.
318. Kreger BE, Craven DE, McCabe WR. Gram-negative bacteremia. Am J Med. 1980;68:344–55.
319. Bryan CS, Reynolds KL, Brenner ER. Analysis of 1,186 episodes of gram-negative bacteremia in non-university hospitals: the effects of antimicrobial therapy. Rev Infect Dis. 1983;5:629–38.
320. Vallo J, Relles J, Ocagavia A. Community-acquired bloodstream infection in critically ill adult patients: impact of shock and inappropriate antibiotic therapy on survival. Chest. 2003;123:1615–23.
321. Diaz-Martin A, Martinez-Gonzalez ML, Ferrer R. Antibiotic prescription patterns in empirical therapy of severe sepsis: combination of antimicrobials with different mechanisms of action reduce mortality. Crit Care. 2012;16:R223.
322. Brunkhorst FM, Oppert M, Marx G. Effect of empirical treatment with moxifloxacin and meropenem vs meropenem on sepsis-related organ dysfunction in patients with severe sepsis. JAMA. 2012;307:2390–9.
323. Craig WA. Pharmacokinetic/pharmacodynamic parameters: rationale antibacterial dosing of mice and men. Clin Infect Dis. 1998;16:1–12.
324. Parrillo JE, Parker MM, Natanson C. Septic shock in humans. Advances in the understanding of pathogenesis, cardiovascular dysfunction, and therapy. Ann Intern Med. 1990;113:227–42.
325. Sharper J, Ahmed J, Schafer T. Volume therapy with colloid solutions preserves intestinal microvascular perfusion in endotoxemia. Resuscitation. 2008;76:120–8.
326. Trzeciak S, McCoy JV, Dellinger PR. Early increases in microcirculatory perfusion during protocol-directed resuscitation are associated with reduced multi-organ failure at 24 h in patients with sepsis. Intensive Care Med. 2008;34:2210–7.
327. Ospina-Tascon G, Neves AP, Occhipinti G. Effects of fluids on microvascular perfusion in patients with severe sepsis. Intensive Care Med. 2010;36:949–55.
328. Al-Khafaji AH, Webb AR. Fluid resuscitation. Anaesth Crit Care Pain. 2004;4:127–31.
329. Ognibene FP. Hemodynamic support during sepsis. Clin Chest Med. 1996;17:279–87.
330. Haupt MT, Gilbert EM, Carlson RW. Fluid loading increases oxygen consumption in septic patients with lactic acidosis. Am Rev Respir Dis. 1985;131:912–6.
331. Marik PE, Monnet X, Teboul JL. Hemodynamic parameters to guide fluid therapy. Ann Intensive Care. 2011;1:1.
332. Rivers E, Nguyen B, Havstad S. Early goal-directed therapy in the treatment of severe sepsis and septic shock. N Engl J Med. 2001;345:1368–77.
333. Wiedemann HP, Wheeler AP, Bernard GR. Comparison of two fluid-management strategies in acute lung injury. N Engl J Med. 2006;354:2564–75.
334. Uchino S, Bellomo R, Morimatsu H. Pulmonary artery catheter versus pulse contour analysis: a prospective epidemiological study. Crit Care. 2006;10:R174.
335. Murphy CV, Schramm GE, Doherty JA. The importance of fluid management in acute lung injury secondary to septic shock. Chest. 2009;136:102–9.
336. Boyd JH, Forbes J, Nakada TA. Fluid resuscitation in septic shock: a positive fluid balance and elevated central venous pressure are associated with increased mortality. Crit Care Med. 2011;39:259–65.
337. Rivers EP. Fluid management strategies in acute lung injury: liberal, conservative, or both? N Engl J Med. 2006;355:2564–75.
338. Levy MM, Macias WL, Vincent J-L. Early changes in organ function predict eventual survival in severe sepsis. Crit Care Med. 2005;33:2194–201.
339. Pottecher J, Deruddre S, Teboul JL. Both passive leg raising and intravascular volume expansion improve sublingual microcirculatory perfusion in severe sepsis and septic shock patients. Intensive Care Med. 2010;36:1867–74.
340. Rosenberg AL, Dechert RE, Park PK. Review on a large clinical series: association of cumulative fluid balance on outcome in acute lung injury: a retrospective review of the ARDSnet tidal volume study cohort. J Intensive Care Med. 2009;24:35–46.

341. Reinhart K, Karzai W. Anti-tumor necrosis factor therapy in sepsis: an update on clinical trials and lessons learned. Crit Care Med. 2001;29(Suppl):S121–5.
342. Bernard GR, Vincent JL, Laterre P-F. Efficacy and safety of recombinant human activated protein C for severe sepsis. N Engl J Med. 2001;344:699–709.
343. Jr Fisher CJ, Agosti JM, Opal SM. Treatment of septic shock with tumor necrosis factor receptor: Fc fusion protein. N Engl J Med. 1996;334:1697–702.
344. Otero RM, Nguyen HB, Huang DT. Early-goal-directed therapy in severe sepsis and septic shock revisited: concepts, controversies, and contemporary findings. Chest. 2006;130:1579–95.
345. He ZY, Gao Y, Wang XR. Clinical evaluation of execution of early goal directed therapy in septic shock. Zhongguo Wei Zhong Bing Ji Jiu Yi Xue. 2007;19:14–6.
346. Kortgen A, Niederprüm P, Bauer M. Implementation of an evidences-based "standard operating procedure" and outcome in septic shock. Crit Care Med. 2006;34:943–9.
347. Shapiro NI, Howell MD, Talmor D. Implementation and outcomes of the Multiple Urgent Sepsis Therapies (MUST) protocol. Crit Care Med. 2006;34:1025–32.
348. Trzeciak S, Dellinger RP, Abate NL. Translating research to clinical practice: a 1-year experience with implementing early goal–directed therapy for septic shock in the emergency department. Chest. 2006;129:225–32.
349. Barochia A, Cui X, Vitberg D. Bundled care for septic shock: an analysis of clinical trials. Crit Care Med. 2010;38:668–78.
350. Hicks P, Cooper S, Webb J. The surviving sepsis campaign: International guidelines for management of severe sepsis and septic shock: 2008. An assessment by the Australian and New Zealand Intensive Care Society. Anaesth Intensive Care. 2008;36:149–51.
351. Perel A. Bench-to-bedside review: the initial hemodynamic resuscitation of the septic patient according to surviving sepsis campaign guidelines–does one size fit all? Crit Care. 2008;12:1.
352. Marik PE, Varon J. Early goal-directed therapy (EGDT): on terminal life support? Am J Emerg Med. 2010;28:243–5.
353. Marik PE, Baram M, Vahid B. Does the central venous pressure predict fluid responsiveness? A systematic review of the literature and the tale of seven mares. Chest. 2008;134:172–8.
354. Mirsky MR, Payen D. Functional hemodynamic monitoring. Crit Care. 2005;9:566.
355. Pinsky MR. Hemodynamic evaluation and monitoring in the ICU. Chest. 2007;132:2020–9.
356. Bafaquech F. CVP and volume responsiveness of cardiac output. Am J Respir Crit Care Med. 2004;169:A344.
357. Blumberg N, Heal JM. Transfusion-induced immunomodulation and its possible role in cancer recurrence and perioperative bacterial infection. Yale J Biol Med. 1990;63:429–33.
358. Shander A, Goodnough LT. Why an alternative to blood transfusion? Crit Care Clin. 2009;25:261–77.
359. Shander A, Javidroozi M, Ozawa S. What is really dangerous: anaemia or transfusion? Br J Anaesth. 2011;107(S 1):i41–59.
360. Barrett NA, Kam PC. Transfusion-related acute lung injury: a literature review. Anaesthesia. 2006;61:777–85.
361. Shander A, Fink A, Javidroozi M. Appropriateness of allogeneic red blood cell transfusion: the international consensus conference on transfusion outcomes. Transfus Med Rev. 2011;25:232–46.
362. Marik PE. Hazards of blood transfusions. Br J Hosp Med. 2009;70:12–5.
363. Napolitano LM, Kurek S, Luchette FA. Clinical practice guideline: red blood cell transfusion in adult trauma and critical care. Crit Care Med. 2009;37:3124–57.
364. Improvement, Joint Commission on Healthcare. Raising the bar with the bundles: improving the quality of care by improving the work environment. Joint Commission Perspectives on Patient Safety. [Online] April 2006. [Zitat vom: 2 Sep 2009]. http://www.ingentaconnect.com/content/jcaho/jcpps/2006/00000006/00000004/art00003.
365. Harden C. What is a bundle? A resource from the Institute for Healthcare Improvement. [Online] [Zitat vom: 27 Feb 2008]. http://www.ihi.org/IHI/Topics/CriticalCare/IntensiveCare/ImprovementStories/WhatIsaBundle.htm.

366. Rabuel C, Mebazaa A. Septic shock: a heart story since the 1960s. Intensive Care Med. 2006;32:799–807.
367. De Backer D, Crteur J, Dubois MJ. The effects of dobutamine on microcirculatory alterations in patients with septic shock are independent of its systemic effects. Crit Care Med. 2006;34:403–8.
368. De Backer D, Donadello K, Taccone FS. Microcirculatory alterations: potential mechanisms and implications for therapy. Ann Intensive Care. 2011;1:27.
369. Sackett DL, Rosenberg WM, Gray JA. Evidence-based medicine: what it is and what it isn't. BMJ. 1996;312:71–2.
370. Delany A, Peake SL, Bellomo R. Australian resuscitation in sepsis evaluation trial statistical plan. Emerg Med Australas. 2013;25:406–15.
371. Pike F, Yealy DM, Kellum J. Protocolized care for early septic shock (ProCESS) statistical analysis plan. Crit Care Resusc. 2013;15:301–10.
372. Power S, Harrison D, Mouncey P. The protocolised management in sepsis (PRoMISE) trial statistical analysis plan. Crit Care Resusc. 2013;15:311–7.
373. Conterno LO, Silva Filho CR, Ruggeberg JU. Conjugate vaccines for preventing meningococcal C meningitis and septicemia. Cochrane Database Syst Rev. 2006;3, CD001834.
374. Swingler G, Fransman D, Hussey G. Conjugate vaccines for preventing Haemophilus influenzae type B infections. Cochrane Database Syst Rev. 2007;2, CD001729.
375. Mykietiuk A, Carratala J, Dominguez A. Effect of prior pneumococcal vaccination on clinical outcome of hospitalized adults with community-acquired pneumococcal pneumonia. Eur J Clin Microbiol Infect Dis. 2006;25:457–62.
376. Plebani A, Soresina A, Rondelli R. Clinical, immunological, and molecular analysis in a large cohort of patients with X-linked agammaglobulinemia: an Italian multicenter study. Clin Immunol. 2002;104:221–30.
377. Dark PM, Dean P, Warhurst G. Bench-to-bedside review: the promise of rapid infection diagnosis during sepsis using polymerase chain reaction-based pathogen detection. Crit Care. 2009;13:217.
378. NISS. National Nosocomial Infections Surveillance (NNIS) system report, data summary from January 1992–April 2000, issued June 2000. Am J Infect Control. 2000;28:429–48.
379. Cunha BA. Antibiotic treatment of sepsis. Med Clin North Am. 1995;79:551–8.
380. Zubert S, Funk DJ, Kumar A. Antibiotics in sepsis and septic shock: like everything else in life, timing is everything. Crit Care Med. 2010;38:1211–2.
381. Torres A, Serra-Batlles J, Ferrer A. Severe community-acquired pneumonia: epidemiology and prognostic factors. Am Rev Respir Dis. 1991;144:312–8.
382. Cunha BA. Antibiotic essentials. 7th ed. Royal Oak: Physicians Press; 2008.
383. Hardawy RM. A review of septic shock. Am Surg. 2000;66:22–9.
384. Lim WS, Baudouin SV, George RC. British Thoracic Society guidelines for the management of community acquired pneumonia in adults: update 2009. Thorax. 2009;64(Suppl III):iii 1–55.
385. Cabarello J, Rello J. Combination antibiotic therapy for community-acquired pneumonia. Ann Intensive Care. 2011;1:48.
386. Lodise TP, Kwa A, Cosler L. Ritel : Comparison of beta-lactam and macrolide combination therapy versus flourquinolone monotherapy in hospitalized Veterans Affair patients with community-acquired pneumonia. Antimicrob Agents Chemother. 2007;51:3977–82.
387. Wunderink RG, Mandell L. Adjunctive therapy in community-acquired. Semin Respir Crit Care Med. 2012;33:311–8.
388. Rodríguez A, Mendia A, Sirvent JM. Combination antibiotic therapy improves survival in patients with community-acquired pneumonia and shock. Crit Care Med. 2007;35:1493–8.
389. Bodí M, Rodríguez A, Solé-Violán J, Community-Acquired Pneumonia Intensive Care Units (Capuci) Study Investigators. Antibiotic prescription for community-acquired pneumonia in the intensive care unit: impact of adherence to Infectious Diseases Society of America guidelines on survival. Clin Infect Dis. 2005;41:1709–16.
390. File TM, Bartlet JG, Romer AR. Treatment of community-acquired pneumonia in patients who require hospitalization. [Online] 14 June 2013. [Zitat vom: 25 July 2013]. http://www.uptodate.com/contents/treatment-of-community-acquired-pneumonia-in-adults-who-require-hospitalization#H5.

391. Marik PE. Assessment of intravascular volume: a comedy of errors. Crit Care Med. 2001;29:1635–6.
392. Marik PE, Varon J. Sepsis: state of the art. Dis Mon. 2001;47:465–532.
393. Monnet X, Teboul JL. Passive leg raising. Intensive Care Med. 2008;34:659–63.
394. Vavallaro F, Sandroni C, Marano C. Diagnostic accuracy of passive leg raising for prediction of fluid responsiveness in adults: systematic review and meta-analysis of clinical studies. Intensive Care Med. 2010;36:1475–83.
395. Lafanechere A, Pene F, Goulenok C. Changes in aortic blood flow induced by passive leg raising predict fluid responsiveness in critically ill patients. Crit Care. 2006;10:R132.
396. Andrijauskas A, Svensen CH, Ivaskevicius J. Goal directed fluid therapy revised: indirect monitoring of interstitial fluid accumulation during mini fluid challenges with crystalloids. Open Conf Proc J. 2012;3:43–51.
397. Durairaj L, Schmidt GA. Fluid therapy in resuscitated sepsis. Chest. 2008;133:252–63.
398. Boulain T, Achard JM, Teboul JL. Changes in blood pressure induced by passive leg raising predict response to fluid loading in critically ill patients. Chest. 2002;121:1245–52.
399. Rackow EC, Falk JL, Fein IA. Fluid resuscitation in circulatory shock: a comparison of cardiorespiratory effects of albumin, hetastarch, and saline solutions in patients with hypovolemic and septic shock. Crit Care Med. 1983;11:839–50.
400. Thooft A, Favory R, Salgado-Ribeiro D. Effects of changes in arterial pressure on organ perfusion during septic shock. Crit Care. 2011;15:R222.
401. Bellomo R, Kellum JA, Wisnieski SR. Effects of norepinephrine on the renal vasculature in normal and endotoxemic dogs. Am J Respir Crit Care Med. 1999;159:1186–92.
402. Guyton AC, Hall JE. Textbook of medical physiology. Philadelphia: Saunders; 2005.
403. Pottecher S, Cavalat H, Dupont J. Hemodynamic management of severe sepsis: recommendations of the French intensive care societies (SFAR/SRLF) consensus conference, 13 October 2005, Paris, France. Crit Care. 2006;10:311.
404. Bourgoin A, Leone M, Delmas A. Increasing mean arterial pressure in patients with septic shock: Effects on oxygen variables and renal function. Crit Care Med. 2005;33:780–6.
405. Dubin A, Pozo MO, Casabella CA. Increasing arterial blood pressure with norepinephrine does not improve microcirculatory blood flow: a prospective study. Crit Care. 2009;13:R92.
406. Perrson PB. Renal blood flow autoregulation in blood pressure control. Curr Opin Nephrol Hypertens. 2002;11:67–72.
407. Correa TD, Vuda M, Takala J. Increasing mean arterial blood pressure in sepsis: effects on fluid balance, vasopressor load and renal function. Crit Care. 2013;17:R21.
408. Badin J, Boulain T, Ehrmann S. Relation between mean arterial pressure and renal function in the early phase of shock: a prospective, explorative cohort study. Crit Care. 2011;15:R135.
409. De Backer D, Biston P, Devriendt J. Comparison of dopamine and norepinephrine in the treatment of shock. N Engl J Med. 2010;362:779–89.
410. Jhanji S, Stirling S, Patel N. The effect of increasing doses of norepinephrine on tissue oxygenation and microvascular flow in patients with septic shock. Crit Care Med. 2009;37:1961–6.
411. De Backer D, Aldecoa C, Njimi H. Dopamine versus norepinephrine in the treatment of septic shock: a meta-analysis. Crit Care Med. 2012;40:725–30.
412. Alsous F, Khamiees M, DeGirolamo A. Negative fluid balance predicts survival in patients with septic shock: a retrospective pilot study. Chest. 2000;117:1749–54.
413. Groeneveld AB, Polderman KH. Acute lung injury, overhydration or both. Crit Care. 2005;9:136–7.
414. Astiz ME, Rachow EC. Crystalloid-colloid controversy revisited. Crit Care Med. 1999;27:34–5.
415. Perel A, Roberts I. Colloids versus crystalloids for fluid resuscitation in critically ill patients. Cochrane Database Syst Rev. 2011;3, CD000567.
416. Choi PT, Yip G, Quinonez LG. Crystalloids vs colloids in fluid resuscitation: a systematic review. Crit Care Med. 1999;27:200–10.
417. Finfer S, Bellomo R, Boyce N. A comparison of albumin and saline for fluid resuscitation in the intensive care unit. N Engl J Med. 2004;250:2247–56.

418. Jungheinrich C, Scharpf R, Wargenau M. The pharmacokinetics and tolerability on an intravenous infusion of the new hydroxyethyl starch 130/0.4 (6%, 500 ml) in mild-to severe renal impairment. Anaesth Analg. 2002;95:544–51.
419. Kato A, Yonemura K, Matsushima H. Complication of oliguric acute renal failure in patients treated with low-molecular weight dextran. Ren Fail. 2001;23:679–84.
420. Finfer S, Chittock DR, Su SY, NICE-SUGAR Study Investigators. Intensive versus conventional glucose control in critically ill patients. N Engl J Med. 2009;360:1283–97.
421. Study, Investigators of the NICE-SUGAR. Hypoglycemia and risk of death in critically ill patients. N Engl J Med. 2012;367:1108–18.
422. Marik PE, Bellomo R. Stress hyperglycemia: an essential survival response! Crit Care. 2013;17:305.
423. Cade JF. High risk of the critically ill for venous thromboembolism. Crit Care Med. 1982;10:448–50.
424. Hirsch DR, Ingenito EP, Goldhaber SZ. Prevalence of deep venous thrombosis among patients in medical intensive care. JAMA. 1995;274:335–7.
425. Garlund B. Randomized, controlled trial of low-dose heparin for prevention of fatal pulmonary embolism in patients with infectious disease: the Heparin Prophylaxis Study Group. Lancet. 1996;347:1357–61.
426. Geerts W, Cook D, Shelby R. Venous thromboembolism and its prevention in critical care. J Crit Care. 2002;17:95–104.
427. Kupfer Y, Anwar J, Seneviratne C. Prophylaxis with subcutaneous heparin significantly reduces the incidence of deep venous thrombophlebitis in the critically ill. Am J Crit Care Med. 1999;159(Suppl):A519.

Chapter 5
Acute Kidney Injury

5.1 Definition

A sudden – minutes to days – deterioration of the kidney function, characterized by a rapid decline in glomerular filtration rate (GFR) resulting in acid–base, fluid and electrolyte disorders, and accumulation of nitrogenous and other metabolic waste products, defines acute renal failure (ARF) [1–6], now termed acute kidney injury (AKI) [1, 2, 5, 7, 8].

In 2004, the Acute Dialysis Quality Initiative (ADQI) inaugurated an integrative and uniform consensus definition of disorders of the renal function, the *RIFLE* criteria [9]. This definition is based on the common and widely used parameters serum creatinine level (SCre) and urinary output (uo) and has been intended to replace the more than 30 definitions of ARF which had been in use until then [10]. Furthermore, the RIFLE criteria classify the severity of renal dysfunction:

RIFLE: criteria and classification [9]

Stage	GFR criteria	Urine output criteria
Risk	Increased SCre × 1.5 or GFR decrease > 25 %	uo < 0.5 ml/kg/h × 6 h
Injury	Increased SCre × 2 or GFR decrease > 50 %	uo < 0.5 ml/kg/h × 12 h
Failure	Increased SCre × 3 or GFR decrease > 75 % or SCre ≥ 4 mg/dl (Acute rise ≥ 0.5 mg/dl)	uo < 0.3 ml/kg/h × 24 h or anuria × 12 h
Loss	Persistent AKI = complete loss of kidney function > 4 weeks	
ESKD	End-stage kidney disease (>3 months)	

In order to overcome a few shortcomings of the RIFLE definition and classification the acute kidney injury network (AKIN), an international collaboration of nephrologists and intensivists, suggested in 2007 some adjustments to the RIFLE criteria [11]: The reduction in renal function occurs abruptly, the *diagnosis should*

W. Krüger, A.J. Ludman, *Core Knowledge in Critical Care Medicine*,
DOI 10.1007/978-3-642-54971-7_5, © Springer-Verlag Berlin Heidelberg 2014

be made within 48 h, an *acute exacerbation of chronic renal failure is not included in RIFLE*, *patients requiring or who are on renal replacement therapy* are automatically assigned to stage III AKIN and the matter that *even small increases in creatinine are substantially and independently affecting the prognosis* [12–16] should be taken into consideration when diagnosing and staging critically ill patients. Of course, there is no place for RIFLE loss and ESKD class in the acute classification scale of AKIN [11]. The AKIN definition and classification are as follows [11]:

Stage	Creatinine criteria		Urine output criteria
Stage I	Increase[a] SCre ×1.5 or increase SCre ≥ 0.3 mg/dl	or	uo < 0.5 ml/kg/h × 6 h
Stage II	Increase[a] SCre ×2	or	uo < 0.5 ml/kg/h × 12 h
Stage III	Increase[a] SCre ×3 or SCre ≥ 4 mg/dl (with acute rise of ≥0.5 mg/dl)	or	uo < 0.3 ml/kg/h × 24 h or anuria × 12 h

Patients requiring/receiving renal replacement therapy (RRT) are assigned to stage III
[a]Increase from baseline

AKIN stage **I** equates RIFLE-**R**, AKIN **II** RIFLE-**I** and AKIN **III** RIFLE-**F** [10].

In case the baseline serum creatinine level is not known, the following simplified and modified formulas are recommended for calculation [17, 18]:

$$\mathrm{GFR} = 186 \times \left(\text{serum creatinine}^{-1.154}\right) \times \left(\text{age}^{-0.0203}\right) \quad -\text{if man}$$

$$\mathrm{GFR} = 186 \times \left(\text{serum creatinine}^{-1.154}\right) \times \left(\text{age}^{-0.0203}\right) \times 0.742 \quad -\text{if female}$$

$$\mathrm{GFR} = 186 \times \left(\text{serum creatinine}^{-1.154}\right) \times \left(\text{age}^{-0.0203}\right) \times 1.210 \quad -\text{if black}$$

$$\left[\mathrm{GFR\ in\ ml/min/1.73\,m^2}\right]$$

For classification in both RIFLE and AKIN, the worst of each criterion has to be used [19].

In the meantime, the RIFLE criteria and classification have been well validated in several studies [20–25]: All demonstrate a clear correlation between RIFLE class and mortality, as well as length of ICU and hospital stay; furthermore, the more severe the disease and thus the higher the RIFLE class, the lower the renal recovery rate [26]. Ricci [20] found in his review that in several studies the relative risk for death was higher when applying the creatinine criterion alone compared to studies where both, uo criterion and creatinine criterion, had been considered. In line with this result, patients classified as being at risk by creatinine criterion were found as being more severely ill when compared with those in the same RIFLE class but considered by the uo criterion alone [27].

The dynamic timeframe over which renal dysfunction develops, and the degree of increase in serum creatinine is relevant for both staging and estimating the patient's prognosis by RIFLE and by AKIN criteria [28]. Misclassifications occur less often in RIFLE than in AKIN with RIFLE showing a higher sensitivity and robustness in predicting outcomes in AKI [28].

The comparison between the original RIFLE definition and classification and AKIN does not demonstrate any advantage of the AKIN modification regarding diagnosing patients with AKI, robustness of the criteria and predictive abilities when applied within the first 24 h after admission to ICU [29]: Kellum [19] states that "*AKI*, as *defined by RIFLE criteria*, is now recognized as an important ICU syndrome alongside other syndromes used in ICU patients for the purpose of epidemiology and trial execution, such as the acute lung injury/acute respiratory distress syndrome consensus criteria".

AKI implies injury or damage, but not necessarily dysfunction [30]. Neither the RIFLE criteria nor the AKIN categories with their modifications are able to exclude transient changes in creatinine or uo as may be found temporary, the so-called "prerenal" azotemic conditions due to volume depletion, elusive decreases in renal blood flow or other easily reversible situations [7, 26]. Furthermore, due to the absence of any relation between the functional degree of the kidney disorder and tissue damage (severe tissue damage may be found in only mild dysfunction and vice versa) [4, 31, 32], an international consensus conference on acute renal failure suggested in order to delineate the syndrome to use the expression "acute kidney insufficiency" rather than acute kidney injury [33].

5.2 Epidemiology

Roughly 1 % of all hospital admissions are caused by acute renal failure [34], and the course of about 5–7 % of all inpatients is complicated by an acute deterioration of kidney function [35, 36]. The prevalence of renal dysfunction is particularly high in patients hospitalized in the ICU, varying between 5–20 % [36] and 1–25 % [37–39], and in the critically ill with multiorgan failure (MOF), where up to 50 % develop AKI [40]. In case of sepsis or even septic shock, depending on the definition used, 11 % [41] to 50 % [37, 42–44] of patients are affected and may develop AKI.

Uchino et al. [45] published a prospective observational study and reported that 1,738 from 29,269 patients (5.9 %) admitted to an ICU unit suffered from AKI. 1,260 or 4.3 % of all ICU admissions or 72.5 % – 1,260 from 1,738 – of those suffering from ARF needed renal replacement therapy (RRT); septic shock was found in 47.5 %, major surgery in 34.3 %, cardiogenic shock in 26.9 %, hypovolaemia in 25.6 % and medication/drug induced in 19 %, emphasizing the very important fact that often multiple reasons are contributing. 30 % of the patients had a known (chronic) renal dysfunction.

When applying the new RIFLE criteria and classification, up to 18 % of all hospitalized patients were found to suffer or to develop AKI classified as AKI-R 9.1 %, AKI-I 5.2 %, AKI-F 3.7 % [46]. An evaluation of ICU patients revealed more than 65 % of patients with compromised renal function when RIFLE criteria and classification were adopted: RIFLE-R 12.4 %, RIFLE-I 26.6 % and RIFLE-F 28.1 % [47]. This is in accordance with the review by Hoste who found in his analysis including 17 epidemiological studies with more than 100,000 patients that more than 2/3 of the patients admitted to an ICU suffered from AKI and that 4–5 % required RRT [48]. These results confirm earlier ones, where AKI severe enough to require RRT was found in about 5 % of general ICU population [45].

If AKI occurs as a single and primary renal disease, mortality rates between 7 and 23 % are reported [36, 40, 49], and in critically ill populations, mortality rates between 50 and 80 % reliant on the severity of comorbidities and the quantity of failing organs are described [36, 49–51]: AKI substantially affects the prognosis and hence the outcome of patients admitted to intensive care unit [13, 39, 45].

Factors that may contribute to increased mortality in ARF include volume overload, acid–base derangements, insulin resistance and enhanced oxidative stress and inflammation [52, 53]. Infections and cardiorespiratory failure are the main causes of death [36]; 15–30 % progress to ESRD [49, 52].

ARF is recognized as an independent risk for death [15], and even mild reductions in kidney function are associated with both increased in-hospital mortality [14, 15, 54] and poor long-term outcome [55, 56].

In RIFLE class F, hospital mortality was found to be as high as approximately 50 % [46]; if the condition is severe enough to require RRT, mortality lies as high as 50–60 % [48] and may even exceed 75 % in patients with underlying sepsis or in patients with MOF [57–59].

However, although the numbers are still severely varying between studies, there is consensus that the problem is big and the mortality high [48].

Acute kidney injury is almost ubiquitous in critically ill patients [60], and sepsis is the predominant reason for AKI in ICU patients [42, 43, 45, 61].

5.3 Aetiology

In the majority of cases, the *aetiology of AKI is multifactorial* [5, 36, 44, 45, 49]. Hemodynamic disorders, largely

- *Low cardiac output* and
- *Hypovolaemia*
  causing hypotension and renal hypoperfusion are together with
- *Sepsis*,
- *Major surgery* and
- *Nephrotoxic agents/drugs*

reported being the five leading sources of AKI [35, 36, 40, 45, 49, 62]; *often two or more are co-existing* [45, 49].

Other frequent conditions are [5]:

- *Intra-abdominal hypertension* including abdominal compartment syndrome [1, 63–66]
- *Venous congestion*/elevated central venous pressure secondary to miscellaneous disorders [67–69]
- Hepatorenal syndrome [5, 70, 71]

- Cardiopulmonary bypass [14]
- Trauma [72]
- Rhabdomyolysis [73]

Medications/drugs may alter autoregulation and/or intra-renal hemodynamics, e.g. NSAIDs and ACE inhibitors [5, 74, 75], exert direct tubular toxic effects, e.g. aminoglycosides and vancomycin [5, 75, 76], or affect the renal structures and function by several other mechanisms like induction of allergic interstitial inflammation, e.g. methicillin [77]; may cause obstructive tubulopathy, e.g. aminoglycosides, vancomycin, cisplatin and amphotericin [74, 78, 79], or vasculopathy, e.g. antineoplastics, immunotherapeutics and antiplatelets [74, 80, 81].

Radiocontrast (by affecting renal blood flow and intra-renal haemodynamics [82], and by direct toxic effects, the latter probably the predominant harmful effect in the pathology of contrast media-associated renal dysfunction [82–84]) is shown to provoke AKI in a substantial number of cases.

Nash reported in his trial [35] that acute radiocontrast nephropathy is the third leading cause of acute tubular necrosis (ATN), and that patients with underlying chronic kidney disease and contrast media administration in the setting of hypovolaemia are at special risk [5, 74, 85, 86].

Special risk factors to develop AKI comprise *diabetes mellitus*, *chronic* (pre-existing) *renal impairment*/kidney disease and *older age* [74, 87, 88] with patients suffering from diabetes with renal failure, and those with diabetic nephropathy are at highest exposure [89].

Other risk factors reported are specific drugs, i.e. radiocontrast media; hypovolaemia; major surgery, in particular cardiopulmonary bypass; heart failure; and ICU patients [90, 91].

Traditionally, the aetiologies of AKI are broadly classified into three categories: prerenal, intra-renal or intrinsic renal and postrenal causes [1, 3, 4, 6, 92]. This grouping is still widely accepted and in use [70, 74] although challenged, and in particular the expression prerenal has become less precise [93].

5.3.1 Pre-renal

The expression "prerenal" implies the cause(s) affecting and compromising the kidney function arises (arise) "outside" the kidneys and outside the urinary excretory pathways [94–96]. Furthermore, prerenal denotes a structurally normal kidney parenchyma and that the condition is transient or functional, implicating a relatively easily reversible dysfunction of the organ [70, 93]. However, a transient, functional reduction in renal performance characterized by an increase in serum creatinine concentration (called prerenal azotemia) and/or diminution in uo attributed to a reduction in GFR may simply be an adequate physiological response to a diminished renal perfusion [4, 19, 89, 92, 97].

The kidney shows remarkable compensatory abilities in case of hypoperfusion (see section on pathophysiology) [70, 89], whereupon *any reduction in renal perfusion can induce prerenal azotemia* [70, 74, 89, 92]. Renal hypoperfusion is mainly determined by (*relative*) *hypotension**, *true hypovolaemia*, *reduction in effective circulating volume* (and thus systemic and renal perfusion pressure [98]), e.g. low cardiac output and systemic vasodilation, or by *intra-renal vasoconstriction* as induced by drugs, and as found in sepsis or in case of inadequate compensatory circulatory reactions to pressure drops [1, 70, 87, 89, 99]. Examples are to follow [1, 70, 87, 89, 99]:

Haemorrhage	Such as trauma, gastrointestinal, renal, *surgery*
Distributive shock	Such as *sepsis*, *hepatorenal syndrome*, vasodilative drugs, anaphylaxis, SIRS
Heart failure	
Cardiogenic shock	
Fluid losses	Such as *bowel obstruction*, pancreatitis, vomiting, burns, *diuretics*, diabetes mellitus and insipidus, *surgery*, *diarrhoea*, *polyuria*
Drugs/medication	Inducing intra-renal vasoconstriction such as *radiocontrast agents* [100] or attenuate prostaglandin-mediated intra-renal vasodilation, i.e. NSAIDs [101, 102]

*Relative hypovolemia may be found in conditions such as chronic hypertension, atherosclerosis, chronic kidney disease and renal artery stenosis, in drug therapy with NSAIDs and ACE inhibitors, and in sepsis, liver cirrhosis or hepatorenal syndrome as well. The kidney is very vulnerable to even transient pressure drops or diminished perfusion, both inducing a temporary hypoperfusion [99, 103–105].

Diminished renal perfusion can even occur if systemic blood pressure is normotensive as in case of increased sympathetic discharge avoiding a significant blood pressure fall as may be seen in low cardiac output, pulmonary oedema, acute myocardial infarction or other maladies where the compensatory circulatory reaction is excessive but nevertheless causes an intra-renal vascular tone imbalance entailing to a drop in GFR [99, 106, 107].

Thus, prerenal azotemia, a very common feature in different clinical settings [60, 70], is defined as "decreased glomerular filtration rate (GFR), resulting from renal hypoperfusion in a structurally intact kidney, which is rapidly reversible when the underlying cause is corrected" [1, 4, 6, 108]. Hence, prerenal azotemia has to be seen as a functional response of the kidneys to a temporary diminished organ perfusion [99].

If the condition causing prerenal azotemia persists long enough and/or is severe enough, tubular cell damage may ensue – this is termed acute tubular necrosis, ATN [6, 109, 110]. Under those circumstances renal ischaemia may become manifest and established, and *intrinsic* ischaemic AKI occurs [3, 4, 6, 70, 87, 89]. However, in humans the magnitude of ischaemia necessary to induce a significant damage and/or substantial functional limitation of kidney performance is not known [87].

ATN, a specific histological feature describes the morphological findings emerging when mimicking renal hypoperfusion by occlusion of the renal artery in animal models [60, 111]. However, we know that such results are not always applicable to humans: The term ATN is somewhat of a misnomer [4], and does not describe and represent the condition and feature correctly. There is plenty of evidence demonstrating that the presence of tubular necrosis in the histological examination is rare, especially in human clinical settings, and frank necrotic tubular cells may even not be detectable [31, 32, 112, 113]. Furthermore, substantial impairment of renal function may not be accompanied by a relevant amount of necrotic tubular cells [114, 115]. Moreover, quite often ATN is even used synonymous with ARF (and hence more recently AKI) in the clinical setting, although their definitions are markedly different [1, 116].

ATN and prerenal azotemia may be considered as different stages on a continuum of the same pathophysiological process of tubular injury, with ATN historically denoting a marker of severe ischaemic or, less common, toxic injury [3, 6, 35, 60, 70, 109]. Recent publications, anyhow, revealed the substantial role of systemic inflammation, endothelial dysfunction, coagulation abnormalities and oxidative stress in the pathology of acute kidney injury [116, 117]. Of course, "true ATN does in fact occur" [19], i.e. in case of arterial catastrophes such as ruptured aneurysms or acute aortic dissections, but the numbers are very small (<1 %) [19].

AKI is entitled to overcome the described shortcomings, particular

(a) When does pre-renal azotemia end and when does pre-renal failure and ATN begin, and how much structural injury is necessary to convert pre-renal azotemia to ATN,
(b) The imbalance between functional deficiency and histological picture, and,
(c) The confusion in the terminology as the expression acute kidney injury encompasses the entire spectrum of the syndrome of disturbed renal function [9, 19, 25, 93].

AKI includes pre-renal azotemia and ATN, but other conditions such as mild and severe states of ARF as well [8, 19, 93].

Compromised renal perfusion is reported as being the *prevailing cause of AKI* [34, 35, 45, 74, 87, 110, 118, 119], although *hypoperfusion may, at least, coexist* with any stage of AKI [89, 97]. Hypoperfused kidneys are highly sensitive to nephrotoxic drug effects [89, 99]: Patients with sustained prerenal azotemia are proned to ischaemic-induced AKI; thereby often multiple factors [44, 87, 120], in particular combinations with drugs, are engaged [119, 121], e.g. IV radiocontrast given to patients on ACE inhibitors [121], aminoglycoside application in septic patients [122] and NSAIDs applied to patients suffering from congestive heart failure [89, 123]. Furthermore, patients with diabetes mellitus and pre-existing renal dysfunction are in particular susceptible to develop AKI (of what degree ever) at times of impaired perfusion even if the event is transient [88].

5.3.2 Intrinsic, Intra-renal Causes

Intrinsic, intra-renal or parenchymal renal failure is defined as a condition where the harmful resource directly injures the kidney structures as (1) glomeruli, (2) renal tubules, (3) interstitium and/or (4) vasculature [1, 3, 60, 116, 124].

Typical examples are

1. Acute or rapidly progressive *glomerulonephritis;*
2. Functional or structural tubular abnormalities of *toxic* (drugs such as intravenous contrast, aminoglycosides, amphotericin B [34, 35, 62, 74, 75, 89]) or *ischaemic genesis* or their combination are reported as being the predominant reasons for “intra-renal” pathology in the *vast majority of* established AKI [45, 49, 74, 87, 89, 119].
3. *Interstitial nephropathy* such as acute interstitial nephritis (e.g. methicillin, beta-lactam antibiotics, NSAIDs), infection (bilateral pyelonephritis) and infiltration by lymphoma or in case of amyloidosis
4. *Vasculitis* secondary to thrombotic thrombocytopenic purpura/haemolytic–uraemic syndrome and disseminated intravascular coagulation [3, 34, 35, 60, 62, 124].

Although “intrinsic” AKI is often multifactorial [87, 116], *ischaemia* is considered as being *most common cause* for (structural and/or functional) *tubular abnormalities*, with prerenal pathology generating hypoperfusion progressing to ischaemia, potentially inducing ATN [1, 87, 89, 116, 124].

Following the new definition and nomenclature, we should term this clinical condition *ischaemic AKI* emphasizing that no correlation and no evidence have been demonstrated in studies between tissue damage and severity of organ dysfunction [114, 115]. But scepticism is advisable as even experimental data cannot provide convincing evidence that ischaemia induces persistent AKI: at least severe, but transient, ischaemia was not sufficient to cause established AKI [125–127].

5.3.3 Postrenal Reasons

Postrenal AKI is attributed to obstructions of the urinary excretory pathways [116, 124]. Typical diseases are *prostatic hypertrophy or carcinoma*, *pelvic tumours*, calculi, strictures, clot, sloughed papillae, retroperitoneal fibrosis, lymphadenopathy and abdominal aortic aneurysm with surrounding inflammation [1, 116, 124]. It is relatively common in the community [60, 128] but accounts in total for less than 5 [35] to 10 % of all cases of AKI [1, 128].

Two special issues of remarkable importance in the aetiology of AKI and in daily clinical practice have just recently attracted our attention: intra-abdominal hypertension and systemic and local venous congestion.

5.3.3.1 Intra-abdominal Hypertension and Abdominal Compartment Syndrome

In recent years, abdominal hypertension and the abdominal compartment syndrome have been recognized as being a clinically relevant reason for AKI [66, 129–132]. Some authors allege that elevated intra-abdominal pressure, aside arterial hypotension, is one of the most substantial threats to renal perfusion in the acute setting [133].

Increased intra-abdominal pressures, in the worst case called abdominal compartment syndrome defined as an intra-abdominal pressure >20 mmHg accompanied by a new organ dysfunction, e.g. renal or liver failure, may decisively affect other organs [64, 134]. Some authors advocate to diagnose an abdominal compartment syndrome in daily practice in case a new organ dysfunction attributed to an elevated abdominal pressure occurs, and thus independently of a special pressure threshold [135–137]. A raised abdominal pressure leads, amongst others, to increased venous and tissue pressures and a decrease in renal blood flow and perfusion pressure comprising cardiac output and systemic oxygen delivery [132, 138, 139]. Reduction in urinary output correlates directly with the intra-abdominal pressure present: Oliguria is observed as the intra-abdominal pressure passes 15 mmHg, and anuria has to be expected when the pressure exceeds 30 mmHg [131, 139].

Several different conditions may increase the pressure in the closed abdominal cavity: *ascites*, *trauma*, colonic dilation, pseudo-obstruction, *mechanical bowel obstruction*, gastric dilation, *peritonitis*, pancreatitis, retroperitoneal haemorrhage, etc. [140–143]. Furthermore, extra-abdominal diseases may decisively influence the degree of the intra-abdominal pressure such as *excessive fluid* resuscitation/administration [64, 134, 144], *chronic obstructive pulmonary disease and obesity* [145]. In most of the cases of abdominal hypertension associated with AKI, intra-abdominal pressures exceed 20–24 mmHg, and the disease onset is acute or hyperacute [66].

Several pathomechanisms are suggested as being responsible for the compromised kidney function. Recent results favour the compression of the renal vein as the primary disorder [131, 146, 147] leading to a reduction in glomerular filtration pressure [148, 149]. Furthermore, reduced cardiac output and elevation of renal vascular resistance, both reducing renal blood flow, and direct compression of renal parenchyma by the elevated intra-abdominal pressure contribute to reduced urine generation [131, 138, 139, 146, 147].

Of special interest, mechanical ventilation with positive end expiratory pressure (PEEP) always decreases uo secondary to decreased renal perfusion pressure resulting from diminished renal blood flow [150].

5.3.3.2 Venous Congestion

In general, central venous pressure has a considerable impact on renal function. Damman et al. [67] demonstrated a substantial effect of the central venous pressure (CVP) on renal function and mortality in patients suffering from wide range of cardiovascular diseases: Elevated CVP may impair renal function and is a strong indicator of all-cause mortality. The association between the height of CVP and compromised kidney performance was independent of the cardiac function; notably, the correlation was even stronger in patients with preserved left ventricular function. This may be in line with results by others reporting the coexistence between diastolic dysfunction (characterized by increased filling pressures) and renal functional disorders, a relationship which turned out to be vice versa [151–153]. Furthermore, Kos found a strong correlation between CVP and renal blood flow in advanced heart failure [154]. Francis describes venous congestion attributed to advanced heart failure may worsen kidney function [155]. In right-sided heart failure, the backup venous pressure in all abdominal organs rises [156]. Conversely, mildly impaired renal function may initiate salt and water retention resulting in increased cardiac filling pressures [157].

Attempts to explain the depicted results always have to take into consideration the activation of the sympathetic and the renin–angiotensin–aldosterone system, the effects of natriuretic peptides and vasopressin and several other hormones and systems involved [158]. The pathophysiology of the contribution of venous congestion to a diminished GFR in heart failure seems to be extremely complex [159], the capillary hydrostatic pressure, the driving force of filtration, is substantially more influenced by changes in renal venous pressure rather than by changes in renal arteriolar pressure [160]. Increases in renal venous pressure reduce the capillary hydrostatic pressure [160]. Furthermore, as net filtration depends on the hydrostatic gradient across the capillary, venous congestion thereby increases interstitial hydrostatic pressure and thus will decrease the hydrostatic pressure gradient resulting in reduced net filtration [161].

Experimental data showing a linear decrease in GFR upon increases in renal venous pressures, in particular during volume expansion, support the considerable role renal venous pressure and renal venous congestion play in maintaining renal function [162].

"In the acute setting, the two most significant *threats to renal perfusion pressure* are *systemic arterial hypotension* and *increased intra-abdominal pressure* (including abdominal compartment syndrome)" [163].

5.3.4 Doubts about Our Traditional Concept

However, Schneider and Bellomo challenge our current concept, differentiating between pre-renal failure reflecting transient kidney malfunction with intact nephrons and easily reversible conditions, and intra-renal failure representing ATN [164]:

I. It is well established that renal tubular cells appear structurally well preserved and normal in the vast majority of histopathological examinations, particularly in patients with sepsis associated acute kidney injury, although the sick person may suffer from severe AKI suggestive for ATN [165].
II. Furthermore, our current urinary biochemistry (urinary indices) is unable to distinguish pre-renal AKI from persistent AKI traditionally indicating ATN [166–169].

Accordingly, AKI seems to reflect more a continuum of the same pathophysiological process rather than functional (pre-renal/transient) and structural (intra-renal) abnormalities [164, 170]. As such, our classical concept to discriminate between the two entities seems to be wrong and pre-renal AKI and ATN in our traditional view do only exist in our minds not reflecting the true renal pathobiology in case of acute kidney injury [164, 170].

5.4 Pathophysiology

Renal physiology and especially pathophysiology are highly complex, sophisticated [118, 171–173] and multifactorial [70, 174–176]. In recent years, the central, dominating and somehow overrated *pathophysiological role of global hypoperfusion* being necessary and almost imperative in the development of particularly ischaemic AKI has been attenuated and modified: A bigger review (analysing the available literature) of renal hemodynamics in sepsis by Langenberg et al. [177] revealing that in about one-third of all studies, renal blood flow (RBF) was preserved or even elevated with the GFR found decreased although CO was reported to be normal or actually raised. Meanwhile more clinically lifelike models [178, 179] and a clutch of other remarkable findings [173, 176, 180–184] on the mechanisms and processes involved in renal haemodynamics confirm Langenberg's results.

Hence, there is no doubt that microhaemodynamic (intra-renal) alterations and their ramifications along with non-hemodynamic, particularly cellular responses and issues, and notably the subsequent *inflammation* meaningfully and decisively contributing in the pathogenesis of acute (ischaemic and non-ischaemic) kidney injury [4, 118, 174, 185–187].

Thus, the primacy of renal blood flow in the pathogenesis of (ischaemic) AKI has been challenged and has to be modified. *Renal hypoperfusion*, at least in all low-flow conditions (i.e. severe heart failure and cardiogenic shock, hemorrhagic shock and other hypovolaemic states such as liver failure and hepatorenal syndrome, during most septic courses and other systemic inflammatory issues) and in case of systemic hypotension secondary to a systemic fall in vascular resistance, is not only the *most common source and contributor* of (ischaemic) AKI but still *a key pathogenetic factor and hallmark in the pathogenesis of acute* (and chronic) kidney injury [70, 87, 98, 118, 173, 185, 188]. By all means, hypoperfusion may coexist with any threat or renal insult [97]. However, "severe, *transient hypoperfusion* is able to reduce GFR and urinary output but not sufficient to induce *persistent* AKI", Legrand and Payen [189] state, following and embedding several study results in the pathogenesis of AKI [125–127, 176].

Glomerular capillary filtration pressure and consecutively glomerular filtration rate, the latter being a hallmark of renal function, considerably depend on renal perfusion/renal blood flow (RBF) [188, 190]. RBF is determined by (a) the effective circulating blood volume, (b) cardiac output and (c) renal perfusion pressure which is defined by and is a result of the relation between the tone of renal vascular resistances (RVR) and the systemic vascular resistance (SVR) [70, 103, 191, 192]; that is, if RVR increases in relation to SVR, RBF and hence renal perfusion will decrease [70, 192]. RBF regulation is very complex and modulated by several features [87, 110, 193]; however, a sophisticated intrinsic (local) system of mechanisms is keeping renal perfusion and thus GFR constant over a wide range of varying blood flow and renal perfusion pressures: *renal autoregulation* [87, 110, 116, 192–195]. Two components – myogenic response of renal vasculature and the tubuloglomerular feedback (TGF) – of the renal autoregulatory setup are approved and established [110, 116, 190, 193, 195]. A third one suggested [196], contributing to a balanced renal perfusion and filtration irrespective of perfusion and perfusion pressure variations. Autoregulation is largely *mediated by adoptions of the tone of the preglomerular arteriole* in relation to changes in renal perfusion pressure [116, 193, 197]: The effective filtration pressure determined by the hydrostatic pressure within the glomerulus and the opposing glomerular and tubular oncotic pressures is mainly affected by the ratio of the tones of afferent and efferent arteriole [70, 116, 198]. The myogenic response basically opposes and compensates fluctuations in perfusion pressure by vasoconstriction (in case of increasing/high perfusion pressure) preventing transmission of the high pressure to the glomerulus, or by vasodilation, in case of easing or low perfusion pressure of the afferent arteriole [191, 193]. On the other hand, TGF fine-tunes renal perfusion and GFR by sensing the sodium chloride load in the distal tubules and adjusting the tone of the afferent arteriole *via local mechanisms and mediators* such as adenosine, NO and charging afferent vasoconstriction in case of increased/high load and vice versa [110, 195, 197]? Concomitantly, the proximal tubular sodium reabsorption is ameliorated (attenuated respectively) and accordingly reduces (increases respectively) distal sodium load, accomplishing afferential vasodilation (vasoconstriction respectively) and thus counteracting and *balancing* the initial reaction [89]. Furthermore, a drop of the sodium chloride amount at the macula densa of the distal tubules stimulates local renin secretion, potentially resulting in a very decided constriction of the efferent arteriole facilitating glomerular filtration pressure [191]. Insofar, TGF keeps the fluid and sodium chloride delivery to that distal part of the nephron (with the macula densa) in certain limits, stabilizing GFR and supporting the kidney's necessary adjustments in reabsorption due to metabolic requirements [110, 199, 200]. The autoregulation of GFR and distal tubular flow, by the help of TGF, contributes to the body's salt and water homeostasis, in particular preventing renal salt and water losses in case of high GFR or tubular dysfunctions [200–202].

Myogenic response and TGF are modulated by a number of paracrine and endocrine agents [203]. The local renin secretion – attuned by the macula densa cells – exerts local vascular effects but may exhibit "*systemic*" *effects* affecting the relation between SVR and RVR and hence influences renal micro- and macrocirculation [98, 110, 193, 204, 205]. Renin participates and may be seen as one of the links to and an element within the extrinsic control mechanisms of renal perfusion engaged in the reflex arc of the neurohormonal renal control and regulation [98, 193, 206].

Preferential increases of vas afferent resistance (vasoconstriction) will diminish RBF into the glomerulus and the glomerular filtration pressure [191, 198], whereas predominant constriction (raise in resistance) of the efferent arterioles accentuates glomerular filtration pressure, preserves or even increases GFR despite a drop in RBF [191, 198]. Thus, the net effect on RBF, filtration pressure and GFR of the altered vascular tones *depends on the relative modification in the tone of afferent arteriole compared to efferent vessel and thus determines the intra-renal vascular tone* [70, 191, 198, 207, 208]. Accordingly, the tones of the arterioles are regulated by circulating hormones, paracrine factors, autoregulation and renal sympathetic nerves [70, 193, 198], consequently show preferentially microcirculatory local, intra-renal effects, but may, in turn (in case of pathological situations), show macro-circulatory effects as well [70, 172].

Of interest to understand the pathophysiology, it is worth to remember the uneven blood distribution within the kidneys: 90 % of the renal blood supply is distributed to the cortex, and 10 % to the medulla [110, 118, 209]. This limited blood flow is appropriate due to the functions necessary to be fulfilled by the medulla, so to preserve osmotic gradient and to optimize urinary concentration [110, 172, 209]. Concurrently, the *medulla – in particular the outer medulla – is posed under constant threat of hypoxic injury even under physiological conditions* [91, 118, 209–212]: Active reabsorption transport (necessary to generate the osmotic gradient [110]) driven by Na–K-ATPase is a highly oxygen-consuming process in the setting of only fair oxygen delivery [110, 118, 172, 209, 212]. Exceptionally the so-called S3 segments of the proximal tubules localized in the outer stripe of the outer medulla [209] are in addition rarely equipped with peritubular capillaries [118] – vessels which supply the tubular cells with oxygen and all vital nutrients. Furthermore, in case of ongoing or substantial hypoperfusion or due to additional threats like drugs or radiocontrast, disproportionally impaired perfusion with considerable regional differences of blood supply can be observed [87, 188]: Intrarenal regulative forces (e.g. sympathetic nerve activation and (local) production and effects of angiotensin II) preferentially lead to vasoconstriction of the cortex [213, 214] and the preglomerular arterioles, subsequently the perfusion-pressure and the blood flow of the post-glomerular (peritubular) capillaries are markedly diminished [1, 4, 173]. Thus, although concomitantly corticomedullary blood redistribution [213, 214] along with local medullary vasodilation and reduction of tubular transport promoted by prostaglandins, NO and adenosine [209, 215] potentially maintaining medullary oxygenation has been demonstrated [201, 202, 215], medullary hypoxia/ischaemia frequently ensues [209]. This may even merely follow disruption or attenuation of the protective forces, e.g. as found in diabetic or aged patients, in case of administration of radiocontrast and other drugs, etc. [70, 118, 216, 217]. Exposure of hypoperfused kidneys to radiocontrast induces hypoxic injury [218–220].

For the sake of completeness, as the reader will already have noticed, RBF and GFR do not always change consecutively or in parallel, rather their regulation is somehow independent under physiological conditions [186, 221–223] as well as in pathological states, particularly in sepsis [150, 177, 189, 224, 225]. This is largely due to local, intra-renal hemodynamic, hormonal and neural [172, 188, 189, 213, 226] and autoregulatory effects [118, 189, 197]. Furthermore, differential neural innervations of the afferent and efferent arterioles are found [198, 226]: The afferent arterioles are three times more densely innervated compared to the efferent ones [227]. Insofar, the sympathetic nervous system may elicit selective changes in pre- and postglomerular resistance [227]. Moreover, differential neural control of the cortical and medullar perfusion – with the medullary to be less sensitive to nerval discharge – qualifies for a very decided and fine-tuned neurohormonal regulation of local vascular resistance and hence regional perfusion [228, 229].

In the setting of mild to moderate reductions in cardiac output, intravascular volume depletion or blood pressure drops, consecutively renal *autoregulatory mechanisms will basically promote a gradual vasodilation of the afferent, preglomerular arterioles* mediated by myogenic response and vasodilating agents such as prostaglandin I_2 and NO generating an increase in RBF [70, 87, 99, 111, 116, 230]. TGF via renin [191] and sympathetic discharge (e.g. due to true or relative hypovolaemia, BP decline, reduced effective circulatory volume as in heart failure or hepatorenal syndrome, etc.) ensues [1, 3, 36, 87, 193]. Subsequently, a "*very mild, minor*" *constriction of the efferent, postglomerular arterioles* is induced, in order to preserve glomerular filtration pressure and thus GFR [87, 111, 116, 193]. The change in the tone of the efferent arterioles is largely mediated via locally generated angiotensin II [116, 193, 197, 198]. Lower concentrations of angiotensin II preferentially lead to constriction of the vas efference, and with higher concentrations, the resistance of the afferent arterioles increases as well [193, 198, 231, 232]. If however, as the result of compensatory response in order to sustain RBF and especially GFR, blood urea nitrogen (urea and creatinine) increases in the presence of a mild/modest decline in GFR, "prerenal azotemia" applies [19, 70, 87, 92].

Although the metabolism of BUN is quite complex, the reabsorption of urea (accompanied by water, as a reduction of the effective circulating blood volume induces a non-osmotic activation of vasopressin antidiuretic hormone [87, 98]) in the setting of prerenal azotemia is found to be intensified in case of any hypovolaemic state, if true or relative [191]. Under the described conditions, the mild accumulation of urea and creatinine has to be seen as an appropriate physiological reaction [92, 191, 201].

Prerenal azotemia may result from any deterioration in renal perfusion and can complicate any malady attributed to true or relative (diminished effective circulating blood volume) *hypovolaemia*, *hypotension* or *intra-renal vasoconstriction* [3, 36, 70, 74, 89, 92, 233].

If the *hypoperfusion* cannot be resolved by the compensatory, basically autoregulatory, mechanisms (and therapeutic measures) and continues or even worsens, renal tissue may be exposed to hypoxia [1, 70, 87, 110, 221]: TGF is highly activated but may contribute to further impairment of RBF and GFR as increasing levels of angiotensin II, due to ongoing decreased renal blood flow, *convey afferent arteriole constriction* [234–236]. The systemic sympathetic nervous system will noticeably be activated by baroreceptors in the afferent arterioles [193] and by those within the systemic circulation, e.g. aortic arch, due to hypovolaemia (true or relative) and/or hypotension. Together with the impelled renin-angiotension-aldosterone system, a *more profound generalized vasoconstrictive environment is engendered* within the renal vascular bed [4, 70, 87, 237]. However, local differences in the intensity of the increase in vascular resistance occur, resulting in *diverse regional perfusion within different kidney areas* [87, 172, 173, 188, 211, 238–240]. Under these circumstances, preferentially *preglomerular resistance is enhanced* since the afferent arterioles are three times more densely innervated by sympathetic nerves than the efferent ones [227]. Furthermore, heightened sensitivity of the vascular endothelium to the renal sympathetic nervous and renin–angiotensin system and their mediators is demonstrated [238–240]. Likewise, vasorelaxation in response to endothelium-dependent vasodilator agents is mitigated [241].

Continuously and increasing afferent vascular tone will attenuate glomerular capillary and postglomerular pressures, hence filtration pressure and glomerular blood flow, resulting in reduced ultrafiltration (GFR). The latter supplementary elicited by boosted oncotic capillary pressure [237]. Moreover, as the postglomerular peritubular capillaries perfuse the tubules, the tubular epithelial cells are subsequently supplied with less blood and oxygen [1, 4, 173]. Beyond that, although preferential cortical vasoconstriction resulting in corticomedullary redistribution has been demonstrated [213, 214], the *medulla*, particularly the outer medulla, *faces a very high risk ending up hypoxic/ischaemic* [118, 173, 209, 210, 221].

The described condition is characterized by a in general more constricted renal vascular bed [87], particularly the afferent arterioles, and the disproportionally modified, diverse local intra-renal blood flows [87, 171, 173, 185, 210, 242, 243]. The latter attributed to a *markedly altered intra-renal microcirculation* [172, 173, 243, 244] responsible for the unbalanced glomerular vascular tones (disturbed intra-glomerular hemodynamics) [193, 215, 245, 246]. The result is a diluted glomerular blood flow and filtration pressure and thus GFR [99], may be called *functional AKI* [185]. Functional AKI is reversible but may proceed to established (intrinsic) renal failure [210, 247].

Although transiently, and especially continuously, impaired *regional* rather than global [118, 173] blood flows may provoke functional AKI [173, 210, 248], renal hypoperfusion alone seems not to be sufficient to establish apparent AKI [125–127, 189]. Typically, multiple nephrotoxic insults are considered necessary to elicit manifest intrinsic AKI [70, 175, 176, 188, 209, 249, 250].

Moderate and especially severe *hypoperfusion*, notably in the presence of blunted autoregulation, of increasing duration and severity and hallmarked by regional diverse, *amended and diminished blood flows (referred to altered renal microcirculation) will expose the kidneys, particularly the outer medulla* (with its special vascular architecture and its very high oxygen demand due to active electrolyte reabsorption [126]) [118, 120, 210, 243, 251] *to hypoxic/ischemic injury* [101, 118, 163, 172, 173, 210, 252]. The medulla may be exposed to hypoxia/ischemia even if the systemic blood pressure is normotensive [99, 101], but autoregulation disrupted [253–255] as found in diabetic patients, chronic hypertension or in arteriosclerosis. The *central role of outer medullary hypoxia* [234, 243] in the development of ischaemic AKI has become more and more apparent in a variety of clinical settings [118, 256]: Medullary oxygen paucity and consecutive hypoxia have been demonstrated in various experimental and clinical conditions leading to ischaemic injury and ischaemic AKI [6, 118, 172, 209], e.g. hemoglobinuria, application of radiocontrast, as mentioned above, in case of disrupted/attenuated "protective mechanisms" found in diabetes, arteriosclerosis, aged people, chronic hypertension, and during reperfusion [118, 242]. Other common situations exposing especially the outer medullary kidney tissue to hypoxic/ischaemic insults are general anaesthesia, intra-operative hypotension and haemodilution-induced anaemia, postoperative low cardiac output, cardiac arrest, cardiac surgery, sepsis and shock [88, 118, 257].

Hypoxic/ischemic events, inevitably following the marked depletion of ATP [1, 87, 210] attributed to the hypoperfusion *and* the consecutive exposure of the outer medulla to hypoxia, will injure endothelial cells causing endothelial activation and dysfunction [4, 120, 243, 258]. Subsequently, swelling of the endothelial cells, even in transient ischaemia [259], ensues [173, 210, 258]. These oedematous endothelial cells are physically narrowing the capillary lumen and thus contributing to the vascular, peritubular congestion found in histological examinations, further deteriorating or abolishing the perfusion in this kidney region [258–261].

Persistent hypoperfusion and hypoxia may induce an accrued expression of a couple of adhesion molecules, e.g. intercellular adhesion molecules 1 (ICAM-1) [262, 263]. Interaction between ICAM-1 with the so-called CD11/CD18 receptor of circulating neutrophils results in adherence of the leukocytes to the vascular wall. Subsequent (inflammatory) infiltration of that tissue area, further injury of endothelial cells, local production of vasoactive metabolites promoting vasoconstriction, and occlusion of the vessel even more decrease the medullary blood flow [6, 262, 264]. Ensuing microcirculatory flow breakdowns deepen hypoxia/ischaemia [6, 265]. Recent evidence revealed and established that the *inflammatory response* (initiated and activated by endothelial cell dysfunction following hypoperfusion and/ or nephrotoxic insults resulting in local hypoxia/ischemia) has an essential and major impact in the genesis of AKI [173, 266–271]. The activated inflammatory cascade can be substantially amplified by the generation of a variety of potent mediators, e.g. pro-inflammatory and chemotactic cytokines, among others addressed to ischaemic tubules [180, 271–273].

Moreover, endothelial cell injury with its affiliated inflammatory reaction will display a significant imbalance between vasodilative and vasoconstrictive mediators decisively responsible for the vascular tone and thus can, in turn, affect microcirculation even more [172, 180, 274, 275]. Although quite a number of mediators are involved, the imbalance between vasoconstrictors, the nitric oxide system, and oxidative stress have reportedly most impact on local vascular tone [276–278]. The nitric oxide system acts as a regulator in a paracrine fashion, affecting surrounding cells, inhibiting platelet aggregation and leukocyte activation, and exerts a direct vasodilatory effect on the vessel contributing to the regulation of intra-renal haemodynamics and renal oxygen consumption but may be severely disturbed in dysfunctional endothelial cells or in AKI [279, 280]. Thus, *renal microcirculatory alterations can promote evolution of AKI through the interlinked occurrence of renal hypoxia and activation of inflammatory pathways* [172].

Furthermore, the intra-renal local haemodynamics may be directly affected by agents like lipopolysaccharide LPS – as found in Gram-negative sepsis – provoking microvascular hypoxia and that independently of macrocirculation [126].

To conclude, altered *regional*, *microcirculatory* intra-renal haemodynamics (widely appreciated and accepted as having substantial pathophysiological impact on the very complex, multifaceted, multi-pathophysiological and multi-pathomechanistic pathology of AKI [87, 118, 171, 172, 185, 188, 210, 214, 281]) and the subsequent (but as well concomitant) endothelial alterations *with the inflammatory response* in the centre of the cellular engagement *dominate the evolution and pathobiology of ischaemic AKI* [172, 244, 265]. The compromised intra-renal haemodynamic interplays with the inflammatory pathways through hypoxic insults resulting in and from microcirculatory flow breakdowns [265], with the structures of outer medulla being affected most by the hypoxic lesions [118, 209, 243]. Moreover, AKI due to local hypoxia/ischaemia may develop without relevant attenuation of global RBF [177, 179, 282, 283].

In the following, a few of the other setups involved in that very complex pathogenesis are worth to be mentioned.

Sublethal injury of the proximal tubular epithelium, particularly the S3 segments (it is not only the scarce supply of capillaries but also the pressure drop along the glomerular capillary net following preferentially preglomeru-lar vasoconstriction [1, 4, 173]) as a result of ATP paucity causes several functional and structural changes. Examples are changes in the cytoskeleton, loss of intercellular junction integrity, confined efficiency of electrolyte reabsorption resulting in high sodium concentrations at the macula densa and therefore lighting the TGF even more, production of autocrine (especially vasoconstrictive) factors and particularly enhanced permeability of the tubules with consecutive so-called tubular back-leak [6, 173, 180, 284–287]. Tubular back-leak is a special phenomenon of tubule cell dysfunction and disrupted cellular integrity. Nearly 60 % of tubular filtrate may move back to circulation, thus contributing to GFR reduction [4, 87, 210]. In consequence, resulting from destroyed tubular epithelial cells, tubular urine flow will be diminished in a situation of tubular cell swelling, tubular obstruction following cell debris secondary to cell deaths, and impaired cellular barrier function [210, 288]. Whether the even more profound activation of TGF resulting from increased sodium chloride delivery to the macula densa (with potentially intensification of afferent arteriolar constriction and hence reduction of GFR) exerts beneficial effects by limiting solute delivery is still an issue of debate [87, 204, 210, 289].

Apoptotic cell deaths are demonstrated in both distal and proximal tubules attributed to ischaemic and non-ischaemic nephrotoxic occurences in human acute renal failure [113, 290]. Under hypoxic conditions, both, the shortage of oxygen itself but also the so-called reactive oxygen species (oxidative stress molecules generated by the processes ushered [278, 291, 292]) may injure the cells [118, 210], while the progress to apoptotic or frank cell necrosis depends on the extent of hypoxic stress [217, 256, 293].

Persistence and/or prolongation of the hypoxic/ischaemic damage will be accompanied by *reperfusion*, a process known to cause harm by provoking cell deaths, resulting in amplification of the inflammatory process and aggravation of the vasoconstriction [87, 116, 173]. Reperfusion itself may induce ischaemia and cellular damage, primarily of the (outer) medulla [173, 242, 294].

The marked *vascular congestion found in the outer medulla* following local hypoperfusion and associated endothelial cell injury is a causal and dominant pathophysiological mechanism resulting in ongoing disruption of perfusion (and hence oxygen delivery) rather than in a no blood flow phenomenon [180, 272, 274, 295]. The activation of the coagulation, not at least due to the loss of antithrombogenic properties of the insulted endothelial cells, is reported to contribute to the medullary vascular congestion of the peritubular vessels and hence the perfusion deficits [173, 243, 255, 296, 297]. Vascular congestion heralds profound depression of GRF and confines RBF and renal vascular as well as tubular cellular functional and morphological abnormalities [87]. Subsequently aggravation of hypoxia and induction of cell death to follow [210, 298].

Disrupted autoregulation is another relevant feature: In case of a marked fall in systemic and/or renal perfusion pressure beyond the autoregulatory threshold, autoregulation will be impaired or lost [5, 189, 253, 299]. Further, "compensatory, autoregulatory mechanisms" may fail with ongoing hypoperfusion [110, 300], especially in risk groups, hypertensive patients, hepatorenal syndrome [101], and/or in case of additional non-ischaemic injuries, such as nephrotoxic drugs [101, 193]. *Even mild, but repetitive hypotensive, episodes exhibit ischaemic detriment* [301]. Disordered intra-renal vascular resistances will deteriorate autoregulation [189, 240, 254, 302], and an impaired autoregulatory capacity secondary to prolonged hypoperfusion was demonstrated [300]. Established acute ischaemic injury has been found to be associated with loss of autoregulation [5, 127, 238].

Agents that interfere with autoregulation may provoke "prerenal" AKI/functional AKI or may worsen renal dysfunction to established AKI, e.g. *NSAIDs* as a typical example for cyclooxygenase inhibition with particular risk of all patients suffering from sodium depletion, liver cirrhosis and congestive heart failure or patients on diuretics [303, 304]. *ACE inhibitors* [1] or even *loop diuretics* may attenuate autoregulation while interfering with sodium chloride reabsorption [70].

Impaired autoregulation makes the kidneys susceptible to recurrent hypoxic/ischaemic insults [305]. Failure of autoregulation will affect endothelial and vascular smooth muscle cells, contributing to the mismatch between vasodilative and vasoconstrictive setups [306].

To address the compromised autoregulation, the underlying factors have to be disengaged, and proper intravascular volume administration is recommended [253, 307, 308]. Unfortunately, almost always, volume replenishment alone is not able to return systemic blood pressure and hence renal perfusion pressure to a level *well above the autoregulatory threshold* of the kidney, thus additional administration of vasoactive drugs is necessary and pivotal [5, 189, 254].

Hypoperfused and hypoxic kidneys are prone to and show an augmented sensitivity to (recurrent) hypoxic and other non-hypoxic insults [89, 99, 188, 305]. Additional jeopardy lurks from the treatment initiated, particularly the drugs administered. Drugs not only exert haemodynamic effects, provoking systemic and intra-renal perfusion disturbances and interfere with autoregulation, but rather perform direct toxicity to renal cells, structures and physiological actions [74, 75]. Antibiotics, for example, are able to induce allergic interstitial inflammation (e.g. penicillins, cephalosporins) ending up with renal tubular epithelial cell injury, or may cause acute tubular necrosis (e.g. aminoglycosides, vancomycin) [75, 77]. Radiocontrast given in such a situation potentially will deteriorate renal function by direct tubular cell injury and alteration of intra-renal circulation, resulting in regional rather than global reno-parenchymal ischaemia, particularly of the medullary area [82, 309].

In the setting of functional AKI, exposure of renal tissue to hypoxia/ischaemia and/or other threats will further *markedly diminish renal function and parenchymal cell injury will inevitably emerge* [1, 4, 163, 173, 210].

However, due to a remarkable potential of the tubule cells to recover [1, 310, 311], AKI may not even be manifested in case of severely impaired function [210, 310] as no correlation has been demonstrated between patho-morphological and functional abnormalities [4, 114, 115]. Appropriate interventions in time, easing the setups running, may restore adequate renal utilization [5, 108, 233, 307]. The balance between cell survival and cell death is a decisive determinant for the course of the syndrome [210], whereupon the repair of both, endothelial and epithelial cells seems to be critical [118, 210]. No doubt, it is quite obvious that *multiple pathophysiological operations are running, interacting and amplifying each other* decisively and fatefully pushing the process in one direction or the other [87, 118, 210, 246, 312, 313].

In summary, renal physiology and especially pathophysiology is highly complex, sophisticated, multifacet [171–173] and multifactorial [174–176, 210]. The kidneys show a remarkable capacity to autoregulate and preserve function over a wide range of haemodynamic variations, and may resolve milder/moderate abnormal circulatory conditions independently [103, 187, 191, 200, 314].

Prerenal azotemia, the most common (40–55 %) cause of AKI [35, 128, 315], is literally a reversible, physiological response to any sort of renal hypoperfusion [4, 92, 163, 201, 255], approached by compensatory, autoregulatory mechanisms [19, 70, 87, 92]. However, hypoperfusion may complicate any malady ascribed to either true hypovolaemia or diminished effective circulating volume as found in low cardiac output, systemic vasodilation (often, but not always indicated by low blood pressure) or intra-renal vasoconstriction [3, 36, 70, 89, 92] and may precede ischaemic AKI [243]: "If pre-renal acute renal failure is not treated in a timely manner or is allowed to get progressively worse, ischaemic injury to renal tubular cells ultimately occurs and precipitates ATN" – now called AKI [163].

Human renal tissue hypoxia/ischaemia is in general attributed to prolonged diminished renal blood flow [118, 163, 172, 173, 209, 316]. Hypoxic renal parenchyma arises in several clinical conditions [6, 118, 172], even under circumstances of direct tubulo-toxicity [118] as hypoperfusion may coexist with any stage of AKI [97]. Renal parenchymal hypoxia is an essential issue of the cascades of repercussions, eliciting cellular injury and tubular and (severe) vascular dysfunction [173, 243, 265, 286, 287].

*Regional micro*circulatory rather than global renal hemodynamic alterations (*resulting in local hypoperfusion* and disproportionate blood flows ascribed to the selective, inappropriate and diverse intra-renal vasoconstriction [172, 173, 188, 212, 213, 244]) *notably affecting the outer medulla* [118, 209, 243] *may promote ischaemic AKI* [172, 188, 258]: *Hypoperfusion* is potentially *exposing kidney parenchyma,* particularly the outer medulla, *to hypoxic/ischaemic insults* [118, 163, 172, 210, 252].

Hypoxia/ischaemia is still a very important and the *most common* pathological condition in AKI [118, 163, 173], and a *hallmark in the pathophysiology of acute renal dysfunction* [118, 163, 185, 188, 221]. Following the local perfusion mismatch, cellular ATP depletion applies [118, 255], *endothelial activation and dysfunction arises with the inflammatory response* being in the centre of the variety of cellular responses, and epithelial cell injury may ensue [173, 180, 267, 268, 304]. Accordingly, elicited by the hypoxic renal insult, a complex interplay between the *altered intra-renal microhaemodynamics and the arising inflammation may lead to established AKI* [172, 265]. Moreover, the afflicted and altered microhemodynamics will be even more influenced and may be markedly disrupted by the reactions and processes associated with and triggered by the abnormal regional renal perfusion and hypoxic/ischemic injury present, thus involving and spreading to other parts of the organ [243, 274].

The reactions and processes include inflammation, blunted autoregulation, coagulopathy, congestion, reperfusion injury, white cell adhesion, and epithelial tubular cell defects (such as capillary leakage and tubular obstruction following epithelial cell damage).

However, due to a high potential to recover, as long as the launched processes can be disengaged, calmed down and balanced in time, restitutio ad integrum is possible [210, 247, 310] (Figs. 5.1, 5.2, and 5.3).

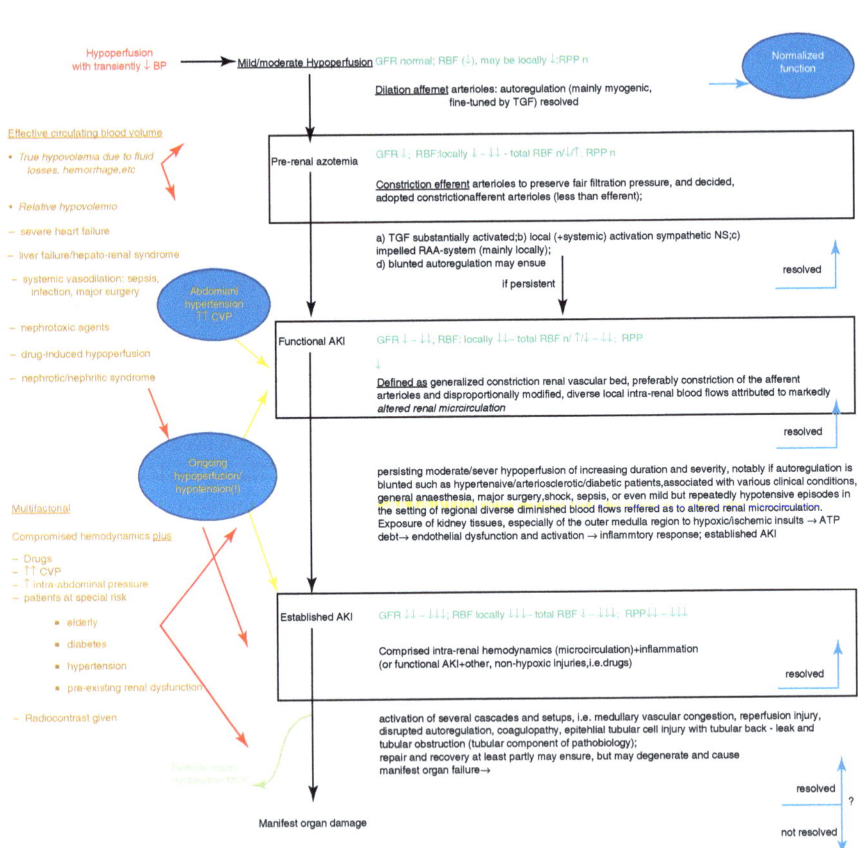

Fig. 5.1 Summary of the pathophysiology of ischaemic AKI

Fig. 5.2 Executive summary of ischaemic AKI by Lameire [163]

(a) $\downarrow$ GFR in case of pre-renal failure is due to abnormalities in glomerular perfusion.Although RBF is $\downarrow$ it remains adequate to provide enough oxygen and metabolic substances to sustain viability of the kidney (163; 280).

(b) If pre-renal failure is not treated timely and adequate $\rightarrow$ gets progressively worse, and ischemic injury to the tubular cells ultimately occurs and precipitates ATN(163).

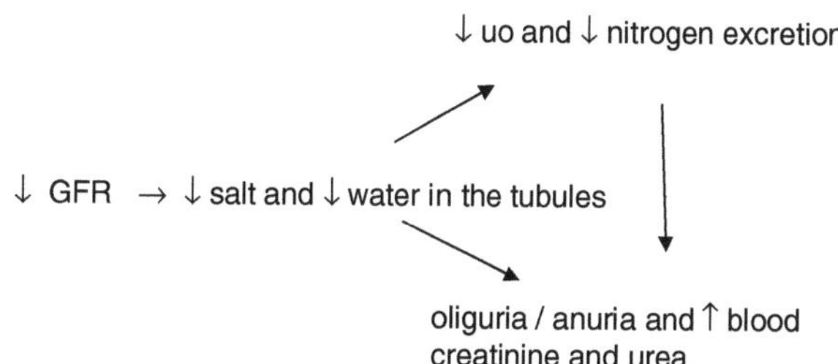

Fig. 5.3 Sequence of results following impaired GFR by Kellum [19]

5.5 Diagnostic Remarks

Azotemia, particularly creatinine serum concentration, reduced urinary output, and *oliguria/anuria*, are *characteristic features* of *impaired renal function and thus AKI* [8, 19, 317]. They are surrogates of renal function and the most common parameters used in daily clinical practice to identify, to diagnose and to classify renal functional disorders [8, 19, 318]. Both have been applied as cardinal items in the newly launched diagnostic and classification systems of renal dysfunction, the RIFFLE classification [9] and its modification, the AKI grading system [11]: Thus, the *diagnosis of AKI should be made on the basis* of these two issues, *creatinine serum levels and urinary output* (uo) [8, 9, 11, 19].

However, both have considerable limitations, and we are far away from a setup that would suit that serious and demanding illness [246, 317]. Creatinine is generally linked to GFR by an inverse correlation, and an increase in serum creatinine level parallels with a decrease in GFR, and vice versa [319]. But serum creatinine levels are affected by muscle mass, tubular secretion, gender, age, metabolism, hydration, medication and nutrition [320, 321] and hence may change disproportional to the extent of the renal insult [321].

Creatinine is insensitive and non-specific [19, 189, 246, 318, 321], and abnormal levels generally occur "too" late in the course to get aware of the disorder and its progression in time. Creatinine accumulation lags behind the threat by up to 48 h (steady-state equilibrium) [36, 318, 322], and significant reductions in GFR may initially cause only small elevations [36, 322]. Hence, repeated measurements of serum creatinine are necessary [321]. Even serious renal diseases (e.g. fibrosis) may apply without relevant alteration in creatinine serum concentration attributed to raised tubular secretion or renal reserves [320, 323]. Furthermore, slight or mild increases in creatinine or decreases in uo may just be related to volume depletion or a transient, milder decline in renal blood flow and/or renal perfusion pressure, conditions which are often referred to as "prerenal" failure which may just represent a physiological response [19, 92, 189, 191, 201, 324]. On the other hand, as demonstrated, even marginal elevations of creatinine have a detrimental impact on the patient's outcome [13–15, 25, 46, 54].

Urinary output is determined by glomerular filtration and tubular function (secretion and reabsorption) [192]. Severe oliguria and anuria indicate either marked reductions in GFR with serious functional or even structural renal tubular cell damage but may be found in case of urinary tract obstruction as well [19, 192, 325]. Renal ultrasound will in the vast majority be able to identify postrenal pathology [108, 325], it might miss up to 5 % of all cases [326]. However, a markedly diminished uo reflects in general a more substantial renal injury or indicates severer AKI [327]. Contrariwise, even in severely compromised renal function with markedly diluted GFR, uo output may be normal or nearly normal. This is due to a disruption of the tubules to concentrate the glomerular filtrate and a high urine osmolality accompanied by a low urine sodium concentration in the setting of oliguria, and azotemia is suggestive for a preserved tubular function [19, 189, 192, 328]. Moreover, particularly a transient decrease in uo is not necessarily associated with a relevant reduction of the GFR as in case of mild to moderate volume depletion or hypotension where oliguria represents a physiological response [19, 189, 201, 324]. Furthermore, neuroendocrine response may be triggered by pain, surgery and trauma, provoking oliguria without any haemodynamic compromise and with normal renal function [324].

Being under sophisticated neurohormonal control, the association between renal function and uo is inconstant and complex [19, 98, 189], and a rise in renal perfusion, e.g. volume replenishment, may boost uo and natriuresis but may not translate into an improvement in GFR [108, 189, 246].

Overall, although some consider uo as being a reasonable sensitive and clinically useful indicator of renal function (GFR) [19, 324], biomarker of tubular injury [19], and one of the parameters determining the recently published consensus definition of AKI, it is after all non-specific and insensitive [189, 318, 324] with limited clinical utility and has only a fair potential to detect and to classify subsequent AKI [318, 324]. Oliguria is frequently observed in ICU patients, but most episodes are not indicative for AKI [192, 324]. Conversely, normal uo and thus apparently intact renal function may be seen in several pathologies, particularly at the onset as in glomerulonephritis or in sepsis, the most common underlying problem of AKI [19, 168, 192, 324]. There is some evidence that oliguria in the presence of haemodynamic dysfunction may be a valuable clinical indicator of patients at risk to develop AKI [150, 189, 324, 329].

Urinary biochemistry and indices of combined blood and urinary analysis such as FE_{Na} or FE_{UREA} are likewise insensitive, unspecific and not indicative for *timely* and *precise* diagnosis [168, 317, 318, 330]: FE_{Na}, for example, is relatively highly specific but has a poor sensitivity [4]; neither is FE_{UREA} although having a higher sensitivity of 85 % and a specificity of 92 % not ample to diagnose *established*, persistent AKI with manifest tubular injury confidently, reliably, and reproducibly [330–333]. Hence urinary indices are unable *to differentiate prerenal and functional AKI from persistent manifest AKI* in critically ill patients [179, 330, 334, 335].

More promising for an early and accurate diagnosis with proper sensitivity and specificity are possibly recently developed biomarkers, foremost the neutrophil gelatinase-associated lipocalin (NGAL) [246, 317, 321]. This protein is secreted by flawed epithelial cells and in case of tubular injury will be secreted into the urine [317, 321]. NGAL could show its potential as being an early marker of AKI in a wide range of clinical conditions [8, 317, 321, 336], but unfortunately is not able to discriminate properly and reliably manifest from transient AKI [337]. However, maybe the traditional paradigm of prerenal AKI representing conditions of oliguria and azotemia in the presence of histologically normal nephrons is, not only due to a lack of a uniform definition, but also particularly due to the absence of bioptic evidence too unrealistic and too theoretical [164], and needs to be desperately adapted when new data, especially histological ones become available forming a conception of the condition in a more appropriate way [164].

It may be of interest to remember that urea expires toxic effects while creatinine does not [318], as this may affect the decision to initiate RRT [338].

A normotensive blood pressure in the patient should not be considered safe, since normal blood pressures may be well beyond renal autoregulatory threshold as found in several patient groups.

The clinical approach to patients suggestive for or suffering from AKI has to include a history and a physical examination [108, 233, 339], although the diagnosis needs evaluation of serum creatinine and uo [9]. Both history and physical examination give substantial information and clues to the cause of AKI [108, 233, 339]. This is of immense importance as the underlying reason needs to be treated properly, e.g. compromised macrohaemodynamics, abdominal compartment syndrome, bacterial infection, mechanical obstruction of urine pathways, etc. [5, 8, 108, 233, 307, 325]. Furthermore, possible consequences of AKI affecting other organs will be recognized by history and physical examination, i.e. pulmonary dysfunction secondary to fluid overload, deranged acid–base homeostasis and electrolyte disorders [233, 340, 341].

Pure "prerenal" problems and postrenal pathology are rare in the hospital setting, but the consequences of a prerenal input are not always benign [8, 19]. Slightly elevated serum creatinine levels and small urine volumes in the presence of some degree of volume depletion are pretty common, a picture often supposed to be not critical and hence not taken serious [8, 93]. Rather, as we know, even mildly abnormal creatinine values are associated with increased mortality [15, 342].

Prerenal azotemia arises in diseases which may be accompanied by diluted effective arterial blood volume, e.g. congestive heart failure, liver failure, hepatorenal syndrome and haemorrhagic and nephritic syndrome [3, 233, 343]. Highly concentrated urine with a low pH in the absence of cellular elements, disproportional rise of urea compared to creatinine and elevated urea acid further support the suggestion of an adaptive response to volume depletion, hence a "prerenal" condition [237]. However, almost always a multifactorial aetiology exists with often concomitant nephrotoxic drug administration [74, 176]. A *special drug history is indispensible* as potentially causal, but also has therapeutic consequences, e.g. adoption of dosage, suspension of significantly nephrotoxic agents, etc. [3, 5, 108, 233]. Haematuria or proteinuria is suspicious for renal parenchymal diseases – "intrinsic renal failure" [108, 233] – while rush, fever and arthralgia are found in acute interstitial nephritis [108, 233]. All patients should undergo ultrasound investigation to rule out postrenal obstructive problems; further, ultrasound may be helpful to recognize chronic renal parenchymal disease (shrunken parenchyma) and acute states with enlarged and big kidneys [108, 233].

The following initial investigations are recommended [108, 233]:

- Urine analysis
- Bloods
  - Urea, creatinine and electrolytes, with creatinine repeatedly
  - Full blood count
  - Blood glucose
  - Coagulation screen
  - Liver enzymes
  - Ca and phosphate
  - Inflammatory markers
  - Urea acid
- Urine specimen for microscopy
- Arterial blood gases
- ECG
- Chest x-ray
- Renal ultrasound

5.6 Management

Treatment strategies of AKI are grossly suggestive, assumptive and based on the premise that an early and immediate approach may be able to reverse kidney dysfunction in the incipient phase of the disease [344–346]. However, the quite substantial body of literature lacks of definite evidence for many of the commonly used daily practice, and our measures are largely supportive rather than causal-therapeutic [308, 327, 343–345, 347].

Conservative prevention and treatment of renal dysfunction comprise:

- Identification of risk factors such as diabetes mellitus, chronic renal insufficiency, hypertension, heart failure, atherosclerosis, hepatic cirrhosis and advanced age

- To rule out or to treat obstructive uropathy
- *Optimalization of haemodynamics aiming to improve renal perfusion*
- *Discontinuation of nephrotoxic drugs* and the *avoidance of exposure to nephrotoxins*, in particular to abandon radiocontrast examinations whenever possible
- Addressing metabolic abnormalities, electrolyte disorders – particularly hyperkalaemia – and remote organ complications such as pulmonary congestion/oedema
- To mitigate progression of renal dysfunction/AKI
- To provide nutritional support where necessary [5, 108, 233, 307, 308, 327, 344, 348]

5.6.1 *Optimalization of Haemodynamics, Fluids and Vasopressive Agents*

The cornerstone in our approach still is to optimize renal perfusion in order to maintain or achieve favourable haemodynamics and haemodynamic stability [5, 307, 308, 346, 349]. Two components are essential to be applied: fluids and vasopressor agents [5, 133, 253, 254, 325, 350]. "Correcting intravascular volume depletion with fluid administration and maintaining adequate renal perfusion pressure with vasopressors in vasoplegic states are both first-line strategies to preserve renal oxygen supply" [118]. Furthermore, improvement of renal haemodynamics prior to application of agents able to compromise renal perfusion will minimize the risk of AKI [351].

Volume status, fluid utilization and fluid balance are central features, but due to most recent publications crucial as well: central, because only fluid therapy gave consistently evidence as being effective to prevent AKI [133, 308, 346, 349], and crucial, because there is no evidence that fluid administration can reverse established AKI and recent data strongly advice to avoid fluid overload as we have substantial evidence of worse outcomes in case of fluid accumulation, often due to iatrogenic genesis [189, 343, 349, 352–355].

Fluid administration is assumed (a) to minimize ischaemia by ameliorating systemic blood pressure (major determinant of renal perfusion pressure) and CO (prerequisite for adequate renal blood flow) secondary to the increase in effective circulating volume *and* by attenuating neuroendocrine reflexes responsible for the rise in renal vascular resistance (particularly vasoconstriction of afferent arterioles) [356] *and* (b) to mitigate nephrotoxic injuries by diluting nephrotoxins.

Both resulting in enhanced renal blood flow, GFR and uo [347, 357–360].

Fluid resuscitation during the *early acute phase* of critical illness is a well-established measure and target of our treatment strategy, based on broad consensus and on some evidence to maintain and support tissue perfusion and oxygen delivery [9, 11, 347, 361–365]. A newly published study by Jhanji is the first one supporting our theory "that cardiac output guided hemodynamic therapy can improve tissue perfusion and oxygenation" [366]; particularly, renal function was found to be protected by optimized haemodynamics with appropriate fluids being an essential element [366].

However, recently more and more evidence emerged revealing that excess fluid and fluid overload may be harmful in several conditions and may contribute to illness deterioration and even increased mortality [343, 346, 349, 352, 353, 355, 367–370]. Patients with oliguria are especially at high risk to develop volume overload [192, 352, 353, 370]. In critically ill paediatric patients with AKI, an association of fluid overload and mortality has been found in several studies [368, 371, 372]. In the PICARD study a positive balance of 1 l within 24 h incremented the mortality risk of about 20 % [373]. If excess fluid directly affects and impairs renal function unfortunately remains unanswered [343].

Thus, a well-balanced concept of fluid application and monitoring is essential: (a) Fluids should be given early on, (b) the (haemodynamic) effects have to be monitored and (c) fluid accumulation has to be avoided:

(a) As we already know from the Rivers study, it is advantageous to apply volume early [364]. The results of more recent trials clearly favour fluid resuscitation within 6 h [363, 365, 374]. What is not given early often cannot be given later in the course; in contrary, volume administration later on may be deleterious [365, 374]. In the early goal-directed trial by Rivers, the study group with the better outcome (mortality rate 30.5 % vs 46.5 %) received even more fluids early on (hour 0–6) than the conventional group; after 72 h, both groups had received the same amount of volume [364].

(b) Clinical signs of hypovolaemia are insensitive and non-specific, and indicators of dehydration are particularly misguiding [375, 376]. However, diminished skin turgor, dry mouth, decreased toe temperature, low blood pressure (hypotension), orthostatic dysregulation and tachycardia are clinically in use and, if present, may suggest hypovolaemia [375, 377].

    Mottled extremities, altered level of consciousness, cold peripheries (forearms and lower leg), moist and clammy skin and oliguria may be clinical signs of organ and tissue hypoperfusion [378].

    Tachycardia is only present in 42 % of all patients with moderate blood loses and seen in only 5–24 % of patients with severe blood loss [376]. Although volume deficits typically induce tachycardia, high heart rates may occur in stress, fever and pain as well as being drug induced [375]. Hypotension may be seen in hypovolaemia but found in heart failure, vascular obstruction (e.g. pulmonary embolism) or vasodilative states also, thus is very unspecific [375, 377]. Therefore, it might be more reliable to assess and to monitor the so-called end point markers such as CO, BP (MAP), CVP and uo and parameters of tissue perfusion (SvO_2 or $ScvO_2$, lactate and base excess, or even sublingual – tissue – pCO_2) in response to volume loading and fluid application [360, 375, 377]. The effects of fluid infusions on the macrohaemodynamic parameters CO, blood pressure, CVP, etc., depend on miscellaneous factors such as myocardial compliance [379] and contractility [380, 381], venous capacitance [382], systemic vascular resistance [383], capillary permeability [382] and regional blood flow distribution [384, 385], making the evaluation of fluid status and the appropriateness of fluid replenishment really challenging [360]. A proper applied and monitored fluid challenge is still a commonly adopted method to test fluid responsiveness in the setting of suggested but uncertain hypovolaemic or haemodynamically unstable patients [375]: Infuse 500–1,000 ml crystalloid fluid (250–500 ml colloidal fluid) within 30 min. A positive response is documented by an increase in SV of ≥10 % or BP rise by ≥10 % with the heart rate being lowered or unchanged [386–388]. Further, CVP should increase by less than 2 mmHg; if the increase is ≥2–5 mmHg, further development of CVP has to be taken into consideration before more volume is given, 5-2 rule [375]. In case of an increase of more than 5 mmHg, no more fluids should be administered at all [375]. Most often a compromised heart function is the limiting element [386, 389].

An international consensus conference from 2006 suggested a rise in CVP of at least 2 mmHg either by 250 ml crystalloid given within 15 min or by passive leg raising has to be interpreted as sign of sufficient fluid administration – defining a positive response if cardiac function and tissue perfusion improve [390]. The classical fluid challenge recommended to administer 100–200 ml crystalloids given every 10 min and monitor blood pressure response and CVP [391]. The surviving sepsis campaign consensus conference currently recommends to apply 500–1,000 ml crystalloid within 30 min in the initial resuscitation phase [362]. Others have modified the expert recommendations and administer 250 ml in 5–7 min or 500 ml crystalloids within 10 min [392].

However, CVP does not reflect and predict preload [389, 393–395] as more than 100 publications state delivering impressive evidence that there is no relationship at all between CVP/change in CVP and fluid responsiveness [396]. As such, the 5–2 rule by Vincent and Weil [375] is not really helpful and needs to be basically challenged as does the 2 mmHg unit of measurement: 2 mmHg is the minimum to confidently detect changes of pressures on currently available monitors apart from imprecisions due to the settings of zero position [392]. A CVP>9–10 mmHg is clearly pathological, putting the patient at risk to ensue diastolic ventricular interaction resulting in even more impaired heart performance [396–399], and affects glomerular filtration pressure and renal function [69, 156, 158]. Fluid loading in excess of a CVP>10 mmHg failed to significant increase CO in most cases [400]. Volume infusion is most effective when a reduced effective circulating volume is replenished [364, 401]. On the other hand, fluid responsiveness (positive effects as described above) does not mean that volume is needed and is per se not an indication to apply volume [327, 353, 370, 402]. Furthermore, quite a substantial number (28–60 %) of haemodynamic unstable critical ill patients do not reply positively to volume application [389, 403], and even in shock only roughly 50 % display an increase in CO when fluids are applied [389]. Regrettably, improvement of macrocirculation may not translate into enhanced organ function [384].

Notably, aside the above-mentioned traditional markers used for fluid management, dynamic markers such as systolic pressure variation, pulse pressure variation and stroke volume variation have emerged [387, 394, 404, 405]. They reflect changes in LV–SV due to heart–lung interactions induced by mechanical positive pressure ventilation [387, 406–408]. All are clearly more sensitive and accurate than the traditional parameters [387, 394, 404, 409–411]. Passive leg raising in spontaneously breathing patients is causing similar variations in LV filling and LV–SV due to a significant change in venous blood return and may currently be the most suitable and reliable clinical way to assess fluid responsiveness and requirements in the setting of suggested reduced effective circulating volume and hypovolaemia [412–416].

Regarding the choice of fluid, crystalloids or colloids, currently there are no substantial differences evident [5, 327, 346, 347, 417]. But large surveys on that issue are lacking [418]. Recently, some concerns have been expressed regarding the synthetic starch HES associated with adverse effects, particularly contributing to the development of AKI in septic patients [419, 420]; however, the data are not delivering convincing evidence [345–347, 421]. Nevertheless, crystalloids are predominantly recommended for volume resuscitation, in particular in septic and kidney-transplanted patients [345, 362, 422, 423]. This approach is supported by the results of the Australian CHEST study and the 6S trial [424, 425].

By the way, don't forget, larger amounts of normal saline may induce hyperchloremic acidosis [426, 427].

Albumin given for resuscitation in the ICU has been assessed in the very large Australian landmark trial, the SAFE study [428]. About 7,000 patients had been included, randomized to saline or 4 % albumin in saline, for fluid resuscitation. Survival and the need for renal replacement therapy (RRT) were identical in both groups; thus, no significant distinction became evident. A subgroup analysis revealed a trend towards a lower mortality rate in sepsis when treated with albumin [428], whereas in traumatic brain injuries the albumin group was associated with a higher mortality [429]. Of note, hypoalbuminemia was found to be associated with a higher mortality rate [430].

Most recently, a study by Bragadottir et al. [260] supports previous suggestions; mannitol may be of some value in advanced but still early and reversible phases of AKI: Administration of mannitol was able to redistribute blood flow to the kidneys and induces renal vasodilation in postoperative AKI and thus enhances RBF [260, 431–433]. Furthermore, mannitol may exert an ebb-away effect on the swollen epithelial cells, thus meliorating the compromised tubular urinary flow in advanced functional AKI [259].

(c) Volume depletion in the setting of hypovolaemia and/or diminished effective circulating volume should be replenished, but no further fluids are necessary; in contrary, they may be deleterious [192, 327, 343, 346, 374]. In a study by Murphy [355], patients were initially given ≥20 ml/kg bolus fluid resuscitation up to a CVP of ≥8 mmHg within the first 6 h after admission. Thereafter, a more restrictive regimen was applied achieving at least on two consecutive days a negative fluid balance in the following 7 days. In Payen's study on the survival of patients with acute renal failure, the survivors received after the initial resuscitation phase 1.06±0.15 l per day and the non-survivors 1.5±0.98 l per day [354]. Thus, it is important and obviously relevant for a positive outcome that after the initial early and aggressive fluid resuscitation phase, we have to aim for a cumulatively mildly negative balance to be attained in the following days [112, 352, 434, 435]. In the study by Alsous in patients with septic shock, all study participants achieving a negative balance of 500 ml within day 3 after ICU admission survived, while in the group of patients where a cumulative negative balance could not be accomplished, the mortality rate reached 80 % [434].

Giving high volumes is not unusual in our ICUs: Often 5–10 % of body weight is infused [352, 436], and, as Pohlman found, just for analgo-sedative purpose, patients may receive up to 1.2–1.3 l per day [437]. Administration of the so-called maintenance fluid is still an issue of practice in several ICUs – maintenance fluids are absolutely unnecessary [346, 360, 374]. ICU's liberal volume application may lead to fluid accumulation, particularly in case of oliguria [192, 327, 353]. Neither oliguria [327, 346] nor a positive fluid responsiveness [352, 353, 370] are per se an indication to apply volume [327, 402]. Fluid overload results in multiple organ complications such as pulmonary oedema or congestive heart failure [192, 327, 343]. Visceral oedema is a risk factor for abdominal hypertension and in consequence for AKI/deterioration of AKI [64, 438, 439]. *Positive fluid balances* are *associated with an increased risk for intra-abdominal hypertension* [64, 134, 144] and hence AKI [134, 144]. Even moderately elevated intra-abdominal pressures may affect glomerular filtration in setting of markedly altered systemic circulation [440], e.g. in patients with elevated right atrial pressures secondary to heart failure, intra-abdominal pressures as low as 8 mmHg are found to compromise renal function [441]. Accumulated fluid will increase CVP and hence central venous congestion, compromising renal perfusion pressure with ensuing renal dysfunction [442].

Replenishing hypovolaemia may ameliorate renal perfusion pressure and renal perfusion but is commonly not sufficient to substantially increase blood pressure, in particular in vasodilative conditions such as inflammatory and septic settings [254, 443]. In patients with sepsis or congestive heart failure, intravenous volume administration might not solve the problem, since low cardiac output and the total vascular resistance impede renal perfusion [97, 174, 444]. Perfusion of organs depends among other factors on CO, representing the flow component, and the height of the MAP, pressure component [390, 445]. Unfortunately, increasing SV due to volume loading (fluid responsive patient) does not always translate into an increase in blood pressure as the relationship between blood pressure and SV/CO is complex and decisively reliant on vascular tone; hence, in case of "vasoplegic" settings, the MAP and thus organ perfusion may not consecutively improve although SV/CO does [446–448]. The vascular tone is determined by both, vascular resistance and compliance [448], and as such, in case of hypotension which may be attributed to either low CO, low systemic vascular resistance (systemic vasodilation) or both [449], vasopressor substances will be causal to solve the problem [450–452]. Hypotension is considered to be a risk factor for the development of AKI [453, 454]; further, an association with the progression of renal dysfunction in septic patients is reported [455].

Renal perfusion is principally determined by CO, MAP, renal blood flow and autoregulation [439]. If the blood pressure falls below autoregulatory range, autoregulation will be disturbed, and renal blood flow becomes directly dependent on the perfusion pressure [253, 254]. Due to the sharp relation beyond autoregulatory threshold, the reduction in RBF is more pronounced for any given pressure drop [299]. Failing renal autoregulation commences early in the course of AKI [87]. Thus, maintenance of an adequate perfusion pressure, basically MAP, well above the autoregulatory threshold is an essential goal to preserve renal function [4, 5, 253, 254, 308, 456].

There is good evidence that restoration of a sufficient blood pressure in case of systemic vasodilation is associated with enhanced RBF and restoration of uo [253, 299, 457, 458]. Although we know from several examinations that in mammalians renal autoregulation works within a range between 80 and 180 mmHg [299, 459, 460], it is still unclear which pressure we shall aim for. In septic shock, an MAP between ≥65 and 75 mmHg (in patients with known atherosclerosis and occlusive arterial disease as well as hypertension) has been and still is (≥65 mmHg in sepsis and septic shock) recommended [362, 364, 457, 461]. Furthermore, Bourgoin could not find further improvements when targeting an MAP of 85 mmHg [457]. More recently, an enhanced renal oxygen delivery and an even-tempered oxygen delivery/consumption balance was achieved when boosting the MAP from 60 to 75 mmHg in cardiosurgical patients with vasodilatory shock [462]. In patients with coronary artery disease or cardiogenic shock, mean arterial pressures between 70 (75) and 80 mmHg are suggested [461, 463, 464]. Of special interest and with consequences for our daily practice are recent publications providing evidence that with MAPs between 75 and 85 mmHg, not only a further enhancement in renal perfusion pressure can be achieved, but renal microcirculatory abnormalities found in septic states could be beneficially addressed with increases in renal vascular conduction, and recoupling the renal circulation with the systemic one attenuates the disperse regional renal vasoconstriction and perfusion [465–467]. Patients with disrupted microcirculation probably will benefit from a higher MAP (75 mmHg), while those with preserved microcirculation, as may be found in early stages, will not [467–469]. A very recent although retrospective study by Lehman including nearly 17,000 patients revealed hypotension as being a significant risk factor for the development of AKI in all ICU patients [470]. It turned out that the cut-off MAP was 80 mmHg below that the risk evolving into AKI increases [470]. All those study outcomes lead to the conclusion that the MAP we are aiming for needs to be individualized and, rather clearly higher than 65, potentially around 80 mmHg [5, 133, 308, 469, 470]. Furthermore, re-establishment of lost autoregulation may need higher MAPs [254, 457]. Notably, in case of intra-abdominal hypertension, the required MAP is suggested to be at least 65 mmHg plus the value of the intra-abdominal pressure [471].

Norepinephrine (NA) is the preferred vasoconstrictor agent to be used: Quite a substantial body of literature indicates NA as being the most suitable drug available, particularly when taking the special renal issues and requirements into consideration [253, 472–475]. In a recent multicentre, randomized trial, De Backer and coworkers revealed NA as being clearly superior to dopamine in the treatment of a variety of shock states and encourage, due to their profound results, to change the recommendations for the treatment of cardiogenic shock in favour of NA [476]. NA not only boosts renal perfusion pressure but also increases renal vascular conductance, both resulting in improved renal blood flow [477, 478]. NA has the ability to increase peripheral vascular tone [479], a crucial issue in organ perfusion and tissue blood flow distribution (autoregulation) [466]. Furthermore, no evidence from neither human nor well-composed animal studies raises concern, NA could be associated with increased risk of AKI [133, 177, 457, 458, 477, 480]. Of course, the indication to apply NA has to be straightforward: to treat arterial hypotension in life-threatening circumstances, shock states and vasodilatory conditions such as severe sepsis and septic shock [350, 443, 475, 476]. Generalized inflammation is present in various systemic diseases and may stimulate the synthesis of inducible nitric oxide synthetase in the vascular endothelium precipitating generalized vasodilation [481]. Moreover, NA administration is only indicated after hypovolaemia is compensated; nevertheless, in common daily practice, NA will be started while fluids are running and stopped as soon as hypovolaemia is balanced and the MAP is in the envisaged range [133, 308, 479].

Dosages used in AKI studies are up to 1.0 μg/kg/min in humans [475] and in well-designed animal studies up to 1.6 μg/kg/min [482], but individual requirements may be even higher.

In the special setting of hepatorenal syndrome, terlipressin in dosages up to 6 mg every 6 h may be preferred to NA if vasopressor therapy is required [483–485]. As the application of albumin in addition to terlipressin (2–12 mg/day, given in equal doses qds) has shown mortality reducing effects in the treatment of hepatorenal syndrome, 1 g of albumin should be added on day 1 of therapy and 20–40 g per day on every following day of management [486].

5.6.2 Loop Diuretics

The application of loop diuretics in the prevention and therapy of AKI, particularly in the setting of oliguria, is a very common tradition [353, 487, 488]. It is based on our belief, non-oliguric AKI has a better prognosis than oliguric one which indeed was found in a study on patients with suggested ATN [489]. However, quite a large bundle of trials, among others [487, 488, 490] a prospective RCT [491], a study using the Bayesian evidence synthesis technique [492] and meta-analyses by Kellum [493] and Bagshaw et al. [494], all provide conflicting results. A recent meta-analysis by Ho could not demonstrate any beneficial effects in the furosemide group, but may be an increased risk of harm [495]. In conclusion, the studies demonstrate a lack of effectiveness of the loop diuretics in prevention and treatment of

AKI. Clear, if any, beneficial effects of furosemide are missing; in contrary, in several publications, loop diuretics were found to be potentially detrimental. Hence, currently there is no place for furosemide in the treatment or prevention of AKI, particularly not to convert oliguric AKI to non-oliguric [5, 133, 327]. To further clarify the situation, high-quality evidence is necessary [327].

Furthermore, dopamine in low dose cannot be recommended any longer in acute renal failure and should be abandoned from use, an evidence of A I-level recommendation [133, 496].

5.6.3 Miscellaneous

1. Hyperkalaemia

   Severe hyperkalaemia (K > 6.5 mol/l) needs to be addressed immediately: A combination of *insulin dextrose* (50 ml 50 % dextrose plus 10 IU fast-acting insulin infused over 10–20 min; in case of a blood glucose concentration >14 mmol/l insulin solely) and β_2-*adrenergic agonists* (i.e. 0.5 mg salbutamol subcutaneously given) has demonstrated to be more efficient than either measure alone [108, 497]. Ion exchange resins are indicative in stable situations of moderately elevated potassium levels (around 6.5 mmol/l) with planned dialysis and where a delay of 2–3 h is tolerable [108].
2. Severe acidosis

   In severe acidosis (pH < 7.2 [108, 233]; <7.1 [498]), commonly sodium bicarbonate IV bolus buffering is recommended [233]. But bicarbonate administration may *deteriorate intracellular acidosis* and worsen the overall conditions by osmotic vasodilation, hypernatraemia and extravasation of fluids, putting the patient even more at risk to organ, particularly pulmonary, oedema [108, 499]. Crucially, there is in general little evidence supporting the application of buffers [108, 500], in particular sodium bicarbonate [500, 501]. Usually, significant acidosis is an indication for renal replacement therapy (RRT) [498].
3. Hyperglycaemia

   Control of hyperglycaemia in critically ill patients is still under controversial debate [502, 503]. With regard to AKI, two larger randomized control trials [503, 504] and one secondary analysis [505] strongly suggest tight blood glucose control as a having a decrease in incidence of AKI development and RRT requirement. In contrary, Finfer recently found no significant difference in the necessity of RRT; unfortunately, the appearance of AKI was not reported [502].
4. Alkalization in rhabdomyolysis is only fairly evidence based [506].
5. Renoprotective measures/prevention of AKI.

   (a) Contrast-induced nephropathy (CIN)

   If the GFR is < 60 ml/min, an increased risk for the development of CIN is displayed [507] and is suggested to increase exponentially with further deterioration of kidney function or additional risk factors [508]. Risk factors include hypotension, diabetes mellitus, anaemia and pre-existing impaired renal function [509–511].

CIN develops within 24–72 h after contrast has been applied [512]. Contrast agents take a direct toxic effect to renal tubule cells and exhibit haemodynamic effects (microthrombi showers, reduction of medullary blood flow) [513, 514].

Hydration is a common practice in the prevention of AKI [5, 108, 133, 307]. Unfortunately, no randomized control trials evaluating volume hydration compared with placebo are available [133, 307, 308, 512]. Nevertheless, quite a substantial number of studies have been done on that issue, and in conclusion, IV isotonic fluids (particularly 0.9 % saline) are largely recommended for prevention of CIN [133, 307, 515–517]. More recently, sodium bicarbonate compared to normal saline showed some advantages; however, study results are conflicting [85, 518, 519]. Furthermore, the use of iodixanol (iso-osmolar, nonionic) agents at the lowest possible volume and temporary cessation of nephrotoxic drugs like metformin or diuretics are pivotal [5, 108, 133, 512, 515].

N-acetylcysteine (NAC) is shown to decrease serum creatinine [520], albeit not affecting GFR and thus renal dysfunction, but it probably increases tubular secretion [133] as well having antioxidant properties [207] (after contrast exposure, renal free radical production is increased). Thus, NAC may be beneficial, but the studies provide conflicting results [133, 308, 512, 521–523].

A possible practical approach to prevent CIN 3–12 h before the radiographic procedure is planned to start with 1–2 ml/kg/h normal saline or Hartman's or isotonic sodium bicarbonate and continue until 6 h postprocedural, aiming for a uo of 150 ml/h [515–517]. If *N*-acetylcysteine is additionally considered, 600 mg twice a day the day before and the day of the procedure is recommended [522].

(b) Regarding all other drugs and procedures suggested for prevention and treatment of AKI, there is no evidence at all; thus, they should not play a role in daily practice [5, 108, 133, 512].

5.6.4 Renal Replacement Therapy

Since the introduction of haemodialysis in the 1940s, and although in particular huge technical steps have been made, many substantial questions and issues of this treatment option still remain unanswered, e.g. indications to initiate RRT, timing, modality, discontinuation of renal support and dosing – regarding the latter, we may have an answer now [36, 524–527].

Initially, haemodialysis was applied to address advanced symptoms, e.g. clinical uraemia, severe academia or hyperkalaemia [528, 529]. It was Teschan who called for a "prophylactic dialysis" with initiation of RRT before "overt" uraemic symptoms are present indicating an advanced disease state, most often ending up lethal [530]. RRT is nonetheless a *supportive* measure for "severer" cases of AKI [531–533]. Unfortunately, we do not know what severer means and are missing consensus and standards for RRTs utilization in AKI [5, 534]. Thus, a wide range of setups are actually in use [1, 535, 536].

Bellomo and Ronco published in 2001 *indications* to initiate RRT which are largely accepted worldwide [498]:

- Nonobstructive oliguria: uo < 200 ml/12 h or anuria
- Severe academia (pH < 7.1) due to metabolic acidosis
- Azotemia (urea > 30 mmol/l)
- Hyperkalaemia (K^+ > 6.5 mmol/l or rapidly rising)
- Suspected uraemic organ involvement (i.e. pericarditis, encephalopathy, neuropathy or myopathy)
- Progressive severe dysnatraemia (Na^+ > 160 mmol/l or < 115 mmol/l)
- Hyperthermia (core temp > 39.5 C°)
- Clinically significant organ oedema, particularly lung
- Drug overdose with dialyzable toxin
- Coagulopathy requiring large amounts of blood products in patients with risk or at risk of pulmonary oedema or ARDS

Lameire published in his review in 2005 very similar indications [1]:

- Oliguria: uo < 200 ml/12 h
- Anuria: uo < 50 ml/12 h
- Hyperkalaemia: K^+ > 6.5 mmol/l
- Severe academia: pH < 7.0
- Azotemia: urea >30 mmol/l
- Uraemic encephalopathy, neuropathy/myopathy and pericarditis
- Plasma sodium abnormalities: concentrations >155 mmol/l or <120 mmol/l
- Hyperthermia
- Drug overdose with dialyzable toxin

Basically, the criteria described are affiliated to the ones conventionally used in chronic renal failure, but as the traditional indications for dialysis in chronic renal failure apply for late complications, in AKI, RRT shall avoid and prevent complications and thus has to commence prior to incipient complications ensuing [532, 537–539]. Additionally, in AKI, RRT is intended to provide control over fluid and electrolyte balance as well as solute clearance subsequently preventing multiorgan dysfunction while waiting for renal recovery [540–542]. Current clinical practice as explored in the BEST study (beginning and ending supportive therapy for the kidney) – an international multicentre, prospective and epidemiological trial – revealed that the most common reasons to initiate RRT are oliguria (70 %), followed by the urea/creatinine levels (53 %), metabolic acidosis (43.6 %), fluid overload (36.7 %), hyperkalaemia (18.6 %) and immunomodulation (13.6 %) [45].

However, *optimal timing* and on which markers/clinical conditions the indication to start RRT should be based on are still under controversial debate: Urea with its toxic–uraemic effects has been in the focus of research for a long time; unfortunately, the results are somewhat inconsistent except that higher values are obviously harmful [543–546]. Regrettably, the cut-off point is nondistinctive: While former studies used cut-offs of 32.1–35.7 mmol/l urea serum concentration [543, 547], investigated more recent studies levels between 21.4 and 35.7 mmol/l [544, 545, 548]. The largest multicenter observational study, the PICARD experience, reported a significant higher mortality rate if RRT was started in patients with a urea concentration of more than 27.1 mmol/l [546].

Observational studies comparing the grade of volume overload at the time RRT was initiated, and the mortality rate reported that relevant overload (>10 % (25 %) [436]) of body weight is associated with poor outcome [372].

Most recently, a systematic review and meta-analysis compared early versus late initiation of RRT in critically ill patients with AKI [549] and included as well two studies using the new RIFLE criteria for starting up with artificial therapy [541, 550]. Both showed a significant advantage for an early initiation as defined by RIFLE-R [541] and RIFLE-R and I, respectively [550]. With urea concentrations used for indication, the results were inconsistent; 4 studies demonstrated early commencement of RRT (defined by urea serum concentration in a range of <21.2 mmol/l [544] and <35.7 mmol/l [545]) being favourable, and one reported late (>23 mmol/l) initiation being in favour over early (<23 mmol/l) [551]. Just to remark, the odds ratio was far less convincing, if at all, using urea as criterion compared to the RIFLE criteria which in both studies clearly undermatched the discriminant – R or I was more convincing than R alone [549], a quite understandable result as with R some less severe cases were probably already selected for therapy. Maybe RIFLE will become a valuable tool for timing as well.

Those studies analysed in that cited review and meta-analysis applying uo as criterion to start RRT observed conflicting results [552–554].

A study by Leite et al. [555] compared early (defined by initiation of RRT within 24 h in patients who achieved stage 3 AKIN) with late initiation (RRT commenced later than 24 h after reaching stage 3 AKIN) of RRT. The early group showed a significant relative mortality reduction by 30.8 relative units (14.4–45.2 $p=0.002$). The special approach of this study includes that both the creatinine and the uo criteria had to be fulfilled for being classified in stage 3 AKIN. Furthermore, the authors separately analysed patients with known chronic kidney disease, and as both criteria of the classification system had to apply, there was some more profound degree of illness. However, more studies are warranted before these criteria for early initiation of RRT may be applied in our daily practice.

In conclusion, we are still awaiting proper evidence regarding the indications for RRT and, closely linked to this, the timing when to start up. Early initiation (but how is early defined?) of RRT is theoretically beneficial, and some trials are supporting this approach – however, clear evidence is lacking [538, 544, 546].

Interestingly, there is no evidence in the literature at all when patients can be removed from RRT; the BEST study could be elucidated a common practice worldwide when to wean a patient from RRT [45].

Different dialysis modalities in RRT are available to manage AKI: *intermittent* hemodialysis, especially its submodality SLED (sustained slow efficiency dialysis), and the *continuously applied modalities* (CRRT) such as arteriovenous or venovenous haemofiltration (CAVHF, CVVHF respectively), hemodialysis (CAVHD and CVVHD, respectively) and haemodiafiltration (CAVHDF and CVVHDF, respectively) [531, 538]. Haemodiafiltration comprises both, convective and diffuse features. *Convection* (and hence haemo*filtration*) is more effective clearing middle and larger weight molecules, while *dialysis* is more efficient removing small molecules [531, 538, 556].

Intermittent haemodialysis (IHD) is usually performed over 4 h, three to four times a week [557], and SLED lasts typically 6–12 h [558].

Venovenous circuits are clearly preferable to arteriovenous ones as with a venous access not only the adverse complications associated with an arterial cannulation can be avoided, but the venous approach provides higher solute clearance as well [559].

Continuous RRT modalities are applied over 24 h with a blood flow between 100 and 200 ml/min and a dialysate flow between 17 and 34 ml/min [560]. SLED is typically running with 200–300 ml/min blood flow and a dialysate flow rate of 300 ml/min [540]. In distinction to classical IHD, the fluid and solute removal is more gradual and the total amount higher [540, 558], compared to CRRT, it can be scheduled and organized between diagnostic procedures and may be performed during daytime [540].

In CRRT, compared to IHD, fluid and solute shift is slower, at least theoretically, and more haemodynamically stable, and due to gradual removal of solute, the occurrence of a cerebral oedema is less likely [561].

Buffer fluids are used to replace ultrafiltrate; in use are lactate, bicarbonate and citrate – lactate and citrate are metabolized by muscles and liver [540], and lactate may in case of liver failure or sepsis contribute to lactic acidosis [562]. Citrate exerts regional anticoagulative effects on the haemofilter and may be preferred in patients at risk of bleeding and in patients with lactate acidosis and liver failure [563].

Pannu and co-workers [538] identified nine randomized control trials directly comparing CRRT and IHD, including nearly 1,000 patients: They neither report any significant difference in outcome, need for chronic dialysis nor incidence of hypotension during the procedure within the two modalities. There are quite a substantial number of other partly observational studies, e.g. [564–566], and partly meta-analyses [538, 567–569]: A significant disparity regarding outcome and renal recovery between the two modalities could not be demonstrated.

Although both modalities are absolute comparable, Hoste and Dhonte [556] conclude, patients with severe haemodynamic instability "are best treated with CRRT in order to allow removal of extravascular fluid and to maintain a stable blood pH".

Otherwise, it does depend on the clinical scenario which modality is most suitable. In patients with significant coagulation abnormalities, IHD may be the modality of choice – easier anticoagulation regime – patients with congestive heart failure may benefit from isolated ultrafiltration [570] (detailed review by Hoste [556]).

The dosing of RRT has been a long-lasting discussion; quite a lot of studies have been performed on this issue, and results and recommendations moved back and forth: Six prospective RCTs using doses between 17 and 48 ml/kg/h have been done – 3 concluded the higher dose to display beneficial effects on the outcome and recovery [571–573], and 3 could not work out any mortality rate difference or a better recovery [552, 574, 575]. In addition, the large international randomized RENAL study (**R**andomized **E**valuation of **N**ormal versus **A**ugmented **L**evel replacement therapy) comparing 20 and 40 ml/kg/h concluded the higher-intensity CRRT was not associated with a lower mortality rate at 90 days [576]. Furthermore, we have to recognize that numerous studies on this issue verified substantial discrepancies between prescribed doses and the actually delivered ones during the replacement procedure [577, 578]. Clotting problems are the most common reason for this deficit in delivery, and the effective dose delivered is on average approximately 20 % beyond the prescribed one [579]. Adjusting these findings, both RENAL study [576] and the so-called ATN study by Palevsky and co-workers [575] we have to conclude that survival is not different when CRRT is on 19–22 ml/kg/h compared to higher doses – but to achieve 19–22 ml/kg/h in daily practice, 25–30 ml/kg/h has to be prescribed [527]. Although Kellum and Ronco, based on the results of RENAL and ATN, recommend to aim for delivery doses between 19 and 45 ml/kg/h [527], why 45 ml/kg/h, the rule in our daily practice should be to run CRRT with 25–30 ml/kg/h, and we might have got a proper answer on that dosing issue.

References

1. Lameire N, Van Biesen W, Vanholder R. Acute renal failure. Lancet. 2005;365:417–30.
2. Finkel KW. Acute tubular necrosis. In: Bellomo R, Kellum JA, Ronco C, editors. Critical care nephrology. 3rd ed. s.l.: Philadelphia: Saunders Elsevier; 2009, chapter 63, p. 351.
3. Singri N, Ahya SN, Levin ML. Acute renal failure. JAMA. 2003;289:747–51.
4. Schrier RW, Wang W, Poole B. Acute renal failure: definitions, diagnosis, pathogenesis, and therapy. J Clin Invest. 2004;114:5–14.
5. Dennen P, Douglas IS, Anderson R. Acute kidney injury in the intensive care unit: an update and primer for the intensivist. Crit Care Med. 2010;38:261–75.
6. Thadani R, Pascual M, Bonventre JV. Acute renal failure. N Engl J Med. 1996;334:1448–60.
7. Lewington A, Kanagasundaram S. Acute kidney injury. 5th ed. Uk Renal Association. 2011. www.renal.org/guidelines. Final version 08.03.11.
8. Ricci Z, Ronco C. Today' s approach to the critically Ill patient with acute kidney injury. Blood Purif. 2009;27:127–34.
9. Bellomo R, Ronco C, Kellum JA. Acute renal failure – definitions, outcome, measures, animal models, fluid therapy and information technology needs: the second international consensus conference of the acute dialysis quality initiative (ADQI) group. Crit Care. 2004;8:R204–12.
10. Cruz DN, Ricci Z, Ronco C. Clinical review: RIFLE and AKIN – time for reappraisal. Crit Care. 2009;13:211–9.
11. Mehta RL, Kellum JA, Shah SV. Acute Kidney Injury Network: report of an initiative to improve outcomes in acute kidney injury. Crit Care. 2007;11:R31.
12. Praught ML, Shlipak ML. Are small changes in serum creatinine an important risk factor. Curr Opin Nephrol Hypertens. 2005;14:265–70.
13. Hoste EA, Clermont G, Kersten A. RIFLE criteria for acute kidney injury are associated with hospital mortality in critically ill patients: a cohort analysis. Crit Care. 2006;10: R73–82.

14. Lassnigg A, Schmidlin D, Mouhieddine M. Minimal changes of serum creatinine predict prognosis in patients after cardiothoracic surgery: a prospective cohort study. J Am Soc Nephrol. 2004;15:1597–605.
15. Chertow GM, Burdick M, Honour M. Acute kidney injury, mortality, length of stay, and costs in hospitalized patients. J Am Soc Nephrol. 2005;16:3365–70.
16. Barrantes F, Tian J, Vazquez R. Acute kidney injury criteria predict outcomes in critically ill patients. Crit Care Med. 2008;31:1397–403.
17. National Kidney Foundation. K/DOQI clinical practice guidelines for chronic kidney disease: evaluation, classification and stratification. Am J Kidney Dis. 2002;39 Suppl 1:S76–92.
18. Levey AS, Bosch JP, Lewis JB. A more accurate method to estimate glomerular filtration rate from serum createnine: a new prediction equation. Modification of diet in renal disease study group. Ann Intern Med. 1999;130:461–70.
19. Kellum JA. Acute kidney injury. Crit Care Med. 2008;36(suppl):S141–5.
20. Ricci Z, Cruz D, Ronco C. The RIFLE criteria and mortality in acute kidney injury: a systematic review. Kidney Int. 2008;73:538–46.
21. Perez Valdivieso JR, Bes-Rastrollo M, Mondreno P. Evaluation of the prognostic value of the risk, injury, failure, loss, and end-stage renal failure (RIFLE) criteria for acute kidney injury. Nephrology (Carlton). 2008;13:361–6.
22. Bagshaw SM, George C, Dinu I. A multi-centre evaluation of the RIFLE criteria for early acute kidney injury in critically ill patients. Nephrol Dial Transplant. 2008;23:1203–10.
23. Chen YC, Jenq CC, Tian YC. RIFLE classification for predicting in-hospital mortality in critically ill sepsis patients. Shock. 2009;31:139–45.
24. Maccariello E, Soares M, Valente C. RIFLE classification in patients with acute kidney injury in need for renal replacement therapy. Intensive Care Med. 2007;33:597–605.
25. Ostermann M, Chang RW. Acute kidney injury in the intensive care unit according to RIFLE. Crit Care Med. 2007;35:1837–43.
26. Perez-Valduvieso JR, Bes-Rastrollo M, Monedero P. Prognosis and serum creatinine levels in acute renal failure at the time of nephrology consultation: an observational cohort study. BMC Nephrol. 2007;8:14.
27. Hoste EA, Kellum JA. Acute kidney injury: epidemiology and diagnostic criteria. Curr Opin Crit Care. 2006;12:531–7.
28. Joannidis M, Metnitz B, Bauer P. Acute kidney injury in critically ill patients classified by AKIN versus RIFLE using the SAPS 3 database. Intensive Care Med. 2009;35:1692–702.
29. Bagshaw SM, George C, Bellomo C. A comparison of the RIFFLE and AKIN criteria for acute kidney injury in critically ill patients. Nephrol Dial Transplant. 2008;23:1569–74.
30. Ronco C, Kellum JA, Haase M. Subclinical AKI is still AKI. Crit Care. 2012;16:313.
31. Hotchkiss RS, Swanson PE, Freeman BD. Apoptotic cell death in patients with sepsis, shock, and multiple organ dysfunction. Crit Care Med. 1999;27:1230–51.
32. Racusen LC, Nast CC. Renal histopathology, urine cytology, and cytopathology in acute renal failure. In: Schrier RW, editor. Atlas of diseases of the kidney. Philadelphia: Blackwell Science; 1999; pp 1–9, 12.
33. Brochard L, Abroug F, Brenner M. An official ATS/ERS/ESCIM/SCCM/SRLF statement: prevention and management of acute renal failure in the ICU patient: an international consensus conference in intensive care medicine. Am J Respir Crit Care Med. 2010;181:1128–55.
34. Kaufman J, Dhakal M, Patel B. Community-acquired acute renal failure. Am J Kidney Dis. 1991;17:191–8.
35. Nash K, Hafeez A, Hou S. Hospital-acquired renal insufficiency. Am J Kidney Dis. 2002;39:930–6.
36. Star RA. Treatment of acute renal failure. Kidney Int. 1998;54:1817–31.
37. Brivet FG, Kleinknecht DJ, Loirat P. Acute renal failure in intensive care units – causes, outcome, and prognostic factors of hospital mortality: a prospective multicenter study. Crit Care Med. 1996;24:192–8.
38. Akposso K, Hertig A, Couprie R. Acute renal failure in patients over 80 years old: 25 years' experience. Intensive Care Med. 2000;26:400–6.

39. De Mendonca A, Vincent JL, Suter P. Acute renal failure in ICU: risk factors and outcome evaluated by the SOFA score. Intensive Care Med. 2000;26:915–21.
40. Schrier RW, Wang W. Acute renal failure and sepsis. N Engl J Med. 2004;351:159–69.
41. Yegenage I, Hoste E, Van Biesen W. Clinical characteristics of patients developing ARF due to sepsis/systemic inflammatory response syndrome: results of a prospective study. Am J Kidney Dis. 2004;43:817–24.
42. Rangel-Fausto MS, Pittet D, Costigan M. The natural history of the systemic inflammatory response syndrome (SIRS): a prospective study. JAMA. 1995;273:117–23.
43. Cole L, Bellomo R, Silvester W. A prospective, multicenter study on the epidemiology, management, and outcome of severe acute renal failure in a 'closed' ICU system. Am J Respir Crit Care Med. 2000;162:191–6.
44. Liano F, Junco E, Pascual J. The spectrum of acute renal failure in the intensive care unit compared with that seen in other settings. Kidney Int. 1998;66(suppl):S16–24.
45. Uchino S, Kellum JA, Bellomo R, Beginning and Ending Supportive Therapy for the Kidney (BEST Kidney) Investigators. Acute renal failure in critically ill patients: a multinational, multicenter study. JAMA. 2005;294:813–8.
46. Uchino S, Bellomo R, Goldsmith D. An assessment of the RIFLE criteria for acute renal failure in hospitalized patients. Crit Care Med. 2006;34:1913–7.
47. Hoste EA, Clermont G, Kersten A. RIFLE criteria for acute kidney injury are associated with hospital mortality in critically ill patients: a cohort analysis. Crit Care. 2006;10:R13.
48. Hoste EA, Schurgers M. Epidemiology of acute kidney injury: how big is the problem. Crit Care Med. 2008;36(suppl):S146–51.
49. Mehta RL, Pascual MT, Soroko S. Spectrum of acute kidney failure in the intensive care unit: the PICARD experience. Kidney Int. 2004;66:1613–21.
50. Bagshaw SM, Laupland KB, Doig CJ. Prognosis for long-term survival and renal recovery in critically ill patients with severe acute renal failure: a population-based study. Crit Care. 2005;9:R700–9.
51. Chen YC, Fang JT, Tien YC. Organ system failures predict prognosis in critically ill patients with acute renal failure requiring dialysis. Chang Gung Med J. 2000;23:8–13.
52. Himmelfarb J, McMonagle E, Freedman S. Oxidative stress is increased in critically ill patients with acute renal failure. J Am Soc Nephrol. 2004;15:2449–56.
53. Perianayapan MC, Liangos D, Kolyala AM. NADPH oxidase p22phox and catalase gene variants are associated with biomarkers of oxidative stress and adverse outcomes in ARF. J Am Soc Nephrol. 2007;18:255–63.
54. Ryckwaert F, Boccara G, Frappier J-M. Incidence, risk factors, and prognosis of a moderate increase in plasma creatinine early after cardiac surgery. Crit Care Med. 2002;30:1495–8.
55. Loef BG, Epema AH, Smilde TD. Immediate postoperative renal function deterioration in cardiac surgical patients predicts in-hospital mortality and long-term survival. J Am Soc Nephrol. 2005;16:195–200.
56. Gruberg L, Mintz GS, Mehran R. The prognostic implications of further renal function deterioration within 48 h of interventional coronary procedures in patients with pre-existent chronic renal insufficiency. J Am Coll Cardiol. 2000;36:1542–8.
57. Hegarty J, Middleton RJ, Krebs M. Severe acute renal failure in adults: place of care, incidence and outcomes. QJM. 2005;98:661–6.
58. Metcalf W, Simpson M, Khan IH. Acute renal failure requiring renal replacement therapy: incidence and outcome. QJM. 2002;95:579–83.
59. Abosaif NY, Tolba YA, Heap M. The outcome of acute renal failure in the intensive care unit according to RIFLE: model, application, sensitivity, and predictability. Am J Kidney Dis. 2005;46:1038–48.
60. Bellomo R, Kellum JA, Ronco C. Basic principles of renal support. In: Ronco C, Bellomo R, Kellum JA, editors. Critical care nephrology. Philadelphia: Saunders Elsevier; 2009. Chapter 12, p 71.
61. Müller M, de la Pena A, Derendorf H. Issues in pharmacokinetics and pharmacodynamics of anti-infective agents: distribution in tissue. Antimicrob Agents Chemother. 2004;48:1441–53.

62. Hou SH, Bushinsky DA, Wish JB. Hospital- acquired renal insufficiency: a prospective study. Am J Med. 1983;74:243–8.
63. Walker J, Criddle LM. Pathophysiology and management of abdominal compartment syndrome. Am J Crit Care. 2003;12:367–71.
64. Malbrain ML, Cheatham ML, Kirkpatrick A, Sugrue M. Results from the international conference of experts on intra-abdominal hypertension and compartment syndrome. I: Definitions. Intensive Care Med. 2006;32:1722–32.
65. De Iaet I, Malbrain ML, Jadoul JL. Renal implications of increased intra-abdominal pressure: are the kidneys the canary for abdominal hypertension? Acta Clin Belg. 2007;(Suppl):119–30.
66. Maerz L, Kaplan LJ. Abdominal compartment syndrome. Crit Care Med. 2008;36(Suppl):S212–5.
67. Damman K, van Deursen VM, Navis G. Increased central venous pressure is associated with impaired renal function and mortality in a broad spectrum of patients with cardiovascular disease. J Am Coll Cardiol. 2009;53:582–8.
68. Mullens W, Abrahams Z, Francis GS. Importance of venous congestion for worsening of renal function in advanced decompensated heart failure. J Am Coll Cardiol. 2009;53:589–96.
69. Joles JA, Bongartz LG, Gaillard CA. Renal venous congestion and renal function in congestive heart failure. J Am Coll Cardiol. 2009;54:1632–3.
70. Macedo E, Mehta RL. Prerenal failure: from old concepts to new paradigms. Curr Opin Crit Care. 2009;15:467–73.
71. Guevara M, Ginès P. Hepatorenal syndrome. Dig Dis. 2005;23:47–55.
72. Viviano G, Antonelli M, Moro ML. Risk factors for acute renal failure in trauma patients. Intensive Care Med. 1998;24:808–14.
73. Sharp LS, Rozycki GS, Feliciano DV. Rhabdomyolysis and secondary renal failure in hospitalized African Americans: comparing community-acquired vs hospital-acquired disease. Am J Surg. 2004;188:801–80.
74. Evenepoel P. Toxic acute renal failure. In: Ronco C, Bellomo R, Kellum JA, editors. Critical care nephrology. 3rd ed. Philadelphia: Saunders Elsevier; 2009. Chapter 30, pp 168.
75. Pannu N, Nadim MR. An overview of drug-induced acute kidney injury. Crit Care Med. 2008;36(suppl):S216–23.
76. Baliga R, Ueda N, Walker PD. Oxidant mechanisms in toxic acute renal failure. Am J Kidney Dis. 1997;29:465–77.
77. Michel DM, Kelly CJ. Acute interstitial nephritis. J Am Soc Nephrol. 1998;9:506–15.
78. Ries F, Klastersky J. Nephrotoxicity induced by cancer chemotherapy with special emphasis on cisplatin toxicity. Am J Kidney Dis. 1986;8:368–79.
79. Izzedine H, Launay-Vacher V, Deray G. Antiviral drug-induced nephrotoxicity. Am J Kidney Dis. 2005;45:804–17.
80. Zakarija A, Bennett C. Drug-induced thrombotic microangiopathy. Semin Thromb Hemost. 2005;31:681–90.
81. Dlott JS, Danielson CF, Blue-Hnidy DE. Drug-induced thrombotic thrombocytopenic purpura/hemolytic uremic syndrome: a concise review. Ther Apher Dial. 2004;8:102–11.
82. Rudnick MR, Berns JS, Cohen RM. Contrast media-associated nephrotoxicity. Curr Opin Nephrol Hypertens. 1996;5:127–33.
83. Stone GW, McCullough PA, Tumlin JA. Fenoldopam mesylate for the prevention of contrast-induced nephropathy. A randomised controlled trail. JAMA. 2003;290:2284–91.
84. Katzberg RW. Contrast medium-induced nephrotoxicity: which pathway? Radiology. 2005;235:752–5.
85. Ho KM, Morgan DJ. Use of isotonic sodium bicarbonate to prevent radiocontrast nephropathy in patients with mild preexisting renal impairment: a meta-analysis. Anaesth Intensive Care. 2008;36:646–53.
86. Meire P, Ko DT, Tamura A. Sodium bicarbonate-based hydration prevents contrast-induced nephropathy: a meta-analysis. BMC Med. 2009;7:23.

87. Garwood S. Ischemic acute renal failure. In: Ronco C, Bellomo R, Kellum JA, editors. Critical care nephrology. 3rd ed. Philadelphia: Saunders Elsevier; 2009. Chapter 28, pp 157.
88. Mangano CM, Diamondstone LS, Ramsey JG. Renal dysfunction after myocardial revascularization: risk factors, adverse outcomes, and hospital resource utilization. Ann Intern Med. 1998;128:194–203.
89. Marquis F, Ahern SP, Leblanc M. Etiology of acute renal failure in the intensive care unit. In: Ronco C, Bellomo R, Kellum JA, editors. Critical care nephrology. 3rd ed. Philadelphia: Saunders Elsevier; 2009. Chapter 17, pp 96.
90. Novis BK, Roizen MF, Aronson S. Association of preoperative risk factors with postoperative acute renal failure. Anesth Analg. 1994;78:143–9.
91. Block CA, Manning HL. Prevention of acute renal failure in the critically ill. Am J Respir Crit Care Med. 2002;165:320–4.
92. Blantz RC. Pathophysiology of pre-renal azotemia. Kidney Int. 1998;53:512–23.
93. Bellomo R, Bagshaw S, Langenberg C. Pre-renal azotemia: a flawed paradigm in critically Ill septic patients. In: Ronco C, Bellomo R, Kellum JA, editors. Acute kidney injury, Contribution to nephrology, vol. 157. Basel: Karger; 2007. p. 1–9.
94. Kasper DL, Braunwald E, Fauci AS, editors. Harrison's principles of internal medicine. 15th ed. New York: McGraw-Hill; 2001.
95. Davidson A, Cameron JS, Grunfeld JP, editors. Oxford textbook of clinical nephrology. 3rd ed. Oxford: Oxford University Press; 2005.
96. Fink MP, Abraham E, Vincent J-L, editors. Textbook of critical care. 5th ed. Philadelphia: Elsevier; 200.
97. Himmelfarb J, Joannides M, Molitoris B. Evaluation and initial management of acute kidney injury. Clin J Am Soc Nephrol. 2008;3:962–7.
98. Rennke HG, Denker BM. Review of renal physiology. In: Renal pathophysiology. 2nd ed. Philadelphia: Lippincott Williams and Wilkins; 2007. Chapter 2, pp 31–67.
99. Abuelo JG. Normotensive ischemic acute renal failure. N Engl J Med. 2007;357:797–805.
100. Oudemans-van-Straaten HM. Contrast nephropathy, pathophysiology and prevention. Int J Artif Organs. 2004;27:1054–65.
101. Abuelo JG. Diagnosing vascular causes of renal failure. Ann Intern Med. 1995;123:601–14, (Erratum, Ann Intern Med 1995; 124: 78).
102. Schlondorff D. Renal complication of nonsteroidal anti-inflammatory drugs. Kidney Int. 1993;44:643–53.
103. Badr KF, Ichikawa I. Prerenal failure: a deleterious shift from renal compensation to decompensation. N Engl J Med. 1988;319:623–9.
104. Brady HR, Clarkson MR, Lieberthal W. Acute renal failure. In: Brenner BM, editor. Brenner and Rector's the kidney. 7th ed. Philadelphia: Saunders; 2004. p. 1215–92.
105. Palmer BF. Renal dysfunction complicating the treatment of hypertension. N Engl J Med. 2002;347:1256–61.
106. Givertz MM, Colucci WS, Braunwald E. Clinical aspects of heart failure: pulmonary edema, high-output heart failure. In: Zipes DP, Libby P, Bonow RO, Braunwald E, editors. Braunwald's heart disease: a textbook of cardiovascular medicine. 7th ed. Philadelphia: Elsevier Saunders; 2005. p. 539–68.
107. Antman EM, Braunwald E. ST-Elevation myocardial infarction: pathology, pathophysiology, and clinical features. In: Zipes DP, Libby P, Bonow RO, Braunwald E, editors. Braunwald's heart disease: a textbook of cardiovascular medicine. Philadelphia: Elsevier Saunders; 2005. p. 1141–65.
108. Fry AC, Farrington K. Management of acute renal failure. Postgrad Med J. 2006;82:106–16.
109. Brady H. Acute renal failure. In: Kasper DL et al., editors. Harrison's principles of internal medicine. 16th ed. New York: McGraw-Hill; 2001. p. 1644.
110. Schetz M. Should we use diuretics in acute renal failure? Best Pract Res Clin Anaesthesiol. 2004;18:75–89.

111. Myers BD, Moran SM. Hemodynamically mediated acute renal failure. N Engl J Med. 1986;314:97–105.
112. Rosen S, Heyman S. Difficulties in understanding human "acute tubular necrosis": limited data and flawed animal models. Kidney Int. 2001;60:1220–4.
113. Castaneda MP, Swiatecka-Urban A, Mitsnefes MM. Activation of mitochondrial apoptotic pathways in human renal allografts after ischemia- reperfusion injury. Transplantation. 2003;76:50–4.
114. Bohle A, Christensen J, Kokot F. Acute renal failure in man: new aspects concerning pathogenesis. A morphometric study. Am J Nephrol. 1990;10:374–88.
115. Racusen L. The morphologic basis of acute renal failure. In: Molitors BA, Finn WF, editors. Acute renal failure. 1st ed. Philadelphia: Saunders; 2001. p. 425–50.
116. Lameire N. The pathophysiology of acute renal failure. Crit Care Clin. 2005;21:197–210.
117. Klenzak J, Himmelfab J. Sepsis and the kidney. Crit Care Clin. 2005;21:211–22.
118. Rosen C, Rosenberger S, Heyman SN. Renal parenchymal oxygenation and hypoxia adaption in acute kidney injury. Clin Exp Pharmacol Physiol. 2006;33:980–8.
119. Santos WJQ, Zanetta DMT, Pires AC. Patients with ischemic, mixed and nephrotoxic acute tubular necrosis in the intensive care unit – a homogeneous population? Crit Care. 2006;10:R68–76.
120. Lameire N, Vanholder R. Pathophysiology of ischaemic acute renal failure. Best Pract Res Clin Anaesthesiol. 2004;18:21–36.
121. Davidman M, Olson P, Kohen J. Iatrogenic renal disease. Arch Intern Med. 1991;151:1809–12.
122. Moore RD, Smith CR, Lipsky JJ. Risk factors for nephrotoxicity in patients treated with aminoglycosides. Ann Intern Med. 1984;100:352–7.
123. Shankel SW, Johnson DC, Clark PS. Acute renal failure and glomerulopathy caused nonsteroidal anti-inflammatory drugs. Arch Intern Med. 1992;152:986–90.
124. Hilton R. Acute renal failure. BMJ. 2006;333:786–90.
125. Saotome T, Ishikawa K, May CN. The impact of experimental hypoperfusion on subsequent kidney function. Intensive Care Med. 2010;36:533–40.
126. Johannes T, Mik EG, Ince C. Nonresuscitated endotoxemia induces microcirculatory hypoxic areas in the renal cortex in the rat. Shock. 2009;31:97–103.
127. Kelleher SP, Robinette JB, Miller F. Effect of hemorrhagic reduction in blood pressure on recovery from acute renal failure. Kidney Int. 1987;31:725–30.
128. Liano F, Pascual J. Epidemiology of acute renal failure: a prospective, multicenter, community-based study. Kidney Int. 1996;50:811–8.
129. Schein M, Wittmann D, Aprahamian C. The abdominal compartment syndrome: the physiological and clinical consequences of elevated intra-abdominal pressure. J Am Coll Surg. 1995;180:745–53.
130. Nebelkopf H. Abdominal compartment syndrome. Am J Nurs. 1999;99:53–6, 58, 60.
131. Richards WO, Scovill W, Shin B. Acute renal failure associated with increased intraabdominal pressure. Ann Surg. 1983;197:183–7.
132. Cheatham ML. Abdominal compartment syndrome: pathophysiology and definitions. Scand J Trauma Resusc Emerg Med. 2009;17:10–27.
133. Venkataraman R, Kellum JA. Prevention of acute renal failure. Chest. 2007;131:300–8.
134. Vidal MG, Weisser JR, Gonzalez F. Incidence and clinical effects of intra-abdominal hypertension in critically ill patients. Crit Care Med. 2008;36:1823–31.
135. Surgrue M. Abdominal compartment syndrome. Curr Opin Crit Care. 2005;11:333–8.
136. Bailey J, Shapiro MJ. Abdominal compartment syndrome. Crit Care. 2000;4:23–9.
137. Mailbrain ML, Deeren D, De Potter TJ. Intra-abdominal hypertension in the critically ill: it is time to pay attention. Curr Opin Crit Care. 2005;11:156–71.
138. Bradley S, Bradley G. The effect of increased intra-abdominal pressure on renal function in man. J Clin Invest. 1947;26:1010–5.
139. Harman PK, Kron IL, McLachlan HD. Elevated intra-abdominal pressure and renal function. Ann Surg. 1982;196:594–7.

140. Malbrain ML, Chiumello D, Pelosi P. Prevalence of intra-abdominal hypertension in critically ill patients: a multicentre epidemiological study. Intensive Care Med. 2004; 30:822.
141. Morken J, West MA. Abdominal compartment syndrome in the intensive care unit. Curr Opin Crit Care. 2001;7:268.
142. Mentula P, Hienonen P, Kemppainen E. Surgical decompression for abdominal compartment syndrome in severe acute pancreatitis. Arch Surg. 2010;145:764–9.
143. Stassen NA, Lukan JK, Dixon MS. Abdominal compartment syndrome. Scand J Surg. 2002;91:104–8.
144. Dalfino L, Tullo L, Donadio I. Intra-abdominal hypertension and acute renal failure in critically ill patients. Intensive Care Med. 2008;34:707–13.
145. Lambert DM, Marceau S, Forse RA. Intra-abdominal pressure in the morbidity obese. Obes Surg. 2005;15:1225–32.
146. Shenasky JH. The renal hemodynamic and functional effects of external counterpressure. Surg Gynecol Obstet. 1972;134:253–8.
147. Bloomfield GL, Blocher CR, Fakhry IF. Elevated intra-abdominal pressure increases plasma renin activity and aldosterone levels. J Trauma. 1997;42:997–1004.
148. Sugrue M, Hallal A, D'Amours S. Intra-abdominal hypertension and the kidney. In: Ivatury RR, Cheatham ML, Malbrain MLNG, Sugrue M, editors. Abdominal compartment syndrome. Georgetown: Landes Biosciences; 2006. p. 119–28.
149. De Iaet I, Malbrain ML, Jadoul J. Renal implications of increased intra-abdominal pressure: are the kidneys the canary for abdominal hypertension? Acta Clin Belg Suppl. 2007;1:119–30.
150. Farge D, De la Coussaye JE, Beloucif S. Interactions between hemodynamic and hormonal modifications during PEEP- induced antidiuresis and antinatriuresis. Chest. 1995;107:1095–100.
151. Bruch C, Rothenburger M, Gotzmann M. Chronic kidney disease in patients with chronic heart failure – impact on intracardiac conduction, diastolic function and prognosis. Int J Cardiol. 2007;118:375–80.
152. Miyazato J, Horio T, Takiuchi S. Left ventricular diastolic dysfunction in patients with chronic renal failure: impact of diabetes mellitus. Diabet Med. 2005;22:730–6.
153. Ogata C, Horio T, Kamide K. Association between left ventricular diastolic dysfunction and renal hemodynamic change in patients with treated essential hypertension. Hypertens Res. 2003;26:971–8.
154. Kos T, Pacher R, Wimmer A. Relationship between kidney function, hemodynamic variables and circulating big endothelin levels in patients with severe refractory heart failure. Wien Klin Wochenschr. 1998;110:89–95.
155. Francis GS. Importance of venous congestion for worsening of renal function in advanced decompensated heart failure. J Am Coll Cardiol. 2009;53:589–96.
156. Damman K, Navis G, Smilde TD. Decreased cardiac output, venous con-gestion and the association with renal impairment in patients with cardiac dysfunction. Eur J Heart Fail. 2007;9:872–8.
157. Schrier RW. Role of diminished renal function in cardiovascular mortality: marker or pathogenetic factor? J Am Coll Cardiol. 2006;47:1–8.
158. Cody RJ. Hormonal alterations in heart failure. In: Hosenpud JB, Greenberg BH, editors. Congestive heart failure: pathophysiology, diagnosis and comprehensive approach to management. Philadelphia: Lippincott, Williams and Wilkins; 2000. p. 199–212.
159. Joles JA, Bongartz LG, Gaillard CA. Letter to the editor "Renal venous congestion and renal function in congestive heart failure". JACC. 2009;54:1632.
160. Klabunde RE. Cardiovascular physiology concepts. Philadelphia: Lippincott, Williams and Wilkins; 2011. ISBN 9781451113846, or www.cvphysiology.com/Microcirculation/M012.htm
161. Boberg U, Persson AE. Tubuglomerular feedback during elevated renal venous pressure. Am J Physiol. 1985;249:F524–31.

162. Burnett Jr JC, Knox FG. Renal interstitial pressure and sodium excretion during renal vein constriction. Am J Physiol. 1980;238:F279–82.
163. Lameire N, van Biesen W, Vanholder R. Acute kidney injury. Lancet. 2008;372:1863–5.
164. Schneider AG, Bellomo R. Urinalysis and pre-renal acute kidney injury: time to move on. Crit Care. 2013;17:141.
165. Takasu O, Gaut JP, Watanabe E. Mechanisms of cardiac and renal dysfunction in patients dying of sepsis. Am J Respir Crit Care Med. 2013;187:509–17.
166. Prowle J, Bagshaw SM, Bellomo R. Renal blood flow, fractional excretion of sodium and acute kidney injury: time for a new paradigm? Curr Opin Crit Care. 2012;18:585–92.
167. Bagshaw SM, Bennett M, Devarajan P. Urine biochemistry in septic and non-septic acute kidney injury: a prospective observational study. J Crit Care. 2013;28:371–8.
168. Bagshaw SM, Langenberg C, Bellomo R. Urinary biochemistry and microscopy in septic acute renal failure. Am J Kidney Dis. 2006;48:695–705.
169. Claure-Del Granado R, Macedo E, Mehta E. Urine microscopy in acute kidney injury: time for a change. Am J Kidney Dis. 2011;57:657–60.
170. Maciel AT, Vitorio D. Urine biochemistry in acute kidney injury: are we moving in the right direction? Crit Care. 2013;17:459.
171. Molitors BA. Renal blood flow in sepsis: a complex issue. Crit Care. 2005;9:327–8.
172. Le Dorze M, Legrand M, Pyaen D. The role of microcirculation in acute kidney injury. Curr Opin Crit Care. 2009;15:503–8.
173. Bonventr JV, Weinberg JM. Recent advances in the pathophysiology of ischemic acute renal failure. J Am Soc Nephrol. 2003;14:2199–210.
174. Wan L, Bagshaw SM, Langenberg C. Pathophysiology of septic acute kidney injury: what do we really know. Crit Care Med. 2008;36(suppl):S198–203.
175. Ljungman S, Laragh JH, Cody RJ. Role of kidney in congestive heart failure – relationship of cardiac index to kidney function. Drugs. 1990;39:10–21.
176. Stam F, van Guldner C, Schalkwijk CG. Impaired renal function is associated with markers of endothelial dysfunction and increased inflammatory activity. Nephrol Dial Transplant. 2003;18:892–8.
177. Langenberg C, Bellomo R, May C. Renal blood flow in sepsis. Crit Care. 2005;9:R363–74.
178. Langenberg C, Wan L, Egi M. Renal blood flow in experimental septic acute renal failure. Kidney Int. 2006;69:1996–2002.
179. Langenberg C, Wan L, Egi M. Renal blood flow and function during recovery from experimental septic acute kidney injury. Intensive Care Med. 2007;33:1614–8.
180. Bonventre JV, Zuk A. Ischemic acute renal failure: an inflammatory disease? Kidney Int. 2004;66:480–5.
181. Marshall JC, Vincent JL, Fink MP. Measures, markers, and mediators: towards a staging system for clinical sepsis. A report of the Fifth Toronto Sepsis Roundtable, Toronto, Ontario, Canada, October 25–26, 2000. Crit Care Med. 2003;31:1560–7.
182. Baud L, Oudinet JP, Bens M. Production of tumor necrosis factor by rat mesangial cells in response to bacterial lipopolysaccharide. Kidney Int. 1989;35:1111–8.
183. Knotek M, Rogachev B, Wang W. Endotoxemic renal failure in mice: role of tumor necrosis factor independent of inducible nitric oxide synthase. Kidney Int. 2001;59:2243–7.
184. Cunningham PN, Dyanov HM, Park P. Acute renal failure in endotoxemia is caused by TNF nacting directly on TNF receptor 1 in kidney. J Immunol. 2002;168:5817–23.
185. Matejovic M, Rademacher P, Joannidis M. Acute kidney injury in sepsis: is renal blood flow more than just an innocent bystander? Intensive Care Med. 2007;33:1498–500.
186. Wan L, Bagshaw SM, Langenberg C. Septic acute renal failure. In: Ronco C, Bellomo R, Kellum JA, editors. Critical care nephrology. 3rd ed. Philadelphia: Saunders Elsevier; 2009. Chapter 29, pp 163.
187. Myers BD, Miller DC, Mehigan JT. Nature of the renal injury following total renal ischemia in man. J Clin Invest. 1984;73:329–41.
188. Tumlin JA. Impaired blood flow in acute kidney injury: pathophysiology and potential efficacy of intrarenal vasodilator therapy. Curr Opin Crit Care. 2009;15:514–9.

189. Legrand M, Payen D. Understanding urine output in critically ill patients. Ann Intensive Care. 2011;1:13–20.
190. Licari E, Calzavacca P, Bellomo R. Renal blood flow and perfusion pressure. In: Ronco C, Bellomo R, Kellum JA, editors. Critical care nephrology. 3rd ed. Philadelphia: Saunders Elsevier; 2009. Chapter 31, pp 172.
191. Renneke HG, Denker BM. Review of renal physiology. In: Renal pathophysiology, the essentials. 2nd ed. Philadelphia: Lippincott, Williams & Wilkins; 2007. Chapter 1, pp 1–30.
192. Rimmele' T, Kellum JA. Oliguria and fluid overload. In: Ronco C, Costanzo MR, Bellomo R, Maisel AS, editors. Contrib nephrol, vol. 164. 2010. p. 39–45.
193. Sladen RN, Landry D. Renal blood flow regulation, autoregulation, and vasomotor nephropathy. Anesthesiol Clin North America. 2000;18:791–807.
194. Shipley RE, Study RS. Changes in renal blood flow, extraction of inulin, glomerular filtration rate, tissue pressure, and urine flow with acute alterations of renal artery pressure. Am J Physiol. 1951;167:676–88.
195. Navar LG, Burke TJ, Robinson RR. Distal tubular feedback in the autoregulation of single nephron glomerular filtration rate. J Clin Invest. 1974;53:516–25.
196. Just A, Ehmke H, Toktomambetowa L. Dynamics characteristics and underlying mechanism of renal blood flow autoregulation in the conscious dog. Am J Physiol. 2001;280:F1062–71.
197. Osswald H, Vallon V. Adenosine and tubuloglomerular feedback in the pathophysiology of acute renal failure. In: Ronco C, Bellomo R, Kellum JA, editors. Critical care nephrology. Philadelphia: Saunders Elsevier; 2009. Chapter 23, pp 128.
198. May CN, Langenberg C, Ronco B, Bellomo R, Kellum JA. The physiology of the afferent and efferent arterioles. In: Critical care nephrology. Philadelphia: Saunders Elsevier; 2009. Chapter 21, pp 118.
199. Thomson S, Bao D, Deng A. Adenosine formed by 5′-nucleotidase mediates tubuloglomerular feedback. J Clin Invest. 2000;106:289–98.
200. Walker 3rd M, Harrison-Bernard LM, Cook AK. Dynamic interaction between myogenic and TGF mechanisms in afferent arteriolar blood flow autoregulation. Am J Physiol. 2000;279:F858–65.
201. Thurau K, Boylan JW. Acute renal success. The unexpected logic of oliguria in acute renal failure. Am J Med. 1976;61:308–14.
202. Schurek HJ, Johns O. Is tubuloglomerular feedback a tool to prevent nephron oxygen deficiency? Kidney Int. 1997;51:386–92.
203. Cupples WA. Modulation by NO of renal myogenic autoregulation: TGF dependency. J Physiol. 2005;569:713.
204. Schnerman J. The juxtaglomerular apparatus: from anatomical peculiarity to physiological relevance. J Am Soc Nephrol. 2003;14:1681–94.
205. Stanton BA, Koeppen BM. Elements of renal function. In: Berne RM, Levy MN, editors. Physiology. 3rd ed. St. Louis: Mosby-Year Book; 1993. p. 719–53.
206. Levens NR, Peach MJ, Carey RM. Role of intra-renal renin- angiotensin system in the control of renal function. Circ Res. 1981;48:157–67.
207. Brady HR, Singer GG. Acute renal failure. Lancet. 1995;346:1533–40.
208. Skaden NS, Landrz D. Renal blood flow regulation, autoregulation, and vasomotor nephropathy. Anesthesiol Clin North America. 2000;18:791–807.
209. Brezis M, Rosen S. Hypoxia of the renal medulla – its implications for disease. N Engl J Med. 1995;332:647–55.
210. Devarjan P. Update on mechanisms of ischemic acute kidney injury. J Am Soc Nephrol. 2006;17:1503–20.
211. Brezis M, Heyman SN, Epstein FH. Determinants of intrarenal oxygenation. II. Hemodynamic effects. Am J Physiol. 1994;267:F1063–8.
212. Sheridan AM, Bonventre JV. Pathophysiology of ischemic acute renal failure. Contrib Nephrol. 2002;132:7–21.
213. Evans RG, Madden AC, Denton KM. Diversity of responses of renal cortical and medullary blood flow to vasoconstrictors in conscious rabbits. Acta Physiol Scand. 2000;169:297–308.

214. Guild SJ, Eppel GA, Malpas SC. Regional responsiveness of renal perfusion to activation of the renal nerves. Am J Physiol Regul Integr Comp Physiol. 2002;283:R1177–86.
215. Pallone TL, Zhang Z, Rhinehart K. Physiology of the renal medullary microcirculation. Am J Physiol Renal Physiol. 2003;284:F253–66.
216. Palm F, Cederberg J, Hansell P. Reactive oxygen species cause diabetes-induced decrease in renal oxygen tension. Diabetologia. 2003;46:1153–60.
217. Heyman SN, Brezis M, Epstein FH. Early renal medullary hypoxic injury from radiocontrast and indomethacin. Kidney Int. 1991;40:632–42.
218. Agmon Y, Peleg H, Greenfield Z. Nitric oxide and prostanoids protect the renal outer medulla from radiocontrast toxicity in the rat. J Clin Invest. 1994;94:1069–75.
219. Brezis M, Heyman SN, Dinour D. Role of nitric oxide in renal medullary oxygen balance. Studies in isolated and intact rat kidneys. J Clin Invest. 1991;88:390–5.
220. Heyman SN, Fuchs S, Yafe R. Renal microcirculation and tissue damage during acute ureteral obstruction: the effect of saline infusion, indomethacin and radiocontrast. Kidney Int. 1997;51:653–63.
221. Evans RG, Gardiner BS, Smith DW. Intrarenal oxygenation: unique challenges and the biophysical basis of homeostasis. Am J Physiol Renal Physiol. 2008;295:F1259–70.
222. Navar LG, Inscho EW, Majid SA. Paracrine regulation of the renal microcirculation. Physiol Rev. 1996;76:425–536.
223. Moran M, Kapsner C. Acute renal failure associated with elevated plasma oncotic pressure. N Engl J Med. 1987;317:150–3.
224. Andersen JL, Andersen LJ, Sandgaard NC. Volume expansion natriuresis during servo control of systemic blood pressure in conscious dogs. Am J Physiol Regul Integr Comp Physiol. 2000;278:R19–27.
225. Langenberg C, Bellomo R, May C. Renal vascular resistance in sepsis. Nephron Physiol. 2006;104:1–11.
226. DiBona GF. Neural control of the kidney: functionally specific renal sympathetic nerve fibers. Am J Physiol Regul Integr Comp Physiol. 2000;279:R1517–24.
227. Denton KM, Luff SE, Shweta A. Differential neural control of glomerular ultrafiltration. Clin Exp Pharmacol Physiol. 2004;31:380–6.
228. Leonhard BL, Evans RG, Navakatikyan MA. Differential neural control of intrarenal blood flow. Am J Physiol Regul Integr Comp Physiol. 2000;279:R907–16.
229. Leonhard BL, Malpas SC, Denton KM. Differential control of intrarenal blood flow during reflex increases in sympathetic nerve activity. Am J Physiol Regul Integr Comp Physiol. 2001;280:R62–8.
230. De Nicola L, Blantz RC, Gabbai FB. Nitric oxide and angiotensin II. Glomerular and tubular interaction in the rat. J Clin Invest. 1992;89:1248–56.
231. Yuan BH, Robinette JB, Conger JD. Effect of angiotensin II and norepinephrine on isolated rat afferent and efferent arterioles. Am J Physiol. 1990;258:F741–50.
232. Denton KM, Fennessy PA, Alcorn D. Morphometric analysis of the actions of angiotensin II on renal arterioles and glomeruli. Am J Physiol. 1992;262:F411–8.
233. Needham E. Management of acute renal failure. Am Fam Physician. 2005;72:1739–46.
234. Bankir L, Kriz W, Goligorsky M. Vascular contributions to pathogenesis of acute renal failure. Ren Fail. 1998;20:663–77.
235. Manson J. Tubulo-glomerular feedback in the early stages of experimental acute renal failure. Kidney Int. 1976;6:S106–14.
236. Kribben A, Edelstein CL, Schrier RW. Pathophysiology of acute renal failure. J Nephrol. 1999;12(supplement 2):S142–51.
237. Agraharkar M, Safirstein RL. Pathophysiology of acute renal failure. In: Primer on kidney disease. 3rd ed. s.l.: The National Kidney Foundation; 2001. Chapter 32, p. 239–45.
238. Conger JD, Robinette JB, Schrier RW. Smooth muscle calcium and endothelium derived relaxing factor in abnormal vascular responses of acute renal failure. J Clin Invest. 1988;82:532–7.

239. Robinette JB, Conger JD. Angiotensin and thromboxane in the enhanced renal adrenergic nerve sensitivity of acute renal failure. J Clin Invest. 1990;86:1532–9.
240. Kelleher SP, Roninette JB, Conger JD. Sympathetic nervous system in the loss of autoregulation in acute renal failure. Am J Physiol. 1984;246:F379–89.
241. Conger JD, Robinette JB, Villar A. Increased nitric oxide synthase activity despite lack of response to endothelium- dependent vasodilators in postischemic acute renal failure in rats. J Clin Invest. 1995;96:631–8.
242. Vetterlein F, Petho A, Schmidt G. Distribution of capillary blood flow in rat kidney during postischemic renal failure. Am J Physiol. 1986;251:H510–9.
243. Sutton TA, Fisher CJ, Molitoris BA. Microvascular injury and dysfunction during ischemic acute renal failure. Kidney Int. 2002;62:1539–49.
244. Molitoris BA, Sandoval RM. Intravital multiphoton microscopy of dynamic renal processes. Am J Physiol Renal Physiol. 2005;288:F1084–9.
245. Yamaguchi N, Jesmin S, Zaedi S. Time-dependent expression of renal vaso-regulatory molecules in LPS-induced endotoxemia in rat. Peptides. 2006;27:2258–70.
246. Lattanzio MR, Kopyt NP. Acute kidney injury: new concepts in definition, diagnosis, pathophysiology, and treatment. J Am Osteopath Assoc. 2009;109:13–9.
247. Wan L, Bellomo R, Giantomasso D. The pathogenesis of septic acute renal failure. Curr Opin Crit Care. 2003;9:496–502.
248. Bonventre JV. Mechanisms of ischemic acute renal failure. Kidney Int. 1993;43:1160–78.
249. Epstein FH, Veves A, Prasad P. Effect of diabetes on renal medullary oxygenation during water diuresis. Diabetes Care. 2002;25:575–8.
250. Losonczy G, Bloch JF, Samsell L. Impact of surgery on nitric oxide in rats: evidence for activation of inducible nitric oxide synthase. Kidney Int. 1997;51:1943–9.
251. Okusa MD, Linden J, MacDonald T. Selective A2a adenosine receptor activation reduces ischemia-reperfusion injury in rat kidney. Am J Physiol. 1999;277:F404–12.
252. Mason J, Torhorst J, Welsch J. Role of the medullary perfusion defect in the pathogenesis of ischemic renal failure. Kidney Int. 1984;26:283–93.
253. Bellomo R, Wan L, May C. Vasoactive drugs and acute kidney injury. Crit Care Med. 2008;36(suppl):S179–86.
254. Bersten AD, Holt AW. Vasoactive drugs and importance of renal perfusion pressure. New Horiz. 1995;3:650–61.
255. Sharfuddin A, Molitoris BA. Pathophysiology of ischemic acute kidney. Nat Rev Nephrol. 2011. doi:10.1038/nrneph.2011.16.
256. Heyman SN, Brezis M, Reubinoff CA. Acute renal failure with selective medullary injury in the rat. J Clin Invest. 1988;82:401–12.
257. Kanji HD, Schulze CJ, Hervas-Malo M. Difference between pre-operative and cardiopulmonary bypass mean arterial pressure is independently associated with early cardiac surgery-associated acute kidney injury. J Cardiothorac Surg. 2010;5:71.
258. Molitoris BA, Sandoval R, Sutton TA. Endothelial injury and dysfunction in ischemic acute renal failure. Crit Care Med. 2002;30:S235–40.
259. Flores J, DiBona DR, Beck CH. The role of cell swelling in ischemic renal damage and the protective effect of hypertonic solute. J Clin Invest. 1972;51:118–26.
260. Bragadottir G, Rdefors B, Ricksten S-E. Mannitol increases renal blood flow and maintains filtration fraction and oxygenation in postoperative acute kidney injury: a prospective interventional study. Crit Care. 2012;16:R159.
261. Redfors B, Sward K, Sellgren J. Effects of mannitol plus frusemide on renal oxygen consumption, blood flow and glomerular filtration after cardiac surgery. Intensive Care Med. 2009;35:115–22.
262. Kelly KJ, Williams WW, Colvin RB. Intercellular adhesion molecule-1-deficient mice are protected against ischemic renal injury. J Clin Invest. 1996;97:1056–63.
263. Jang HR, Rabb H. The innate immune response in ischemic acute kidney injury. Clin Immunol. 2009;130:41–50.

264. Sheridan AM, Bonventre JV. Cell biology and molecular mechanisms of injury in ischemic acute renal failure. Curr Opin Nephrol Hypertens. 2000;9:427–34.
265. Cao CC, Ding XQ, Zhu CF. In vivo transfection NF-kappaB decoy oligodeoxynucleotides attenuates renal ischemia /reperfusion injury in rats. Kidney Int. 2004;65:834–45.
266. Benes J, Chvojka J, Sykora R. Searching for mechanisms that matter in early septic acute kidney injury: an experimental study. Crit Care. 2011;15:R256.
267. Kinsey GR, Li L, Okusa MD. Inflammation in acute kidney injury. Nephron Exp Nephrol. 2008;109:e102–7.
268. Wen X, Murugan R, Peng Z. Pathophysiology of acute kidney injury: a new perspective. Contrib Nephrol. 2010;165:39–45.
269. Holly MK, Dear JW, Hu X. Biomarker and drug-target discovery using proteomics in a new rat model of sepsis-induced acute renal failure. Kidney Int. 2006;70:496–506.
270. Kwon O, Nelson WJ, Sibley R. Backleak, light junctions, and cell-cell adhesion in postischemic injury to the renal allograft. J Clin Invest. 1998;101:2054–64.
271. Ramesh G, Reeves WB. Inflammatory cytokines in acute renal failure. Kidney Int Suppl. 2004;91:S56–61.
272. Friedewald JJ, Rabb H. Inflammatory cells in ischemic acute renal failure. Kidney Int. 2004;66:486–90.
273. Simmons EM, Himmelfarb J, Sezer MT. PICARD Study Group: plasma cytokine levels predict mortality in patients with acute renal failure. Kidney Int. 2004;65:1357–65.
274. Molitoris B, Sutton TA. Endothelial injury and dysfunction: role in the extension phase of acute renal failure. Kidney Int. 2004;66:496–9.
275. Dagher PC, Herget-Rosenthal S, Ruehm SG. Newly developed techniques to study and diagnose acute renal failure. J Am Soc Nephrol. 2003;14:2188–98.
276. Legrand M, Mik EG, Johannes T. Hypoxia and dysoxia after reperfusion of ischemic kidney. Mol Med. 2008;14:502–16.
277. Miloradovic Z, Jerkic M, Jovovic D. Bosentan and losartan ameloriate acute renal failure associated with mild but not strong NO blockage. Nephrol Dial Transplant. 2007;22:2276–84.
278. Gulmen S, Kiris I, Narin C. Tezosentan reduces the renal injury induced by abdominal aortic ischemia-reperfusion in rats. J Surg Res. 2009;22:4804–9.
279. Guan Z, Gobe G, Willgoss D. Renal endothelial dysfunction and impaired autoregulation after ischemia-reperfusion injury result from excess nitric oxide. Am J Physiol Renal Physiol. 2006;291:F619–28.
280. Legrand M, Payen D, Ince C. L-NIL prevents microvascular hypoxia and increase of renal oxygen consumption after ischemia-reperfusion in rats. Am J Physiol Renal Physiol. 2009;296:1109–17.
281. Besso J, Blanco R, Gonzalez R. Acute renal failure. In: Gullo A, Besso J, Lumb PD, Williams GF, editors. Intensive and critical care medicine. s.l.: Springer-Italia; 2009. Chapter 16, p. 175.
282. Corrigan G, Ramaswamy D, Kwon O. PAH extraction and estimation of plasma flow in human postischemic acute renal failure. Am J Physiol. 1999;277:F312–8.
283. Ramaswamy D, Corrigan G, Polhemus C. The maintenance and recovery stages of postischemic acute renal failure in humans: a study of cadaveric renal allografts. Am J Physiol Renal Physiol. 2001;282:F271–80.
284. Molitoris BA. Ischemia-induced loss of epithelial polarity: potential role of the actin cytoskeleton. Am J Physiol. 1991;260:F769–78.
285. Glaumann G, Glaumann H, Trump BF. Studies of cell recovery from injury. Ultrastructural studies of the recovery of the pars recta of the proximal tubule (P3) segment of the rat kidney from temporary ischemia. Virchows Arch. 1977;25:281–308.
286. Safirstein RL, Bonventre JV. Molecular response to ischemic and nephrotoxic acute renal failure. In: Bonventre D, Schlöndorff JV, editors. Molecular nephrology. New York: Marcel Dekker; 1995. p. 839–54.
287. Donohoe JK, Venkatachalam MA, Bernard DB. Tubular leakage and obstruction after renal ischemia: structural-functional correlations. Kidney Int. 1978;13:208–22.

288. Benad H-M. Acute renal failure in intensive care medicine. Anästh Intensivmed. 2011;52:757–70.
289. Lee HT, Xu H, Nasr SH. A1 adenosine receptor knockout mice exhibit increased renal injury following ischemia and reperfusion. Am J Physiol Renal Physiol. 2004;286:F298–306.
290. Safirstein RL. Acute renal failure: from renal physiology to the renal transcriptome. Kidney Int. 2004;66:S62–6.
291. Wu L, Tiwari MM, Messer KJ. Peritubular capillary dysfunction and renal tubular epithelial cell stress following lipopolysaccharide administration in mice. Am J Physiol Renal Physiol. 2007;292:F261–8.
292. Miloradovic Z, Jerkic M, Jovovic D. Bosentan and losartan ameloirate acute renal failure associated with mild but not strong NO blockage. Nephrol Dial Transplant. 2007;22:2476–84.
293. Beeri R, Symon Z, Brezis M. Rapid DNA fragmentation from hypoxia along the thick ascending limb of rat kidneys. Kidney Int. 1995;47:1806–10.
294. Brodsky SV, Yamamoto T, Tada T. Endothelial dysfunction in ischemic acute renal failure: rescue by transplanted endothelial cells. Am J Physiol Renal Physiol. 2002;282: F1140–9.
295. Solez K, Kramer EC, Fox JA. Medullary plasma flow and intravascular leukocyte accumulation in acute renal failure. Kidney Int. 1974;6:24–37.
296. Goligorsky MS, Brodsky SV, Noiri E. NO bioavailability, endothelial dysfunction, and acute renal failure: new insights into pathophysiology. Semin Nephrol. 2004;24:316–23.
297. Druid H, Enestrom S, Rammer L. Effect of anticoagulation upon nephron obstruction in experimental acute ischaemic renal failure. A morphological study. Int J Exp Pathol. 1998;79:55–66.
298. Goligorsky MS. Whispers and shouts in the pathogenesis of acute renal ischemia. Nephrol Dial Transplant. 2005;20:261–6.
299. Bellomo R, Di Giantomasso D. Noradrenaline and the kidney: friends or foes? Crit Care. 2001;5:294–8.
300. Conger JD, Schultz MF, Miller F. Responses to hemorrhagic arterial pressure reductions in different ischemic renal failure models. Kidney Int. 1994;46:318–23.
301. Conger JD, Hammond WS. Renal vasculature and ischemic injury. Ren Fail. 1992;14:307–10.
302. Adams PL, Adams FF, Bell PD. Impaired renal blood flow autoregulation in ischemic acute renal failure. Kidney Int. 1980;18:68–76.
303. Brater DC. Anti-inflammatory agents and renal function. Semin Arthritis Rheum. 2002;32 suppl 1:33–42.
304. Gambro G, Perazella MA. Adverse renal effects of anti-inflammatory agents: evaluation of selective and nonselective cyclooxygenase inhibitors. J Intern Med. 2003;253:643–52.
305. Conger JD. Vascular alterations in acute renal failure: roles in initiation and maintenance. In: Molitoris BA, Finn WF, editors. Acute renal failure—A companion to Brenner and Rector's the kidney. Philadelphia: Saunders; 2001. p. 13–29.
306. Kelly KJ, Tolkoff-Rubin NE, Rubin RH. An oral platelet-activating factor antagonist, Ro-24-4736, protects the rat kidney from ischemic injury. Am J Physiol. 1996;271:F1061–7.
307. Karajala-Subramanyam V, Venkataraman R, Kellum JA. Expert Consult Nephrology, section XIII. Nonpharmacological management of acute renal failure. 2009. Philadelphia: Elsevier; Chapter 76.
308. Venkataraman R. Can we prevent acute kidney injury? Crit Care Med. 2008;36(suppl):S166–71.
309. Humes HD, Hunt DA, White MD. Direct toxic effect of the radiocontrast agent diatrizoate on renal proximal tubule cells. Am J Physiol. 1987;252:F246–55.
310. Deverajan P, Mishra J, Supavekin S. Gene expression in early ischemic renal injury: clues towards pathogenesis, biomarkers discovery, and novel therapeutics. Mol Genet Metab. 2003;80:365–76.
311. Nadasdy T, Laszik Z, Blick KE. Human acute tubular necrosis: a lectin and immunohistochemical study. Hum Pathol. 1995;26:230–9.

312. Molitoris BA. Transitioning to therapy in ischemic acute renal failure. J Am Soc Nephrol. 2003;14:265–7.
313. Schrier RW. Need to intervene in established acute renal failure. J Am Soc Nephrol. 2004;15:2756–8.
314. Dworkin LD, Brenner RM. The renal circulations. In: Brenner BM, editor. Brenner and Rector's the kidney. 7th ed. Philadelphia: Saunders; 2004. p. 307–52.
315. Himmelfarb J. Acute kidney injury in the elderly: problems and prospects. Semin Nephrol. 2009;29:658–64.
316. Kelly KJ, Molitors BA. Acute renal failure in the new millennium: time to consider combination therapy? Semin Nephrol. 2000;20:4–19.
317. Parik CR, Devarajan P. New markers of acute kidney injury. Crit Care Med. 2008;36(suppl):S159–65.
318. Bagshaw SM, Gibney RTN. Conventional markers of kidney function. Crit Care Med. 2008;36(suppl):S152–8.
319. Stevens LA, Levey AS. Measurement of kidney function. Med Clin North Am. 2005;89:457–73.
320. Tomlanovich S, Golbertz H, Perlroth M. Limitations of createnine in quantifying the severity of cyclosporine-induced chronic nephropathy. Am J Kidney Dis. 1986;8:332–7.
321. Nikolas TL, O'Rourke MJ, Yang J. Sensitivity and specificity of a single emergency department measurement of urinary neutrophil gelatinase-associated lipocalin for diagnosing acute kidney injury. Ann Intern Med. 2008;148:810–9.
322. Lameire N, Hoste EA. Reflections on the definition, classification, and diagnostic evaluation of acute renal failure. Curr Opin Crit Care. 2004;10:468–75.
323. Baboolal K, Jones GA, Janezic A. Molecular and structural consequences of early renal allograft injury. Kidney Int. 2002;61:686–96.
324. Prowle JR, Liu Y-L, Bagshaw SM. Oliguria as predictive biomarker of acute kidney injury in critically ill patients. Crit Care. 2011;15:R172–81.
325. Subramanian S, Kellum JA, Ronco C. Oliguria. In: Ronco C, Bellomo R, Kellum JA, editors. Critical care nephrology. Philadelphia: Saunders Elsevier; 2009. Chapter 61, pp 341.
326. Spital A, Valvo JR, Segal AJ. Nondilated obstructive uropathy. Urology. 1988;31:478–82.
327. Bagshaw SM, Bellomo R, Kellum JA. Oliguria, volume overload, and loop diuretics. Crit Care Med. 2008;36:S172–8.
328. Anderson RJ, Linas SL, Berns AS. Nonoliguric acute renal failure. N Engl J Med. 1977;296:1134–8.
329. Prowle JR, Ishikawa K, May CN. Renal plasma flow and glomerular filtration rate during acute kidney injury in man. Ren Fail. 2010;32:349–55.
330. Pons B, Lautrette A, Oziel J. Diagnostic accuracy of early urinary index changes in differentiating transient from persistent acute kidney injury in critically ill patients: multicenter cohort study. Crit Care. 2013;17:R56.
331. Kaplan AA, Kohn OF. Fractional excretion of urea as a guide to renal dysfunction. Am J Nephrol. 1992;12:49–54.
332. Fang L, Sirota R, Ebert T. Low fractional excretion of sodium with contrast media-induced acute renal failure. Arch Intern Med. 1980;140:531–3.
333. Vaz AJ. Low fractional excretion of urine sodium in acute renal failure due to sepsis. Arch Intern Med. 1983;143:738–9.
334. Darmon M, Vincent F, Dellamonica J. Diagnostic performance of fractional excretion of urea in the evaluation of critically ill patients with acute kidney injury: a multicenter cohort study. Crit Care. 2011;15:R178.
335. Darmon M, Schortgen F, Vargas F. Diagnostic accuracy of Doppler renal resistive index for reversibility of acute kidney injury in critically ill patients. Intensive Care Med. 2011;37:68–76.
336. Devarajan P. Neutrophil gelatinase-associated lipocalin – an emerging troponin for kidney injury. Nephrol Dial Transplant. 2008;23:3737–43.

337. Nickolas TL, Schmidt-Ott KM, Canetta P. Diagnostic and prognostic stratification in the emergency department using urinary biomarkers of nephron damage: a multicenter prospective cohort study. J Am Coll Cardiol. 2012;59:246–55.
338. Rona R, Fumagalli R. Indications for renal replacement therapy in the critically ill. In: Ronco C, Bellomo R, Kellum JA, editors. Critical care nephrology. Philadelphia: Saunders Elsevier; 2009. Chapter 241, pp 1328.
339. Best HA, Counselman FL. Evaluating and managing acute renal failure. Emerg Med. 2008;40:16–36.
340. Hoste EA, De Waele JJ. Physiologic consequences of acute renal failure on the critically ill. Crit Care Clin. 2005;21:251–60.
341. Golab F, Kadkhodaee M, Zahmatkesh M. Ischemic and non – ischemic acute kidney injury cause hepatic damage. Kidney Int. 2009;75:783–92.
342. Uchino S, Bellomo R, Bagshaw SM. Transient azotemia is associated with a high risk of death in hospitalized patients. Nephrol Dial Transplant. 2010;25:1833–9.
343. Bouchard J, Mehta RL. Fluid accumulation and acute kidney injury: consequence or cause? Curr Opin Crit Care. 2009;15:509–13.
344. Hinds CJ, Watson JD. Acute renal failure: clinical course and management. In: Hinds CJ, Watson JD, editors. Intensive care. A concise textbook. 3rd ed. Philadelphia: Saunders Elsevier; 2008. p. 365–70.
345. Roche AM, James MFM. Colloids and crystalloids: does it matter to the kidney? Curr Opin Crit Care. 2009;15:520–4.
346. Townsend DR, Bagshaw SM. New insights on intravenous fluids, diuretics and acute kidney injury. Nephron Clin Pract. 2008;109:c206–16.
347. Kellum JA, Cerda J, Kaplan LJ. Fluids for the prevention and management of acute kidney injury. Int J Artif Organs. 2008;31:96–110.
348. Agrawal M, Swartz A. Acute renal failure. Am Fam Physician. 2000;61:2077–88, (Erratum 2001; 63: 445).
349. Kellum JA, Ronco C, Mehta RL. Fluid management in acute kidney injury. Int J Artif Organs. 2008;31:94–5.
350. Matejovic M, Träger K, De Backer D. Noradrenaline and the kidney: foe, friend, or both? Intensive Care Med. 2005;31:1476–8.
351. Molitoris BA. Contrast nephropathy: are short-term outcome measures adequate for quantification of long-term renal risk? Nat Clin Pract Nephrol. 2008;4:594–5.
352. Wiedemann HP, Wheeler AP, Bernard GR. Comparison of two fluid-management strategies in acute lung injury. N Engl J Med. 2006;354:2564–75.
353. Van Biesen W, Yegenaga L, Vanholder R. Relationship between fluid status and its management on acute renal failure (ARF) in intensive care unit (ICU) patients with sepsis: a prospective analysis. J Nephrol. 2005;18:54–60.
354. Payen D, de Pont AC, Sakr Y. A positive fluid balance is associated with a worse outcome in patients with acute renal failure. Crit Care. 2008;12:R74.
355. Murphy CV, Schramm GE, Doherty JA. The importance of fluid management in acute lung injury secondary to septic shock. Chest. 2009;136:102–9.
356. Schrier RW. Body fluid volume regulation in health and disease: a unifying hypothesis. Ann Intern Med. 1990;113:155–9.
357. Pearse RM, Belsey JD, Cole J. Effect of dopexamins infusion on mortality following major surgery: individual patient data meta-regression analysis of published clinical trails. Crit Care Med. 2008;36:1323–9.
358. McKendry M, McGloin H, Saberi D. Randomised controlled trial assessing the impact of a nurse delivered, flow monitored protocol for optimisation of circulatory status after cardiac surgery. BMJ. 2004;329:258–60.
359. Boyd O, Grounds RM, Bennett ED. A randomized clinical trial of the effect of deliberate perioperative increase of oxygen delivery on mortality in high-risk surgical patients. JAMA. 1993;270:2699–707.

360. Prowle JR, Echeverri JE, Ligabo V. Fluid balance and acute kidney injury. Nat Rev Nephrol. 2010;6:107–15.
361. Lewington A, Kanagasundaram S. Acute kidney injury. Guidelines by the Renal association of UK. 2011. http://www.renal.org/clinical/guidelinessection/AcuteKidneyInjury.aspx
362. Dellinger RP, Levy MM, Carlet JM. Surviving sepsis campaign: international guidelines for management of severe sepsis and septic shock. Crit Care Med. 2008;36:296–327.
363. Lin SM, Huang CD, Lin HC. A modified goal-directed protocol improves clinical outcomes in intensive care unit patients with septic shock: a randomized controlled trail. Shock. 2006;26:551–7.
364. Rivers E, Nguyen B, Hawstad S. Early goal-directed therapy in the treatment of severe sepsis and septic shock. N Engl J Med. 2001;345:1368–77.
365. Pinsky MR, Brophy P, Padilla J. Fluid and volume monitoring. Int J Artif Organs. 2008;31:111–26.
366. Jhanji S, Vivian-Smith A, Lucena-Amaro S. Hemodynamic optimalization improves tissue microvascular flow and oxygenation after major surgery: a randomised control trail. Crit Care. 2010;14:R151.
367. Schuller D, Mitchell JP, Calandrino FS. Fluid balance during pulmonary oedema: is fluid gain a marker or cause of poor outcome? Chest. 1991;100:1068–75.
368. Gillespie RS, Seidel K, Symons JM. Effect of fluid overload and dose of replacement fluid on survival in hemofiltration. Pediatr Nephrol. 2004;19:1394–9.
369. Vincent JL, Sakr Y, Sprung CL. Sepsis in European intensive care units: results of the SOAP study. Crit Care Med. 2006;34:344–53.
370. Sakr Y, Vincent JL, Reinhart K. High tidak volume and positive fluid balance are associated with worse outcome in acute lung injury. Chest. 2005;128:3098–108.
371. Foland JA, Fortenberry JD, Warshaw BL. Fluid overload before continuous hemofiltration and survival in critically ill patients: a retrospective analysis. Crit Care Med. 2004;32:1771–6.
372. Goldstein SL, Somers MJ, Baum MA. Pediatric patients with multi-organ dysfunction syndrome receiving continuous renal replacement therapy. Kidney Int. 2005;67:653–8.
373. Bouchard J, Soroko S, Chertow CM. Program to improve care in acute renal disease (PICARD) study group: fluid accumulation, survival, and recovery of kidney function in critically ill patients with acute kidney injury. Kidney Int. 2009;76:422–7.
374. Durairaj L, Schmidt GA. Fluid therapy in resuscitated sepsis: less is more. Chest. 2008;133:252–63.
375. Vincent JL, Weil MH. Fluid challenge revisited. Crit Care Med. 2006;34:1333–7.
376. McGee S. Is the patient hypovolemic? JAMA. 1999;281:1022–4.
377. Stevens WJ. Fluid balance and resuscitation. Nurs Crit Care. 2008;3:11.
378. Raper R, Sibbald WJ. Misled by the wedge? The Swan-Ganz catheter and left ventricular preload. Chest. 1986;89:427–34.
379. Bouhemad B. Isolated and reversible impairment of ventricular relaxation in patients with septic shock. Crit Care Med. 2008;36:766–74.
380. Bouhemad B. Acute left ventricular dilatation and shock-induced myocardial dysfunction. Crit Care Med. 2009;37:441–7.
381. Rüdiger A, Singer M. Mechanisms of sepsis-induced cardiac dysfunction. Crit Care. 2007;35:1599–608.
382. Fleck A. Increased vascular permeability: a major cause of hypalbuminaemina in disease and injury. Lancet. 1985;325:781–4.
383. Conference, American College of Chest Physicians/Society of Critical Care Medicine Consensus. Definition for sepsis and organ failure and guidelines for the use of innovative therapies in sepsis. Crit Care Med. 1992;20:864–74.
384. Di Giantomasso D, May CN, Bellomo R. Vital organ blood flow during hyperdynamic sepsis. Chest. 2003;124:1053–9.
385. Ruokonen E. Regional blood flow and oxygen transport in septic shock. Crit Care Med. 1993;21:1296–303.

386. Michard F, Alaya S, Zarka V. Global end-diastolic volume as an indicator of cardiac preload in patients with septic shock. Chest. 2003;124:1900–8.
387. Michard F, Boussat S, Chemla D. Relation between respiratory changes in arterial pulse pressure and fluid responsiveness in septic patients with acute circulatory failure. Am J Respir Crit Care Med. 2000;162:134–8.
388. Kirov MY, Kuzkov VV, Kuklin VN. Extravascular lung water assessed by transpulmonary single thermodilution and postmorten gravity in sheep. Crit Care. 2004;8:R451–8.
389. Michard F, Teboul JL. Predicting fluid responsiveness in ICU patients: a critical analysis of the evidence. Chest. 2002;121:2000–8.
390. Antonelli M, Levy M, Andrews PJ. Hemodynamic monitoring in shock and implications for management. International Consensus Conference, Paris, France, 27–28, April 2006. Intensive Care Med. 2007;33:575–90.
391. Weil MH, Henning RJ. New concepts in the diagnosis and fluid treatment of circulatory shock. Anesth Analg. 1979;58:124–32.
392. Magder S. Central venou pressure monitoring. Curr Opin Crit Care. 2006;12:219–27.
393. Michard F. Changes in arterial pressure during mechanical ventilation. Anesthesiology. 2005;103:419–28.
394. Feissel M, Michard F, Mangin I. Respiratory changes in aortic blood velocity as an indicator of fluid responsiveness in ventilated patients with septic shock. Chest. 2001;119:867–73.
395. Tavernier B, Makhotine O, Lebuffe G. Systolic pressure variation as a guide to fluid therapy in patients with sepsis-induced hypotension. Anesthesiology. 1998;89:1313–21.
396. Marik PE, Baram M, Vahid B. Does the central venous pressure predict fluid responsiveness? A systematic review of the literature and the tale of seven mares. Chest. 2008;134:172–8.
397. Pinsky MR. Hemodynamic evaluation and monitoring in the ICU. Chest. 2007;132:2020–9.
398. Mirsky MR, Payen D. Functional hemodynamic monitoring. Crit Care. 2005;9:566–72.
399. Applegate RJ, Johnston WE, Vinten-Johansen J. Restraining effect of intact pericardium during acute volume loading. Am J Physiol Heart Circ Physiol. 1992;262:H1725–33.
400. Magder S, Bafaqeeh F. The clinical role of central venous pressure measurements. Intensive Care Med. 2007;22:44–51.
401. Carcillo JA, Davis AL, Zaritsky A. Role of early fluid resuscitation in pediatric septic shock. JAMA. 1991;266:1242–5.
402. Michard F. Volume management using dynamic parameters: the good, the bad, and the ugly. Chest. 2005;128:1902–3.
403. Thiel SW, Kollef MH, Isakow W. Non-invasive stroke volume measurement and passive leg raising predict fluid responsiveness in medical ICU patients: an observational cohort study. Crit Care. 2009;13:R111.
404. Preisman S, Kogan S, Berkenstadt H. Predicting fluid responsiveness in patients undergoing cardiac surgery: functional haemodynamic parameters including the Respiratory Systolic Variation Test and static preload indicators. Br J Anaesth. 2005;95:746–55.
405. Reuter DA, Kirchner A, Felbinger TW. Usefullness of left ventricular stroke volume variation to assess fluid responsiveness in patients with reduced cardiac function. Crit Care Med. 2003;31:1399–404.
406. Perel A. The physiological basis of arterial pressure variation during positive-pressure ventilation. Reanimation. 2005;14:162–71.
407. Fessler HE, Brower RG, Wise RA. Mechanism of reduced LV afterload by systolic and diastolic positive pleural pressure. J Appl Physiol. 1988;65:1244–50.
408. Brower R, Wise RA, Hassapoyannes C. Effect of lung inflation on lung blood volume and pulmonary venous flow. J Appl Physiol. 1985;58:954–63.
409. Deuter DA, Felbinger TW, Schmidt C. Stroke volume variation for the assessment of cardiac responsiveness to volume loading in mechanically ventilated patients after cardiac surgery. Intensive Care Med. 2002;28:392–8.
410. Wyler von Ballmoos M, Takahla J, Roeck M. Pulse-pressure variation and hemodynamic response in patients with elevated pulmonary artery pressure: a clinical study. Crit Care. 2010;14:R111.

411. Marik PE, Cavallazzi R, Veron T. Dynamic changes in arterial waveform derived variables and fluid responsiveness in mechanically ventilated patients: a systematic review of literature. Crit Care Med. 2009;37:2642–7.
412. Monnet X, Rienzo M, Osman D. Passive leg raising predicts fluid responsiveness in the critically ill. Crit Care Med. 2006;34:1402–7.
413. Boulain T, Achard J-M, Teboul J-L. Changes in BP induced by passive leg raising predict response to fluid loading in critically Ill patients. Chest. 2002;121:1245–52.
414. Lafaneclere A, Pène F, Goulenok C. Changes in aortic blood flow induced by passive leg raising predict fluid responsiveness in critically ill patients. Crit Care. 2006;10:R132.
415. Lamia B, Ochagavia A, Monnet X. Echocardiographic prediction of volume responsiveness in critically ill patients with spontaneous breathing activity. Intensive Care Med. 2007;33:1125–32.
416. Chemla D, Hébert J-L, Coirault C. Total arterial compliance estimated by stroke volume-to-aortic pulse pressure ratio in humans. Am J Physiol Heart Circ Physiol. 1998;274:H500–5.
417. Perel P, Roberts I. Colloids versus crystalloids for fluid resuscitation in critically ill patients. Cochrane Database Syst Rev. 2007;(4):CD000567.
418. McIntyre LA, Hebert PC, Fergusson D. A survey of Canadian intensivists' resuscitation practices in early sepsis: a Canadian multicenter observational study. Crit Care. 2007;11:R74.
419. Schortgen F, Lacherade JC, Bruneel F. Effects of hydroxyethylstarch and gelatin on renal function in severe sepsis: a multicenter randomised study. Lancet. 2001;357:911–6.
420. Brunkhorst FM, Engel C, Bloos F. Intensive insulin therapy and pentastarch resuscitation in severe sepsis. N Engl J Med. 2008;358:125–39.
421. Wiesen P, Canivet JL, Ledoux D. Effect of hxdroxyethylstarch on renal function in cardiac surgery: a large scale retrospective study. Acta Anaesthesiol Belg. 2005;56:257–63.
422. Brunkhorst FM, Reinhart K. Diagnose und kausale Therapie der Sepsis. Internist. 2009;50:810–6.
423. Pinsky MR. Goals of resuscitation from circulatory shock. Contrib Nephrol. 2004;144:94–104.
424. Myburgh JA, Finfer S, Bellomo R. Hydroxyethylstarch or saline for fluid resuscitation in intensive care. N Engl J Med. 2012;367:1901–11.
425. Perner A, Haase N, Guttormsn AB. Hydroxyethylstarch 130/0.42 versus Ringer's acetate in severe sepsis. N Engl J Med. 2012;367:124–34.
426. Fencl V, Leith DE. Stewart's quantitative acid–base chemistry: applications in biology and medicine. Respir Physiol. 1993;91:1–16.
427. Stewart PA. Modern quantitative acid–base chemistry. Can J Physiol Pharmacol. 1983;61:1444–61.
428. Finfer S, Bellomo R, Boyce N. A comparison of albumin and saline for fluid resuscitation in the intensive care unit. N Engl J Med. 2004;350:2247–56.
429. Myburgh J, Cooper DJ, Finfer S. Saline or albumin for fluid resuscitation in patients with traumatic brain injury. N Engl J Med. 2007;357:874–84.
430. Vincent JL, Dubois MJ, Navickis RJ. Hypalbuminemia in acute illness: is there a rationale for intervention? A meta-analysis of cohort studies and controlled trails. Ann Surg. 2003;237:319–34.
431. Velasquez MT, Notargiacomo AV, Cohn JN. Comparative effects of saline and mannitol on renal cortical blood flow and volume in the dog. Am J Physiol. 1973;224:322–7.
432. Selkurt E. Changes in renal clearance following complete ischemia of kidney. Am J Physiol. 1945;144:395–403.
433. Morris CR, Alexander EA, Bruns FJ. Restoration and maintenance of glomerular filtration by mannitol during hypoperfusion of the kidney. J Clin Invest. 1972;51:1555–64.
434. Alsous F, Khamiees M, DeGirolamo A. Negative fluid balance predicts survival in patients with septic shock: a retrospective pilot study. Chest. 2000;117:1749–54.
435. Uchino S, Bellomo R, Morimatsu H. Pulmonary artery catheter versus pulse contour analysis: a prospective epidemiological study. Crit Care. 2006;10:R174.
436. Bouchard J. Fluid accumulation, survival, and recovery of kidney function in critically ill patients with acute kidney injury. Kidney Int. 2009;76:422–7.

437. Pohlman AS, Simpson KP, Hall JB. Continuous intravenous infusions of lorazepam versus midazolam for sedation during mechanical ventilatory support: a prospective randomized study. Crit Care Med. 1994;149:1241–7.
438. McNelis J, Marini CP, Jurkiewicz A. Predictive factors associated with the development of abdominal compartment syndrome in the surgical intensive care unit. Arch Surg. 2002;137:133–6.
439. Parik A, Moitra VK. Hepatorenal syndrome: one size does not fit all. Crit Care. 2012;16:421–2.
440. Jacobs FM, Pinot G, Pilorge C. Intra abdominal pressure elevation and its link with renal failure in medical intensive care patients. Intensive Care Med. 2009;35(suppl):223.
441. Mullens W, Abrahams Z, Skouri HN. Elevated intra-abdominal pressure in acute decompensated heart failure: a potential contributor to worsening renal function. J Am Coll Cardiol. 2008;51:300–6.
442. Kastner PR, Hall JE, Guyton AC. Renal hemodynamic responses to increased renal venous pressure: role of angiotensin II. Am J Physiol. 1982;243:260–4.
443. Losser MR, Forget AP, Payen D. Nitric oxide involvement in the hemodynamic response to fluid resuscitation in endotoxic shock in rats. Crit Care Med. 2006;34:2426–31.
444. Wencker D. Acute cardio-renal syndrome: progression from congestive heart failure to congestive kidney failure. Curr Heart Fail Rep. 2007;4:134–8.
445. Pinsky MR. Both perfusion pressure and flow are essential for adequate resuscitation. Sepsis. 2000;4:143–6.
446. Pinsky MR. Functional hemodynamic monitoring: applied physiology at the bedside. In: Vincent JL, editor. Yearbook of intensive care and emergency medicine. Heidelberg: Springer; 2002. p. 534–51.
447. Gracia MIM, Cano AG, Romero MG. Dynamic arterial elastance to predict arterial pressure response to volume loading in preload-dependent patients. Crit Care. 2011;15:R15.
448. Pinsky MR. Defining the boundaries of bedside pulse contour analysis: dynamic arterial elastance. Crit Care. 2011;15:120.
449. Prowle JR, Chua H-R, Bagshaw SM. Clinical review: volume of fluid resuscitation and the incidence of acute kidney injury – a systematic review. Crit Care. 2012;16:230.
450. Guarracino F, Baldassari R, Pinskc MR. Ventriculo-arterial decoupling in acutely altered hemodynamic states. Crit Care. 2013;17:213.
451. Zarbock A, Ertmer C, Van Aken H. Current controversies in volume therapy in critically ill patients. Adv Anaesthesiol Crit Care. 2009;1:3–10.
452. Ellender TJ, Skinner JC. The use of vasopressors and inotropes in the emergency medical treatment of shock. Emerg Med Clin North Am. 2008;26:759–86.
453. Rasmussen HH, Ibels LS. Acute renal failure. Multivariante analysis of causes and risk factors. Am J Med. 1982;73:211–8.
454. Liu YL, Prowle J, Licari E. Changes in blood pressure before the development of nosocomial acute kidney injury. Nephrol Dial Transplant. 2009;24:504–11.
455. Badin J, Boulain T, Ehrmann S. Relation between mean arterial pressure and renal function in the early phase of shock: a prospective, explorative cohort study. Crit Care. 2011; 15:R135.
456. Esson ML, Schrier RW. Diagnosis and treatment of acute tubular necrosis. Ann Intern Med. 2002;137:744–52.
457. Bourgoin A, Leone M, Delmas A. Increasing mean arterial pressure in patients with septic shock: effects on oxygen variables and renal function. Crit Care Med. 2005;33:780–6.
458. Albanese J. Renal effects of norepinephrine in septic and nonseptic patients. Chest. 2004;126:534–9.
459. Persson PB. Renal blood flow autoregulation in blood pressure control. Curr Opin Nephrol Hypertens. 2002;11:67–72.
460. Navar LG. Integrating multiple paracrine regulators of renal microvascular dynamics. Am J Physiol Renal Physiol. 1998;274:F433–44.
461. LeDoux D, Astiz ME, Carpati CM. Effects of perfusion pressure on tissue perfusion in septic shock. Crit Care Med. 2000;28:2729–32.

462. Redfors B, Bragadottir G, Sellgren J. Effects of norephinephrine on renal perfusion, filtration and oxygenation in vasodilatory shock and acute kidney injury. Intensive Care Med. 2010;37:60–7.
463. Pijls NHJ, Van Gelder B, Van der Voort P. Fractional flow reserve: a useful index to evaluate the influence of an epicardial coronary stenosis on myocardial blood flow. Circulation. 1995;92:3183–93.
464. Bourdarias JP. Coronary reserve: concept and physiological variations. Eur Heart J. 1995;16(suppl I):2–6.
465. Deruddre S, Cheisson G, Mazoit JX. Renal arterial resistance in septic shock: effects of increasing mean arterial pressure with norephnephrine on renal resistive index assessed by Doppler ultrasonography. Intensive Care Med. 2007;33:1557–62.
466. Thooft A, Favory R, Salgado- Ribeiro D. Effects of changes in arterial pressure on organ perfusion during septic shock. Crit Care. 2011;15:R222.
467. Trzeciak S, Dellinger RP, Parrillo JE, Microcirculatory alterations in resuscitation and shock (MARS) Investigators. Early microcirculatory perfusion derangements in patients with severe sepsis and septic shock: relationship to hemodynamics, oxygen transport, and survival. Ann Emerg Med. 2007;49:88–98.
468. Pottecher J, Deruddre S, Teboul JL. Both passive leg raising and intravascular volume expansion improve sublingual microcirculatory perfusion in severe sepsis and septic shock patients. Intensive Care Med. 2010;26:1867–74.
469. Silva S, Teboul JL. Defining the adequate arterial pressure target during septic shock: not a "micro" issue but the microcirculation can help. Crit Care. 2011;15:1004.
470. Lehman L, Saeed M, Moody G. Hypotension as a risk factor for acute kidney injury in ICU patients. Comput Cardiol. 2010;37:1095–7.
471. Wauters J, Claus P, Brosens N. Pathophysiology of renal hemodynamics and renal cortical microcirculation in a porcine model of elevated intra-abdominal pressure. J Trauma. 2009;66:713–9.
472. Martin C, Eon B, Saux P. Renal effects of norepinephrine used to treat septic shock patients. Crit Care Med. 1990;18:282–5.
473. Desjars P, Pinaud M, Bugnon D. Norepinephrine therapy has no deleterious renal effects in human septic shock. Crit Care Med. 1990;18:1048–9.
474. Rudis MI, Basha MA, Zarowitz BJ. Is it time to reposition vasopressors and inotropes in sepsis? Crit Care Med. 1996;24:525–37.
475. Martin C, Viviand X, Leone M. Effect of norepinephrine on the outcome of septic shock. Crit Care Med. 2008;28:2758–65.
476. De Backer D, Biston P, Devriendt J. Comparison of dopamine and norepinephrine in the treatment of shock. N Engl J Med. 2010;362:779–89.
477. Bellomo R, Kellum JA, Wisniewski SR, Pinsky MR. Effects of norepinephrine on the renal vasculature in normal and endotoxemic dogs. Am J Respir Crit Care Med. 1999;159:1186–92.
478. Albert M, Losser MR, Hayon D, Faivre V, Payen D. Systemic and renal macro- and microcirculatory responses to arginine vasopressin in endotoxic rabbits. Crit Care Med. 2004;32:1891–8.
479. Kellum JA, Pinsky MR. Use of vasopressor agents in critically ill patients. Curr Opin Crit Care. 2002;8:236–41.
480. Levy B, Bollaert PE, Charpentier C. Comparison of norepinephrine and dobutamine to epinephrine for hemodynamics, lactate metabolism, and gastric tonometric variables in septic shock: a prospective, randomized study. Intensive Care Med. 1997;23:282–7.
481. Moncada S, Palmer RM, Higgs EA. Nitric oxide: physiology, pathophysiology, and pharmacology. Pharmacol Rev. 1991;43:109–42.
482. Schaer GL, Fink MP, Parrillo JE. Norepinephrine alone versus norepinephrine plus low-dose dopamine: enhanced renal blood flow with combination pressor therapy. Crit Care Med. 1985;13:492–6.

483. Alessandria C, Ottobrelli A, Debernardi-Vernon W. Noradrenaline vs terlipressin in patients with hepatorenal syndrome: a prospective randomised unblinded pilot study. J Hepatol. 2007;47:499–505.
484. Therapondos G, Stanley AJ, Hayes PC. Systemic, portal and renal effects of terlipressin in patients with cirrhotic acsites: pilot study. J Gastroenterol Hepatol. 2004;19:73–7.
485. Gluud LL, Kjaer MS, Christensen E. Terlipressin for hepatorenal syndrome. Cochrane Database Syst Rev. 2006;(4):CD005162.
486. Salero F, Gerbes A, Ginès P. Diagnosis, prevention and treatment of hepatorenal syndrome in cirrhosis. Gut. 2007;56:1310–8.
487. Mehta RL, Pascual MT, Soroko S. Diuretics, mortality, and nonrecovery of renal function in acute renal failure. JAMA. 2002;288:2547–53.
488. Uchino S, Doig GS, Bellomo R. Diuretics and mortality in acute renal failure. Crit Care Med. 2004;32:1669–77.
489. Chertow GM, Lazarus JM, Paganini EP. Predictors of mortality and the provision of dialysis in patients with acute tubular necrosis: The auriculin Anaritride Acute Renal Failure Study Group. J Am Soc Nephrol. 1998;9:692–8.
490. Brown CB, Ogg CS, Cameron JS. High dose frusemide in acute renal failure: a controlled trail. Clin Nephrol. 1981;15:90–6.
491. Cantarovich F, Rangoonwala B, Lorenz H. High-dose frusemide for established ARF: a prospective, randomised, double-blind, placebo-controlled, multicenter trail. Am J Kidney Dis. 2004;44:402–9.
492. Sampath S, Moran JL, Graham PL. The efficacy of loop diuretics in acute renal failure: assessment using Bayesian evidence synthesis techniques. Crit Care Med. 2007;35:2516–24.
493. Kellum JA. The use of diuretics and dopamine in acute renal failure: a systematic review of evidence. Crit Care. 1997;1:53–9.
494. Bagshaw SM, Delany A, Haase M. Loop diuretics in the management of acute renal failure: a systematic review and meta-analysis. Crit Care Resusc. 2007;9:60–8.
495. Ho KM, Sheridan DJ. Meta-analysis of frusemide to prevent or treat acute renal failure. BMJ. 2006;333:420.
496. Kellum JA, Decker JM. Use of dopamine in acute renal failure: a meta-analysis. Crit Care Med. 2001;29:1526–31.
497. Ahee P, Crowe AV. The management of hyperkalemia in the emergency department. J Accid Emerg Med. 2000;17:188–91.
498. Bellomo R, Ronco C. An introduction to continuous renal replacement therapy. In: Baldwin R, Ronco I, Golper C, Bellomo G, editors. Atlas of hemofiltration. London: Bailliere Tindal; 2001. p. 1–9.
499. Kraut JA, Kurtz I. Use of base in the treatment of severe acidaemic states. Am J Kidney Dis. 2001;38:703–21.
500. Nahas GG. Guidelines for the treatment of acidaemia with THAM. Drugs. 1998;55:191–224.
501. Robertson C, Steen P, Adgey J. The 1998 European Resuscitation Council guidelines for adult advanced life support: a statement from the Working Group on Advanced Life Support, and approved by the executive committee of the European Resuscitation Council. Resuscitation. 1998;37:81–90.
502. Finfer S, Chittock DR, Su SY. Intensive versus conventionell glucose control in critically ill patients. N Engl J Med. 2009;360:1283–97.
503. van den Berghe G, Wilmer A, Hermans G. Intensive insulin therapy in the medical ICU. N Engl J Med. 2006;354:449–61.
504. van den Berghe G, Wouters P, Weekers F. Intensive insulin therapy in the critically ill patients. N Engl J Med. 2001;345:1359–67.
505. Schetz M, Vanhorebeek I, Wouters PJ. Tight glucose control is renoprotective in critically ill patients. J Am Soc Nephrol. 2008;19:571–8.

506. Abassi ZA, Ellahham S, Winaver J. Acute renal failure complicating muscle crush injury. Semin Nephrol. 1998;18:558–65.
507. McCullough PA, Adams A, Becker CR. Risk prediction of contrast – induced nephropathy. Am J Cardiol. 2006;98:27K–36.
508. Meier M, Bongartz G, Prauhauser C. Niere als Opfer: Kontrastmittelschäden. info@herz+gefäß_06_2011. 2011, Vol. 06, 23–23.
509. Lautin EM, Freeman NJ, Schoenfeld AH. Radiocontrast-associated renal dysfunction: incidence and risk factors. AJR Am J Roentgenol. 1991;157:49–58.
510. Brown JR, DeVries JT, Piper WD. Serious renal dysfunction after percutaneous coronary interventions can be predicted. Am Heart J. 2008;155:260–6.
511. Mehran R, Aymong ED, Nikolsky E. A simple risk score for prediction of contrast-induced nephropathy after percutaneous coronary intervention: development and initial validation. J Am Coll Cardiol. 2004;44:1393–9.
512. McCulllough PA, Soman SS. Contrast-induced nephropathy. Crit Care Clin. 2005;21:261–80.
513. Solomon R. Preventing contrast-induced nephropathy: problems, challenges, and future directions. BMC Med. 2009;7:24.
514. Margulies K, Schirger J, Brunett Jr J. Radiocontrast-induced nephropathy: current status and future prospects. Int Angiol. 1992;11:20–5.
515. Solomon R, Werner C, Mann D. Effects of saline, mannitol, and frusemide to prevent acute decreases in renal function induced by radiocontrast agents. N Engl J Med. 1994;331:416–20.
516. Mueller C, Buerkle G, Buettner HJ. Prevention of contrast media – associated nephropathy: randomized comparison of 2 hydration regimens in 1620 patients undergoing coronary angioplasty. Arch Intern Med. 2002;162:329–36.
517. Stevens MA, McCullough PA, Tobin KJ. A prospective randomized trail of prevention measures in patients at high risk for contrast nephropathy: results of the P.R.I.N.C.E. study. J Am Coll Cardiol. 1999;33:403–11.
518. Joannidis M, Schmidt M, Wiedemann CJ. Prevention of contrast media-induced nephropathy by isotonic sodium bicarbonate: a meta-analysis. Wien Klin Wochenschr. 2008;120:742–8.
519. Hogan SE, L'Allier P, Chetcuti S. Current role of sodium bicarbonate-based preprocedural hydration for the prevention of contrast-induced acute kidney injury: a meta-analysis. Am Heart J. 2008;156:414–21.
520. Hoffmann U, Fischereder M, Kruger B. The value of N-acetylcysteine in the prevention of radiocontrast agent-induced nephropathy seems questionable. J Am Soc Nephrol. 2004;15:407–10.
521. Kshirsagar AV, Poole C, Mottl A. N-acetylcysteine for the prevention of radiocontrast induced nephropathy: a meta-analysis of prospective controlled trails. J Am Soc Nephrol. 2004;15:761–9.
522. Tepel M, van der Giet M, Schwarzfeld C. Prevention of radiographic-contrast-agent-induced reductions in renal function by acetylcysteine. N Engl J Med. 2000;343:180–4.
523. Diaz-Sandoval LJ, Kosowsky BD, Losordo DW. Acetylcysteine to prevent angiography-related renal tissue injury (the APART trail). Am J Cardiol. 2002;898:356–8.
524. Paganini EP, Tapolyai M, Goormastic M. Establishing a dialysis therapy/patient outcome link in intensive care unit acute dialysis for patients with acute renal failure. Am J Kidney Dis. 1996;28 Suppl 3:S81–9.
525. Chertow GM, Christiansen CL, Cleary PD. Prognostic risk stratification in critically ill patients with acute renal failure requiring dialysis. Arch Intern Med. 1995;155:1505–11.
526. Mehta R, McDonald B, Gabbai F. A randomised clinic trail of continuous versus intermittent dialysis for acute renal failure. Kidney Int. 2001;60:1154–63.
527. Kellum JA, Ronco C. Results of RENAL – what is the optimal CRRT target dose? Nat Rev Nephrol. 2010;6:191–2.
528. Kolf WJ. First clinical experience with arteficial kidney. Ann Intern Med. 1965;62:608–19.

529. Merrill JP, Smith S, Callahan EJ. Use of artificial kidney. II. Clinical experience. J Clin Invest. 1950;29:425–38.
530. Teschan PE, Baxter CR, O'Brien TE. Prophylactic hemodialysis in the treatment of acute renal failure. Ann Intern Med. 1960;53:992–1016.
531. Palevsky PM, Berns JS, Post TW. Renal replacement therapy (dialysis) in acute kidney injury (acute renal failure): indications, timing, and dialysis dose. UpToDate. 2010 www.uptodate.com
532. Palevsky PM. Renal replacement therapy I: Indications and timing. Crit Care Clin. 2005;21:347–56.
533. Mehta RL. Indications for dialysis in the ICU: renal replacement vs renal support. Blood Purif. 2001;19:227–32.
534. Uchino S. Beginning and ending continuous therapies in the intensive care unit, chapter 242. In: Ronco C, Bellomo R, Kellum JA, editors. Critical care nephrology. 3rd ed. Philadelphia: Saunders Elsevier; 2009.
535. Rizzi Z, Picardo S, Ronco C. Results from international questionnaires. Contrib Nephrol. 2007;156:297–303.
536. Ricci Z, Ronco C, D'Amico G. Practice patterns in the management of acute renal failure in the critically ill patient: an international survey. Nephrol Dial Transplant. 2006;21:690–6.
537. Kellum JA, Mehta RL, Angus DC, ADQI Workgroup. The first international consensus conference on continuous renal replacement therapy. Kidney Int. 2002;62:1855–63.
538. Pannu N, Klarenbach S, Wiebe N. Renal replacement therapy in patients with acute renal failure. JAMA. 2008;299:793–805.
539. Seabra VF, Balk EM, Liangos O. Timing of renal replacement therapy initiation in acute renal failure: a meta- analysis. Am J Kidney Dis. 2008;52:272–84.
540. Riviello ED, Christopher KB. Critical care nephrology: acute renal failure in the intensive care unit. Nephrol Rounds. 2006;4(10):1–6.
541. Shiao C-C, Wu V-C, Li W-Y. Late initiation of renal replacement therapy is associated with worse outcomes in acute kidney injury after major abdominal surgery. Crit Care. 2009;13:R171.
542. Traynor JP, Simpson K, Geddes CC. Early initiation of dialysis fails to prolong survival in patients with end-stage failure. J Am Soc Nephrol. 2002;13:2125–32.
543. Fischer RP, Griffen WOJ, Reiser M. Early dialysis in the treatment of acute renal failure. Surg Gynecol Obstet. 1966;123:1019.
544. Gettings LG, Reynolds HN, Scalea T. Outcome in posttraumatic acute renal failure when continuous renal replacement therapy is applied early vs late. Intensive Care Med. 1999;25:805–13.
545. Carl DE, Grossman C, Behnke M. Effect of timing of dialysis on mortality in critically ill, septic patients with acute renal failure. Hemodial Int. 2010;14:11–7.
546. Liu KD, Himmelfarb J, Paganini E. Timing of initiation of dialysis in critically ill patients with acute kidney injury. Clin J Am Soc Nephrol. 2006;1:915–9.
547. Kleinknecht D, Jungers P, Chanard J. Uremic and nonuremic complications in acute renal failure: evaluation of early and frequent dialysis on prognosis. Kidney Int. 1972;1:190.
548. Bagshaw SM, Uchino S, Bellomo R. Timing or renal replacement therapy and clinical outcomes in critically ill patients with severe acute kidney injury. J Crit Care. 2009;24:129–40.
549. Karvellas CJ, Farhat MR, Sajjad I. A comparison of early versus late initiation of renal replacement therapy in critically ill patients with acute kidney injury: a systematic review and meta-analysis. Crit Care. 2011;15:R72.
550. Sabater J, Perez XL, Albertos R. Acute renal failure in septic shock. Should we consider different continuous renal replacement therapies on each RIFLE score stage? Intensive Care Med. 2009;35:S239.
551. Bagshaw SM, Wald R, Barton J. Initiation of renal replacement therapy in critically ill patients with acute kidney injury – a prospective multi-centre observational study. In: 2010 AKI Critical Care Nephrology Conference, Edmonton; 2010

552. Bouman CS, Oudermans-Van Straaten HM, Tijssen JG. Effects of early high-volume continuous venovenous hemofiltration on survival and recovery of renal function in intensive care patients with acute renal failure. Crit Care Med. 2002;30:2205–11.
553. Sugahara S, Suzuki H. Early start on continuous hemodialysis therapy improves survival rate in patients with acute renal failure following coronary bypass surgery. Hemodial Int. 2004;8:320–5.
554. Elahi MM, Lim MY, Joseph RN. Early hemofiltration improves survival in post-cardiotomy patients with acute renal failure. Eur J Cardiothorac Surg. 2004;26:1027–31.
555. Leite TT, Macedo E, Pereira SM. Timing of renal replacement therapy initiation by AKIN classification system. Crit Care. 2013;17:R62.
556. Hoste EAJ, Dhonte A. Clinical review: use of renal replacement therapies in special groups of ICU patients. Crit Care. 2012;16:201.
557. Pannu N, Gibney R. Renal replacement therapy. Ther Clin Risk Manag. 2005;1:141–50.
558. Kihara M, Ikeda Y, Shibata K. Slow hemodialysis performed during the day in managing renal failure in critically ill patients. Nephron. 1994;67:36–41.
559. Bellomo R, Parkin G, Love J. A prospective comparative study of continuous arteriovenous hemodiafiltration and continuous venovenous hemnodiafiltration in critically ill patients. Am J Kidney Dis. 1993;21:400.
560. O'Reilly P, Tolwani A. Renal replacement therapy III: IHD; CRRT; SLED. Crit Care Clin. 2005;21:367–78.
561. Ronco C, Bellomo R, Brendolan A. Brain density changes during renal replacement in critically ill patients with acute renal failure. Continuous hemofiltration versus intermittenz hemodialysis. J Nephrol. 1999;12:173–8.
562. Heering P, Ivens K, Thumer O. The use of different buffers during continuous hemofiltration in critically ill patients with acute renal failure. Intensive Care Med. 1999;25:1244–51.
563. Palsson R, Laliberte KA, Niles JL. Choice of replacement solution and anticoagulant in continuous venovenous hemofiltration. Clin Nephrol. 2006;65:34–42.
564. Van Bommel EFH, Ponssen HH. Intermittent versus continuous treatment for acute renal failure: where do we stand? Am J Kidney Dis. 1997;30 Suppl 4:S72.
565. Lameire N, Van Biesen W, Vanholder R. Dialysing the patient with acute renal failure in the ICU: the emperor's clothes? Nephrol Dial Transplant. 1999;14:2570.
566. Bellomo R, Farmer M, Parkin G. Severe acute renal failure: a comparison of acute continuous hemodiafiltration and conventionel dialytic therapy. Nephron. 1995;71:59.
567. Tonelli M, Manns B, Feller-Kopman D. Acute renal failure in the intensive care unit: a systematic review on the impact of dialytic modality on mortality and renal recovery. Am J Kidney Dis. 2002;40:875.
568. Rabindranath K, Adams J, Macleod AM. Intermittent versus continuous renal replacement therapy for acute renal failure in adults. Cochrane Database Syst Rev. 2007;(3):CD003773.
569. Bagshaw SM, Berthiaume LR, Delany A. Continuous versus intermittent renal replacement therapy for critically ill patients with acute kidney injury: a meta-analysis. Crit Care Med. 2008;36:610.
570. Constanzo MR, Ronco C. Extracorporal fluid removal in heart failure patients. Contrib Nephrol. 2010;165:236–43.
571. Schiffl H, Lang SM, Fischer R. Daily hemodialysis and the outcome of acute renal failure. N Engl J Med. 2002;346:305–10.
572. Ronco C, Bellomo R, Homel P. Effects of different doses in continuous venovenous haemofiltration on outcomes of acute renal failure. Lancet. 2000;356:26–30.
573. Saudan P, Niederberger M, De Seigneux S. Adding a dialysis dose to continuous hemofiltration increases survival in patients with acute renal failure. Kidney Int. 2006;70:1312–7.
574. Tolwani AJ, Campbell RC, Stofan BS. Standard versus high dose CVVHDF for ICU – related acute renal failure. J Am Soc Nephrol. 2008;19:1233–8.
575. Palevsky PM, Zhang JH, O'Connor TZ. Intensity of renal support in critically ill patients with acute kidney injury. N Engl J Med. 2008;359:7–20.

576. Bellomo R, Cass A, Cole L. Intensity of continuous renal-replacement therapy in critically ill patients. N Engl J Med. 2009;361:1627–38.
577. Evanson JA, Himmelfarb R, Wingard S. Prescribed versus delivered dialysis in acute renal failure patients. Am J Kidney Dis. 1998;32:731–8.
578. Venkataraman R, Kellum JA, Palevsky P. Dosing patterns for continuous renal replacement therapy at a large academic medical center in the United States. J Crit Care. 2002;17:246–50.
579. Vesconi S, Cruz DN, Fumagalli R. Delivered dose of renal replacement therapy and mortality in critically ill patients with acute kidney injury. Crit Care. 2009;13:R57.

Chapter 6
Nutrition in Critical Illness

Appropriate feeding has become an integral part of the management of critically ill patients [1–5]. Moreover, Wischmeyer has postulated a key role for nutrition in survival after critical illness that extends for months or even years [6]. Disease-related malnutrition in critically ill patients is associated with increased morbidity and mortality [4, 7–12]. Compromised immune function, blunted respiratory drive and respiratory muscle weakness due to energy deficit may predispose the patients to further complications, the so-called second hit of nosocomial infections, and prolonged dependence on respiratory support with its inherent complications [13–16]. Associated with the inflammatory response triggered in critical illness, there is hypermetabolism due to metabolic stress and bed rest, all of which may promote protein catabolism [4, 17]. Thus, the critically ill may quickly develop malnutrition [12].

Malnutrition is common in acutely severely ill patients with a reported occurrence of 30–50 % [14, 18, 19]. In the critically ill, this overall weakness is termed ICU-acquired critical illness weakness and is the result of substantial loss of lean body mass, inadequate physical activity and marked catabolic metabolism [20–22]. Feeding patients has been shown to be beneficial, improving wound healing, mitigating the catabolic response, maintaining gastrointestinal barrier function preventing bacterial translocation, attenuating disease severity, modulating the immune system, enhancing insulin sensitivity and improving the overall outcome [2, 23]. However, feeding may result in harmful side effects such as gastric volume retention (gastroparesis) due to disturbed gut motility, bacterial colonization of the stomach and potentially an increased risk of aspiration and nosocomial pneumonia [24, 25]. Overfeeding can also result in significant metabolic stress aggravating the metabolic response to stress [26, 27] and is associated with harmful effects such as disturbed glycaemic control, deranged liver function, increased infections and even poor outcomes [28–31]. Furthermore, some authors argue that enteral feeding is contraindicated in unstable patients in shock requiring vasoactive medications [32, 33] as pyloric dysfunction is common in cardiogenic shock [34] and often combined with intestinal atony [35]. As such, after cardiac surgery, acute mesenteric ischaemia resulting from hypoperfusion – often related to hypotension – is seen in 0.5 % [36] to 1.4 % [37] of patients and is associated with substantial mortality rates ranging

W. Krüger, A.J. Ludman, *Core Knowledge in Critical Care Medicine*,
DOI 10.1007/978-3-642-54971-7_6, © Springer-Verlag Berlin Heidelberg 2014

between 11 and 27 % [36, 37]. Enteral feeding is shown to enhance the local mesenteric oxygen demand at the expense of vital organs [35]. Publications have suggested that enteral feeding may be generally deleterious in the critically ill unstable patient resulting in disturbed gastrointestinal motility, poor glycaemic control and worsened clinical outcome [26, 29, 38]. However, other studies show that enteral feeding may be beneficial even in haemodynamically compromised critically ill patients if applied with caution [39–43], and malnutrition is found to be associated with reduced LV systolic function, lower contractility and compliance [44]. There are ongoing controversies as to the amount of enteral feed and which formulation should be offered [4–6, 28]. The timing is not conclusively answered yet, and the role of parenteral nutrition, particularly in the case of compromised enteral function, is not determined [3, 12, 45, 46], though recent evidence has helped to clarify some of these issues.

The typical metabolic feature of the acute phase of critical illness is rapid catabolism [6]. The cytokines, with their central role in the systemic inflammatory response [47, 48], are key in affecting the metabolic pathways [49]. There is good evidence that the proinflammatory agents are the predominant reason for the energy deficit and cachexia occurring in severe acute illness [50]. Critically ill patients may not be able to utilize the substrates available due to metabolic abnormalities compounding the metabolic problem further [51]. Cytokines are shown to activate transcription factors which reduce muscle protein synthesis [52, 53], interfere with proteins involved in muscular development and stimulate myosin proteolysis [54]. They drive the ubiquitin-mediated proteolytic system resulting in muscle proteolysis [55] and trigger the release of cortisol and catecholamines [56, 57], the latter causing a rise in resting metabolic rate [49]. Boosted lipolysis and beta-oxidation [58] as well as reduced lipoprotein lipase activity and thus hypertriglyceridaemia are to follow [59]. Amino acids and other substances are mobilized for rapid energy supply triggering the body's stress system and are involved in the activation of survival enzymes [60].

A summary of the pathophysiologic effects of the cytokines is illustrated in Fig. 6.1.

Exhaustion of the body's metabolic reserve is suggested to occur in most patients and under most clinical circumstances after 24–48 h of illness [61]. Accordingly, it is recommended to initiate feeding within this period [2, 28] based on the belief that enteral nutrition support should start early [3, 45]. In the sickest patients, early enteral feeding showed the most pronounced effect on reducing mortality, when feeding was initiated within 48 h of ICU admission [41]. Delaying beyond this time frame may result in gross loss of proteins as demonstrated in patients staying for more than 3 days in the ICU [61]. In long-stay patients (≥7–10 days), the cumulative protein balance is reported to be severely negatively deranged and associated with substantial morbidity and mortality [4, 12, 62, 63].

There is no doubt and considerable evidence that appropriate feeding of critically ill patients has to be an essential element of the treatment regimen as prolonged starvation is associated with and may even be causal for adverse outcome and death [5, 46, 64–66]. However, the amount of calories to be supplied and when to reach the caloric target are not so clear [28, 46].The nutritional requirements of critically

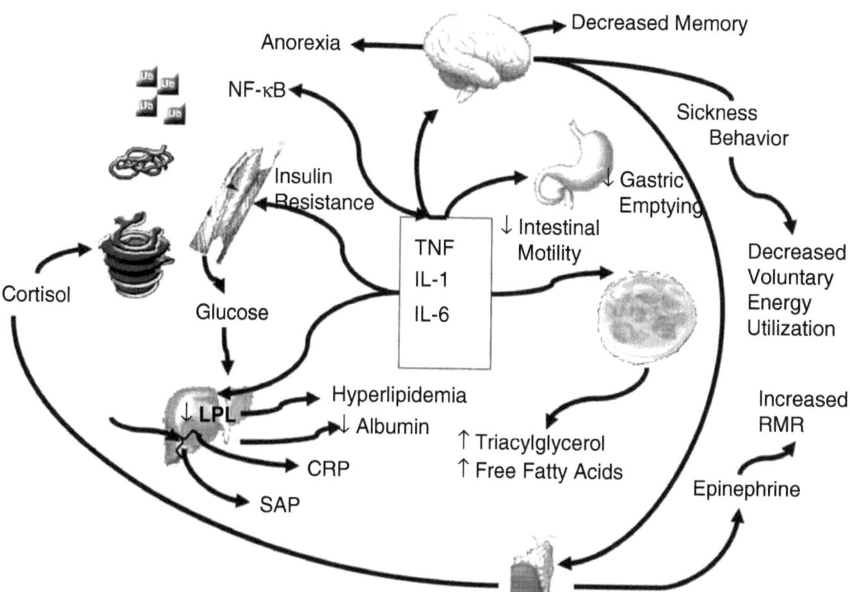

Fig. 6.1 The impact of the cytokines in the pathophysiology of cachexia, with permission from Morley et al. [49]. As obvious, cytokines exert multifaceted effects on several organs and metabolic pathways, notice particularly the effects on the brain. Cytokines: *IL-1* interleukin 1, *IL-6* interleukin 6, *TNF* tumor necrosis factor, *CRP* C-reactive protein, *SAP* serum amyloid protein, *LPL* lipoprotein lipase, *RMR* resting metabolic rate

ill patients are highly variable [35] and will increase with expending activity during recovery [27].

The resting energy demand will stay elevated for several weeks after overcoming critical illness [67]. The results of two recent randomized controlled trials suggest that hypocaloric enteral feeding during the first days does not have any deleterious impact on outcome [68, 69]. Moreover, as both the TICACOS and the EPaNIC trial revealed, reduced energy intake early on may even have a beneficial effect on morbidity [70, 71].

Enteral feeding is, without dispute, the clearly preferred approach and essential to improve outcomes in critically ill patients [2, 4, 65, 66, 72–74]. Most recent study results have shown that late initiation (day 4 [28], in the *EPaNIC study* day 7–10 [71]) of additional parenteral nutrition (PN) in the case of failure to meet caloric goals by enteral nutrition (due to intolerance) is reasonable. Early supplemental PN is reported not to be clinically beneficial [75] or may even be deleterious [71]. Moreover, PN alone as a single measure instead of enteral nutrition has turned out to just cause hypergylcaemia and immunosuppression, and further hepatocellular injury may result [76].

Gastric intolerance commonly complicates enteral feeding and may lead to undernutrition. Altered gastric emptying is found in about 50 % of mechanically ventilated patients and in 80 % of patients suffering from increased intracranial pressure following head injury [77, 78]. Disturbed gastric motility does not only cause discomfort, but is associated with a higher risk of bacterial translocation, bacterial overgrowth, infections, aspiration and ventilator-associated pneumonia [79–82]. Several mechanisms including the release of neuroendocrine peptides (cholecystokinin and serotonin) in response to nutrients resulting in vagal and spinal afferent neuronal activation subsequently inhibiting gastric emptying are proposed [83]. Furthermore, hyperglycaemia reduces vagal efferent activity and the release of NO from the myenteric plexus [84], hindering gastric emptying and interacting with erythromycin reducing its pro-kinetic effects [85]. The small bowel and colon are affected differently [86]. As such, in the case of reduced gastric emptying not sufficiently responding to pro-kinetic drugs (metoclopramide/erythromycin [87]), tube placement in the small bowel and post-pyloric feeding is recommended although no significant benefit is described in studies [88, 89]. Hence, additional parenteral feeding may be required, commencing after 7–10 [4] days of failing to meet caloric needs via the enteral route due to compromised gut function or other complicating factors [28, 71]. If less than 50–65 % of calculated and currently recommended calorie intake during the first week of hospitalization cannot be administered, the clinical benefit of EN will be blunted or even grinded, and supplemental PN is indicated [64, 90–92].

6.1 Practical Aspects

6.1.1 Enteral Nutrition Versus Parenteral Nutrition

EN is clearly preferred over PN as the beneficial effects of EN over PN are well demonstrated in numerous studies [93–97]. It also has a grade B recommendation in the American Society for Parenteral and Enteral Nutrition (ASPEN) [2] and a grade C recommendation in the European Society for Parenteral and Enteral Nutrition (ESPEN) guidelines, respectively [1].

PN alone may even be harmful [76]. If the caloric target for the patient cannot be achieved within 7–10 [4] days or EN is not possible, PN should be initiated or added, a grade C ASPEN recommendation [2]. ASPEN [2] further distinguishes between suitably nourished and malnourished patients prior to disease onset. They advise no PN within the first 7 days (grade C) and to commence PN from day 8 (grade E) in well-nourished patients. In those, malnourished patients on admission, PN should be initiated as soon as possible if EN is not available or contraindicated (grade C). The ESPEN on the contrary recommends to add or start PN in such cases of insufficient caloric delivery straight away, a grade C recommendation [1].

However, initial intentional underfeeding may even be beneficial as long as the overall amount of calories supplied during the first 7–10 [71] [4 [28]] days is not below 50–65 % of the calculated caloric demands [64, 90–92] and the target amount of calories per day estimated is achieved within a period of 3–4 days [35]. Both the EPaNIC and the TICACOS trial revealed benefits in terms of reduced morbidity if early hypocaloric enteral feeding was accepted [70, 71].

6.1.2 Timing of Initiation of Enteral Nutrition

The initiation of enteral feeding within 24–48 h after ICU admission, a grade C recommendation of both ESPEN and ASPEN guidelines [1, 2], is essential. A recently published meta-analysis grades early enteral feeding as a level A recommendation [96].

6.1.3 Dosing of Enteral Nutrition

The amount of calories to be supplied is not very well investigated; most authors advocate 20–25 kcal/kg IBW/day [1, 2, 42, 70, 98]; this is a grade C recommendation of both the ASPEN [2] and ESPEN guidelines [1]. During the acute stress response phase, 10–20 kcal/kg/day ideal or adjusted body weight (BW) may be sufficient and adequate [99]. In the later recovery phase of critical illness, the ESPEN guidelines recommend the calories may be increased to 25–30 kcal/kg/day [1].

Anyhow, several controversies exist, for example, in the early phase of critical illness, 20–25 kcal/kg/day may be too high for women [1, 100] and that the caloric supply has to be different for those who are obese (BMI $\geq$ 35) and for those with a BMI < 25. A multicenter observational study has suggested that those with a BMI < 25 or $\geq$ 35 may benefit from more calories than those with a BMI of 25–34.9, although this should not exceed 25 kcal/kg/day early, and 30 kcal/kg/day later in the disease course [4]. ASPEN suggests less calories if the BMI > 30, defined as 60–70 % of target calories or 11–14 kcal/kg actual BW per day or 22–25 kcal/kg ideal BW per day, a grade D recommendation [2]. This directly opposes the findings by Alberda et al. who advocate more energy provision to substantially obese patients (BMI > 35) rather than less [4]. Berger et al. [43] acknowledged that an average delivery of 1,250 and 1,500 kcal/day via the gastric or post-pyloric routes, respectively, is realistic practice in haemodynamically compromised patients. That implies 15–20 kcal/kg/day and is more than acceptable in the early illness period [43]. The latter is indirectly confirmed by a most recently published study in critically ill septic patients showing a beneficial impact on outcome (60-day mortality rate significantly lower) and respirator days (significantly less days on mechanical ventilation) if 1,300 kcal/day (14.5 $\pm$ 7.2 kcal/kg IBW/day) – and on average $\geq$ 58.9 g/day protein – can been applied in the early days of disease [101].

The gold standard for the assessment of metabolism and caloric requirement and hence the calculation of calories needed to be applied, as being the only valid and most accurate method, is *indirect calorimetry* [27, 35, 101, 102]. This analyses the oxygen consumption (VO_2) and the carbon dioxide expired (VCO_2), calculating the respiratory quotient (RQ) (see Table 6.1) [103, 104]. For determination of accurate values, the patient, in fact, needs to be intubated with a stable respiratory pattern, no air leak in the respiratory system, and the FiO_2 provided has to be less than 0.6 [103, 104].

However, unfortunately, indirect calorimetry is not used in the vast majority of hospitals in daily practice, although recommended by the societies, since it is too

Table 6.1 Respiratory quotient related to metabolized substrate

Substrate metabolized	Respiratory quotient (RQ)
Lipogenesis (overfeeding)	1.00–1.20
Carbohydrate oxidation	1.00
Mixed substrate oxidation (appropriate feeding)	0.85
Protein oxidation	0.82
Fat oxidation (underfeeding)	0.71
Ketosis	0.67–0.70

Data from Hester [103] and French [104]. The different RQs between the individual items protein, fat and carbohydrate are reflecting net substrate utilization at the time of assessment

complex and intricate – instead, simplistic formulas and algorithms are used in real life [105, 106].

The formulas and equations to estimate caloric demand are decisively based on body weight [107] and (occasionally) biometric parameters, such as the Harris–Benedict (H-B) equation [108], the most widely used formula [109]. To particularly overcome the quite divergent opinions on which body weight (result of H-B, ideal, adjusted, actual or anamnestic ((pre-)admission) body weight) has to be applied for the most valid and accurate calculation of energy requirement, recent publications seem to prefer the simple-to-ascertain pre-admission or so-called *anamnestic* body weight [28, 106]: anamnestic or (pre-) admission body weight is not influenced by the fluid accumulation resulting from initial volume resuscitation as does the actual body weight (untruly larger than true body weight), and is insofar generally considered to be inaccurate [28, 106]; further, actual weights are found to be frequently augmented by excess fat mass which should not to be fed at all [28]. Berger and Pichard [28] proposed a simple-to-adopt scheme in daily practice employing anamnestic weight for lean (BMI < 20)- and normal (BMI 20–25)-body-weight patients to estimate BMI, but advocate ideal-body-weight-based BMI calculations in overweight (BMI > 20–30) and obese ones (BMI > 30 resp.); see Fig. 6.2.

As the actual body weight in that patient group is generally falsified by fluid loading usually required by these patients increasing their weight, it is advocated to apply ideal body weight for overweight and obese patients to calculate their energy requirements, while the anamnestic body weight should be the basis for computation in case of lean- or normal-weight patients.

Body mass index [BMI] = [body weight/(body height)^2]

Anamnestic BW (AnBW): BMI calculation based on pre-admission/pre-anamnestic body weight and height

Ideal-body-weight (IBW) calculations [110]: BMI calculation based on ideal body weight and known height

Men = 50 kg + 2.3 kg for each inch (2.54 cm) above 5 ft (152.4 cm)
Women = 45.5 kg + 2.3 kg for each inch (2.54 cm) above 5 ft (152.4 cm) or
Men = 50 kg + 0.91 (height − 152.4 cm) kg
Women = 45.5 kg + 0.91 (height − 152.4 cm) kg
(Ideal body weight is also called predicted body weight!)

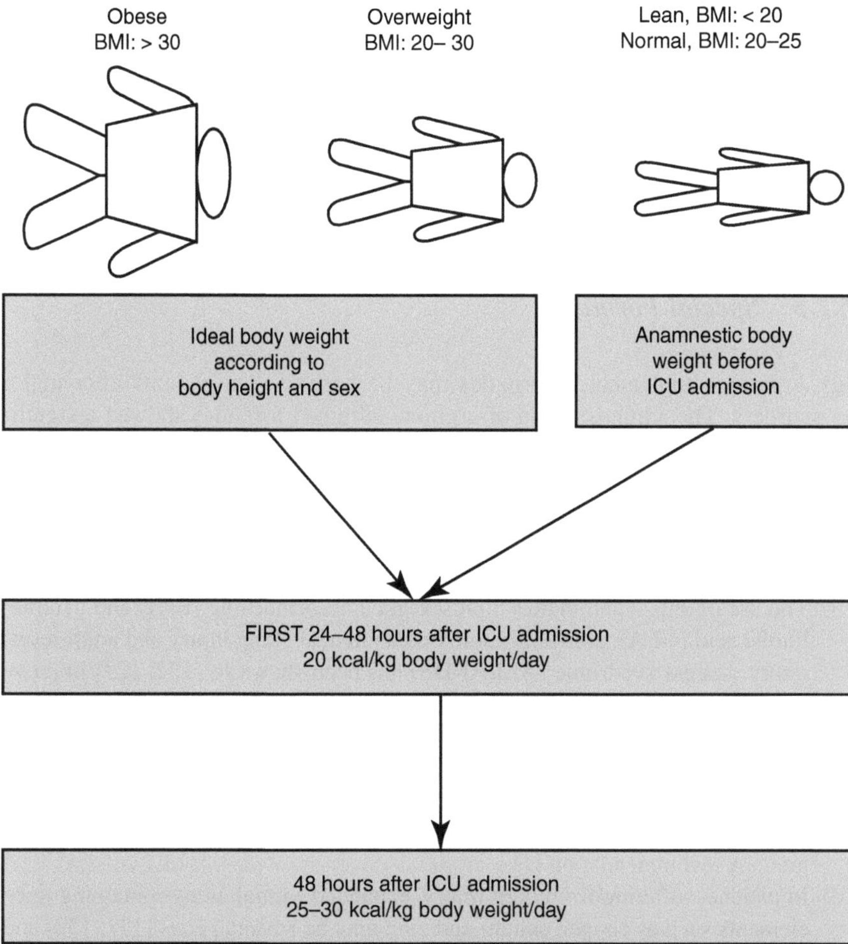

Fig. 6.2 Recommended amount of calories required by critically ill patients, with permission by Berger and Pichard [28]

Just for the sake of completeness – Adjusted body weight: AdjBW = [(Actual, anamnestic weight – IBW) × 0.25] + IBW [105, 111, 112])

Patients at risk for refeeding should be fed with "just" 15–20 kcal/kg AnBW [28].

6.1.4 Protein Requirements

The recommended protein requirements are also rather arbitrary. Between 1.2 and 2.0 g/kg BW/day has been suggested [28], with the ESPEN guidelines 1.3–1.5 g/kg BW/day [1] and the ASPEN guidelines 1.2–2.0 g/kg BW/day [2]. Weijs reported

1.2 g/kg BW/day [113] and Bonaito 1.5–2.0 g/kg a day [99]) of actual body weight. As has been suggested previously, anamnestic body weight may provide more accurateness than actual body weight in the resuscitated critically ill patients as actual body weight is in general artificially increased [28] by body water [28, 113]. There is no need for special protein composition and/or restrictions in case of considerable renal or liver malfunction any longer [105, 114–118].

6.1.5 Special Formulas

(a) Arginine-supplemented formulas may be beneficial in patients after major surgery. The administration of arginine-enriched formulas showed a significantly reduced risk of infections in the perioperative phase and an overall shorter length of hospital stay [119, 120]. This is an ASPEN grade A [2] and ESPEN grade B [1] recommendation. The caveat is that in septic patients, arginine-supplemented formulas did not show any relevant benefit; on the contrary, administration may in fact be harmful [120, 121].
(b) The use of anti-inflammatory lipids, e.g. eicosapentaenoic (EPA) and gamma-linolic acid (GLA), alongside antioxidants in acute lung injury and adult respiratory distress syndrome (ALI/ARDS) has been shown to [122, 123] improve ICU length of stay, duration of mechanical ventilation, organ failure and mortality compared to standard formulations [124–127], grade A ASPEN [2] and grade B ESPEN [1] recommendation.
(c) Immunomodulating formulas supplemented with arginine, nucleotides and ω-3 fatty acids are advocated in trauma patients [128–130] by the ESPEN with a grade A recommendation [1].
(d) In patients suffering from burn injury, expanded formulations containing trace elements such as copper, selium and zinc may be advantageous [131, 132] and are recommended by ASPEN (grade B) [2] and ESPEN (grade A) [1].

6.1.6 Management of Aspiration Risk

Factors associated with an increased special risk to aspirate [133] are listed below:

- Positive-pressure mechanical ventilation with endotracheal tube in situ
- Older than 70 years.
- Reduced level of consciousness.
- Poor oral health.
- Intermittent bolus feeding is provided.
- Supine position.
- Needing transport out of the ICU for diagnostic reasons, e.g. CT scanning.
- Poor nursing care.

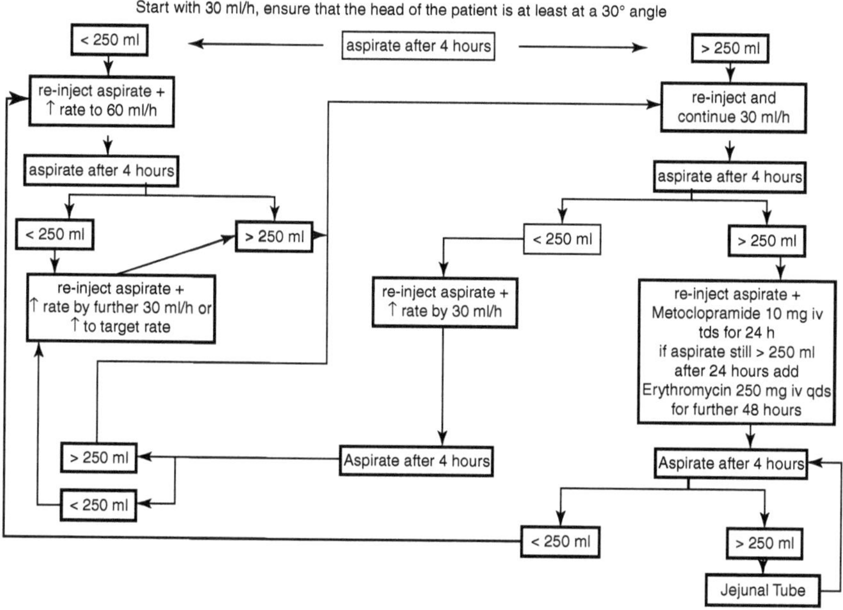

Fig. 6.3 Proposed Enteral Feeding Protocol used at Queen Elizabeth Hospital, London (Adapted and modified from Raper and Maynard [137], and adjusted to ASPEN [2], ESPEN [1], the Canadian guidelines from 2003 [45] and to PE Marik (Australia) [138])

The risk for aspiration can be reduced by:

- Positioning the head of bed and the patient upright to 30–45° [134], grade C recommendation by ASPEN
- Use post-pyloric feeding route [135, 136], grade C recommendation by ASPEN
- Applying continuous instead of bolus feeding, grade D recommendation by ASPEN
- Administration of pro-kinetic drugs (metoclopramide and/or erythromycin) where gastric motility is an issue, grade C recommendation by ASPEN [2]

The initiation of enteral feeding is variable, both, in dose and timing across institutions. A proposed Enteral Feeding Protocol used at Queen Elizabeth Hospital, London, is cited as one potential guideline (Fig. 6.3):

References

1. Kreymann KG, Berger MW, Deutz NE. ESPEN guidelines on enteral nutrition: intensive care. Clin Nutr. 2006;25:210–33.
2. McClave SA, Martindale RG, Vanek VW. Guidelines for the provision and assessment of nutrition support therapy in the adult critically ill patient: Society of Critical Care Medicine

(SCCM) and American Society for parenteral and enteral nutrition (A.S.P.E.N.). JPEN J Parenter Enteral Nutr. 2009;33:277–316.
3. Singer P, Berger MW, Van den Berghe G. ESPEN guidelines on parenteral nutrition: intensive care. Clin Nutr. 2009;28:387–400.
4. Alberda C, Gramlich L, Jones N. The relationship between nutritional intake and clinical outcomes in critically ill patients: results of an international multicenter observational study. Intensive Care Med. 2009;35:1728–37.
5. Wischmeyer PE, Heyland DK. The future of critical care nutrition therapy. Crit Care Clin. 2010;26:433–41.
6. Wischmeyer PE. The evolution of nutrition in critical care: how much, how soon? Crit Care. 2013;17(suppl1):S7.
7. Stratton RJ, Green CJ, Elia M. Disease-related malnutrition: an evidence-based approach to treatment. Oxon: CAB International; 2003. p. 3.
8. Correia MI, Waitzberg DL. The impact of malnutrition on morbidity, mortality, length of hospital stay and costs evaluated through a multivariate model analysis. Clin Nutr. 2003;22:235–9.
9. Lim SL, Ong KC, Chan YH. Malnutrition and its impact on cost of hospitalization, length of stay, readmission and 3-year mortality. Clin Nutr. 2012;31:345–50.
10. Giner M, Laviano A, Meguid MM. In 1995 a correlation between malnutrition and poor outcomes in critically ill patients still exists. Nutrition. 1996;12:23–9.
11. Petros S, Engelmann L. Enteral nutrition delivery and energy expenditure in medical intensive care patients. Clin Nutr. 2006;25:51–9.
12. Villet S, Chiolero RL, Bollmann MD. Negative impact of hypocaloric feeding and energy balance on clinical outcome in ICU patients. Clin Nutr. 2005;24:502–9.
13. Chandra R. Nutrition, immunity and infection: present knowledge and future direction. Lancet. 1983;1:688–91.
14. Norman K, Pichard C, Lochs H. Prognostic impact of disease-related malnutrition. Clin Nutr. 2008;27:5–15.
15. Chandra RA. 1990 McCollum Award lecture. Nutrition and immunity: lessons from the past and new insights into the future. Am J Clin Nutr. 1991;53:1087–101.
16. Rubinson L, Diette GB, Song X. Low caloric intake is associated with nosocomial bloodstream infections in patients in the medical intensive care unit. Crit Care Med. 2004;32:350–7.
17. Thibault R, Chikhi M, Clerc A. Assessment of food intake in hospitalized patients: a 10-year comparative study of a prospective hospital survey. Clin Nutr. 2011;30:289–96.
18. Baccaro F, Moreno JB, Borlenghi C. Subjective global assessment in the clinical setting. JPEN J Parenter Enteral Nutr. 2007;31:406–9.
19. Singh H, Watt K, Veitch R. Malnutrition is prevalent in hospitalized medical patients: are housestaff identifying the malnourished patient? Nutrition. 2006;22:350–4.
20. Griffiths RD, Hall JB. Intensive care unit acquired weakness (review). Crit Care Med. 2010;38:779–87.
21. Kaplan V, Clermont G, Griffin MF. Pneumonia: still the old man's friend? Arch Intern Med. 2003;163:317–23.
22. Herridge MS, Cheung AM, Tansey CM. One-year outcomes in survivors of acute respiratory distress syndrome. N Engl J Med. 2003;348:683–93.
23. Heyland DK. Nutritional support in the critically ill patient: a critical review of the evidence. Crit Care Clin. 1998;14:423–40.
24. Heyland DK, Cook DJ, Schoenfeld PS. The effect of acidified enteral feeds on gastric colonization in the critically ill patient: results of a multicenter randomized trial. Crit Care Med. 1999;27:2399–406.
25. Mentec H, Dupont H, Bocchetti M. Upper digestive intolerance during enteral nutrition in critically ill patients: frequency, risk factors, and complications. Crit Care Med. 2001;29:1955–61.
26. Krishnan JA, Parce PB, Martinez S. Caloric intake in medical ICU patients: consistency of care with guidelines and relationship to clinical outcomes. Chest. 2003;124:297–305.

27. Loh N-HW, Griffiths RD. The curse of overfeeding and the blight of underfeeding. [Buchverf.]. In: Vincent J-L, editor. Yearbook of intensive care and emergency medicine. Berlin/Heidelberg/New York: Springer; 2009. p. 681.
28. Berger MM, Pichard C. Best timing for energy provision during critical illness. Crit Care. 2012;16:215.
29. Dissanaike S, Shelton M, Warner K. The risk of bloodstream infections is associated with increased parenteral caloric intake in patients receiving parenteral nutrition. Crit Care. 2007;11:R114.
30. Ahrens CL, Barletta JF, Kanji S. Effect of low-calorie parenteral nutrition on the incidence and severity of hyperglycemia in surgical patients: a randomized, controlled trial. Crit Care Med. 2005;33:2507–12.
31. Grau T, Bonet A, Rubio M. Liver dysfunction associated with artificial nutrition in critically ill patients. Crit Care. 2007;11:R10.
32. Dellinger RP, Carlet JM, Masur H, Surviving Sepsis Campaign Management Guidelines Committee. Surviving Sepsis campaign guidelines for the management of severe sepsis and septic shock. Crit Care Med. 2004;32:858–73.
33. force, ASPEN Board of directors and the clinical guidelines task. Guidelines for the use of parenteral and enteral nutrition in adult and pediatric patients. JPEN J Parenter Enteral Nutr. 2002;26 Suppl 1:1SA–38.
34. Goldhill DR, Whelpton R, Winyard JA. Gastric emptying in patients the day after cardiac surgery. Anaesthesia. 1995;50:122–5.
35. Berger MM, Chiolero RL. Enteral feeding during circulatory failure: myths and reality. [Buchverf.]. In: Vincent J-L, editor. Yearbook of intensive care and emergency medicine. Berlin/Heidelberg/New York: Springer; 2009. p. 683.
36. Venkateswaran RV, Charman SC, Goddard M. Lethal mesenteric ischemia after cardiopulmonary bypass: a common complication? Eur J Cardiothorac Surg. 2002;22:534–8.
37. Lazar HL, Hudson H, McCann J. Gastrointestinal complications following cardiac surgery. Cardiovasc Surg. 1995;3:341–4.
38. McCowen KC, Friel C, Sternberg J. Hypocaloric total parenteral nutrition: effectiveness in prevention of hypoglycemia and infectious complications – a randomized clinical trial. Crit Care Med. 2000;28:3606–11.
39. Lasierra JLF, Perez-Vela J, Makikado LDU. Early enteral nutrition in patients with hemodynamic failure following cardiac surgery. JPEN J Parenter Enteral Nutr. 2013. doi:10.1177/0148607113504219.
40. Zaloga GP, Roberts PR, Marik PE. Feeding the hemodynamically unstable-patient: a critical evaluation of the evidence. Nutr Clin Pract. 2003;18:285–93.
41. Doshi P, DiGiovine B. Effects of early enteral feeding on the outcome of critically ill mechanically ventilated medical patients on vasopressors. Chest. 2006;130(Suppl):101s.
42. Revelly JP, Tappy L, Berger MM. Metabolic, systemic and splanchnic hemodynamic responses to early enteral nutrition in postoperative patients treated for circulatory compromise. Intensive Care Med. 2001;27:540–7.
43. Berger MM, Revelly JP, Cayeux MC. Enteral nutrition in critically ill patients with severe hemodynamic failure after cardiopulmonary bypass. Clin Nutr. 2005;24:124–32.
44. Okoshi K, Matsubara LS, Okoshi MP. Food restriction induced myocardial dysfunction demonstrated by the combination of in vivo and in vitro studies. Nutr Res. 2002;22:1353–64.
45. Heyland DK, Dhaliwal R, Drover JW. Canadian clinical practice guidelines for nutrition support in mechanically ventilated, critically ill patients. JPEN J Parenter Enteral Nutr. 2003;27:355–73.
46. Schetz M, Casaer MP, Van den Berghe G. Does artificial nutrition improve outcome of critical illness? Crit Care. 2013;17:302.
47. Bonanno FG. Clinical pathology of shock syndromes. J Emerg Trauma Shock. 2011;4:233–43.
48. Brame AL, Singer M. Stressing the obvious? An allostatic look at critical illness. Crit Care Med. 2010;38(suppl):S600–7.

49. Morley JE, Thomas DR, Wilson M-MG. Cachexia: pathophysiology and clinical relevance. Am J Clin Nutr. 2006;83:735–43.
50. Kotler DP. Cachexia. Ann Intern Med. 2000;133:622–34.
51. Rice TW, Mogan S, Hays MA. Randomized trial of initial trophic versus full-energy enteral nutrition in mechanically ventilated patients with acute respiratory failure. Crit Care Med. 2011;39:967–74.
52. Guttridge DC, Mayo MW, Madrid LV. NF-kB-induced loss of MyoD messenger RNA; possible role in muscle decay and cachexia. Science. 2000;289:2363–6.
53. Wheeler MT, Snyder EC, Patterson MN. An E-box within the MHC IIB gene is bound by MyoD and is required for gene expression in fast muscle. Am J Physiol. 1999; 276:C1069–76.
54. Acharyya S, Ladner KJ, Nelsen LL. Cancer cachexia is regulated by selective targeting of skeletal muscle gene products. J Clin Invest. 2004;114:370–8.
55. Mitch WE, Goldberg AL. Mechanisms of muscle wasting. The role of the ubiquitin-proteasome pathway. N Engl J Med. 1996;335:1897–905.
56. Shintani F, Nakaki T, Kanba S. Role of interleukin-1 in stress responses. A putative neurotransmitter. Mol Neurobiol. 1995;10:47–71.
57. Pende A, Musso NR, Vergassola C. Neuroendocrine effects of interferon alpha 2-a in healthy human subjects. J Biol Regul Homeost Agents. 1990;4:67–72.
58. Reyden M, Arvidsson E, Blomqvist L. Targets for TNF-alpha-induced lipolysis in human adipocytes. Biochem Biophys Res Commun. 2004;318:168–75.
59. Grunfeld C, Feingold KR. Regulation of lipid metabolism by cytokines during host defense. Nutrition. 1996;12(suppl):S24–6.
60. Wischmeyer PE. Glutamine: the first clinically relevant pharmacological regulator of heat shock protein expression? Curr Opin Clin Nutr Metab Care. 2006;9:201–6.
61. Larsson J, Lennmarken C, Martensson J. Nitrogen requirements in severely injured patients. Br J Surg. 1990;77:413–6.
62. Bartlett RH, Dechert RE, Mault JR. Measurement of metabolism in multiple organ failure. Surgery. 1982;92:771–9.
63. Faisy C, Lerolle N, Dachraoui F. Impact of energy deficit calculated by a predictive method on outcome in medical patients requiring prolonged acute mechanical ventilation. Br J Nutr. 2009;101:1079–87.
64. Barr J, Hecht M, Flavin KE. Outcomes in critically ill patients before and after the implementation of an evidence-based nutritional management protocol. Chest. 2004;125: 1446–57.
65. Artinian V, Krayem H, DiGiovine B. Effects of early enteral feeding on the outcome of critically ill mechanically ventilated patients. Chest. 2006;129:960–7.
66. Khalid I, Doshi P, DiGiovine B. Early enteral nutrition and outcomes of critically ill patients treated with vasopressors and mechanical ventilation. Am J Crit Care. 2010;19:261–8.
67. Uehara M, Plank LD, Hill GL. Components of energy expenditure in patients with severe sepsis and major trauma: a basis for clinical care. Crit Care Med. 1999;27:1295–302.
68. Arabi YM, Tamim HM, Dhar GS. Permissive underfeeding and intensive insulin therapy in critically ill patients: a randomized controlled trial. Am J Clin Nutr. 2011;93:569–77.
69. Rice TW, Wheeler AP, Thompson BT, National Heart, Lung, and Blood Institute Acute Respiratory Distress Syndrome (ARDS) Clinical Trial Network. Initial trophic vs full enteral feeding in patients with acute lung injury: the EDEN randomized trial. JAMA. 2012;307:569–77.
70. Singer P, Cohen J, Shapiro H. The tight calorie control study (TICACOS): a prospective, randomized, controlled pilot study of nutritional support in critically ill patients. Intensive Care Med. 2011;37:601–9.
71. Casear MP, Mesotten D, Hermans G. Early versus late parenteral nutrition in critically ill adults. N Engl J Med. 2011;365:506–17.
72. Heyland DK, Wischmeyer PE. Does artificial nutrition improve outcome of critical illness? An alternative viewpoint! Crit Care. 2013;17:324.
73. Heyland DK, Cahill N, Day AG. Optimal amount of calories for critically ill patients: depends on how you slice the cake! Crit Care Med. 2011;39:2619–26.

74. Gramlich L, Kichian K, Pinilla J. Does enteral nutrition compared to parenteral nutrition result in better outcomes in critically ill adult patients? A systematic review of literature. Nutrition. 2004;20:843–8.
75. Vanderheyden S, Casaer MP, Kestelooz K. Early versus late parenteral nutrition in ICU patients: cost analysis of the EPaNIC trial. Crit Care. 2012;16:R96.
76. Marik PE, Pinsky MR. Death by parenteral nutrition. Intensive Care Med. 2003;29:867–9.
77. Tarling MM, Toner CC, Withington PS. A model of gastric emptying using paracetamol absorption in intensive care patients. Intensive Care Med. 1997;23:256–60.
78. Kao CH, ChangLai SP, Chieng PU. Gastric emptying in head-injury patients. Am J Gastroenterol. 1998;93:1108–12.
79. Ritz MA, Fraser R, Tam W. Impacts and patterns of disturbed gastrointestinal function in critically ill patients. Am J Gastroenterol. 2000;95:3044–52.
80. Herbert MK. Impairment of intestinal motility in the critically ill patient. Clinical implications and contribution of drugs, and mediators. [Buchverf.]. In: Holzer MK, Roewer P, Herbert N, editors. Problems of the gastrointestinal tract in anesthesia, the perioperative period and intensive care. Berlin/Heidelberg/New York: Springer; 1999. p. 28–38.
81. Adam S, Batson S. A study of problems with delivery of enteral feed in critically ill patients in five ICUs in the UK. Intensive Care Med. 1997;23:261–6.
82. Chapman M, Fraser R, Vozzo R. Antro-pyloro-duodenal motor responses to gastric and duodenal nutrient in critically ill patients. Gut. 2005;54:1384–90.
83. Raybould HE, Glatzle J, Robin C. Expression of 5-HT3 receptors by extrinsic duodenal afferents contribute to intestinal inhibition of gastric emptying. Am J Physiol Gastrointest Liver Physiol. 2003;284:G367–72.
84. Spallone V. Glycaemic control and gastric emptying. Diabetes Nutr Metab. 2004;17:47–55.
85. Spallone und Deehan S, Dobb GJ. Metoclopramide-induced raised intracranial pressure after head injury. J Neurosurg Anesthesiol. 2002;14:157–60.
86. Frühwald S, Holzer P, Metzler H. Intestinal motility disturbances in intensive care patients pathogenesis and clinical impact. Intensive Care Med. 2007;33:36–44.
87. Booth CM, Heyland DK, Paterson WG. Gastrointestinal promotility drugs in the critical care setting: a systematic review of the evidence. Crit Care Med. 2002;30:1429–35.
88. Kearns PJ, Chin D, Mueller L. The incidence of ventilator-associated pneumonia and success in nutrient delivery with gastric versus small intestinal feeding: a randomized clinical trial. Crit Care Med. 2000;28:1742–6.
89. Neumann DA, Delegge MH. Gastric versus small-bowel tube feeding in the intensive care unit: a prospective comparison of efficacy. Crit Care Med. 2002;30:1436–8.
90. Taylor SJ, Fettes SB, Jewkes C. Prospective randomized controlled trial to determine the effect of early enhanced enteral nutrition on clinical outcome in mechanically ventilated patients suffering head injury. Crit Care Med. 1999;27:2525–31.
91. Ziegler TR, Smith RJ, O'Dwyer ST. Increased intestinal permeability associated with infection in burn patients. Arch Surg. 1988;123:1313–9.
92. Jabbar A, Chang WK, Dryden GW. Gut immunology and the differential response to feeding and starvation. Nutr Clin Pract. 2003;18:461–82.
93. Windsor AC, Kanwar S, Li AG. Compared with parenteral nutrition, enteral feeding attenuates the acute phase response and improves disease severity in acute pancreatitis. Gut. 1998;42:431–5.
94. Kudsk KA, Croce MA, Fabian TC. Enteral versus parenteral feeding: effects on septic morbidity after blunt and penetrating abdominal trauma. Ann Surg. 1992;215:503–13.
95. Braunschweig CL, Levy P, Sheean PM. Enteral compared with parenteral nutrition: a meta-analysis. Am J Clin Nutr. 2001;74:542.
96. Simpson F, Doig GS. Parenteral vs enteral nutrition in the critically ill patient: a meta-analysis of trials using the intention to treat principle. Intensive Care Med. 2005;31:12–23.
97. Peter JV, Moran JL, Phillips-Highes JA. A metaanalysis of treatment outcomes of early enteral versus early parenteral nutrition in hospitalized patients. Crit Care Med. 2005;33:213–20.
98. Foster GD, Knox LS, Dempsey DT. Caloric requirements in total parenteral nutrition. J Am Coll Nutr. 1987;6:231–53.

99. Boitano M. Hypocaloric feeding of the critically ill. Nutr Clin Pract. 2006;21:617–22.
100. Kreymann und Krishnan JA, Parce PB, Martinez A. Caloric intake in medical ICU patients: consistency of care with guidelines and relationship to clinical outcomes. Chest. 2003;124:297–305.
101. Harris JS, Benedict FG. A biometric study of basal metabolism in man. Washington: Carnegie Institute of Washington; 1919.
102. Elamin EM. Nutritional care of the obese intensive care unit patient. Curr Opin Crit Care. 2005;11:300–3.
103. Hester DD, Lawson K. Suggested guidelines for use by dietitians in the interpretation of indirect calorimetry data. J Am Diet Assoc. 1989;89:100–1.
104. French SN. Nutritional assessment via indirect calorimetry. Tech Notes. Medical Graphics Corporation. St. Paul, MN. 1987; 1–12.
105. McClave SA, Martindale RG, Vanek VW. Guidelines for the provision and assessment of nutrition support therapy in the adult critically ill patient: Society of Critical Care Medicine (SCCM) and American Society for Parenteral and Enteral Nutrition (A.S.P.E.N.). J Parenter Enteral Nutr. 2009;33:277–316.
106. Elamin EW, Camporesi E. Evidenced – based nutritional support in the intensive care unit. Inter Anaesth Clin. 2009;47:212–138.
107. Miles JM. Energy expenditure in hospitalized patients: implications for nutritional support. Mayo Clin Proc. 2006;81:809–16.
108. Harris A, Benedict FG. A biometric study of human basal metabolism. Proc Natl Acad Sci U S A. 1918;4:370–3.
109. Blackburn GL, Bistrian BR, Maini BS. Nutritional and metabolic assessment of the hospitalized patient. JPEN J Parenter Enteral Nutr. 1991;1:11–22.
110. Robinson RD, Lupkiewicz SM, Palenik L. Determination of ideal body weight for drug dosage calculations. Am J Hosp Pharm. 1983;40:1016–9.
111. Baumgartner, TG. Parenteral macronutrition. [Buchverf.] Clinical guide to parenteral micronutrition. Baumgartner: Fujisawa; 1997. p. 41.
112. Krenitsky J. Adjusted body weight, pro: evidence to support the use of adjusted body weight in calculating calorie requirements. Nutr Clin Pract. 2005;20:468–73.
113. Weijs PJM, Stapel SN, de Groot SDW. Optimal protein and energy nutrition decreases mortality in mechanically ventilated, critically ill patients: a prospective observational cohort study. JPEN J Parenter Enteral Nutr. 2012;36:60–8.
114. Bozfakioglu S. Nutrition in patients with acute renal failure. Nephrol Dial Transplant. 2001;16 suppl 6:21–2.
115. Cano N, Fiaccadori E, Tesinsky P. ESPEN guidelines on enteral nutrition: adult renal failure. Clin Nutr. 2006;25:295–310.
116. Florez DA, Aranda-Michel J. Nutritional management of acute and chronic liver disease. Semin Gastrointest Dis. 2002;13:169–78.
117. Schutz T, Bechstein WO, Neuhaus P. Clinical practice of nutrition in acute liver failure: a European survey. Clin Nutr. 2004;23:975–82.
118. Henkel AS, Buchman AL. Nutritional support in patients with chronic liver disease. Nat Clin Pract Gastroenterol Hepatol. 2006;3:202–9.
119. Drover JW, Dhaliwal R, Weitzel L. Perioperative use of arginine-supplemented diets: a systematic review of the evidence. J Am Coll Surg. 2011;212:385–99, e381.
120. Heyland DK, Novak F, Drover JW. Should immunonutrition become routine in critically ill patients? A systematic review of the evidence. JAMA. 2001;286:944–53.
121. Waitzberg DL, Saito H, Plank LD. Postsurgical infections are reduced with specialized nutrition support. World J Surg. 2006;30:1592–604.
122. Bringham KL. Metabolites of arachidonic acid in experimental lung vascular injury. Fed Proc. 1985;44:43–5.
123. Henderson Jr WR. Lipid-derived and other chemical mediators of inflammation in the lung. J Allergy Clin Immunol. 1987;79:543–53.

124. Mizock BA. Nutritional support in acute lung injury and acute respiratory distress syndrome. Nutr Clin Pract. 2001;16:319–28.
125. Mizock BA, DeMichele SJ. The acute respiratory distress syndrome: role of nutritional modulation of inflammation through dietary lipids. Nutr Clin Pract. 2004;19:563–74.
126. Pontes-Arruda A. The use of special lipids in the treatment of inflammatory lung disease. Clin Nutr Insight. 2007;33:1–4.
127. Singer P, Theilla M, Fisher H. Benefit of an enteral diet enriched with eicosapentaenoic acid and gamma-linolenic acid in ventilated patients with acute lung injury. Crit Care Med. 2006;34:1033–8.
128. Chuntrasakul C, Siltham S, Sarasombath S. Comparison of a immunonutrition formula enriched arginine, glutamine and omega-3 fatty acid with a currently high-enriched enteral nutrition for trauma patients. J Med Assoc Thai. 2003;86:552–61.
129. Mendez C, Jurkovich GJ, Garcia I. Effects of an immune-enhancing diet in critically injured patients. J Trauma. 1997;42:933–40.
130. Brown RO, Hunt H, Mowatt-Larssen CA. Comparison of specialized and standard enteral formulas in trauma patients. Pharmacotherapy. 1994;14:314–20.
131. Berger MM, Spertini F, Shenkin A. Trace element supplementation modulates pulmonary infection rates after major burns: a double-blind, placebo-controlled trial. Am J Clin Nutr. 1998;68:365–71.
132. Heyland DK, Dhaliwal R, Suchner U. Antioxidant nutrients: a systematic review of trace elements and vitamins in the critically ill patient. Intensive Care Med. 2005;31:327–37.
133. McClave SA, DeMeo MT, DeLegge MH. North American summit on aspiration in the critically ill patient: consensus statement. JPEN J Parenter Enteral Nutr. 2002;26(6 Suppl):S80–5.
134. Drakulovic MB, Torres A, Bauer TT. Supine body position as a risk factor for nosocomial pneumonia in mechanically ventilated patients: a randomised trial. Lancet. 1999;354:1851–8.
135. Lien HC, Chang CS, Chen GH. Can percutaneous endoscopic jejunostomy prevent gastroesophageal reflux in patients with preexisting esophagitis? Am J Gastroenterol. 2000;95:3439–43.
136. Heyland DK, Drover JW, MacDonald S. Effect of postpyloric feeding on gastroesophageal regurgitation and pulmonary microaspiration: results of a randomized controlled trial. Crit Care Med. 2001;29:1495–501.
137. Raper S, Maynard N. Feeding the critically ill. Br J Nurs. 1992;1:273–80.
138. Marik PE, Zaloga GP. Early enteral nutrition in acutely ill patients: a systematic review. Crit Care Med. 2001;29:2264–70.
139. Association, American Dietetic. American Dietetic Association Evidence Analysis Library. Update 1 Apr 2011 [Online]. [Zitat vom: 25. 11 2013.]

Appendix: Analgesia and Sedation in the Critically Ill Patients

Medication with analgesic, anxiolytic and sedative properties are an essential and integral part of critical care conduct, as critically ill patients typically feel uncomfortable displaying stress, anxiety and pain [1–4]. For patients under mechanical ventilation, such drugs are administered in order to reduce discomfort associated with intubation and mechanical ventilation and pain [4, 5] and to minimize oxygen consumption [4, 6]. Although a standard practice and employed for decades, our epidemiologic knowledge about the administration of such drugs under critical illness conditions and particularly our evidence-based insights are marginal [4, 5, 7]. Furthermore, assured scientific findings are often not implemented in daily practice [8]. Indeed, a striking evolution in this area of critical care management took place within the last decade [5]; nevertheless, substantial and high quality trials to safely guide therapy are still lacking. Strongly recommended monitoring issues (e.g. sedation scores) and protocols are not widely established. Changes in the use of specific drugs, such as fentanyl replacing morphine (for assumed clinical benefit) as an analgesic drug in many centres are still not sufficiently tested [5, 9]. These practices in the absence of a strong evidence base are surprising; as well, the outcome of the patient obviously is dependent on patient discomfort/comfort, over- or under-sedation, poor pain control, anxiety and agitation, the presence of delirium and the type of drugs and drug combinations applied [10–15]. In particular, the occurrence of delirium requires a high index of suspicion [4, 7], as the vast majority, potentially up to 87 % [16], of patients are affected [17–19], with an effective appropriate treatment remaining unknown [7, 9].

In the 1980s and 1990s, analgesic and sedative practice on our units were largely just transferred techniques from general anaesthesia [5]. The clinical practice guideline by the Society of Critical Care Medicine (SCCM) and the American College of

W. Krüger, A.J. Ludman, *Core Knowledge in Critical Care Medicine*,
DOI 10.1007/978-3-642-54971-7, © Springer-Verlag Berlin Heidelberg 2014

Critical Care Medicine (ACCM) published in 2013 is the first comprehensive document based on an extensive literature review and database, with considerable evidence on how to approach and maintain analgesia and sedation in critically ill adults [7]. Mainstay concepts are:

1. Analgesia first or analgesia-first sedation [9], implying a light and controlled sedation
2. To recognize and deal with delirium as promptly as possible [7]

Analgesia

The vast majority, 90 % or more [20, 21], of critically ill patients suffer from or develop pain even at rest [7, 22, 23]. Routine intensive care procedures are reported to be some of the most painful and stressful events for patients [20, 24]; simply moving the patient in bed, inserting lines and tubes or suctioning may cause pain [20, 25]. Importantly, however, not all ventilated patients exhibit pain [22]; consequently, universal administration of analgesic agents is not required [26]. When present, pain should be adequately addressed, even pre-emptively (a level C, grade I recommendation [7]), in order to facilitate patient's comfort, to avoid stress response and other possible adverse effects related to pain [27]. Of note, in the case of effectual pain control, the patient may not require any additional sedatives [28], and noticeable evidence is available emphasizing that suitable pain control applied before sedation may reduce analgesia/sedation requirements and shorten ventilator time, supporting the analgesia-first concept [29–33].

Attaching, a strong importance on pain awareness and treatment is pivotal [7, 34]. Routine assessment and monitoring of pain is highly recommended (level I, grade B [7]) and can be appropriately assessed nonverbally, too [35, 36]. To communicate directly with the patient is optimal [4], applying the numeric rating scale using a 0–10 scale, with a zero for no pain and pain as bad as it could a 10 on the scale [4, 37]. The Behavioural Pain Scale (BPS) and the Critical Care Pain Observation Tool (CPOT) can be applied in ventilated patients with no or significantly limited communication [38], both based on the clinician's observation of the patient's response to pain [4, 7]. The latter two are not validated in and applicable to deeply sedated patients [4]. The numeric rating scale is quite well validated in critically ill adults and is advocated even in delirious patients as long as communication by speaking or pointing is possible [39, 40]. BPS and CPOT are the most reliable and best-validated tests performing very well for pain assessment and monitoring, with the exception of brain trauma patients and patients unable to communicate adequately but with preserved motor functions [35, 36].

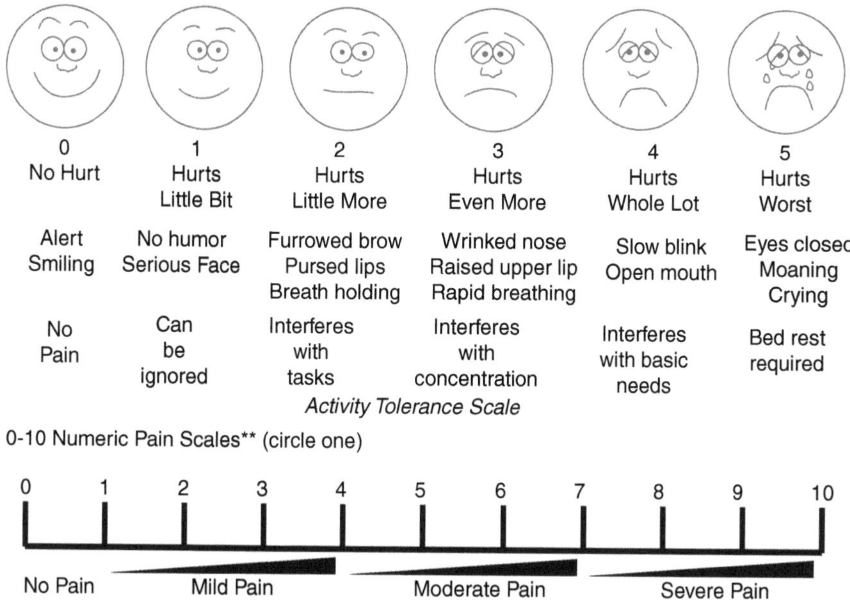

The combined verbal and mimic/facial expression pain scale
The awake, but unable to verbally express herself/himself, patient should point on the face best describing her/his pain. The points (1–5) on this "face scale" have to be doubled to be transferred into our commonly and worldwide-used numeric, 0–10 points comprising a pain scale
Adapted and modified from Hockenberg MJ, Wong D. *Wong's Essentials of Pediatric Nursing*, 8th ed. St Louis: Mosby; 2009, and *Acute Pain Management: Operative or Medical Procedures and Trauma*, clinical practice guidelines No 1. AHCPR Publication No 92–0032, Feb 1992, Agency for Healthcare and research and Quality, Rockville. p. 116–7, and published by the San Diego Patient safety Council, 2009; with permission

Treatment Issues

Opioids, applied intravenously, are considered the first-line drugs to address non-neuropathic pain in critically ill patients (level I, grade C recommendation [7]) [4, 9, 30, 41–44]. Although only a few randomized trials comparing the different drugs are available, remifentanil may afford favourable effects compared to morphine and fentanyl, the agents predominantly in use until now [8]. Remifentanil is reported to show a shorter offset of analgesic effect when discontinued compared to morphine and fentanyl allowing faster weaning from ventilation, extubation and consecutively shorter ICU stay. Furthermore, due to its sedative properties, less additional sedatives are required if any at all [29, 30, 32, 33, 45].

Morphine is metabolized to active products cumulating in the case of kidney failure; the lipophilic fentanyl may exert prolonged effects after offset due to

Table 1 Dose examples of commonly used opioid analgesics

Substance	Single bolus	Infusion	Onset	Elimination half-life
Fentanyl	50–200 μg	0.7–10 μg/kg/h	50–200 1–2 min	2–4 h
Morphine	5–10 mg	2–30 mg/h	5–10 min	3–4 h
Remifentanil	N/A	0.5–15 μg/kg/h[a]	1–3 min	3–10 min

[a]A loading dose of 1.5 μg/kg is necessary [7, 60, 61]

deposition into fat tissues; remifentanil is broken by plasma enzymes to inactive metabolites and thus is independent of kidney or liver function, and its context-sensitive half-time is independent of the length of time of the infusion [46, 47] (see Table 1). All opioids potentially develop tolerance; thus, escalating doses may be required over time to maintain the same analgesic effect [48]. Furthermore, particularly in short-acting opioids like remifentanil, hyperalgesia or paradoxical increases in the sensitivity to pain may be observed [49, 50], the short offset after discontinuation of IV remifentanil may leave the patient without analgesia cover [4]. The typical side effects of opioids like diminished respiratory drive and decreased gastroenteric motility, the latter potentially affecting enteral nutrition [51, 52], may be reduced by complementary administration of non-opioid analgesics (level C, grade II [7]), e.g. ketamine [53, 54], cyclooxygenase inhibitors [55, 56] or paracetamol [41, 57].

Alternate agents should be considered depending on the patient, the operation and the underlying pathophysiology. In the case of neuropathic pain, gabapentin or carbamazepine should be added to opioids [58, 59], a level A, grade I recommendation [7]. Patients undergoing abdominal aortic aneurysm operation may benefit from a thoracic epidural catheter (level B, grade I), and in patients suffering from rib fractures, a thoracic epidural for analgesia should be considered (level B, grade II) [7].

Sedation

Although the analgesia-first concept is reasonable and clearly a step forward, the majority of patients will still require supplemental sedatives [28, 29, 32, 45]. Anxiety and agitation are common in critical illness and are related to poorer outcomes [15, 62, 63]. With sedation quite often a crucial obstacle for extubation [64], a trend to lighter sedation has already been observed in recent years [11, 12]. Study results support and guidelines endorse monitoring sedation and sedation depth and to carefully select the drugs and doses applied [1, 7, 13, 65, 66]. A clear relationship has been demonstrated between the depth of sedation and clinical outcomes, and deeper sedation is associated with longer durations of mechanical ventilation and ICU and hospital stay [11, 12, 67, 68]. Importantly, before the initiation of sedatives, possible underlying reasons for agitation should be questioned for and addressed such as pain, delirium, hypoxemia, hypogylcaemia, hypotension, alcohol and other drug abuse withdrawal [7]. Sleep deficits are a significant problem in ICUs, and sleep should be promoted at all times, a level C, grade I recommendation [7]. The use of sedative agents has increased the recognition and understanding of the associated serious adverse effects. Examples include the prolonged sedative properties of midazolam due to the accumulation of its metabolites [69, 70], the neuro- and nephrotoxicity of the lorazepam solvents propylene and polyethylene glycol if given continuously IV over a long duration and in higher doses [71, 72], the potential development of delirium associated with benzodiazepines in general [14, 73, 74] and their respiratory depressive quality [4], the propofol infusion syndrome [75–77] and the potential to induce bradycardia when dexmedetomidine is administered [14, 78].

Several studies comparing benzodiazepines, the traditionally most common sedatives in US ICUs, with newer agents, particularly propofol, confirm benzodiazepines [79] as having a worse outcome [4]. Propofol-treated patients show a shorter mental recovery time and a faster liberation from ventilation with propofol also proving to be more cost-effective [80–84]. However, the study results have been inconsistent, and a meta-analysis including six moderate- to high-quality studies revealed that benzodiazepines might increase the days of ICU stay by 0.5 compared to nonbenzodiazepine sedation [7, 13, 14, 78, 81, 85–87]. No differences in mortality have been found when comparing the different sedative drugs in use in our ICUs [14, 78, 81, 88].

Adverse effects from propofol may include the quite common occurrence of hypotension due to altered venous and arterial tone with consecutively diminished cardiac output (patients should be well volume loaded to attenuate this effect) [89], increases in triglycerides [90], respiratory depression [8] and the propofol infusion syndrome (PIS). Although PIS rarely is a very serious and complex disorder, circulatory failure may result from bradycardia/asystole and myocardial failure in the setting of metabolic acidosis, rhabdomyolysis and hyperkalaemia [4, 91–94]. Other features observed in PIS are acute kidney injury, hepatomegaly and hypertriglyceridaemia [76]. Typically, long-term infusion (>24 h) with higher doses (>4 mg/kg/h) is a risk factor for provoking this dangerous complication [76].

Dexmedetomidine (DMDT), an alpha-2 agonist with a central action inhibiting the release of norepinephrine [4], has been compared to propofol and lorazepam as well as to midazolam, respectively [14, 78, 88, 95]. Benefits reported include lower incidence of delirium, tachycardia and hypertension and a reduction in ventilator days and faster ICU discharge [14, 73, 78, 96–98]. DMDT has sedative, analgesic and sympathicolytic properties, and its administration will spare opioids [99, 100]. There is no respiratory depression described and the sedation level is lighter; the patient can be more awake and cooperative [14, 78, 101, 102]. Relevant side effects are bradycardia and hypotension; a slow infusion rate and adequate fluid loading prior to application may blunt those responses [78, 103].

Either daily interruptions of IV continuous drug application with renewal of the infusion at a lower rate if indicated or goal-directed sedation algorithms are advocated [4, 7, 12, 67]. A proper assessment is indispensable, and daily interruption allows for a better neurologic evaluation and assessment of the patient's analgesic and sedative requirements [4, 12]. There is ample evidence that daily discontinuation significantly diminishes the time of mechanical ventilation and ICU and hospital stay [12, 13, 67]. Further benefits noted are reduced incidence of critical illness [104, 105] and posttraumatic stress disorders [104] and potentially less risk

Table 2 The Richmond Agitation and Sedation Scale (RASS) [112, 114]

Score	Term	Description
+4	Combative	Overtly combative, violent, immediate danger to staff
+3	Very agitated	Pulls or removes tube(s) or catheter(s), aggressive
+2	Agitated	Frequent non-purposeful movement, fights ventilator
+1	Restless	Anxious, but movements not aggressive or vigorous
0	Alert and calm	
−1	Drowsy	Not fully alert, but has sustained awakening Drowsy (eye opening/eye contact to voice for ≥10 s)
−2	Light sedation	Briefly awakens with eye contact to voice <10 s
−3	Moderate sedation	Movement or eye opening to voice but no eye contact
−4	Deep sedation	No response to voice, but movement or eye opening to physical stimulation
−5	Unaroused	No response to voice or physical stimulation

of myocardial ischemia [106]. An approach using specific sedation protocols may be an equipotent alternative to daily interruption but may not be ideal for all patients (e.g. chronic alcohol and drug abuse [107]) [11, 107, 108]. However, daily interruption is the most widely applied tool and endorsed by several societies [7, 109, 110], a level B, grade I recommendation of the SCCM and ACCM guidelines [7].

Regardless of sedation protocol, a valid, reliable and feasible bedside assessment is strongly recommended [7]. An example of a tool to evaluate the patient's sedative requirements is the Richmond Agitation and Sedation Scale (Table 2); an alternative, equally valid and reliable tool, is the Sedation–Agitation Scale (SAS) which, together with the RASS, performs best of all available tests [7, 111–115].

The target is to keep the patient light to moderately sedated, unless clinically contraindicated [12, 116, 117] and adjust drug infusions of analgesics and sedatives as appropriate. Closely monitoring for pain, sedation depth and delirium (see below) and responding by clearly constituted orders on how to address assessment results receives a level B, grade I recommendation [7]. Nonbenzodiazepine agents, thus either propofol or dexmedetomidine, are advocated over benzodiazepines in order "to improve clinical outcomes in mechanically ventilated patients", level B grade II [7].

Examples of commonly used sedatives

Substance	Loading dose	Maintenance dose	Onset after IV loading	Elimination half-life
Midazolam	0.01–0.05 mg/kg for several min	0.02–0.1 mg/kg/h	2–5 min	3–11 h
Lorazepam	0.02–0.04 mg/kg in total ≤ 2 mg	0.02–0.1 mg/kg/h in total ≤ 10 mg/h	15–20 min	8–15 h
Propofol	5 μg/kg/min over 5 min	5–50 μg/kg/min	1–2 min	3–12 h/50 ± 19 h[a]
DMTD	1 μg/kg over 10 min	0.2–0.7 μg/kg/h	5–10 min	1.8–3.1 h

DMTD dexmedetomidine [7, 8, 44, 118, 119]
[a]Short-term application/long-term application

Presently, neuromuscular blocking is no longer routinely advocated [4, 120, 121], except in the early phase, for a maximum period of 48 h, of acute respiratory distress syndrome or in the case of mechanical ventilation with high-frequency oscillation [122]. The application of neuromuscular-blocking agents is confirmed to substantially contribute to prolonged muscle weakness and thus extended mechanical ventilation [121–123]. Furthermore, the development of myopathy is described [124], unfortunately even in cisatracurium, the recommended preferred drug [125].

Delirium

Diagnosis

Delirium is characterized by the acute onset of cerebral dysfunction described as:

- Change or fluctuation of baseline mental status (feature 1)
- Inattention (feature 2)
- Either altered level of consciousness (feature 3) or disorganized thinking (feature 4) [19, 120–123].

Typical clinical symptoms [7] are:

Reduced clarity/awareness of the environment	Hallucinations
Memory deficits	Language disturbances
Reduced ability to focus, sustain and shift attention	Delusions
Sleep disturbances	Abnormal psychomotor activity
Disorientation	Emotional expressions such as fear, anger, depression, apathy and euphoria

Two subtypes may be distinguished depending on the prevailing clinical features [7]:

- The agitated, *hyperactive* type of patient
- The calm or lethargic, *hypoactive* type of patient
  (some patients may exhibit both hyperactive and hypoactive elements)

The hypoactive patient is far more common, but due to the difficulty of recognition, the hypoactive delirium often is not diagnosed [124, 125]. Hallucinations and delusions are more often associated with the hyperactive type; confusion and calmness/appearing sedated are more associated with hypoactivity [7]. It is important to notice that the delirious patient may describe hallucinations and/or delusions, but to diagnose delirium, these features, as per the definition, are not required [7]. Unfortunately, the occurrence of delirium is associated with worse outcomes including higher mortality and long-term cognitive deficiencies [126–129]. Moreover, the behaviour and attitude of the ICU team is shown to affect the incidence of delirium and the consecutive aftermath [116, 130–132].

Underlying reasons for the development of delirium are either disease-associated issues, e.g. sepsis, or iatrogenic/environmental issues, e.g. medication and immobilization [7].

Risk factors identified to promote delirium development are:

- Pre-existing dementia [132, 133]
- Arterial hypertension prior to ICU admission [133–135]
- Coma [134, 136]
- Severe illness on admission [132, 137]

ICU patients should be routinely screened for delirium development [7], and the tools available provide the potential to confidently identify delirium [18, 19, 123]. It is recommended to use specific psychometric assessment tools such as the confusion assessment method for the ICU (CAM-ICU) or the intensive care delirium screening checklist (ICDSC); both tests are equally efficacious and performed the best compared to others [7, 17, 19, 123, 124, 138, 139]. CAM-ICU assesses the cardinal features of delirium (changes in baseline cerebral constitution during the course, attention, level of consciousness and thinking abilities; see above), and the presence of delirium is diagnosed, if feature 1 *plus* feature 2 *and* either feature 3 or feature 4 are experienced.

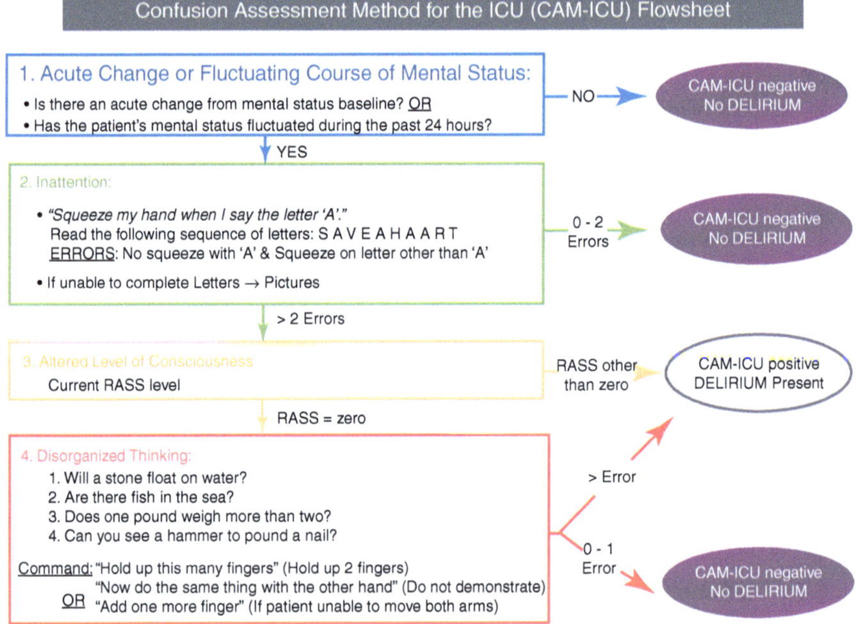

The CAM-ICU confusion assessment

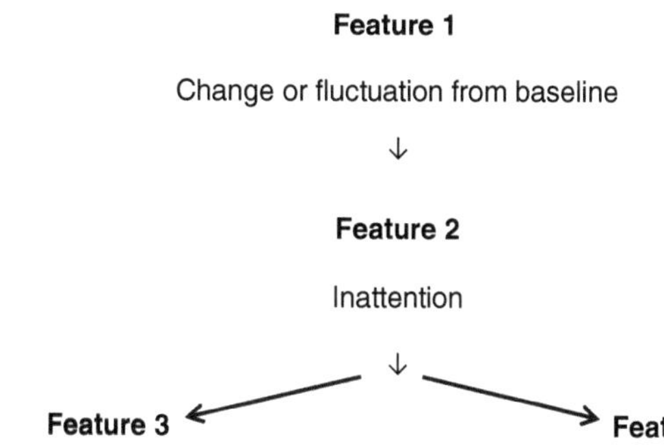

Diagnosis of delirium

Delirium: feature 1 plus feature 2 and either feature 3 or feature 4

Treatment Issues

We neither properly understand the pathophysiology of delirium [126–128] nor do we have relevant studies investigating treatment options [4, 7]. However, the application of antipsychotic agents is endorsed by several international guidelines [140–143], and many intensivists use those drugs to treat delirious patients. Haloperidol is the most commonly used drug [144, 145], especially when administered for psychotic symptoms [4, 118]. Unfortunately, no evidence exists to adopt haloperidol as the first-line agent [7], and the newest guidelines of the SCCM and the ACCM [7] advocate dexmedetomidine rather than haloperidol or benzodiazepines (grade II, level B recommendation). Nonetheless, atypical antipsychotics (e.g. quetiapine [146]) may reduce the duration of delirium, a level C recommendation; no level of evidence is given [7]. Moreover, dexmedetomidine is suggested to reduce the incidence of delirium [78, 95] and its duration [78, 147, 148], further may shorten the ventilator time [78, 88]. Advocated dosing of dexmedetomidine ranges from max 0.7 μg/kg/h for short-term sedation as approved in the United States to infusions of up to 1.5 μg/kg/h for long-term sedation for up to 28 days, the latter demonstrated to be safe and efficient in several studies [78, 96, 103, 149].

Interestingly, early mobilization is the only feature demonstrating preventive properties (level B, grade I [7]) and thus should be considered immediately in every patient [116, 130, 150]. Poor and scarce sleep may contribute to delirium development. Hence, measures to facilitate sleep include a calm environment (low noise and light), the choice of hypnosedatives, the avoidance of medical procedures at night, and between decline of stimuli conserving the sleep cycles [151–155], a level C, grade I recommendation [7].

A subcategory of delirium are those delirious states provoked by patient-inherent factors such as abrupt alcohol withdrawal in chronic alcohol abuse or abrupt drug reduction in patients on high-dose opioids or sedatives, where a physiologic dependence exists [7]. Alcohol-withdrawal-associated delirium is present in 15–20 % of all hospitalized patients [156]. Usually, the symptoms are mild to moderate [157], but some patients might develop a life-threatening combination of central nervous system excitation (agitation, delirium, seizures) and hyperadrenergic symptoms (hypertension, tachycardia, arrhythmias, etc.) [158]. Central in the treatment of alcohol-withdrawal-associated symptoms are still benzodiazepines [126]. In the case of drug dependency, no specific prophylactic or treatment recommendation can be made due to lack of data and studies; weaning over several days is the measure advocated [7].

References

1. Jacobi J, Fraser GL, Coursin DB, Task force of the American College of Critical care medicine (ACCM) of the Society of Critical Care Medicine. Clinical practice guidelines for the sustained use of sedatives and analgesics in the critically ill adult: sedation and analgesia. Crit Care Med. 2002;30:119–41.
2. Ostermann ME, Keenan SP, Seiferling RA. Sedation in the intensive care unit: a systematic review. JAMA. 2000;283:1451–9.
3. Aitkenhead AR. Analgesia and sedation in intensive care. Br J Anaesth. 1989;63:196–206.
4. Patel SB, Kress JP. Sedation and analgesia in the mechanically ventilated patient. Am J Respir Crit Care Med. 2012;185:486–97.
5. Shehabi Y, Bellomo R, Mehta S. Intensive care sedation: the past, present and the future. Crit Care. 2013;17:322.
6. Cheung BMY. Sedation in the intensive -care unit: good and bad. Lancet. 2008;371:95–7.
7. Barr J, Fraser GL, Puntillo K. Clinical practice guidelines for the management of pain, agitation, and delirium in adult patients in the intensive care unit. Crit Care Med. 2013;41:263–306.
8. Mehta S, McCullagh I, Burry L. Current sedation practices: lessons learned from international surveys. Crit Care Clin. 2009;25:471–88.
9. Riker RR, Fraser GL. Altering intensive care sedation paradigms to improve patient outcomes. Crit Care Clin. 2009;25:527–38.
10. Arroliga A, Frutos-Vivar F, Hall J, International Mechanical Ventilation Study Group. Use of sedatives and neuromuscular blockers in a cohort of patients receiving mechanical ventilation. Chest. 2005;128:496–506.
11. Brook AD, Ahrens TS, Schaiff R. Effect of a nursing implemented sedation protocol on the duration of mechanical ventilation. Crit Care Med. 1999;27:2609–15.
12. Kress JP, Pohlman AS, O'Connor MF. Daily interruption of sedative infusions in critically ill patients undergoing mechanical ventilation. N Engl J Med. 2000;342:1471–7.
13. Carson SS, Kress JP, Rodgers JE. A randomized trail comparing intermittent bolus lorazepam and continuous infusion of propofol with daily interruption in mechanically ventilated patients. Crit Care Med. 2006;34:1326–32.
14. Pandharipande P, Pun BT, Herr DL. Effect of sedation with dexmedetomidine vs lorazepam on acute brain dysfunction in mechanically ventilated patients: the MENDS randomised controlled trail. JAMA. 2007;298:2644–53.

15. Fraser GL, Prato BS, Riker RR. Frequency, severity, and treatment of agitation in young versus elderly patients in the ICU. Pharmacotherapy. 2000;20:75–82.
16. Van Rompaey B, Schuurmans MJ, Shortridge-Baggett LM. Risk factors for intensive care delirium: a systematic review. Intensive Crit Care Nurs. 2008;24:98–107.
17. Ely EW, Margolin R, Francis J. Evolution of delirum in critically ill patients: validation of the confusion assessment method for the intensive care unit (CAM-ICU). Crit Care Med. 2001;29:1370–9.
18. Devlin JW, Fong JJ, Schumaker G. Use of a validated delirum assessment tool improves the ability of physicians to identify delirium in medical intensive care unit patients. Crit Care Med. 2007;35:2721–4.
19. Bergeron N, Dubois MJ, Dumont M. Intensive care delirium screening checklist: evaluation of a new screening tool. Intensive Care Med. 2001;27:859–64.
20. Puntillo KA, White C, Morris AB. Patients' perceptions and responses to procedural pain: results from Thunder Project II. Am J Crit Care. 2001;10:238–51.
21. Bergom- Engberg I, Haljamäe H. Assessment of patients' experience of discomfort during respiratory therapy. Crit Care Med. 1989;17:1068–72.
22. Puntillo KA, Arai S, Cohen NH. Symptoms experienced by intensive care unit patients at high risk of dying. Crit Care Med. 2010;38:2155–60.
23. Desbiens NA, Wu AW, Broste SK. Pain and satisfaction with pain control in seriously ill hospitalized adults: findings from the SUPPORT research investigations. Crit Care Med. 1996;24:1953–61.
24. Gelinas C. Management of pain in cardiac surgery ICU patients: have we improved over time? Intensive Crit Care Nurs. 2007;23:298–303.
25. Novaes MA, Aronovich A, Ferraz MB. Stressors in ICU: patients' evaluation. Intensive Care Med. 1997;23:1282–5.
26. Kress JP. New insights into symptoms experienced by high-acuity intensive care unit patients. Crit Care Med. 2010;38:2262–3.
27. Lewis KS, Whipple JK, Michael KA. Effect of analgesic treatment on the physiological consequences of acute pain. Am J Hosp Pharm. 1994;51:1539–54.
28. Strom T, Martinussen T, Toft P. A protocol of no sedation for critically ill patients receiving mechanical ventilation: a randomised trail. Lancet. 2010;375:475–80.
29. Breen D, Karabinis A, Malbrain M. Decreased duration of mechanical ventilation when comparing analgesia-based sedation using remifentanil with standard hypnotic-based sedation for up to 10 days in intensive care unit patients: a randomised trail. Crit Care. 2005;9:R200–10.
30. Dahaba AA, Grabner T, Rehak PH. Remifentanyl versus morphine analgesia and sedation for mechanically ventilated critically ill patients: a randomized double blind study. Anesthesiology. 2004;101:640–6.
31. Muellejans B, Lopez A, Cross MH. Remifentanil versus fentanyl for analgesia based sedation to provide patient comfort in the intensive care unit: a randomized double-blind controlled trail. Crit Care. 2004;8:R1–11.
32. Rozendaal FW, Spronk PE, Snellen FF. Remifentanil-propofol analgo-sedation shortens duration of ventilation and length of ICU stay compared to a conventional regimen: a centre randomised, cross-over, open-label study in the Netherlands. Intensive Care Med. 2008;35:291–8.
33. Muellejans B, Matthey T, Scholpp J. Sedation in the intensive care unit with remifentanil/propofol versus midazolam/fentanyl: a randomized, open-label, pharmacoeconomic trail. Crit Care. 2006;10:R91.
34. Puntillo KA. Pain experiences of intensive care unit patients. Heart Lung. 1990;19:526–33.
35. Payen JF, Bru O, Bosson JL. Assessing pain in critically ill sedated patients by using a behavioral pain scale. Crit Care Med. 2001;29:2258–63.
36. Gelinas C, Fillion L, Puntillo KA. Validation of the critical-care pain observation tool in adult patients. Am J Crit Care. 2006;15:420–7.

37. Williamson A, Hoggart B. Pain: a review of three commonly used pain rating scales. J Clin Nurs. 2005;14:798–804.
38. Odhner M, Wegman D, Freeland N. Assessing pain control in nonverbal critically ill adults. Dimens Crit Care Nurs. 2003;22:260–7.
39. Ahlers SJ, van Gulik L, van der Veen AM. Comparison of different pain scoring systems in critically ill patients in a general ICU. Crit Care. 2008;12:R15.
40. Puntillo K, Pasero C, Li D. Evaluation of pain in ICU patients. Chest. 2009;135:1069–74.
41. Memis D, Inal MT, Kavalci G. Intravenous paracetamol reduced the use of opioids, extubation time, and opioid-related adverse effects after major surgery in intensive care unit. J Crit Care. 2010;25:458–62.
42. Chinachoti T, Kessler P, Kirkham A. Remifentanil vs morphine for patients in intensive care unit who need short-term mechanical ventilation. J Med Assoc Thai. 2002;85 Suppl 3:S848–57.
43. Maddali MM, Kurian E, Fahr J. Extubation time, hemodynamic stability, and postoperative pain control in patients undergoing coronary artery bypass surgery: an evaluation of fentanyl, remifentanil, and nonsteroidal antiinflammatory drugs with propofol for perioperative and postoperative management. J Clin Anesth. 2006;18:605–10.
44. Payen JF, Chanques G, Mantz J. Current practices in sedation and analgesia for mechanically ventilated critically ill patients: a prospective multicenter patient-based study. Anesthesiology. 2007;106:687–95, quiz 891.
45. Park G, Lane M, Rogers S. A comparison of hypnotic and analgesic based sedation in a general intensive care unit. Br J Anaesth. 2007;98:76–82.
46. Sessler CN, Varney K. Patient-focused sedation and analgesia in the ICU. Chest. 2008;133:552–65.
47. Westmoreland CL, Hoke JF, Sebel PS. Pharmacokinetics of remifentanil (GI87084B) and its major metabolite (GI90291) in patients undergoing elective inpatient surgery. Anesthesiology. 1993;79:893–903.
48. Lee M, Silverman SM, Hansen H. A comprehensive review of opioid-induced hyperalgesia. Pain Physician. 2011;14:145–61.
49. Vinik HR, Kissin I. Rapid development of tolerance to analgesia during remifentanil infusion in humans. Anesth Analg. 1998;86:1307–11.
50. Schmidt S, Bethge C, Forster MH. Enhanced postoperative sensitivity to painful pressure stimulation after intraoperative high dose remifentanil in patients without significant surgical site pain. Clin J Pain. 2007;23:605–11.
51. Carrer S, Bocchi A, Candini M. Short term analgesia based on sedation in the intensive care unit: morphine vs remifentanil plus morphine. Minerva Anestesiol. 2007;73:327–32.
52. Nguyen NQ, Chapman MJ, Fraser RJ. The effects of sedation on gastric emptying and intragastric meal distribution in critical illness. Intensive Care Med. 2008;34:454–60.
53. Guillou N, Tanguy M, Seguin P. The effects of small-dose ketamine on morphine consumption in surgical intensive care unit patients after major abdominal surgery. Anesth Analg. 2003;97:843–7.
54. Schmittner MD, Vajkoczy SL, Horn P. Effects of fentanyl and S(+)-ketamine on cerebral hemodynamics, gastrointestinal motility, and need of vasopressors in patients with intracranial pathologies: a pilot study. J Neurosurg Anesthesiol. 2007;19:257–62.
55. Rapanos T, Murphy P, Szalai JP. Rectal indomethacin reduces postoperative pain and morphine use after cardiac surgery. Can J Anaesth. 1999;46:725–30.
56. Hynninen MS, Cheng DC, Hossain I. Non-steroidal anti-inflammatory drugs in the treatment of postoperative pain after cardiac surgery. Can J Anaesth. 2000;47:1182–7.
57. Simmons DL. Cyclooxygenase isozyms: the biology of prostaglandin synthesis and inhibition. Pharmacol Rev. 2004;56:387–437.
58. Pandey CK, Bose N, Garg G. Gabapentin for the treatment of pain in guillain-barre syndrome: a double-blinded. Placebo-controlled, crossover study. Anesth Analg. 2002;95:1719–23.

59. Pandey CK, Raza M, Tripathi M. The comparative evolution of gabapentin and carbamazepine for pain management in Guillain-Barre syndrome patients in the intensive care unit. Anesth Analg. 2005;101:220–5.
60. Truog RD, Campbell ML, Curtis J. Recommendations for end-of-life care in the intensive care unit: a consensus statement by the American College of Critical Care Medicine. Crit Care Med. 2008;36:953–63.
61. Jacobi J, Fraser GL, Coursin DB. Clinical practice guidelines for the sustained use of sedatives and analgesics in the critically ill adult. Crit Care Med. 2002;30:119–41.
62. Atkins PM, Mion LC, Mendelson W. Characteristics and outcomes of patients who self-extubate from ventilatory support: a case-control study. Chest. 1997;112:1317–23.
63. Conti J, Smith D. Hemodynamic responses to extubation after cardiac surgery with and without continued sedation. Br J Anaesth. 1998;80:834–6.
64. Devaquet J, Jonson B, Nikolason L. Effect of inspiratory pause on CO2 elimination and arterial PCO2 in acute lung injury. J Appl Physiol. 2008;105:1944–9.
65. Martin J, Basell K, Burkle H. Analgesie und Sedierung in der Intensivmedizi- S 2- Leitlinie der Deutschen Gesellschaft für Anästehesiologie und Intensivmedizin. Anästh Intensivmed. 2005;46:1–20.
66. Lui LL, Gropper MA. Postoperative analgesia and sedation in the adult intensive care unit: a guide to drug selection. Drugs. 2003;63:755–67.
67. Girard TD, Kress JP, Fuchs BD. Efficacy and safety of a paired sedation and ventilator weaning protocol for mechanically ventilated patients in the intensive care (awakening and breathing controlled trail): a randomised controlled trail. Lancet. 2008;371:126–34.
68. Kollef MH, Levy NT, Ahrens TS. The use of continuous i.v. sedation is associated with prolongation of mechanical ventilation. Chest. 1998;114:541–8.
69. Bauer TM, Ritz R, Haberthür C. Prolonged sedation due to accumulation of conjugated metabolites of midazolam. Lancet. 1995;346:145–7.
70. Spina SP, Ensom MH. Clinical pharmacokinetic monitoring of midazolam in critically ill patients. Pharmacotherapy. 2007;27:389–98.
71. Yahwak JA, Riker RR, Fraser GL. Determination of a lorazepam dose threshold for using the osmol gap to monitor for propylene glycol toxicity. Pharmacotherapy. 2008;28: 984–91.
72. Nelsen JL, Haas CE, Habtemariam B. A prospective evaluation of propylene glycol clearance and accumulation during continuous infusion lorazepam in critically ill patients. J Intensive Care Med. 2008;23:184–94.
73. Pandharipande P, Shintani A, Peterson J. Lorazepam as an independent risk factor for transitioning to delirium in intensive care unit patients. Anesthesiology. 2006;104:21–6.
74. Pandharipande P, Cotton BA, Shintani A. Prevalence and risk factors for development of delirium in surgical and trauma intensive care unit patients. J Trauma. 2008;65:34–41.
75. Mallow Corbet S, Montoya ID, Moore FA. Propofol-related infusion syndrome in intensive care patients. Pharmacotherapy. 2008;28:250–8.
76. Kam PC, Cardone D. Propofol infusion syndrome. Anaesthesia. 2007;62:690–701.
77. Fong JJ, Sylvia L, Ruthazer R. Predictors of mortality in patients with suspected propofol infusion syndrome. Crit Care Med. 2008;36:2281–7.
78. Riker RR, Shebabi Y, Bokesch PM. Dexmedetomidine vs midazolam for sedation of critically ill patients: a randomized trail. JAMA. 2009;30:489–99.
79. Rhoney DH, Murry KR. National survey of the use of sedating drugs, neuromuscular blocking agents, and reversal agents in the intensive care unit. J Intensive Care Med. 2003;18:139–45.
80. Kress JP, O'Connor MF, Pohlman AS. Sedation of critically ill patients during mechanical ventilation: a comparison of propofol and midazolam. Am J Respir Crit Care Med. 1996;153:1012–8.
81. Hall RI, Sandham D, Cardinal P. Propofol vs midazolam for ICU sedation: a Canadian multicenter randomized trail. Chest. 2001;119:1151–9.

82. Chamorro C, de Latorre FJ, Montero A. Comparative study of propofol versus midazolam in the sedation of critically ill patients: results of a prospective, randomized, multicenter trail. Crit Care Med. 1996;24:932–9.
83. Aitkenhead AR, Pepperman ML, Willatts SM. Comparison of propofol and midazolam for sedation in critically ill patients. Lancet. 1989;334:704–9.
84. Barrientos-Vega R, Mar sanchez-Soria M, Morales-Garcia C. Prolonged sedation of critically ill patients with midazolam or propofol: impact on weaning and costs. Crit Care Med. 1997;25:33–40.
85. Weinbroum AA, Halpern P, Rudick V. Midazolam versus propofol for long-term sedation in the ICU: a randomized prospective comparison. Intensive Care Med. 1997;23:1258–63.
86. Fong JJ, Kanji S, Dasta JF. Propofol associated with a shorter duration of mechanical ventilation than scheduled intermittent lorazepam: a database analysis using project IMPACT. Ann Pharmacother. 2007;41:1986–91.
87. Huey-Ling L, Chun-Che S, Jen-Jen T. Comparison of the effect of protocol-directed sedation with propofol vs midazolam by nurses in intensive care: efficacy, hemodynamic stability and patient satisfaction. J Clin Nurs. 2008;17:1510–7.
88. Ruokonen E, Parviainen I, Jakob SM. Dexmedetomidine versus propofol/midazolam for long-term sedation during mechanical ventilation. Intensive Care Med. 2009;35:282–90.
89. Nimmo GR, Mackenzie SJ, Grant IS. Hemodynamic and oxygen transport effects of propofol infusion in critically ill adults. Anaesthesia. 1994;49:485–9.
90. Lowrey TS, Dunlap AW, Brown RO. Pharmacologic influence on nutrition support therapy: use of propofol in a patient receiving combined enteral and parenteral nutrition support. Nutr Clin Pract. 1996;11:147–9.
91. Parke TJ, Stevens JE, Rice AS. Metabolic acidosis and fatal myocardial failure after propofol infusion in children: five case reports. BMJ. 1992;305:613–6.
92. Vasile B, Rasulo F, Candiani A. The pathophysiology of propofol infusion syndrome: a simple name for a complex syndrome. Intensive Care Med. 2003;29:1417–25.
93. Zaccheo MM, Bucher DH. Propofol infusion syndrome: a rare complication with potentially fatal results. Crit Care Nurse. 2008;28:18–25.
94. Sharshar T. ICU-acquired neuromyopathy, delirium and sedation in intensive care unit. Ann Fr Anesth Reanim. 2008;27:617–22.
95. Maldonado J, Wysong A, van der Starre P. Dexmedetomidine and the reduction of postoperative delirum after cardiac surgery. Psychosomatics. 2009;50:206–17.
96. Venn RM, Grounds RM. Comparison between dexmedetomidine and propofol for sedation in the intensive care unit: patient and clinician perceptions. Br J Anaesth. 2001;87:684–90.
97. Dasta JF, Kane-Gill SL, Pencina M. A cost – minimization analysis of dexmedetomidine compared with midazolam for long-term sedation in the intensive care unit. Crit Care Med. 2010;38:497–503.
98. Martin E, Ramsey G, Mantz J. The role of the alpha-2-adrenoceptor agfonist dexmedetomidine in postsurgical sedation in the intensive care unit. J Intensive Care Med. 2003;18:29–41.
99. Bhana N, Goa KL, McClellan KJ. Dexmedetomidine. Drugs. 2000;59:263–8, discussion 269.
100. Khan ZP, Ferguson CN, Jones RM. Alpha-2 and imidazoline receptor agonists. Their pharmacology and therapeutic role. Anaesthesia. 1999;54:146–65.
101. Venn RM, Hell J, Grounds RM. Respiratory effects of dexmedetomidine in the surgical patient requiring intensive care. Crit Care. 2004;4:302–8.
102. Reade MC, O'Sullivan K, Bates S. Dexmedetomidine vs haloperidol in delirious, agitated, intubated patients: a randomized open-label trail. Crit Care. 2009;13:R75.
103. Shehabi Y, Ruettimann U, Adamson H. Dexmedetomidine infusion for more than 24 hours in critically ill patients: sedative and cardiovascular effects. Intensive Care Med. 2004;30:2188–96.
104. Kress JP, Gehlbach B, Lacy M. The long term psychological effects of daily sedative interruption on critical ill patients. Am J Respir Crit Care Med. 2003;168:1457–61.

105. Schweickert WD, Gehlbach BK, Pohlman AS. Daily interruption of sedative infusions and complications of critical illness in mechanically ventilated patients. Crit Care Med. 2004;32:1272–6.
106. Kress JP, Vinayak AG, Levitt J. Daily sedative interruption in mechanically ventilated patients at risk for coronary artery disease. Crit Care Med. 2007;35:365–71.
107. De Wit M, Gennings C, Jenvey WI. Randomised trail comparing daily interrution of sedation and nursing-implemented sedation algorithm in medical intensive care unit patients. Crit Care. 2008;12:R70.
108. De Jonghe B, Bastuji-Garin S, Fangio P. Sedation algorithm in critically ill patients without acute brain injury. Crit Care Med. 2005;33:120–7.
109. Dellinger RP, Levy MM, Carlet JM. Surviving sepsis campaign: international guidelines for the management of severe sepsis and septic shock 2008. Intensive Care Med. 2008;36:296–327.
110. Department of Health. The Health Act 2006-code of practice for the prevention and control of health care associated infections. 2009. www.dh.gov.uk/assetRoot/04/13/93/37/04139337.pdf
111. Riker RR, Fraser GL, Simmons LE. Validating the sedation-agitation scale with bispectral index and visual analog scale in adult ICU patients after cardiac surgery. Intensive Care Med. 2001;27:853–8.
112. Ely EW, Truman B, Shintani A. Monitoring sedation status over time in ICU patients. Reliability and validity of the Richmond Agitation-Sedation Scale (RASS). JAMA. 2003;289:2983–91.
113. Brandl KM, Langley KA, Riker RR. Confirming the reliability of the sedation-agitation scale administered by ICU nurses without experience in its use. Pharmacotherapy. 2001;21:431–6.
114. Sessler CN, Gosnell MS, Grap MJ. The Richmond Agitation–Sedation Scale. Validity and reliability in adult intensive care unit patients. Am J Respir Crit Care Med. 2002;166:1338–44.
115. Masica AL, Girard TD, Wilkinson GR. Clinical sedation scores as indicators of sedative and analgesic drug exposure in intensive care unit patients. Am J Geriatr Pharmacother. 2007;5:218–31.
116. Schweickert WD, Pohlman MC, Pohlman AS. Early physical and occupational therapy in mechanically ventilated, critically ill patients: a randomised controlled trail. Lancet. 2009;373:1874–82.
117. Mehta S, Burry L, Martinez-Motta JC, Canadian Critical Care Trails Group. A randomized trial of daily awakening in critically ill patients managed with a sedation protocol: a pilot trail. Crit Care Med. 2008;36:2092–9.
118. Patel RP, Gambrell M, Speroff T. Delirium and sedation in the intensive care unit: survey of behaviors and attitudes of 1384 healthcare professionals. Crit Care Med. 2009;37:825–32.
119. Tan JA, Ho KM. Use of dexmedetomidine as a sedative and analgesic agent in critically ill adult patients: a meta-analysis. Intensive Care Med. 2010;36:926–39.
120. Meagher DJ, Leonhard M, Donnelly S. A longitudinal study of motor subtypes in delrium. Relationship with other phenomenology, etiology, medications exposure and prognosis. J Psychosom Res. 2011;71:395–403.
121. Gupta N, de Jonge J, Schieveld J. Delirium, dementia and anamnestic and other cognitive disorders. In: American Psychiatric Association, editor. Diagnostic and statistical manual of mental disorders. 4th ed. Washington, DC: s.n.; 1994. p. 123–33.
122. American Psychiatric Association. Practice guidelines for the treatment of patients with delirium. Am J Psychiatry. 1999;156:1–20.
123. Ely EW, Inouye SK, Bernard GR. Delirium in mechanically ventilated patients: validity and reliability of the confusion assessment method for the intensive care unit (CAM-ICU). JAMA. 2001;286:2703–10.

124. Guenther U, Popp J, Koecher L. Validity and reliability of the CAM-ICU flowsheet to diagnose delirium in surgical ICU patients. J Crit Care. 2010;25:144–51.
125. Peterson JF, Pun BT, Dittus RS. Delirium and its motoric subtypes: a study of 614 critically ill patients. J Am Geriatr Soc. 2006;54:479–84.
126. Shehabi Y, Roker RR, Bokesch PM, SEDCOM (Safety and Efficacy of Dexmedetomidine Compared With Midazolam) Study Group. Delirium duration and mortality in lightly sedated, mechanically ventilated intensive care patients. Crit Care Med. 2010;38:2311–8.
127. Pisani MA, Kong SY, Kasl SV. Days of delirium are associated with 1-year mortality in an older intensive care unit population. Am J Respir Crit Care Med. 2009;180:1092–7.
128. Ely EW, Shintani A, Truman B. Delirium as a predictor of mortality in mechanically ventilated patients in the intensive care unit. JAMA. 2004;291:1753–62.
129. Girard TD, Jackson JC, Pandharipande PP. Delirium as a predictor of long-term cognitive impairment in survivors of critical illness. Crit Care Med. 2010;38:1513–20.
130. Needham DM, Korupolu R, Zanni JM. Early physical medicine and rehabilitation for patients with acute respiratory failure: a quality improvement project. Arch Phys Med Rehabil. 2010;91:536–42.
131. Skrobik Y, Ahern S, Leblanc M. Protocolized intensive care unit management of analgesia, sedation, and delirium improves analgesia and subsyndromal delirium rates. Anesth Analg. 2010;111:451–63.
132. Pisani MA, Murphy TE, Araujo KL. Benzodiazepine and opioid use and the duration of intensive care unit delirium in an older population. Crit Care Med. 2009;37:177–83.
133. Van Rompaey B, Elseviers MM, Schuurmans MJ. Risk factors for delirium in intensive care patients: a prospective cohort study. Crit Care. 2009;13:R77.
134. Ouimet S, Kavanagh BP, Gottfried SB. Incidence, risk factors and consequences of ICU delirium. Intensive Care Med. 2007;33:66–73.
135. Dubois MJ, Bergerson N, Dumont M. Delirium in an intensive care unit: a study of risk factors. Intensive Care Med. 2001;27:1297–304.
136. Ely EW, Girard TD, Shintani AK. Apolipoprotein E4 polymorphism as a genetic predisposition to delirium in critically ill patients. Crit Care Med. 2007;35:112–7.
137. Pandharipande PP, Morandi A, Adams JR. Plasma tryptophan and tyrosine levels are independent risk factors for delirium in criticallly ill patients. Intensive Care Med. 2009;35:1886–92.
138. van Eijk MM, van Marum RJ, Klijn IA. Comparison of delirium assessment tools in a mixed intensive care unit. Crit Care Med. 2009;37:1881–5.
139. Tobar E, Romero C, Galleguillos T. Confusion assessment method for diagnosing delirium in ICU patients: cultural adaptation and validation of the Spanish version. Med Intensiva. 2010;34:4–13.
140. Martin J, Heymann A, Bäsell K. Evidence and consensus-based German guidelines for the management of analgesia, sedation and delirium in intensive care – short version. Ger Med Sci. 2010;8:Doc02.
141. Silva JA, Romera MA, Chamorro C. Sedo-analgesia in neurologically ill patients: guidelines revisited. Stroke. 2008;39:e67; author reply e68.
142. O'Mahony R, Murthy L, Akunne A, Development Group. Synopsis of the National Institute for Health and Clinical Excellence guideline for prevention of delirium. Ann Intern Med. 2011;154:746–51.
143. Tropea J, Slee JA, Brand CA. Clinical practice guidelines for the management of delirium in older people in Australia. Australas J Ageing. 2008;27:150–6.
144. Soliman HM, Melot C, Vincent JL. Sedative and analgetic practice in the intensive care unit: the results of a European survey. Br J Anaesth. 2001;87:186–92.
145. Burry L, Perreault M, Williamson D. A prospective evaluation of sedative, analgesic, antipsychotic and paralytic practices in Canadian mechanically ventilated adults. Proc Am Thorac Soc. 2009;179:A5492.

146. Devlin JW, Roberts RJ, Fong JJ. Efficacy and safety of quetiapine in critically ill patients with delirium: a prospective, multicenter, randomized, double-blind, placebo-controlled pilot study. Crit Care Med. 2010;38:419–27.
147. Girard TD, Pandharipande PP, Carson SS, MIND Trail Investigators. Feasibility, efficacy, and safety of antipsychotics for intensive care unit delirium: the MIND randomized, placebo-controlled trail. Crit Care Med. 2010;38:428–37.
148. Pandharipande PP, Girard T, Sanders RD. Comparison of sedation with dexmedetomidine versus lorazepam in septic ICU patients. Crit Care. 2008;12:P275.
149. Dasta JF, Kane-Gill SL, Durtschi AJ. Comparing dexmedetomidine prescribing patterns and safety in the naturalistic setting versus published data. Ann Pharmacother. 2004;38:1130–5.
150. Inouye SK, Bogardus Jr ST, Charpentire PA. A multicomponent intervention to prevent delirium in hospitalized older patients. N Engl J Med. 1999;340:669–76.
151. Redeker NS. Sleep in acute care settings: an integrative review. J Nurs Scholarsh. 2000;32:31–8.
152. Li SY, Wang TJ, Vivienne WU SF. Efficacy of controlling night-time noise and activities to improve patients' sleep quality in a surgical intensive care unit. J Clin Nurs. 2011;20:396–407.
153. Dennis CM, Lee R, Woodward EK. Benefits of quiet time for neuro-intensive care patients. J Neurosci Nurs. 2010;42:217–24.
154. Tamburri LM, DiBrienza R, Zozula R. Nocturnal care interactions with patients in critical care units. Am J Crit Care. 2004;13:102–12, quiz 114.
155. Cooper AB, Thornley KS, Young GB. Sleep in critically ill patients requiring mechanical ventilation. Chest. 2000;117:809–18.
156. Moore RD, Bone LR, Geller G. Prevalence, detection, and treatment of alcoholism in hospitalized patients. JAMA. 1989;261:403–7.
157. Bayard M, McIntyre J, Hill KR. Alcohol withdrawal syndrome. Am Fam Physician. 2004;69:1443–50.
158. Turner RC, Lichstein PR, Peden JG. Alcohol withdrawal syndromes: a review of pathophysiology, clinical presentation, and treatment. J Gen Intern Med. 1989;4:432–44.

Index

W. Krüger, A.J. Ludman, *Core Knowledge in Critical Care Medicine*,
DOI 10.1007/978-3-642-54971-7, © Springer-Verlag Berlin Heidelberg 2014

I

J

L

R